VLSI CAD TOOLS AND APPLICATIONS

THE KLUWER INTERNATIONAL SERIES
IN ENGINEERING AND COMPUTER SCIENCE

VLSI, COMPUTER ARCHITECTURE AND
DIGITAL SIGNAL PROCESSING

Consulting Editor

Jonathan Allen

Other books in the series:

Logic Minimization Algorithms for VLSI Synthesis, R.K. Brayton, G.D. Hachtel, C.T. McMullen, and A.L. Sangiovanni-Vincentelli. ISBN 0–89838–164–9.

Adaptive Filters: Structures, Algorithms, and Applications, M.L. Honig and D.G. Messerschmitt. ISBN: 0–89838–163–0.

Computer-Aided Design and VLSI Device Development, K.M. Cham, S.-Y. Oh, D. Chin and J.L. Moll. ISBN 0–89838–204–1.

Introduction to VLSI Silicon Devices: Physics, Technology and Characterization, B. El-Kareh and R.J. Bombard. ISBN 0–89838–210–6.

Latchup in CMOS Technology: The Problem and Its Cure, R.R. Troutman. ISBN 0–89838–215–7.

Digital CMOS Circuit Design, M. Annaratone. ISBN 0–89838–224–6.

The Bounding Approach to VLSI Circuit Simulation, C.A. Zukowski. ISBN 0–89838–176–2.

Multi-Level Simulation for VLSI Design, D.D. Hill, D.R. Coelho. ISBN 0–89838–184–3.

Relaxation Techniques for the Simulation of VLSI Circuits, J. White and A. Sangiovanni-Vincentelli. ISBN 0–89838–186–X.

VLSI CAD TOOLS
AND APPLICATIONS

edited by

Wolfgang Fichtner
Martin Morf

Institute for Integrated Systems
Swiss Federal Institute of Technology
Zurich, Switzerland

KLUWER ACADEMIC PUBLISHERS
Boston/Dordrecht/Lancaster

Distributors for North America:
Kluwer Academic Publishers
101 Philip Drive
Assinippi Park
Norwell, MA 02061, USA

Distributors for the UK and Ireland:
Kluwer Academic Publishers
MTP Press Limited
Falcon House, Queen Square
Lancaster LA1 1RN, UNITED KINGDOM

Distributors for all other countries:
Kluwer Academic Publishers Group
Distribution Centre
Post Office Box 322
3300 AH Dordrecht, THE NETHERLANDS

Library of Congress Cataloging-in-Publication Data

VLSI CAD tools and applications.

(The Kluwer international series in engineering and
computer science ; SECS 24)
Papers presented at a summer school held from July
21 to Aug. 1, 1986 at Beatenberg, Switzerland.
Includes bibliographies.
1. Integrated circuits—Very large scale integration
—Design and construction—Data processing.
2. Computer-aided design. I. Fichtner, Wolfgang.
II. Morf, Martin. III. Series
TK7874.V5572   1987   621.395   86–21148
ISBN-13: 978-1-4612-9186-2   e-ISBN-13:978-1-4613-1985-6
DOI:10.1007/978-1-4613-1985-6

TABLE OF CONTENTS

Preface   xi

1 **VLSI Design Strategies**   1
   C. Séquin

   1.1 Introduction   1
   1.2 VLSI Complexity   1
   1.3 The Design Spectrum   3
   1.4 The Role of CAD Tools   5
   1.5 The Role of the Designer   8
   1.6 The Synthesis Project at Berkeley   11
   1.7 Conclusions   15
   1.8 References   16

2 **Introduction to VLSI Design**   19
   J. Allen

   2.1 Introduction   19
   2.2 The MOS Transistor   20
   2.3 Inverter Circuits   26
   2.4 Generalized Inverter Circuits   31
   2.5 Transmission Gates   33
   2.6 Full Adders   36
   2.7 Programmed Logic Arrays   38
   2.8 Clocked Circuits   39
   2.9 Finite-State Machines   45
   2.10 Integrated Circuit Fabrication   47
   2.11 Design Rules   52
   2.12 References   54

3 **Simulation Tools for VLSI**   57
   C. J. Terman

   3.1 Introduction   57
   3.2 Circuit-Level Simulation   60
   3.3 A Linear Model for MOS Networks   67
   3.4 A Switch Model for MOS Networks   76
   3.5 Gate-Level Simulation   97
   3.6 References   100

4   Aspects of Computational Circuit Analysis   105
W. M. Coughran Jr., E. Grosse and D. J. Rose

4.1   Introduction   105
4.2   Formulation of the Circuit Equations   106
4.3   Table Representations of Devices   110
4.4   Linear Algebra Techniques   112
4.5   Newton-Like Methods   113
4.6   Continuation Methods   116
4.7   Time-integration Schemes   118
4.8   Macromodeling of Circuits   121
4.9   References   125

**5   VLSI Circuit Analysis,
     Timing Verification and Optimization   129**
A. E. Ruehli and D. L. Ostapko

5.1   Introduction   130
5.2   Circuit Analysis   131
5.3   Timing Verification   134
5.4   Circuit Optimization   136
5.5   References   141

6   CAD Tools for Mask Generation   147
J. B. Rosenberg

6.1   Introduction   148
6.2   Advantages   153
6.3   Mechanisms   155
6.4   ABCD   160
6.5   Design Capture   182
6.6   Compaction   198
6.7   Technology Encapsulation   206
6.8   Software Engineering   210

**7   Design and Layout Generation
     at the Symbolic Level   213**
C. Séquin

7.1   Introduction   213
7.2   The Role of Symbolic Representation   214
7.3   EDISTIX   216
7.4   TOPOGEN   221

7.5   ZORRO   224
7.6   Conclusions   230
7.7   Acknowledgements   230
7.8   References   230

8   Overview of the IDA System:
      A Toolset for VLSI Layout Synthesis   **233**
      D. D. Hill, K. Keutzer and W. Wolf

8.1   Introduction and Background   233
8.2   Key Ideas in IDA   234
8.3   IMAGES: a Symbolic, Constraint-Based
            Generator Design Language   235
8.4   Compaction and Assembly   244
8.5   Layout Synthesis   248
8.6   Other Tools and Features of IDA   255
8.7   Summary   260
8.8   References   261

9   CAD Programming in an Object Oriented
      Programming Environment   **265**
      J. Cherry

9.1   Introduction   265
9.2   The Programming Environment   268
9.3   The Organization of NS   275
9.4   Design Verification in NS   279
9.5   Physical Design in NS   282
9.6   History and Results   292
9.7   References   292

10   Trends in Commercial VLSI
      Microprocessor Design   **295**
      N. Tredennick

10.1 Introduction   296
10.2 Historical Design Practice   296
10.3 Current Design Practice Details   306
10.4 Future Designs   311
10.5 Forecast   313
10.6 Summary   322
10.7 References   323

11   Experience with CAD Tools
for a 32-Bit VLSI Microprocessor   **327**
D. R. Ditzel and A. D. Berenbaum

11.1   Introduction   327
11.2   Top Down Simulation   327
11.3   The Interpreter   328
11.4   Architectural Simulator   328
11.5   Functional Simulator   328
11.6   Schematic Logic Drawings: Draw   331
11.7   Switch Level Simulation: Soisim   335
11.8   Backporting: Switch Level Simulation
          Without Vector Files   336
11.9   Layout Tools: Mulga   337
11.10 Circuit Extraction: Goalie   340
11.11 Netlist Comparison: Gemini   341
11.12 Timing Analysis: ADVICE and Leadout   342
11.13 Naming Conventions   344
11.14 Control the Source Code   346
11.15 UNIX   346
11.16 Some Statistics   347
11.17 Weak Spots   348
11.18 Results   348
11.19 Conclusion   349
11.20 References   350

12   Overview of a 32-Bit Microprocessor
Design Project   **351**
P. Bosshart

12.1   Introduction   351
12.2   Processor Description   352
12.3   High Level Decisions and Influences   354
12.4   Design Tools   355
12.5   Division of Labor   370
12.6   Problems   375
12.7   Important Results and Conclusions   378
12.8   References   380

13   Architecture of Modern VLSI Processors   381
P. M. Lu, D. E. Blahut and K. S. Grant

13.1   Introduction   381
13.2   Microprocessor Architecture Design Considerations   382
13.3   Memory Management Architectures   397
13.4   Memory Interfacing Peripherals   400
13.5   Summary   400
13.6   Acknowledgments   400
13.7   References   401

14   A Comparison of Microprocessor
Architectures in View of
Code Generation by a Compiler   407
N. Wirth

14.1   Introduction   407
14.2   The Target Architectures
         and Their Instruction Formats   409
14.3   Code Generation   413
14.4   Measurements   421
14.5   Conclusions   425
14.6   References   427

15   Fault Tolerant VLSI Multicomputers   429
C. Séquin and Y. Tamir

15.1   Introduction   429
15.2   Fault Tolerance   431
15.3   Self-Checking Nodes   432
15.4   Defects and Faults in VLSI   434
15.5   Self-Testing Comparators in VLSI   435
15.6   Implementation Issues   439
15.7   System Level Protocols   443
15.8   Conclusions   446
15.9   References   447

16   The VLSI Design Automation Assistant:
An IBM System/370 Design   451
T. J. Kowalski

16.0   Introduction   451
16.1   Conception   452
16.2   Birth   454
16.3   First Steps   457

16.4   The IBM System/370 Experiment   461
16.5   Summary   473
16.6   References   473

17   Higher Level Simulation and CHDLs   475
R. W. Hartenstein and U. Welters

17.1   Introduction: Why CHDLs?   476
17.2   Early Phases of the Design Process   477
17.3   Introducing a CHDL and Its Use   480
17.4   CHDL-based Design Environments   489
17.5   Conclusions   496
17.6   Literature   497

18   New Trends in VLSI Testing   501
G. Saucier, C. Bellon and M. Crastes de Paulet

18.1   Introduction   501
18.2   The Test Assembler   505
18.3   The Test Advisor   510
18.4   Interface with the General Data Base   511
18.5   Conclusion   512
18.6   References   512

**19   VLSI Testing:
     DFT Strategies and CAD Tools   515**
M. Gerner and M. Johansson

19.1   Introduction   515
19.2   DFT Strategies   516
19.3   CAD-Tools   531
19.4   Conclusions   546
19.5   Literature   547

PREFACE

The summer school on *VLSI CAD Tools and Applications* was held from July 21 through August 1, 1986 at Beatenberg in the beautiful Bernese Oberland in Switzerland. The meeting was given under the auspices of IFIP WG 10.6 VLSI, and it was sponsored by the Swiss Federal Institute of Technology Zurich, Switzerland. Eighty-one professionals were invited to participate in the summer school, including 18 lecturers. The 81 participants came from the following countries: Australia (1), Denmark (1), Federal Republic of Germany (12), France (3), Italy (4), Norway (1), South Korea (1), Sweden (5), United Kingdom (1), United States of America (13), and Switzerland (39).

Our goal in the planning for the summer school was to introduce the audience into the realities of CAD tools and their applications to VLSI design. This book contains articles by all 18 invited speakers that lectured at the summer school. The reader should realize that it was not intended to publish a textbook. However, the chapters in this book are more or less self-contained treatments of the particular subjects. Chapters 1 and 2 give a broad introduction to VLSI Design. Simulation tools and their algorithmic foundations are treated in Chapters 3 to 5 and 17. Chapters 6 to 9 provide an excellent treatment of modern layout tools. The use of CAD tools and trends in the design of 32-bit microprocessors are the topics of Chapters 10 through 16. Important aspects in VLSI testing and testing strategies are given in Chapters 18 and 19.

We would like to thank all of the invited speakers for the time and effort they had to invest into the preparation of their talks and papers. It would have been impossible to organize the summer school and to edit this book without the help of the members of the Institute of Integrated Systems at the Swiss Federal Institute of Technology Zurich. In particular, we would like to thank L. Heusler, Dr. H. Kaeslin, P. Lamb, R. Meyer, and M. Raths for their assistance. Dr. A. Aemmer was especially helpful in the planning and organization. He deserves special credit for the success of the meeting and the existence of this book. Mrs. Bourquin was a perfect summer school secretary. It was a special pleasure to work with Carl Harris from Kluwer Publishing in the preparation of this book.

W. Fichtner M. Morf

VLSI CAD TOOLS AND APPLICATIONS

1

V L S I   DESIGN   STRATEGIES

Carlo H. Séquin

Computer Science Division
Electrical Engineering and Computer Sciences
University of California, Berkeley, CA 94720

ABSTRACT

The growing complexity of VLSI chips creates a need for better CAD tools and data management techniques. The rapidly changing nature of the field requires a modular toolbox approach — rather than a fixed monolithic design system — and the involvement of the designer in the tool-building process. A short overview over the Berkeley design environment and our recent Synthesis Project is also given.

1. INTRODUCTION

Very large scale integration (VLSI) has made it economically viable to place several hundred thousand devices on a single chip, and the technological evolution will continue to increase this number by more than an order of magnitude within a decade. While the limits on chip growth imposed by technology and materials are still another three orders of magnitude away,[1] the design of the present-day chips already causes tremendous problems. G. Moore coined the term "complexity barrier".[2] This is the major hurdle faced today in the construction of ever larger integrated systems. In Section 2 the nature of the VLSI complexity problem will be discussed.

In order to deal with this complexity and to exploit fully the technological potential of VLSI, some structure has to be introduced into the design process; the resulting design styles are reviewed in Section 3, and the general nature of the VLSI design process is discussed. The size of the task is such that it cannot be done without tools; new tools and new ways of managing the information associated with the design of a VLSI chip must be developed (Section 4). This changes the role of the designer (Section 5).

Section 6 illustrates with the example of the Berkeley Synthesis Project how we think the art of VLSI design is going to evolve.

2. VLSI COMPLEXITY

In the early 1980's. VLSI complexity became a hot topic for concern and discussion.[3] This may appear surprising if one compares the complexity of the chips of that period with other technological structures that mankind has built in the past. Certainly the number of components on a VLSI chip does not exceed the number of parts in a telephone switching station or in the space shuttle, and mainframe computers with an even larger number of transistors have been built for at least a decade before they were integrated onto a chip. System complexity should not differ markedly whether a circuit is contained within a cabinet, on a printed circuit board, or on a single silicon chip.

It is the "large", potentially unstructured space of the VLSI chip that causes the concern. Nobody would dare to insert a million discrete devices into a large chassis using discrete point-to-point wiring. Large systems built from discrete devices are broken down into sub-chassis, mother-boards, and module-boards carrying the actual components. This physical partitioning encourages careful consideration of the logical partitioning and of the interfaces between the modules at all levels of the hierarchy. Since such systems are typically designed by large teams, early top-down decisions concerning the partitioning and the interfaces must be made and enforced rather rigidly — for better or for worse. This keeps the total complexity in the scope of each individual designer limited in magnitude, and thus manageable.

VLSI permits the whole system to be concentrated in a basically unstructured domain of a single silicon chip which does not *a priori* force any partitioning or compartmentalization. On the positive side, this freedom may be exploited for significant performance advantages. On the negative side, it may result in a dangerous situation where the complexity within a large, unstructured domain simply overwhelms the designer.

A similar crisis was faced by software engineers when unstructured programs started to grow to lengths in excess of 10,000 lines of code. The crisis was alleviated by the development of suitable design methodologies, structuring techniques, and documentation styles. Many of the lessons learned in the software domain are also applicable to the design of VLSI systems.[3]

Furthermore, the field of VLSI is rather interdisciplinary in nature. To achieve optimal results, we need a tight interaction of algorithms, architecture, circuit design, IC technology, etc. However, designers who are experts in all these fields are rarely found. How can ordinary mortals attempt to do a reasonable VLSI design? Here again, suitable abstractions have to be found, so that the details of processing are hidden from the layout designer, and the details of the circuit implementation are hidden from the microarchitect. Models that are accurate enough to permit sound decisions based on them need to be created, and clean interfaces between the various domains of responsibility need to be defined. For instance, the semantic meaning of the geometry specified in a layout has to be defined carefully: Is this the geometry of the fabrication masks? Is this the desired pattern on the silicon chip? Or is this a symbolic representation of some of the desired device parameters? These questions still lead to much discussion and often to bad chips. The emergence of silicon brokerage services such as MOSIS[4] has forced clarification of many of these issues.

3. THE DESIGN SPECTRUM

To make the task of filling the void on a VLSI chip manageable, some widely accepted abstractions have emerged that lead to a hierarchy (or rather a multidimensional space) of *views* of a particular design. The different representations generally address different concerns. A typical list of design levels and of the concerns they address is shown in Table 1.

Design Level	Concerns Addressed
Behavior	Functionality
Functional blocks, Linked module abstraction[5]	Resource allocation sequencing, causality
Register-transfer level Clocked register and logic[5]	Testability timing, synchronization
Gate Level, Clocked primitive switches[5]	Implementation with proper digital behavior
Circuit Level	Performance, noise margins
Sticks Level	Layout topology
Mask Geometry	Implementation, yield

Table 1. *Levels of abstraction in chip design.*

The other saving notion is that of prefabricated parts. The same functions at various levels of the design space are needed again and again. Successful designs of frequently used parts can thus be saved in libraries for the reuse by many customers. The nature of these parts and the level to which they are predefined or even prefabricated leads to a variety of different design styles.

3.1. VLSI Design Styles

A large spectrum of possibilities for the design of a VLSI chip has evolved, offering wide ranges of expected turn-around time, resulting performance, and required design effort. Table 2 gives a strongly simplified view of the spectrum of possibilities.

On one end of the spectrum are the Gate Array and Standard Cell technologies. Predesigned logic cells at the SSI and MSI level permit the engineers to use functional blocks that they are already familiar with from TTL breadboard designs. The abstraction and prefabrication of these cells lead to minimal design effort and faster turn-around time but at the price of less functionality per chip and less performance for a given technology.

Method	Complexity	Effort	Main Strength	Automation
Gate Array	20,000	4-8 weeks	Fast changes	Yes
Standard Cell	40,000	4-8 weeks	Resue of logic	Yes
Macro Cell	100,000	1-2 years	Good area use	Almost
Flexible Modules	200,000	1-2 years	High density	Almost
Standard Functions	200,000	2-8 years	Testability	Not yet
Optimized Layout	400,000	$\geq$8 years	High performance	Not so soon

Table 2. *Styles of IC chip design.*

At the other end of the spectrum is the full custom chip in which all modules have been hand-designed with the utmost care for performance and density and have been integrated and packed onto the chip in a tailor-made fashion. This design style can lead to spectacular results in terms of functionality and performance of an individual chip, but it comes at the price of an exorbitant design effort.

Somewhere in the middle between these two extremes are mixed approaches in which the crucial cells have been hand-designed with great care — particularly the cells that are in the critical path determining performance and the cells that are used in large arrays, as they will make the dominant contribution to the size of the chip. Uncritical "glue" logic that is used only once may be generated by a program either in the form of a PLA or as a string of standard cells. These macro cells of varying sizes and shapes are then placed and wired by hand or by emerging CAD programs.[6] This approach leads to higher densities than standard cells, since the degree of integration in the macro cells is typically higher and since a smaller amount of area is wasted in partly filled wiring channels. If the macro cells are procedurally generated and suitably parameterized so that they can be adjusted to the available space, even higher densities can be achieved. When properly used, these intermediate approaches can compete with a full custom design in terms of performance but typically result in a somewhat larger chip size.

Concerns of modularity and testability may outweigh aims for density and performance; functional modules designed for testability with clean interfaces are then used. This is in analogy to the use of properly abstracted and encapsulated software modules. This approach has started to gain acceptance also in VLSI. A good VLSI design environment will permit the designer to mix these various design styles in appropriate ways.

3.2. The Design Process

Even with an agreed-upon set of hierarchical levels, an extensive library of predefined parts, and a chosen design style, the design process can still be rather involved. It is rarely a single forward pass through all the transformation steps that takes a high-level behavioral description through register-transfer and logic level descriptions into a symbolic representation capturing the topology and finally into a dense layout suitable for implementation with a particular technology (Table 1). The overall problem may be structured in a top down manner into simpler subtasks with clearly defined functions. But in parallel, designers intimately familiar with the implementation technology will explore good solutions for generic functions in the given technology in a bottom-up fashion. This effort will result in an understanding of what functions can best be implemented in this environment and produce a set of efficient building blocks.

Hopefully, the top-down decomposition and the bottom-up provision of solutions will meet in the middle and permit completion of the design. However, for a new technology, it is unlikely that this will happen on the first try. The natural building blocks must first be discovered; only then can the architectures be modified and partitioned appropriately. Thus there is an iteration of top-down and bottom-up moves in a Yo-Yo like fashion until the optimal path linking architecture to technology has been found.

It should also be pointed out that the design process is often a mixture of solid established procedures and of free associations and 'trial-and-error'. The guessing part plays a role in finding good partitioning schemes as well as in the definition of generic functions that might constitute worthwhile building block in the given technology. Proven checking methods are then used to evaluate objectively whether the guesses made are indeed usable: Is the decomposition functionally correct? Is it appropriate — or does it cut through some inner loop, causing unnecessary communications penalties? Are the building blocks of general use? How many algorithms, tasks, or architectures can actually make use of them? Is their performance reasonable?

In the next section we explore to what extent this design process can be supported by the computer, and for which part the human intelligence might be hard to replace.

4. THE ROLE OF CAD TOOLS

Good tools help man to achieve more, to obtain better results, or to reach given goals more effortlessly. VLSI design is no exception. I like to split the CAD tools useful in the design of ICs into five classes:

1) *Checking and Verification Tools* typically answer questions such as: Are there any errors? Are the connections between blocks consistent? Does this function behave as specified?

2) *Analysis Tools* tell the designer: How well does a particular approach work? How much power does this circuit consume? What is the worst case settling time?

3) *Optimization Tools* can help the designer to vary component values to achieve a specific performance goal, or they can find "optimal" module placements within given constraints.

4) *Synthesis Tools* combine construction procedures and optimization algorithms. They may decompose a logic function into a minimum number of gates, or they may find a good floor plan from a connectivity diagram.

5) *High-level Decision Tools* support the designer in the "guessing phase" of the design process. These tools try to suggest particular solutions, i.e., partitioning schemes, micro architectures, or network topologies.

4.1. The CAD Wave

Building tools in the above classes 1) through 5) gets progressively more difficult. Typically, checking and verification tools are the first to become viable, helping to eliminate well-defined mistakes. Next, analysis tools permit the designer to find out how good a solution he has chosen and whether the design meets specifications. Based on the analysis algorithms, optimization tools emerge, assisting the designer in fine-tuning a design and in optimizing particular aspects of it. Gradually, these tools evolve into self-reliant synthesis tools; these may use heuristic methods or simulated annealing techniques to find solutions that are becoming competitive in quality with the work of human designers. Finally the tools will invade the areas where it is most difficult to replace the human mind — the high-level decision making process. Here tools from all the previously mentioned classes need to be employed in an iterative way; often techniques from the field of *Artificial Intelligence* are used.

Sweeping through the various levels of the design hierarchy in a bottom-up manner, tools will start to take over the function of the human designer. This general trend has started many years ago. Historically, the first tools to be developed for IC designers were circuit analysis tools such as SPICE.[7] There was a real need for such tools, since calculating the performance even of small integrated circuits would have been too tedious, and including actual fabrication of the chip in every design iteration would have been too slow and costly. At that time, the circuits were small enough so that most checking tasks could be performed without computer assistance. Optimization was done by hand with the help of the available analysis tools. Design decisions were largely based on the intuition or experience of the designer.

In the meantime, tools have matured at the layout level. Circuit extractors and design rule checkers are relied upon by every designer of large ICs. Without timing verification and circuit simulation, it would be impossible to obtain chips that meet performance specifications. Circuit optimization, however, is still largely done by the designer, using analysis tools in the "feedback loop", and synthesis tools are being investigated in the research laboratories.

At the higher levels of the design hierarchy tools have not claimed as much ground yet. Functional simulators are used to verify the correctness of the functional behavior and to obtain some crude idea of the expected performance. Optimization and synthesis tools are the subject of active research. High-level decision tools are being contemplated.

Tools are important at all the levels of the design hierarchy introduced in Section 3. The development of CAD tools started at the circuit level, because there the need was most urgent. This was the level of abstraction that could not easily be breadboarded and evaluated by measurement. As larger and larger systems get integrated onto a single chip, we will need better tools also at the higher levels in the design hierarchy.

4.2. Design Representation

Traditionally, many design systems for custom circuitry have used the geometrical layout information to "glue" everything together. From this low-level description that other representations are derived, and many of the analysis start from this level, e.g. circuit extraction and design rule checking. This is an unsatisfactory approach. Too much of the designer's intent has been lost in that low-level representation and has to be rediscovered by the analysis tools.

If there is to be a "core" description from which other representations are derived, it has to be at a higher level. The trend is to move to a symbolic description[8-10] that is still close enough to the actual geometry, so that ambiguities in the layout specification can be avoided. Yet at the same time, this description must have provisions to specify symbolically the electrical connections and functional models of subcircuits.[11]

In the long run, there is no way that a proper, integrated data management system can be avoided. Such a system can capture the design at various levels of the design hierarchy and, with the help of various tools, ensure consistency between the various representations. An integrated tool system will have to support the mentioned Yo-Yo design process in order to be effective.

4.3. Tool Integration

The art of VLSI design is not yet fully understood, and new methodologies are still evolving. It is thus too early to specify a rigid design system that performs the complete design task; quite likely, such a system would be obsolete by the time it becomes available to the user. It is more desirable to create a framework that permits the usage of many common tools in different approaches and that supports a variety of different design styles and methods. In short, the environment should provide *mechanisms* and *primitives* rather than *policies* and *solutions*.

Intricate interaction between the various tools must be avoided; every tool should do one task well and with reasonable efficiency.[12] The tools are coupled through compatible data formats or a joint data base to which they all interface in a procedural manner. The former solution causes less overhead in the early development phase of a new tool and makes it easy for workers in different locations to share data and test examples since ASCII text files can easily be transmitted over electronic networks.

The data base approach leads to a more tightly coupled system. It has the advantage (or disadvantage) that all data is in one central location. Interfacing a tool to this data base is normally more involved and costly than to simply read and write ASCII files. Unless the data management system is properly constructed and supported, the access to the data base can also get painfully slow. A practical solution is to use a combination of both: a data base that also has proper ASCII representations for each view of the design.

At Berkeley such a collection of tools[13] has been under development since the late 1970s. All tools are embedded in the UNIX[14] operating system. UNIX already provides many of the facilities needed in such an environment: a suitable hierarchical *file structure*, a powerful monitor program in the form of the UNIX *shell*,[15] and convenient mechanisms for *piping* the output of one program directly into the input of a successor program. An example of a newer, object oriented data base[16] will be discussed briefly in Section 6. The corresponding ASCII representation and interchange format is EDIF.[17]

Regardless of the exact structure of the data base, the various different representations of a design should be at the fingertips of the designer, so that he can readily choose the one representation that best captures the problem formulation with which he is grappling at the moment.

5. THE ROLE OF THE DESIGNER

The wave of emerging CAD tools at all levels of the design hierarchy is changing the role of the designer.

5.1. The CAD System Virtuoso

Designers of solid-state systems will spend an ever smaller fraction of their time designing at the solid-state level. More and more technical tasks, particularly at the lower levels of the design hierarchy, can be left to computer-based tools. Systems designers will rely increasingly on tools and on prototype modules generated by expert designers. They will thus change from being technical designers to being players of a sophisticated and rich CAD system.

It will take effort to learn the new skills. The essential experience no longer consists in knowing how to best lay out a Schottky-clamped bipolar gate, but rather in choosing the right tool, setting the right parameters and constraints, using a reasonable number of iterations, or knowing what to look for in a simulation producing a wealth of raw data.

The results obtainable with any CAD system depend to a large extent on the skill with which the designer moves through the maze of options. Furthermore, many of our design tools are still in the state corresponding to the early days of the automobile, where the driver also had to be a mechanic and be prepared to take care of frequent breakdowns.

5.2. The Designer as a Tool Builder

Good tools cannot be constructed in an isolated CAD department. They must be built in close relationship with the user. Who is better qualified than the actual user to understand the needs for a tool and to test whether a new tool really meets expectations? Further, a good CAD tool cannot be built in a single try. Only after the designers have a prototype to play with, they can decide what they really need and provide more accurate specifications for the new tool. The emergence of a tool often changes the nature of the job enough to shift the emphasis to a different bottle neck, thus altering the requirements for the tool. This in turn may necessitate a revision of the user interface or the performance targets. This iterative process to arrive at the proper specifications leads to the tool development spiral shown in Figure 1.

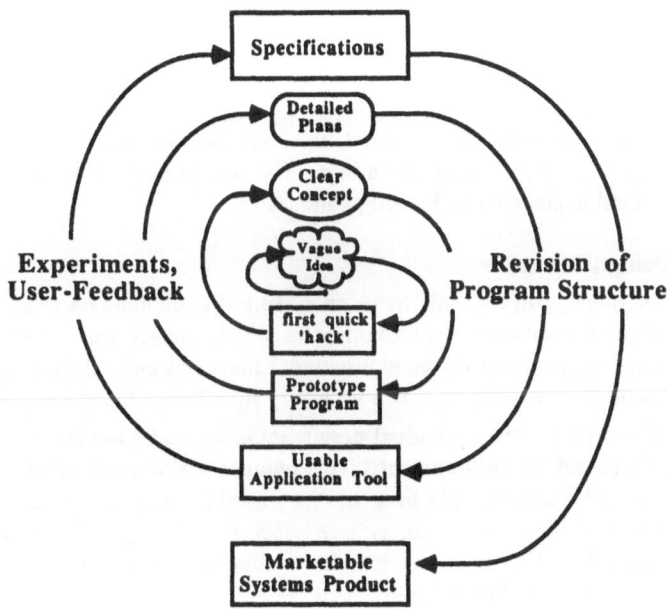

Figure 1. *The spiral of tool development.*

Each implementation serves as the basis for clearer specifications for the next round. The more rounds one can make around this spiral, the better the tool will get. In going from one round to the next, one should not be afraid to start completely from scratch, to throw out the old code, and to keep nothing but the experience and plans for an improved approach. The temptation to just patch up the old code can be reduced if the implementation language is switched. Many tool developers have found it productive to create early prototypes in LISP, SmallTalk, or Prolog, and to code later versions in an efficient procedural language providing some control over the machine resources.

The first one or two turns on the development spiral are crucial. This is where the general directions of a new tool are determined. On later turns it is much harder to make significant conceptual changes. Thus, on the first turn(s) it is particularly important that the development is done in close contact with the designers actually using the tool. How much closer can you get than having the designer himself do the first "quick hack"? Nowaday, more and more engineers receive a good education in programming, and it is thus easier to find persons with the right combination of skills.

Once the framework of the tool is well established and the user interface defined, a formal CAD group could take over to recode the tool, modularize the program, look at efficiency issues, and provide decent documentation. In the production of a good manual, the designers must again be strongly involved, as they understand the needs of the users.

The most leverage out of human ingenuity can be obtained if the latter is used to build new and better tools, which then can help many other designers to do the job better or faster. Using the designers as tool-builders, the impact of the work of individual engineers can be compounded.

5.3. The Design Manager

A good manager, will not only focus on getting the job done on time, but will also concentrate on creating an environment in which the job can get done most efficiently. The improvement of the environment must not be neglected under the pressures of immediate deadlines; it is a necessary investment for the future.

This is also true for the individual design engineer, as he too is a manager of his task, his time, and his environment. This requires a change of attitude on the part of the typical engineer. He may have to spend a larger amount of time, learning about available tools, acquiring new tools, or building tools himself, than working on 'the job'. But experience has shown that, amortized over two or three jobs, this investment into the environment pays off.

6. THE SYNTHESIS PROJECT AT BERKELEY

In Spring 1986, an ambitious project concerning the automatic synthesis of integrated circuits was undertaken in the Department of EECS at Berkeley. The "official" goal of the project was to integrate and enhance our various CAD tools to create a suit of tools that could synthesize a complex microcomputer from a behavioral-level description to the mask-level output with as little manual intervention as possible. As a "fringe benefit" we expected to gain a thorough understanding of the major issues in IC synthesis and to find out where our CAD tool design efforts need to be focussed.

6.1. Project Organization

The Synthesis Project was led by Professors Newton, Sangiovanni-Vincentelli and Séquin together with seven visiting Industrial Fellows. Following a tradition in our department, the project was tightly integrated with our graduate instruction. During the Spring term of 1986, the project was carried by two graduate courses, a design-oriented class (CS 292H) and a CAD tool-oriented class (EECS 290H), both of which had to be taken by all 35 participating students.

The tool development was tied to the SPUR (Symbolic Processing Using RISCs) project[18] which had been in progress for about a year. The main focus of the SPUR project was the development of a set of three chips: the central RISC processor (CPU), a cache controller (CCU), and a floating point coprocessor (FPU), for use in a multiprocessor workstation. We planned to use the architects and original implementors of these chips as consultants and hoped to obtain large parts of chip descriptions in machine-readable form.

A matrix organization was adopted for the graduate-student design teams. Each student, as a participant in the *design class*, was involved in the design of one of the three chips and was responsible for the generation of at least one specific module. As a participant in the *CAD class*, each student was a member of one of several tool development groups (e.g., logic synthesis, place and route, module generation) and was working towards developing a tool suite that would be usable for all three chips.

6.2. Resources and Infrastructure

Resources available to the Synthesis Project included a dozen DEC VAXstationII workstations and seven color VAXstationII/GPX machines. The backbone was a VAX8650 CPU with 500 Mbytes of disc storage dedicated to the course. This machine acted as the central database and as the repository for all the existing and emerging CAD tools. All these machines, as well as all the other computing resources in the department, were coupled through an ethernet, creating a tightly coupled, highly interactive computing and communications environment.

Important software support was provided by the Digital Equipment Corporation in the form of the DECSIM mixed-level simulator and its associated behavioral design language, BDS. This software package was chosen primarily because of its availability and because DEC personnel were on site to provide support for its application. DECSIM also offered the possibility of using mixed-level simulation at the behavior, register-transfer, gate-logic, or switch level — even though during the course we did not get far enough to use all these options.

For the integration of our tools we chose to use a single object-oriented data management system, OCT,[16] the development of which had started some time ago. OCT has as its basic unit the 'cell' which can have many 'views' — physical, logical, symbolic, geometrical. A cell is a portion of a chip that a designer wishes to abstract; it can vary in size from a simple transistor to the entire floorplan of a CPU. The system is hierarchical, i.e., cells can contain instances of other cells. Moreover, cells can have different abstract representations depending on the intended application, and these are represented in OCT by 'facets', which are the accessible units that can be edited. OCT provides powerful constructs for complex data structures but manages this complexity unseen by the user.

A graphical CAD shell, VEM, was developed that permits the user to inspect and alter the contents of the various cells in the data base in a natural manner. OCT also provides project management support in the form of change-lists, time stamps, and search paths. All evolving synthesis tools were provided with interfaces to the OCT data manager.

6.3. Module Generation Tools

One major effort during the Synthesis Project concerned the creation of a module generator that transforms logic equations at the behavioral level into a final mask layout. The important representation levels and the tools that perform the transformations between them are shown in Table 3.

Module generation starts from a DEC BDS behavioral description which is converted with the help of a language translator into BDSYN, a subset of BDS, developed to represent logic partitioned into combinational blocks and latches. From there, another translator maps the BDSYN description into BLIF, the Berkeley Logic Intermediate Format, by expanding high-level constructs into Boolean equations.

MIS, a multilevel interactive logic synthesis program, then restructures the equations to minimize area and to attempt to satisfy timing constraints. MIS first implements global optimization steps that involve the factoring of Boolean equations and multiple-level minimization. Local optimization is then performed to transform locally each function into a set of implementable gates. Finally, MIS includes a timing-optimization phase that includes delay approximation based on technology data and critical-path analysis.[19]

Design Function	Representation Level	Program Name
	Logic / Behavior	
Logic Synthesis		BDSYN MIS
	Logic / Gate	
Topology Optimization		TOPOGEN / EDISTIX GENIE / MKARRAY
	Symbolic / Graphic	
Layout Generation		SPARCS ZORRO
	Layout / Geometry	

Table 3. *Transformations in the module generation process.*

Once the logic equations have been optimized the module generators are responsible for optimal packing of the logic into regular or irregular array-based structures.[20] Some of these tools also consider slack times for critical paths.

TOPOGEN generates a standard-cell-like layout at the symbolic level from a description of a Boolean function in the form of nested AND, OR, INVERT expressions. A complex static CMOS gate is produced in which first the transistors and then the gates have been arranged so as to minimize the module area. The output from TOPOGEN can be inspected and modified with EDISTIX, a graphic editor using a symbolic description on a virtual grid.[10] The symbolic layout can then be sent to one of the compactors mentioned below.

A more sophisticated module generator is the combination of GENIE and MKARRAY. GENIE is a fairly general software package using simulated annealing to optimize the topology of a wide range of array design styles, including PLAs, SLAs, Gate Matrix, and Weinberger arrays. It handles nonuniform transistor dimensions, allows a variety of pin-position constraints, approximates desired aspect ratios by controlling the degree of column folding, and performs delay optimization. Its output is sent to the array composition tool, MKARRAY, which takes specifications of arrays of cells at the topological level. It then places the cells and aligns and interconnects all the terminals.

The modules at the symbolic level have to be spaced or compacted to a dense layout obeying a particular set of design rules.[21] SPARCS is a new constraint-based IC compaction tool that provides an efficient graph-based solution to the spacing problem. It can deal with upper bounds, user constraints, even symmetry requirements. It detects of over-constrained elements, and permits adjustable positioning of noncritical path elements

Another compactor under development, ZORRO,[22] works in two dimensions and is derived from the concept of *zone refining* used in the purification of crystal ingots. ZORRO passes an *open zone* across a precompacted layout. Circuit elements are taken from one side of this zone and are then reassembled at the other side in a denser layout.[10] This compactor gives denser layouts at the cost of longer run times.

6.4. Chip Assembly tools

All the tools described above are employed in the automatic synthesis of modules that are to be used in the design of an entire chip. Various tools have been developed to perform module placement, channel definition and ordering, global routing, and finally detailed routing.[6] These tools handle routing on multiple layers as well as over-the-cell wiring. Table 4 shows the sequence of transformations carried out on the representations in the OCT database from the original tentative floor plan to the final placement of all the modules and of the wiring in-between.

Layout Function		OCT Symbolic View
Floorplanning & Placement	--->	Placed
Channel Definition and Ordering	--->	Channel Defined
Global Router	--->	Routed
Detailed Router	--->	Unspaced
Spacing-Compaction	--->	Spaced

Table 4. *Functions of the chip composition tools.*

The TIMBERWOLF-MC[23] package performs the placement function using simulated annealing techniques. This program handles cells of arbitrary rectilinear shape; it accommodates fixed or variable shapes with optional bounds on aspect ratio, and accepts fixed, constrained, or freely variable pin locations.

CHAMELEON[24] is a new multi-level channel router that allows the specification of layer-dependent pitch and wire widths. It has as its primary objective the minimization of channel area and as its secondary objective the minimization of the number of vias and the length of each net. On two-layer problems it performs as well or better than traditional channel routers.

MIGHTY[25] is a 'rip-up and reroute' two-layer detailed switch-box router that can handle any rectagon-shaped routing region with obstructions and pins positioned on the boundary as well as inside the routing region. It outperforms all the known switch-box routers and even performs well as a channel router on problems with a simple rectangular routing region.

6.5. Results

Fifteen weeks is not enough time to build a complete synthesis system — thus we could not "press the button" on the last day of class and watch the layouts for the three SPUR chips pop out of the computer.

After the fifteen-week course period, all three chip designs had been converted from their original descriptions in 'N.2' or SLANG formats to BDS and inserted into our data management system. In the last few weeks of the course, these descriptions were then used to exercise the pipeline of tools that had been created in parallel. Major parts of these designs have run through various tool groups and produced results of widely varying quality. Improvements were quite visible as the tools were debugged and improved.

The major benefit of this course is a very good understanding of the bottlenecks and missing links in our system and concrete plans to overcome these deficiencies. Over all, the Synthesis Project of Spring 1986 must have been a positive experience; the students polled at the end of the term voted strongly in favor of continuing the Synthesis Project in the Fall term.

7. CONCLUSIONS

There is a broad spectrum of design styles that have proven successful for the construction of VLSI circuits and systems. For all these styles and for all the levels in the design hierarchy, good computer aided tools and data management techniques are indispensable. The emerging wave of CAD tools shows a trend to start at the lower hierarchical levels and to move upwards and to sweep the verification and analysis tools before the synthesis and high-level decision making tools. There is no doubt that eventually the whole design spectrum will be covered.

To make the emerging tools truly useful, the new tools should be developed in close cooperation with the user, or even by the user himself. Several iterations are normally needed to produce a good tool. The development of tools should be planned with this in mind.

Due to the changing nature of VLSI design, a design system will never be "finished". In order to keep up with the needs of the chip designers, the environment and the data representations must be kept flexible and extensible. A modular set of tools coupled to an object-oriented, integrated data base is a good solution.

Finally, we believe that the most effective tool development takes place under the forcing function of actual designs. In a recent push to integrate and complete our synthesis tools at Berkeley, we have used the chip set of an emerging VLSI-based multiprocessor workstation. This effort has given us a clear understanding of the tools that we are still missing. It has charted out enough work to keep us busy for several more years.

ACKNOWLEDGEMENTS

The development of our CAD tools is strongly supported by the Semiconductor Research Corporation.

References

1. J.D. Meindl, "Limits on ULSI," *Keynote Speech at VLSI'85, Tokyo*, Aug. 26, 1985.

2. G.E. Moore, *Quote at the First Caltech Conf. VLSI*, Pasadena, CA, Jan. 1979.

3. C.H. Séquin, "Managing VLSI Complexity: An Outlook," *Proc. IEEE*, vol. 71, no. 1, 1983.

4. MOSIS, "MOS Implementation Service," DARPA-sponsored 'Silicon Brokerage Service' at the University of Southern California's Information Sciences Institute , since 1981.

5. M. Stefik, D.G. Bobrow, A. Bell, H. Brown, L. Conway, and C. Tong, "The Partitioning of Concerns in Digital Systems Design," *Proc. Conf. on Adv. Research in VLSI*, pp. 43-52, M.I.T., Cambridge, MA, Jan. 1982.

6. J. Burns, A. Casotto, G. Cheng, W. Dai, M. Igusa, M. Kubota, U. Lauther, F. Marron, F. Romeo, C. Sechen, H. Shin, G. Srinath, H. Yaghutiel, A.R. Newton, A.L. Sangiovanni-Vincentelli, and C.H. Séquin, "MOSAICO: An Integrated Macrocell Layout System," *submitted to ICCAD-86*, Santa Clara, CA, Nov. 1986.

7. L.W. Nagel and D.O. Pederson, "Simulation Program with Integrated Circuit Emphasis," *Proc. 16th Midwest Symp. Circ. Theory*, Waterloo, Canada, April 1973.

8. J.D. Williams, "STICKS - A Graphical Compiler for High Level LSI Design," *AFIPS Conf. Proc., NCC*, vol. 47, pp. 289-295, 1978.

9. N. Weste and B. Ackland, "A Pragmatic Approach to Topological Symbolic IC Design," *Proc. VLSI 81*, pp. 117-129, Academic Press, Edinburgh, Aug. 18-21, 1981.

10. C.H. Séquin, "Design and Layout Generation at the Symbolic Level," in *Proceedings of the Summer School on VLSI Tools and Applications*, ed. W. Fichtner and M. Morf, Kluwer Acadmic Publishers, 1986.

11. S.A. Ellis, K.H. Keller, A.R. Newton, D.O. Pederson, A.L. Sangiovanni-Vincentelli, and C.H. Séquin, "A Symbolic Layout Design System," *Int. Symp. on Circuits and Systems*, Rome, Italy, May 1982.

12. S. Gutz, A.I. Wasserman, and M.J. Spier, "Personal Development Systems for the Professional Programmer," *Computer*, vol. 14, no. 4, pp. 45-53, April 1981.

13. A.R. Newton, D.O. Pederson, A.L. Sangiovanni-Vincentelli, and C.H. Séquin, "Design Aids for VLSI: The Berkeley Perspective," *IEEE Trans. on Circuits and Systems*, vol. CAS-28, no. 7, pp. 666-680, July 1981.

14. B.W. Kernighan and J.R. Mashey, "The Unix Programming Environment," *Computer*, vol. 14, no. 4, pp. 12-22, April 1981.

15. S.R. Bourne, "UNIX Time-sharing System: The UNIX Shell," *Bell Syst. Tech. J.*, vol. 57, no. 6, pp. 1971-1990, Jul.-Aug. 1978.

16. D. Harrison, P. Moore, A.R. Newton, A.L. Sangiovanni-Vincentelli, and C.H. Séquin, "Data Management in the Berkeley Design Environment," *submitted to ICCAD-86*, Santa Clara, CA, Nov. 1986.

17. The EDIF User's Group, "EDIF, Electronic Design Interchange Format, Version 1.0," Technical Report, Design Automation Department, Texas instruments, Dallas, TX, 1985.

18. M.D. Hill, S.J. Eggers, J.R. Larus, G.S. Taylor, G. Adams, B.K. Bose, G.A. Gibson, P.M. Hansen, J. Keller, S.I. Kong, C.G. Lee, D. Lee, J.M. Pendleton, S.A. Ritchie, D.A. Wood, B.G. Zorn, P.N. Hilfinger, D.A. Hodges, R.H. Katz, J.K. Ousterhout, and D.A. Patterson, "SPUR: A VLSI Multiprocessor Workstation," CS Division Report. No. UCB/CSD 86/273, University of California, Berkeley, CA, 1986.

19. R. Brayton, A. Cagnola, E. Detjens, K. Eberhard, S. Krishna , P. McGeer, L.F. Pei, N. Phillips, R. Rudell, R. Segal, A. Wang, R. Yung, T. Villa, A.R. Newton, A.L. Sangiovanni-Vincentelli, and C.H. Séquin, "Multiple-Level Logic Optimization System," *submitted to ICCAD-86*, Santa Clara, CA, Nov.1986.

20. G. Adams, S. Devadas, K. Eberhard, C. Kring, F. Obermeier, P.S. Tzeng, A.R. Newton, A.L. Sangiovanni-Vincentelli, and C.H. Séquin, "Module Generation Systems," *submitted to ICCAD-86*, Santa Clara, CA, Nov. 1986.

21. J.L. Burns, T. Laidig, B. Lin, H. Shin, P.S. Tzeng, A.R. Newton, A.L. Sangiovanni-Vincentelli, and C.H. Séquin, "Symbolic Design Using the Berkeley Design Environment," *submitted to ICCAD-86*, Santa Clara, CA, Nov. 1986.

22. H. Shin and C.H. Séquin, "Two-Dimensional Compaction by Zone Refining," *Proc. Design Autom. Conf., Paper 7.3*, Las Vegas, July 1986.

23. C. Sechen and A. Sangiovanni-Vincentelli, "TIMBERWOLF 3.2: A New Standard Cell Placement and Global Routing Package," *Proc. Design Autom. Conf., Paper 26.1*, Las Vegas, July 1986.

24. A. Sangiovanni-Vincentelli, D. Braun, J. Burns, S. Devadas, H.K. Ma, K. Mayaram, and F. Romeo, "CHAMELEON: A New Multi-Layer Channel Router," *Proc. Design Autom. Conf., Paper 28.4*, Las Vegas, July 1986.

25. H. Shin and A. Sangiovanni-Vincentelli, "MIGHTY: A 'Rip-up and Reroute' Detailed Router," *submitted to ICCAD-86*, Santa Clara, CA, Nov. 1986.

2

INTRODUCTION TO VLSI DESIGN
by
Jonathan Allen
Research Laboratory of Electronics
and
Department of Electrical Engineering and Computer Science
Massachusetts of Institute of Technology
Cambridge, Massachusetts 02139

INTRODUCTION

In recent years, modern VLSI technology has made possible the realization of complex digital systems on a single silicon chip. The ability to compress much digital logic complexity onto a single chip has provided the means to achieve substantial cost reductions, making this technology very attractive for designers of custom systems.

However, there have been two obstacles in achieving this goal. One has been the ability to obtain mask and wafer fabrication services in standard NMOS and CMOS technologies. The other obstacle has been the understanding of and access to modern design techniques which can effectively exploit this highly volatile technology. In recent years, "foundries" have appeared which have converted the designer's geometric mask representation into finished and packaged chips. Thus, the first obstacle has been effectively removed. At the same time, in light of this new availability of foundry services and the need to cope with increasingly complex designs, new techniques for circuit realization have evolved. These techniques have vastly increased the number of VLSI designers who can create working chips with acceptable performance.

In this chapter, an introduction to VLSI design is presented
by initially describing the basic devices, and then building
up several circuit forms that can be combined to provide
complete systems.   In this way, the designer is provided
with a basic circuit vocabulary which is needed to compose
complex systems.   A brief introduction to MOS fabrication
techniques is also provided, leading to an understanding of
lithography   and   the   design   rule   constraints   which
characterize   acceptable   geometric   mask   representations
("layouts").   These representations will then lead to wafer
processing and the final intended circuits.   Subsequent
chapters in this book build on this foundation, and present
design   tools   and   strategies   to   create   state-of-the-art
chips.

THE MOS TRANSISTOR

     In this introduction, unipolar transistors (commonly
known as MOS transistors) will be the central topic. These
transistors are four-terminal devices, although one terminal
is the connection to the bulk substrate which is often
implied in many designs.   The three remaining terminals are
used to realize a switch which connects two terminals (known
as the source and the drain) controlled by a third terminal
(known as the gate).   The cross-section of a typical MOS
transistor, with its analogy to a switch, is illustrated in
Figure 1.

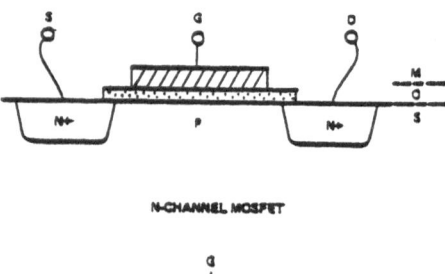

N-CHANNEL MOSFET

Figure 1.   An N-Channel MOS Transistor

The label "MOS" is utilized because of the three layers (indicated in Figure 1), and historically corresponds to a metal gate laying over a thin layer of silicon dioxide which, in turn, rests on the silicon substrate, or wafer. In contemporary practice, gate material is usually fabricated from polycrystalline silicon, and yet, the name "MOS" is still retained.

There are two basic types of MOS transistors. The transistor shown in Figure 1 is an n-channel MOS transistor, sometimes known as a MOSFET (MOS field-effect transistor). In this transistor, the source and drain connections are realized by n-type regions with many excess electrons, but the semiconductor substrate is "doped" in order to provide for a lack of electrons needed to satisfy the crystalline bonds, designated as "holes." A p-channel MOSFET can be obtained in a similar way by providing p-type regions (doped in order to provide many holes) situated in an n-type substrate. In order to understand the physical scale, the source and drain regions are typically several microns deep, the silicon dioxide is several hundred angstroms thick (1 Å $= 10^{-4}\mu$), and the gate material is approximately one micron thick. The distance from the source to the drain varies greatly according to the technology utilized. Usually, it is less than three microns, and in very aggressive practice, it may approach one micron or less.

There are two physical principles which lead to the switching action associated with this transistor structure. The first of these phenomena is the *pn junction* which is achieved whenever n-type material is juxtaposed with p-type material. An understanding of the basic physics of these junctions is assumed.[1] For the purpose of this discussion, it is sufficient to note that the junction provides a diode circuit action across its boundary, and hence, it may be either forward-biased or back-biased. In the normal use of these transistors, the junctions are kept back-biased, so that the source and drain regions are electrically isolated from the underlying crystalline substrate. In other words, for an n-channel transistor, the voltage of the source and drain regions is maintained equal to or greater than that of the substrate. Although there are many variations, it will be assumed that the substrate for n-channel MOSFETs is always maintained at ground, and that the source and drain voltages can range from ground to +5 volts. The gate voltage will also vary between ground and five volts.

The second physical principle used to provide the switching action of MOSFETs is the notion of *accumulation, depletion*, and *inversion*. The gate, together with the silicon dioxide and underlying substrate, form a capacitive structure which effectively allows the changing of the doping's polarity locally underneath the oxide under direct control of the gate voltage. With the substrate at ground and the gate negative, the majority carriers of the substrate (holes) are attracted, or "accumulated," at the substrate's surface directly adjacent to the silicon dioxide. As the gate voltage rises above ground, these holes are electrostatically pushed away from the surface, and hence, the substrate majority carrier population is "depleted" at the surface. This action continues until the gate voltage reaches a *threshold* value, where the holes have not only been pushed away from the surface, but where sufficient electrons have also been attracted to the surface by the positive gate potential to form a thin strip of n-type material below the oxide. At this point, an electrical connection has formed between the source-drain regions, since an n-type path exists between the two. It is important to note that there is still a distributed pn junction surrounding the source, drain, and the n-type channel under the gate. This means that the conducting region is still isolated from the underlying substrate, an important feature of the MOSFET. Typically, the source and drain are physically symmetrical, and conventionally in an n-channel MOSFET, the source is the end of the transistor with the lowest voltage.

A p-channel MOSFET works in a similar way. P-type source and drain regions are formed in an n-type substrate, and the substrate is connected to the highest available voltage (typically five volts). The source, drain, and gate voltages range between ground and five volts, as is the case with an n-channel MOSFET. For a p-channel device, the transistor is off (no conducting channel between source and drain) when the gate voltage is high or at five volts, since high voltage will attract the substrate majority carriers (electrons) to the surface, corresponding to accumulation. As the gate voltage is lowered, the electrons are pushed back, and holes are once again attracted until a thin channel region is created under the silicon dioxide between source and drain which is p-type, corresponding to inversion in that region. At this point, a conducting strip has formed between source and drain, and the corresponding switch is closed.

Both n- and p-channel MOSFETs are easily built, but their speed is constrained by the mobility of the majority carrier, which is substantially higher for electrons than for holes. Thus, when only a single type of MOSFET is used, n-channel technology is preferred since it is faster. CMOS (Complementary MOS) utilizes both n- and p-channel devices in a complementary way to achieve a variety of desirable circuit properties.

From the above discussion, it should be appreciated how the pn junction at either end of the device channel is dynamically altered through gate voltage action to provide either an open circuit between the source and drain, or a thin conducting path. Thus, the interplay of pn junction physics with electrically alterable regions of accumulation and inversion provide the essential device action. The gate voltage at which substantial inversion is obtained is called *the device threshold voltage*. In modern practice, this voltage can be set at any value. When the threshold is positive with respect to the source for n-channel devices (negative with respect to the source for p-channel devices), the device is called *an enhancement-mode transistor*, whereas in the opposite case, it is called *a depletion-mode transistor*. All MOS circuit technologies provide devices with at least one enhancement-mode threshold, typically in the neighborhood of one volt with respect to the source.

In most of this discussion, these MOSFETs will be regarded as either on or off. But, it is important to understand that there are typically three regions of recognized "on" behavior. When the channel is inverted, and the source-drain voltage is small, then the thickness of the inversion region is approximately uniform, and the current between the source and the drain varies *linearly* with the source-drain voltage using the gate voltage as a parameter. However, as the source-drain voltage rises, the thickness of the inversion region at the drain end diminishes with respect to the source end (due to less available vertical electric field). Increases in source-drain voltage lead to correspondingly smaller increases in source-drain current for a given gate voltage. In turn, this leads to a *transition* region which evolves into a *saturation* region where the inversion layer is completely pinched off at the drain end, and increased drain-source voltage does not yield any additional source-drain current for a given gate voltage. Keeping these three regions in mind, the overall current voltage characteristic of an individual MOSFET appears in Figure 2.

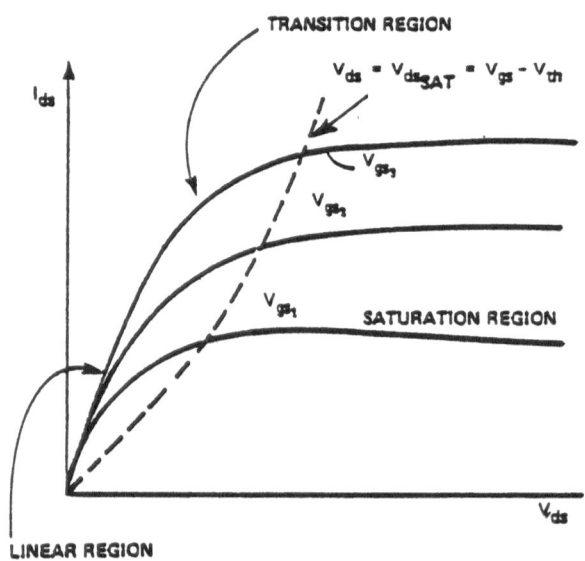

Figure 2. I_{ds} versus V_{ds} for an N-Channel MOSFET, with V_{gs} as a parameter.

These characteristics can be rigorously derived, and there are several books available on semiconductor device physics that can be referenced for a rigorous derivation.

At this point, a brief introduction to the individual MOSFET has been provided. In today's technology, these are extremely small switches which can be interconnected to provide a variety of circuit forms that will be discussed below. Certainly, these are not ideal switches. When the devices are off, the impedance between source and drain is very high, and is commonly several megohms. On the other hand, when the device is on, a minimal-sized device (where the channel's width equals its length) has an impedance of approximately 10,000 ohms. Nevertheless, these switches are used to manage charge distribution on the circuit nodes, and this substantial resistance is tolerable. It should be remembered that the gate threshold voltage can easily be set in the manufacturing process at any desired voltage. It is not unusual for four different thresholds to be used in contemporary circuits. In this discussion, it will always be assumed that the supply voltage for the circuit is five volts, and that the enhancement threshold is nominally one volt. Additionally, in NMOS circuits, a negative four-volt depletion-mode threshold is also assumed which is useful for load devices. These self-isolating devices, whose active switching channel area is only several square microns, can be used in large numbers to provide very complex circuits. Indeed, circuits containing over a million transistors are routinely fabricated. As a result, we look for regular and repeatable circuit structures that can be combined easily in order to minimize the design effort for these large circuits.

While there are many definitions of the term "VLSI," its most important connotation is the use of vast numbers of very small devices, together with the corresponding problems of fabrication and design. Effective custom design is possible only because general circuit and layout design techniques that provide dependable circuits with acceptable performance have evolved. In the sequel, many of these basic circuits will be shown, and how they can be easily expanded and generalized to provide all desired logical functions.

INVERTER CIRCUITS

Now that an understanding has been reached regarding
the switching action of a single MOS transistor, the ways in
which these transistors can be combined to provide desirable
circuit action can be considered. For digital circuits, the
simplest structure is the inverter which simply inverts the
input logic level. Thus, in MOS circuits where voltage can
range between ground and +5 volts, an input high level of
five volts should lead to a low output value of ground, and
vice versa. Furthermore, the circuits should switch rapidly
at an "inverter threshold" between these two configurations.
This inverter threshold should also be ideally halfway
between ground and five volts to provide the best noise
margins.

First, consider NMOS circuits. A simple inverter can
be obtained by using one MOSFET controlled by the input
voltage, as illustrated in Figure 3.

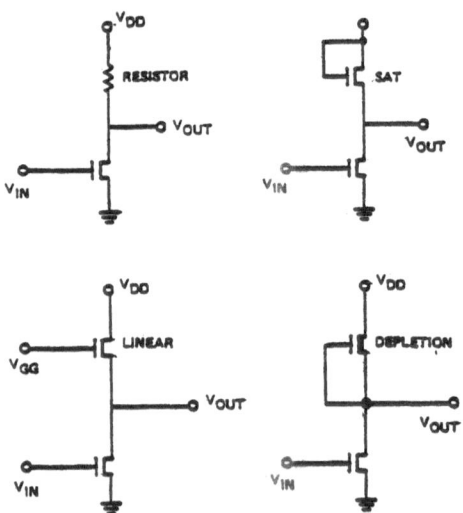

Figure 3.   Four NMOS inverter circuits.

A load resistor is connected to the high-voltage V_{dd}, commonly five volts. The device threshold for the enhancement MOSFET is typically one volt, so when V_{in} is less than one volt with respect to ground, the MOSFET is off and the output voltage rises to V_{dd}, since it is always assumed that the output drives a pure capacitive load with with no dc path to ground (unless stated otherwise). On the other hand, when V_{in} rises above the device threshold, the MOSFET turns on, and the output voltage is pulled down to a level determined by the ratio of the device on resistance, and the resistive load. For this reason, such a circuit is called *a ratioed circuit*. It is important to note that the low output voltage never reaches ground, since it is constrained by this ratio. Furthermore, it is essential that this low-output voltage be less than the device threshold voltage. Otherwise, the output is incapable of turning off a succeeding inverter pull-down device. Several other load structures have been proposed, and three of them are illustrated in Figure 3. They all replace the resistor load by a MOSFET pull-up in order to save space in the circuit, and to provide desirable circuit action. Usually, the least desirable load device is the so-called *saturated load*, where the gate of the pull-up is permanently connected to its drain. This pull-up device will certainly provide current to the output as desired, but as the output voltage rises, the gate-to-source voltage diminishes which provides less current as the output rises. Finally, the device is cut-off when the gate-to-source voltage reaches the device threshold, typically at one volt. This means that the output cannot rise above 4 volts, and it is very slow in approaching this value.

The next alternative load structure to be considered is the so-called *linear load*, where the load MOSFET has its gate connected to an additional supply voltage, V_{gg}, which is higher than five volts, with eight volts a typical value. The linear load performs well, and provides strong drive (high gate-to-source voltage) when the output is low, and allows the output to rise up to V_{dd}. On the other hand, a power supply is needed for V_{gg}, and wiring between the supply and the individual circuits must also be provided. Usually, these disadvantages are sufficient to rule out the use of a linear load.

Finally, the *depletion load* is observed, where a device with a deep depletion threshold is provided, and its gate is connected to its source. Since the deep depletion threshold is typically between -3 and -4 volts, and the gate-to-source voltage is constrained to be zero, this load device is always on. It provides constant drive into the output. It is still a ratioed circuit, but it provides the best rising output transient of these circuit alternatives. It is compact in its layout, and readily supported by the available technology. For this reason, the basic NMOS circuit style is referred as *enhancement/depletion ratioed NMOS*, or simply E/D NMOS. It is essential to remember that the enhancement/depletion inverter just described is still a ratioed circuit. The size of the pull-down enhancement MOSFET, as well as the size of the depletion-mode pull-up, must be appropriately sized so that the low output voltage is appreciably below the enhancement-mode threshold, providing a successful cut-off of succeeding inverter-like circuits. This sizing depends on the actual device threshold values provided by the technology, and hence, it will vary. But, a common value of the ratio of the channel length divided by the channel width of the pull-up, divided by the corresponding ratio for the pull-down, is four. This leads to the frequently mentioned "four-to-one" rule. In many NMOS circuits, however, the input to the E/D inverter is provided through a series ("pass") transistor. As a result, the input to the inverter never rises above four volts. When this occurs, the available gate-to-source drive voltage on the pull-down is diminished, and the sizing ratio previously mentioned must be doubled.

In CMOS technology, the presence of both n- and p-channel devices permits a more desirable inverter characteristic without the need for concern over device sizing which is dictated by ratioing considerations. The two inverter styles are contrasted in Figure 4.

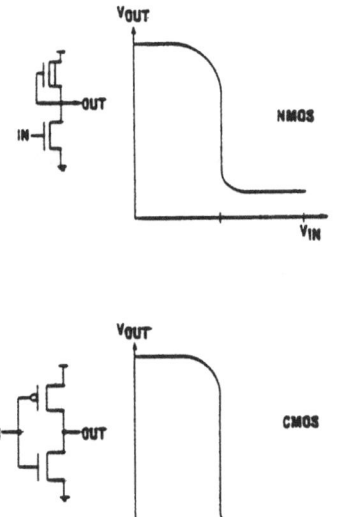

Figure 4.   NMOS and CMOS Inverter Characteristics.

By definition, in CMOS circuits, all input and output
signals vary between five volts and ground, or "from
rail-to-rail." Thus, when the input to a CMOS inverter is
high (five volts), the NMOS pull-down device is turned on
hard, and the p-channel pull-up device is turned off
completely. This causes the output to fall to ground. On
the other hand, when the input is low (at ground), the
p-channel device is turned on hard, and the n-channel device
is turned off completely. This causes the output to be
driven high to five volts, and leads to a symmetrical
inverter characteristic with very good noise margins. These
two characteristics are responsible for much of CMOS'
popularity. In addition, the two devices are statically
complementary, so that only one or the other is on at a
specific time, except during transitions. This minimizes
power consumption, and contrasts with the NMOS E/D inverter
where there is continuous static conduction from five volts
to ground when the pull-down device is on.

Both CMOS and NMOS circuits are highly utilized, although there is an increasing trend towards CMOS design styles. Generally, NMOS circuits are denser than CMOS circuits, and are often faster. CMOS circuits operate at much lower power levels, and provide superior noise immunity. Like all generalizations, however, these observations have exceptions, and continuing circuit innovation has led to the utilization of the best features of both approaches.

GENERALIZED INVERTER CIRCUITS

NMOS and CMOS inverters can be easily generalized to provide universal logic families. NAND and NOR circuits are illustrated in Figures 5 and 6.

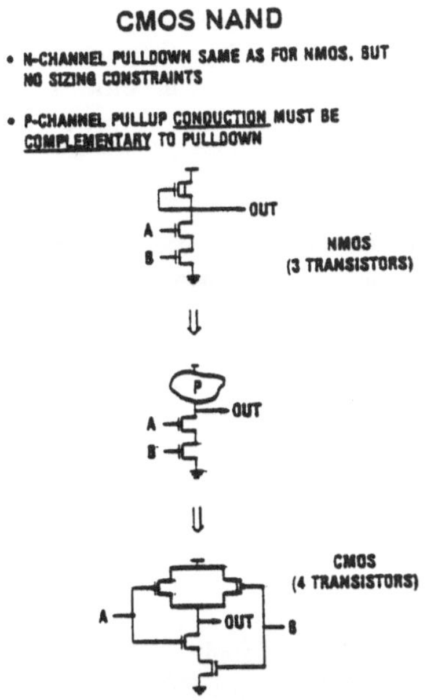
CMOS NAND
• N-CHANNEL PULLDOWN SAME AS FOR NMOS, BUT NO SIZING CONSTRAINTS
• P-CHANNEL PULLUP CONDUCTION MUST BE COMPLEMENTARY TO PULLDOWN
OUT
A
B
NMOS (3 TRANSISTORS)
P
OUT
A
B
CMOS (4 TRANSISTORS)
A
OUT
B

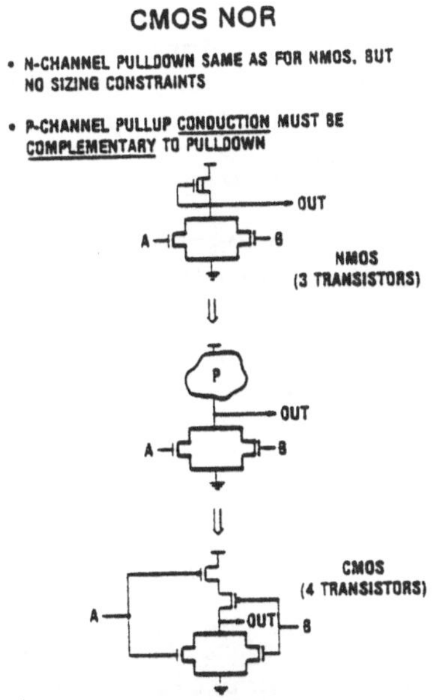
CMOS NOR
• N-CHANNEL PULLDOWN SAME AS FOR NMOS, BUT NO SIZING CONSTRAINTS
• P-CHANNEL PULLUP CONDUCTION MUST BE COMPLEMENTARY TO PULLDOWN
OUT
A
B
NMOS (3 TRANSISTORS)
P
OUT
A
B
CMOS (4 TRANSISTORS)
A
OUT
B

Figure 5. Relation of NMOS and CMOS NAND gates.
Figure 6. Relation of NMOS and CMOS NOR gates.

It can easily be verified that in the case of NMOS, two pull-down devices in series provide the two-input NAND circuit function, whereas two pull-down devices in parallel provide the two-input NOR function. It must not be forgotten that these are still ratioed circuits, and that all of the devices involved in these gates must be appropriately sized. This leads to an important *modular* property of the NOR gate, because once an NMOS inverter is properly sized, the addition of further pull-downs in

parallel cannot upset the ratio constraint. On the other
hand, if a ratioed NMOS inverter is extended to a NAND gate,
then sizing must be varied according to the number of
pull-down devices in series. One might be tempted to argue
for making the ratio sufficiently high that these sizing
conditions would not have to be varied with the number of
pull-down devices. But, this would lead to exceedingly slow
circuits. It is essential to remember that ratioed NMOS
circuits are inherently slower in the pull-up transient than
in the pull-down transient. Hence, care must be taken to
ensure that the pull-up transient is kept as short as
possible.

The NMOS NOR and NAND circuits be easily extended to
complementary (or classical) CMOS forms. One merely takes
the NMOS pull-down structure, and composes it with an
appropriate p-channel pull-up structure so that the total
pull-up structure is off when the pull-down is on, and vice
versa. This leads to a circuit form which satisfies the
criteria for CMOS, and is illustrated in Figures 5 and 6.
All input and output signals vary between ground and five
volts, and there is no static power dissipation between the
five-volt supply and ground.

The augmentation of inverter circuits to provide NAND
and NOR capability is easily extended to provide more
general logic capability. One simply builds a network of
pull-down transistors that will realize the desired logic
function, keeping in mind that all devices must be sized in
NMOS to provide the appropriate ratioing. For CMOS, the
NMOS pull-down structure is retained, and the corresponding
conduction complement pull-up (composed of p-channel
devices) is provided to achieve the desired complementary
circuit properties, but sizing is not critical, except for
speed.

The generalized inverter circuits described above are
widely utilized, and form the basis for much combinational
logic design. The augmentation of the basic inverter is
responsible for nearly all forms of combinational and
sequential logic designed in MOS technologies. Hence, it is
the fundamental, canonical form for both NMOS and CMOS
circuit design.

TRANSMISSION GATES

In the previous section, it has been shown that networks of transistors can be composed in such a way to realize a broad variety of logic functions in both NMOS and CMOS. These transistors provide a network of switches controlled by appropriate gate signals that either discharge the output node toward ground, or charge it toward the V_{dd} voltage which is typically five volts. Hence, each transistor participates in paths between the output node and either ground or V_{dd} which can be selected by appropriate gate signals. It is also possible to introduce transistors as switching elements in *series* with the output node. Then, they are considered *transmission gates* between the output of one logical gate and the input of other gate structures. Thus, in NMOS, it is possible to insert an n-channel MOSFET in series between the output of one inverter and the input of another. This is commonly done to clock signals from one stage to another, as in a shift register. More complicated networks of NMOS transistors can also be used as transmission gates between output and input nodes of logic gates, and hence, they can provide additional combinational logic function to that provided by generalized inverter structures. Typically, these networks provide selection or routing capability between logic gates, but the gate signals are frequently ANDed with clock signals in order to provide an overall timing discipline.

The CMOS transmission gate is inherently more complicated than the NMOS single-transistor transmission gate, since it must retain its rail-to-rail signal swing. As illustrated in Figure 7, the CMOS transmission gate is realized by connecting an n-channel and p-channel transistor back-to-back, and making their corresponding gate signals complementary.

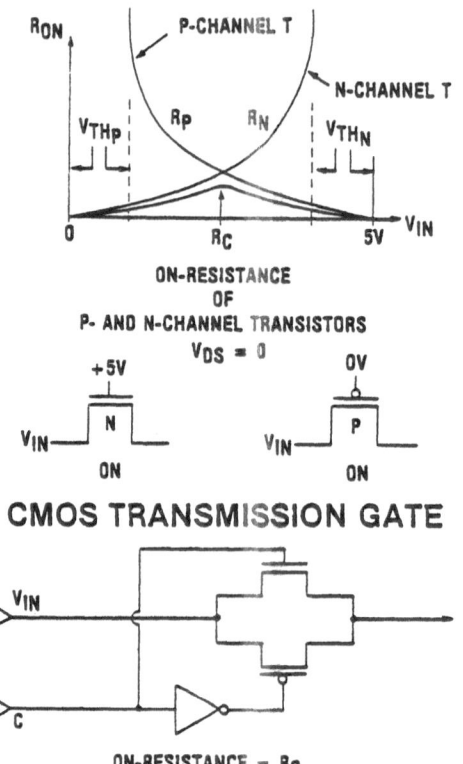

CMOS TRANSMISSION GATE

Figure 7.   The CMOS Transmission Gate.

One of the two devices is always on when the control signal
is high, and both devices are off when the control signal is
low.   In this way, the complete logic swing of the input is
conveyed to the output.   This property is indispensable for
many CMOS logic forms.

One of the most clever of these circuits is the six-transistor CMOS-exclusive OR gate, shown in Figure 8.

**EXCLUSIVE OR
WITH
TRANSMISSION GATE
AND
"PROGRAMMABLE INVERTER"**

	A	B	A⊕B
①	0	0	0
	0	1	1
②	1	0	1
	1	1	0

Put the Circuits for Case 1 and Case 2 together:

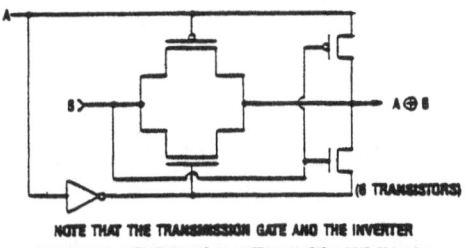

NOTE THAT THE TRANSMISSION GATE AND THE INVERTER
SUPPLY THE OUTPUT VALUE IN MUTUALLY EXCLUSIVE FASHION.

Figure 8. A CMOS Exclusive OR Circuit.

This circuit realizes the four possibilities indicated in its truth table through the complementary selection of two separate subcircuits. When the input A is low, then the output is just the value of the alternate input B, as illustrated in the truth table, and hence, this value can be routed to the output of the circuit by means of a transmission gate. On the other hand, when the input A is high, then the output is the inverse of B. This must be realized by an inverter. The circuit is ingenious because when the transmission gate action is desired, the inverter structure is effectively decoupled from the output, whereas when the inverter action is needed, the transmission is simply turned off. The ability to electronically decouple an inverter in a circuit is peculiar to CMOS, and has no corresponding NMOS analogy.

FULL ADDERS

The use of generalized inverter structures, as well as transmission gates is amply illustrated by the design of full adder circuits. These are complex modules in an overall system, and indeed, the basic building blocks of MOS circuits are rarely more complicated than these examples.

The NMOS full adder circuit is interesting because the carry output is generated first, and then used as partial input to the output sum bit circuit, as shown in Figure 9.

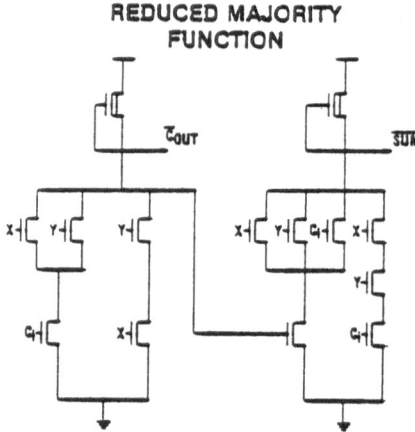

Figure 9.   NMOS Full Adder.

It is easy to understand the carry out circuit, since it must only test all possible pairs of the three input signals to determine if any two are simultaneously true. If the carry out signal is false, then the sum circuit merely checks to see if any one of the three inputs is true. In addition, it must also check to see if all three inputs are simultaneously true. It should be noticed that this circuit is easy to "read," since the way in which the individual transistors contribute to the logic switching function of the overall circuit can be readily determined. This circuit also has the useful property of generating the carry out before the sum bit, which helps to speed up parallel adders.

A CMOS full adder could be obtained from the previously described NMOS full adder by retaining the NMOS pull-down structures, and generating the needed PMOS conduction complements as pull-up structures. If this is done, the result is a rather cumbersome circuit which is both large and slow. A better circuit is illustrated in Figure 10.

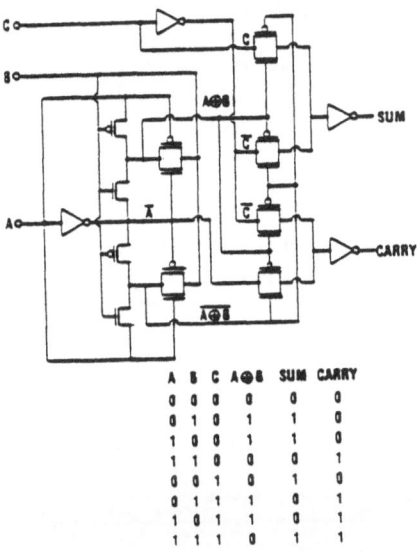

A	B	C	A⊕B	SUM	CARRY
0	0	0	0	0	0
0	1	0	1	1	0
1	0	0	1	1	0
1	1	0	0	0	1
0	0	1	0	1	0
0	1	1	1	0	1
1	0	1	1	0	1
1	1	1	0	1	1

Figure 10.   CMOS Full Adder Circuit.

Here the previously described CMOS exclusive OR gate is utilized as a basic building block for the full adder. It is combined with a succession of transmission gates to achieve the output sum and carry signals. Both true and complement forms of the exclusive OR of inputs A and B are generated in order to improve the circuit's speed. It can be readily appreciated that without previous understanding of the properties of both CMOS transmission gates and the CMOS exclusive OR circuits, this circuit would be exceedingly difficult to understand.

PROGRAMMED LOGIC ARRAYS

It has been shown that both generalized inverter structures and networks of transmission gates can be utilized to provide combinational logic capability. It has also been mentioned that in many designs, several hundred transistors may be utilized, and hence, a design strategy must be available to permit the designer to readily implement large amounts of combinational logic in a straightforward manner. The solution to this problem is provided, in part, by *program logic arrays* (PLAs) which are highly regular in repeatable structures, and can be generated automatically from a logic specification by appropriate CAD tools. The basic building block for an NMOS PLA is the NOR gate, which has already been shown to have an admirable modular property whereby changing the number of pull-down devices in parallel does not change the requirement for overall gate sizing. In a PLA, logic inputs are provided to a so-called "AND" plane which is realized through an array of NOR gates. This AND plane generates a group of product terms which are then combined in an "OR" plane, which in turn is realized by a regular array of NOR gates. The overall structure is shown for one example in Figure 11.

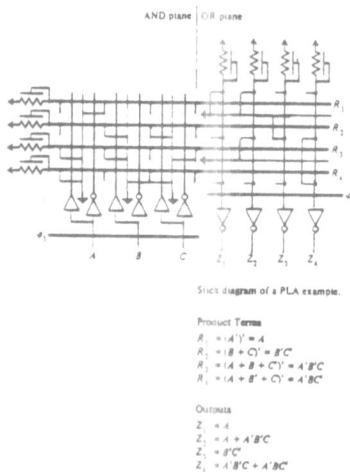

Figure 11.   An NMOS PLA Circuit.

Both the true and complement vales of all logic inputs are routed vertically across the AND plane consisting of a set of distributed NOR gates, one for each desired product term. These product terms are then bussed horizontally into the OR plane, where they control another set of distributed NOR gates that feed a set of output inverters, providing the correct polarity for the output logic signals. It cannot be overemphasized that the regular layout strategy afforded by NMOS NOR-NOR PLAs is due to the modular sizing property of NOR gates. Notice, however, that while these NMOS NOR gates can all be sized by using fixed load devices, a CMOS PLA which is constructed by using the normal complement techniques would not be feasible because the size of the pull-up structures would be logic-dependent. For many years, this was a substantial obstacle in using regular PLA circuit forms. It has only been recently overcome through the introduction of so-called "precharge-evaluate" circuit techniques which will be described later in this chapter.

NMOS PLAs are not only highly *regular*, but they are also *universal* since any logic function can be realized as a sum of products. In addition, recent research has provided several invaluable CAD tools, including facilities for logic optimization and array size minimization through techniques of input encoding,[2] and line folding,[3] both of which are beyond the scope of this chapter. It is sufficient to say that dense and highly efficient PLA structures can be readily obtained from an input logic specification, leading to the broad use of PLAs in a wide variety of circuits. Thus, good performance is achieved with minimum design effort together with the ability to delay binding many logic decisions until very late in the design process of the overall chip.

CLOCKED CIRCUITS

Both NMOS and CMOS clocked circuits are widely used in order to provide not only sequential circuits, but also

combinational circuits through the use of complex timing
disciplines and precharging techniques. A simple example of
an NMOS sequential circuit is provided by a series of
inverters connected by clocked pass transistors. A
two-phased nonoverlapping clock discipline, as illustrated
in Figure 12, is frequently used.

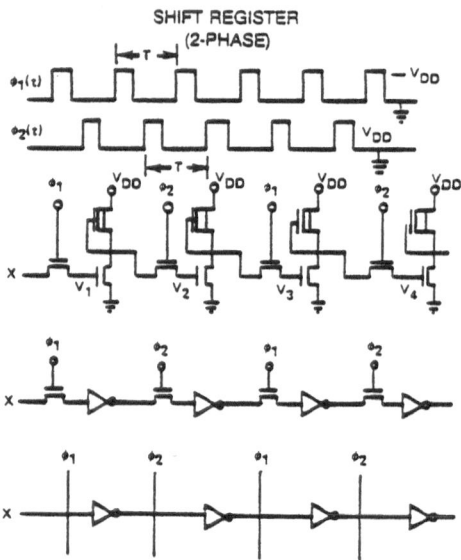

Figure 12.   An NMOS Shift Register Circuit.

This leads to shift register capability as shown in the
figure. This circuit can easily be generalized to provide
shifting in both directions, as well as many other variants
including push-down stack operation. It should be verified
that a complete clock cycle (including both phases) succeeds
in shifting an input bit in Figure 12 through two inverters,
and hence, propagates that bit one bit position in the
overall shift register. This circuit provides memory at
each bit position, and the corresponding values are
maintained by isolating charge at the input of

inverter pull-down gates. This isolation is clearly
achieved by the use of the clocked pass transistors, and
such stored charge values can be safely maintained for
intervals as long as several milliseconds. This ability to
achieve dynamic memory through charged storage on isolated
nodes is a dominant feature of MOS circuit design. Of
course, fully static designs are easily realized by using
generalized inverter structures. An example of a simple
set-reset latch is illustrated in Figure 13, and is
completely static.

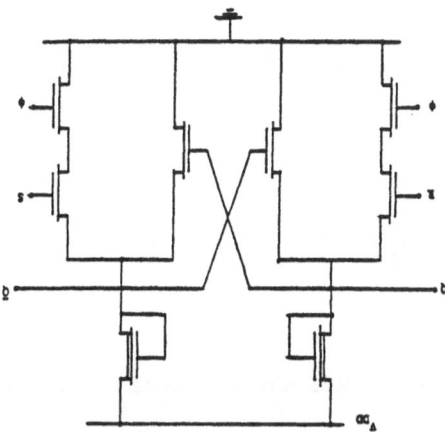

Figure 13. An NMOS Clocked SR Flip-Flop.

Such designs can also be clocked, and are useful when the
clock frequency is very low, or when the clock is
occasionally turned off.

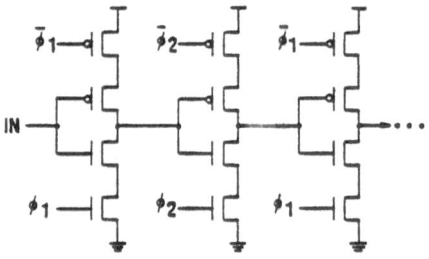

ϕ_1 AND ϕ_2 NON-OVERLAPPING

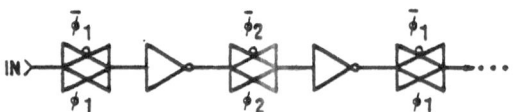

Figure 14.   CMOS Shift Register Circuits.

CMOS inverters can also be augmented by either transmission
gates or series clock devices in both the pull-up and
pull-down to provide for shift registers, as seen in the
NMOS case.   Examples of such circuits are illustrated in
Figure 14, where the transmission gate case is represented
symbolically in the lower part of the circuit, including the
commonly used symbol for a CMOS transmission gate.

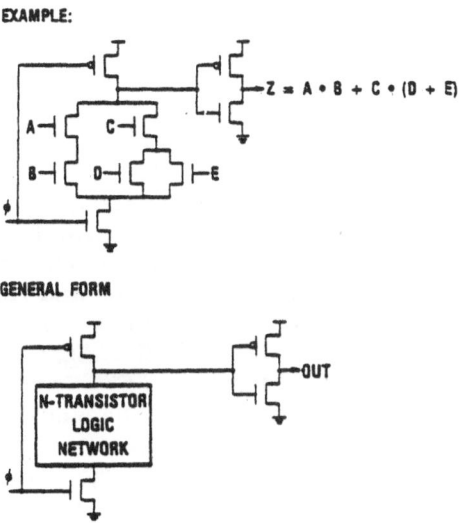

Figure 15.   "Domino" CMOS Circuits.

Perhaps the most prevalent use of clocking in CMOS is provided by the so-called "precharge-evaluate" circuits. Of the many forms of such logic, the "domino" form is widely utilized.   The idea is to build a CMOS circuit that approaches the density of an NMOS circuit by avoiding the use of a complementary PMOS pull-up structure through the introduction of a precharge-evaluate timing discipline.   An example of such a circuit is illustrated in Figure 15.

The basic logic network is achieved through the interconnection of n-channel MOSFETs in the pull-down structure. Just below this logic network, however, an evaluate transistor is connected between the logic network and ground. A static p-channel pull-up transistor is provided, and both this transistor (called the "precharge" transistor) and the evaluate transistor (connected from the logic network to ground) are controlled by a clock signal. Additionally, the output of the precharge-evaluate network is run through an inverter to provide the final output. The circuit action is as follows. While the clock signal is low, the precharge transistor is on and charging up the input to the output inverter. Hence, it is holding the output signal low. When the clock signal goes high, the precharge transistor is turned off, and the pull-down evaluate transistor is turned on. Depending on the values of the input logic signals, the logic network together with the evaluate transistor may discharge the input to the output inverter. This causes output Z to go high. This circuit has the advantage of using an entirely n-channel logic network, which is fast, and minimizing the number of p-channel devices. Thus, for complicated logic functions, the total number of transistors is minimized, and the circuit action is very fast. The output inverter is provided to ensure that all output signals in domino CMOS are low prior to the evaluate phase. If this was not the case, race conditions between circuits could easily arise, thus destroying the utility of this circuit form. CMOS domino circuits, together with other variants of this form, are widely used in order to retain the rail-to-rail and low-power advantages of CMOS, while obtaining the high density and fast circuit speed of NMOS. Hence, this form should be viewed as an innovative compromise motivated by performance considerations. Undoubtedly, many other schemes of this sort will appear in the future. Since these circuits employ a simple one-transistor pull-up, and since sizing considerations are not necessary, both NAND and NOR precharge-evaluate circuits can be combined to provide efficient CMOS PLA structures.

FINITE-STATE MACHINES

In light of both the combinational logic and the memory capability developed in previous sections, it is a simple matter to construct finite-state machines. Many different forms are possible, but PLAs are widely augmented for this purpose. In a finite-state machine, combinational logic is needed to generate the primary outputs from the primary inputs and the present state. The next-state information is also derived from the primary inputs and the present state. This set of combinational logic can be readily realized by a PLA, as illustrated in Figure 16, and the state memory is easily obtained through the clocking of register cells realized with inverters and pass transistors.

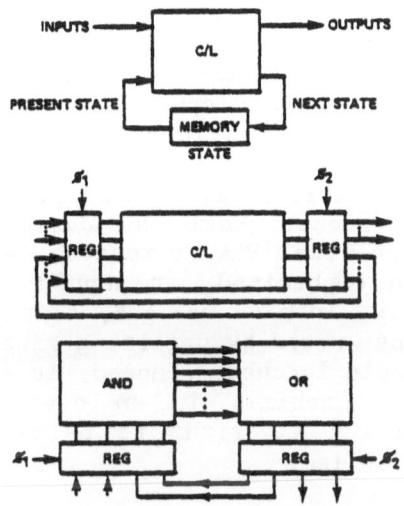

Figure 16. Finite State Machine Forms.

In this way, it is a simple matter to add clocking to a PLA, together with feedback connections needed to propagate the next-state information to the updated present state. As is the case with any finite-state machine, the designer starts by constructing a transition table which shows all possible sets of inputs together with present states, and the consequent next state and output signals that must be generated in each case. From this information, product terms are determined that will be generated in the AND plane of the corresponding PLA, together with the way in which these are combined in the OR plane to provide the next-state information and output logic signals.

Like PLA design, a variety of optimizations can be applied to finite-state machine design. Perhaps the most significant is the optimization that allows the selection of codes for the set of states within the machine. It is often possible to perform state assignment in a way that eliminates rows of the PLA, thus reducing the circuit's area and increasing its speed. Input logic lines can often be multiplexed if they are used in a state-dependent way, and often the PLA can be further reduced by providing some external logic, particularly when output signals are strictly state-dependent. While the design of finite-state machines based on the PLA structure with appropriate feedback has been emphasized, it should be realized that this is unnecessary. In many designs, simple state machines can be readily constructed by utilizing generalized inverter structures and simple latches. Indeed, it is common to have many simple state machines in an overall chip design, particularly when the algorithm to be implemented admits considerable parallelism.

At this point, several methods for the design of combinational logic together with effective techniques for sequential circuit design, including finite-state machines, have been introduced. This is by no means an exhaustive treatment of useful MOS circuit forms, but it does provide the most basic and useful circuit forms in current practice.

Using these circuits, a designer can readily build
complete chips of considerable complexity, aided by
comprehensive logic simulation and timing verification CAD
tools which are currently available. A design is conceived
as an overall architecture comprised of several blocks,
which in turn can be successively broken down into a set of
basic building blocks similar to those considered in this
discussion. The designer then estimates the size and
performance of these circuits by using layout techniques not
yet discussed, in order to provide a floor plan for the
overall design. Different circuit styles and aspect ratios
can be *explored* with a view toward an effective, overall
layout that will fit within the chip size constraints.
Thus, it is important to consider how the circuit designs
previously described can be transformed into mask
specifications for a particular technology which serves as
the fabrication "foundry" to produce final, packaged chips.
In order to appreciate the nature of mask specifications, it
is important to have an understanding of the wafer and chip
fabrication process, including lithographic techniques.
These can then be used to motivate a set of geometric design
rules which restrict the class of all possible mask
specifications to those which can yield correctly working
circuits. Once the design rules are understood, then the
mask layout can be generated, keeping in mind that there are
many degrees of freedom afforded by the design rules which
can lead to many possible mask layouts corresponding to a
given circuit form. With final mask layout as the goal
(considered the designer's interface to the fabrication
foundry), a discussion of the integrated circuit fabrication
process follows.

INTEGRATED CIRCUIT FABRICATION

The common meeting point between design and fabrication
is the specification of the integrated circuit masks. On
one hand, these masks are interpreted by the integrated
circuit fabrication process which actually forms the
physical circuits themselves in and on a silicon wafer. On
the other hand, the designer must transform the circuit
representation of a design to a set of closed shapes

on the several masks, which obey the fabrication design
rules, but will also lead to the needed structures in the
final physical circuit.   There are several programs that
will transform special circuit forms or logic specifications
into mask specifications.   But, in the most general case,
the designer must accomplish this transformation.   The
design rules which constrain the mask representation, or
layout, are a set of simultaneous linear inequalities, and
it is certainly possible to approach mask layout without any
concept of the inequalities' origins.   Nevertheless, a
knowledge of lithography and fabrication is highly useful in
terms of being able to relate the mask set to the final
physical circuit, and in terms of providing justification
for the substantial variety of design rules used in
contemporary practice.   Modern silicon integrated circuit
fabrication is an example of planar fabrication because all
operations are performed at the surface of a silicon wafer.
The starting silicon wafers are purified monocrystalline
silicon doped with a p-type dopant (e. g., boron) for
n-channel MOSFETs, or an n-type dopant (e. g., phosphorus)
for p-channel MOSFETs.   These wafers are usually round and
are several inches in diameter, and several hundred microns
thick.   The main thickness requirement is to permit handling
during manufacturing as opposed to any needed electrical
properties.   Through a variety of processes such as
diffusion, oxide growth, ion implantation, deposition, and
other techniques, regions of doped silicon, thin and thick
oxides, polysilicon, metal, and interlayer contacts are
readily achieved through the utilization of many processing
steps modulated and controlled by the features of the
individual masks.

     The masks are typically realized in glass with some
type of opaque material (e. g., chromium) that serves to
define the shapes on the mask.   For each step in which a
pattern on the mask must be utilized in processing on the
wafer, a material known as photoresist is deposited on the
wafer's surface and then baked.   Ultraviolet light is then
shone through the mask onto the resist on top of the wafer,
exposing the polymer bonds of the resist material.   In the
case of "positive" resist, the bonds in the polymer break
down where the light strikes the resist, and these regions
can   be   easily   etched   away.   In   "negative"   resist,

the complementary regions are etched away. In this way, the patterns contained on the masks (which have been generated by a designer or a CAD program) are transferred to the surface of the wafer, where additional processing can then occur.

In the overwhelming majority of cases, processing proceeds according to the so-called *silicon gate self-aligned process*. A detailed description of this technology is beyond the scope of this chapter,[4] but its essential features can be mentioned. Thin oxide is grown wherever a channel region is needed, and thick oxide (or field oxide) is used in all other places except where diffusion into the substrate will occur. Even before the thin oxide is grown, ion implantation is used to adjust thresholds in those channel regions where tailoring is required, such as in depletion loads. Following the threshold adjustments and the growing of oxide, the polysilicon layer (used for both interconnect and as a gate material) is laid down *before* the actual diffusions for the source-drain regions are made. This sequencing is critical because after the polysilicon is patterned, and only then, is the diffusion into the substrate performed. This results in a doping of the polysilicon material itself, but the dopant will not penetrate the thin oxide in channel regions underneath the gate polysilicon. For this reason, the process is called "*self-aligned*," and earlier problems involving mask registration between the diffusion layer and polysilicon layer are avoided. This leads to lowered capacitances and vastly improved performance. Once the diffused regions are formed, then contact cuts are formed through thick oxide between the metal, polysilicon, and diffusion layers. Finally, metal is deposited over the entire wafer and patterned using a mask that characterizes the metal interconnect.

The process sequence described above is for silicon gate self-aligned NMOS, and is widely used. Obviously, only the briefest outline of this process has been given since a typical fabrication sequence will involve several hundred steps in a carefully controlled procedure.

Furthermore, there are countless variations, and although
one layer of metal interconnect has been suggested, in fact,
two levels of metal interconnect are now commonplace. Some
bipolar technologies currently available use four levels of
metal interconnect. The increased numbers in the levels of
metal provide a highly desirable interconnect material, and
vastly improved ease of placement and routing of logic
signals, clock signals, and supply power. By the end of
this decade, it can be expected that three levels of metal
interconnect will be common.

The CMOS fabrication sequence is complicated by the
need to provide both an n-type substrate for p-channel
devices, as well as a p-type substrate for n-channel
devices. This need has been satisfied in many ways, but a
common technique is to provide a *well* of diffusion by means
of diffusion or ion implantation within the main wafer
substrate for the alternate polarity of doping as opposed to
the one provided by the main substrate. Thus, CMOS is often
built in an n-type wafer with a p-type well formed by deep
diffusion. On the other hand, an n-well in a p-substrate
can also be utilized, or both kinds of wells can be built in
a lightly doped epitaxial substrate built on top of a common
wafer substrate. These choices are illustrated in Figure
17.

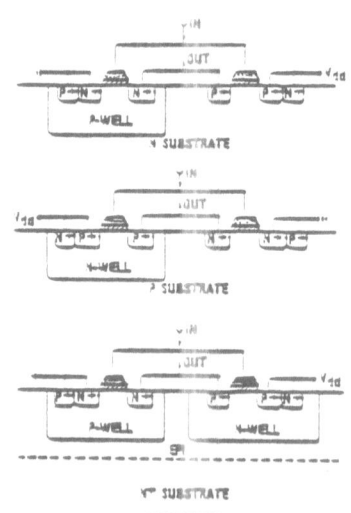

Figure 17.   CMOS Inverter Cross-Sections.

In the case of the p-well approach, the performance of the n-channel devices is somewhat diminished due to the fact that p-type dopants are introduced into the n-doped substrate, leading to diminished mobility in contrast to the corresponding NMOS case. This corresponds to "dragging down" the performance of the n-channel devices to that of the p-channel devices. On the other hand, if an n-well is used in a p-type substrate, then the p-type devices (already slow due to the hole mobility) are made even slower. But, the good performance of the n-channel devices is preserved. In many designs, particularly those where p-channel devices are minimized, as seen in "domino" CMOS, this approach is highly desirable. The last approach uses two wells, each optimized for one device or another, and is highly desirable but more expensive to produce. It also avoids the introduction of parasitic silicon-controlled rectifiers which are common in the single-well processes, and lead to a phenomenon known as *latch-up* once the silicon-controlled rectifier fires. The characterization and control of latch-up are beyond the scope of this chapter, but comprehensive treatments of this phenomenon are available.[5] It is clear, that in years to come, processing innovations will continue to lead to fabrication techniques that either minimize or eliminate latch-up. It is important to note that the latch-up problem is unique to CMOS, and is not found in the normal NMOS processes.

Figure 17 also illustrates another need of CMOS design, namely the use of *body ties* or *plugs* to establish the appropriate substrate potential. It should be noted that diffusions of like-type dopant are made into each substrate in order to make a good contact into both the well and base substrate. Thus, in the case of p-well CMOS, a p-diffusion is made into the p-well in order to establish its potential at ground. On the other hand, in the main wafer substrate, an n-diffusion must be made into the n-type substrate in order to provide an ohmic contact from the power supply, V_{dd}, into the substrate. Many descriptions of CMOS circuit design omit these body plugs, but they are indispensable to proper circuit action, and lead to the introduction of one additional mask in CMOS that provides for the correct contact diffusions. The problems introduced by the placement of body plugs constitute another difficulty associated with CMOS that is not present in NMOS.

DESIGN RULES

Once the decision is made to adopt a particular
fabrication technology, then the designer must utilize a set
of *design rule* constraints peculiar to that process.   As
mentioned above, these constraints are a set of simultaneous
linear inequalities which cover the minimum size of mask
shapes, their separation (both on one mask and between two
masks), and a variety of constraints concerning overlap and
surrounding borders.   A set of design rules will not be
presented here, since they can be readily appreciated from
the particular process chosen.   When custom design is
undertaken, the designer must specify those forms on the
relevant masks that lead to instantiation of transistors,
interconnect layers, and contact interconnections between
layers in a way that is consistent with the design rules,
and which will lead to desirable electrical properties in
the final circuit.   Because there are so many possible mask
layout geometries corresponding to a given circuit design,
designers are often frustrated by the wide variety of
choices available for the layout specification.   This
problem is usually solved by adopting some *layout discipline*
or strategy which introduces conventions to simplify the
layout process without reducing the layout efficiency.   This
is a good example of where circuit area and performance is
sometimes compromised in the name of design efficiency, and
is a common result of the increasing complexity of
integrated circuits.   It is simply impossible to lavish
attention on the individual transistors of a design, and so
there is a great need for repeatable and regular structures
which can serve as the basic building blocks in designs.
Program logic arrays are good examples of such structures,
but register arrays, bit-slice arithmetic logic units, and a
variety of memory structures are other examples which are
frequently utilized.   As CAD tools progress, a variety of
programs have been produced which generate highly optimized
versions of the various building blocks described in this
chapter.   These can be called *macro generators*, and can be
effectively utilized by the designer to quickly generate
mask specifications which correspond to the basic circuit
forms   discussed.   These   are   available   even   for

complex structures such as array multipliers and floating-point units, and can provide both logic optimization as well as layout efficiency. A natural extension of this macrogeneration process is to procedural means for the generation of layout of an *entire* chip from an input functional specification. This process is sometimes called *silicon compilation*, and while it is in its very early stages, it can be expected to lead to acceptable circuit performance with minimal human design time in some restricted, yet important, cases. Nevertheless, it is essential that the designer understand the basic aspects of MOS circuit design, fabrication, and layout at a detailed level in order to critically evaluate and appreciate the growing capability of these CAD programs.

REFERENCES

1. For a basic treatment of device physics in general, and
    pn junctions in particular, see:

    a. Streetman, B. G., Solid State Electronic Devices,
        2nd edition (Prentice-Hall, Englewood Cliffs,
        New Jersey, 1980).

    b. Neudeck, G. W., The PN Junction Diode. Modular
        Series on Solid State Devices, Volume II
        (Addison-Wesley, Reading, Massachusetts, 1983).

2. The basic idea of input encoding for logic arrays has
    been extensively developed at IBM. See:

    Fleisher, H., and Maissel, L. I., "An Intro-
    duction to Array Logic," *IBM Journal of Research
    and Development*, March 1975, 98-109.

3. Optimization of PLA area by means of "folding" has been
    extensively studied. See:

    a. Hachtel, G. D., Newton, A. R., and
        Sangiovanni-Vincentelli, A., "Techniques for
        Programmable Logic Array Folding," *Pro-
        ceedings of the 19th Design Automation Con-
        ference*, 1982, 147-155.

    b. Wong, D. F., Leong, H. W., and Liu, C. L.,
        "Multiple PLA Folding by the Method of Simu-
        lated Annealing," *Proceedings of the Custom
        Integrated Circuits Conference*, 1986, 351-355.

4. The silicon gate self-aligned process is described in
    numerous books on integrated circuit technology:

    a. El-Kareh, B., and Bombard, R. J.,   Introduction
        to VLSI Silicon Devices. (Kluwer Academic
        Publishers, Boston, Massachusetts, 1986).

b. Ghandhi, S. K., VLSI Fabrication Principles. (John Wiley & Sons, New York, New York, 1983).

c. Sze, S. M. (Ed.), VLSI Technology. (McGraw-Hill, New York, New York, 1983).

5. CMOS latch-up is studied comprehensively in:

Troutman, R. R., Latchup in CMOS Technology. Kluwer Academic Publishers, Boston, Massachusetts, 1986 .

General References on VLSI Design.

These books cover much of the material presented in this chapter. The study guide by Allen is comprehensive over these topics.

Allen, J., Introduction to VLSI Design. Video Course Study Guide. (MIT Video Course, Cambridge, Massachusetts, 1985).

Annaratone, M., Digital CMOS Circuit Design. (Kluwer Academic Publishers, Boston, Massachusetts, 1986).

Glasser, L. A., and Dobberpuhl, D. W., The Design and Analysis of VLSI Circuits. (Addison-Wesley, Reading, Massachusetts, 1985).

Mavor, J., Jack, M. A., and Denyer, P. B., Introduction to VLSI Design. (Addison-Wesley, Reading, Massachusetts, 1983).

Mead, C., and Conway, L., Introduction to VLSI Systems. (Addison-Wesley, Reading, Massachusetts, 1980).

Weste, N., and Eshraghian, K., Principles of CMOS VLSI Design. (Addison-Wesley, Reading, Massachusetts, 1985).

3

Simulation Tools for VLSI

Christopher J. Terman

Symbolics, Inc.
Cambridge, Massachusetts

1. Introduction

Simulation plays an important role in the design of integrated circuits. Using simulation, a designer can determine both the functionality and the performance of a design before the expensive and time-consuming step of manufacture. The ability to discover errors early in the design cycle is especially important for MOS circuits, where recent advances in manufacturing technology permit the designer to build a single circuit that is considerably larger than ever before possible. This paper reviews the simulation techniques which are commonly used for the simulation of large digital MOS circuits.

Simulation is more than a mere convenience—it allows a designer to explore his circuit in ways which may be otherwise impractical or impossible. The effects of manufacturing and environmental parameters can be investigated without actually having to create the required conditions; the ability to detect manufacturing errors can be evaluated beforehand; voltages and currents can be determined without the difficulties associated with attaching a probe to a wire 500 times smaller than the period at the end of this sentence; and so on. To paraphrase a popular corporate slogan: without simulation, VLSI itself would be impossible!

To use a simulator, the designer enters a design into the computer, typically in the form of a list of circuit components where each component connects to one or more *nodes*. A node serves as a wire, transmitting the output of one circuit component to other components connected to the same node. The designer then specifies the voltages or logic levels of particular nodes, and calls upon the simulator to predict the voltages or logic levels of other nodes in the circuit. The simulator bases its predictions on models that describe the operation of the components. To be successful, a simulator requires the following characteristics of its models:

- The underlying model must not be too computationally expensive since the empirical nature of the verification provided by simulation suggests that it must be applied extensively if the results are to be useful.

- Component-level simulation is necessary to accurately model the circuit structures found in MOS designs. This allows the designer to simulate what was designed—an advantage, since requiring separate specification of a design for simulation purposes only introduces another opportunity for error.†

- The results must be correct, or at least conservative; a misleading simulation that results in unfounded confidence in a design is probably worse than no simulation at all. Here, we must trade off the conflicting desires of accuracy and efficiency.

Three of the more popular approaches to modeling are:

- component models based on the actual physics of the component; for example, a transistor model that relates current flow through the transistor to the terminal voltages, device topology, and manufacturing parameters of the actual device.

- component models based on a description of the logic operation performed by the component, *e.g.*, NAND and NOR gates.

- component models based on hybrid approaches which aim to approximate the predictions made by physical models, at a computational cost equal to that of gate-level models.

The first type of model is found in circuit analysis programs such as ASTAP [Weeks73] or SPICE [Nagel75] which try to predict the actual behavior of each component with a high degree of accuracy. Current circuit analysis programs do the job well, perhaps too well; at no small cost, they provide

† This is not a strict requirement; simulators which employ higher-level models often provide a "circuit compilation" phase to translate the component-level circuit description into (hopefully) equivalent high-level elements—the circuit compiler essentially automates the construction of a separate specification for simulation.

a wealth of detail, at sub-nanosecond resolution, about the voltage of each node and the amount of current through each device. (For example, a properly calibrated circuit analysis program is able to predict, within a few per cent, the amount of current that flows through an actual transistor.) This level of detail would swamp the designer if collected for the entire circuit while simulating, say, a microprocessor. Fortunately, the designer is spared this fate, since the computational cost of circuit analysis restricts its applicability to circuits with no more than a few thousand devices.

One solution to the problem of simulator performance is to adopt a simpler component model, such as the gate-level model introduced above. This approach works well when dealing with implementation technologies that adhere to gate-level semantics (*e.g.*, bipolar gate arrays). However, MOS circuits contain bidirectional switching elements that cannot be modeled by the simple composition of Boolean gates. Since many of the circuit techniques that make MOS attractive for LSI and VLSI applications take advantage of this non-gatelike behavior, it is important to model such circuits accurately.

Hybrid simulators provide the essential information (functionality and comparative timing) for large digital circuits by using models that bridge the gap between the gate-level and detailed models discussed above. Two hybrid models are examined in detail:

- a *linear model* in which a transistor is modeled by a resistance in series with a voltage-controlled switch. The state of the switch is controlled by the voltage of the transistor's gate node.

- a *switch model*, similar to the linear model, except that a resistance value is limited to one of two quantities: 0 for n- and p-channel devices, and 1 for depletion devices.

There are numerous simulation tools, usually called *functional* or *behavioral* simulators, which support design at higher levels of abstraction. Many of the tools are based on some type of hardware description language (HDL) and provide a catalog of high-level building blocks such as registers, memories, busses, combinational logic elements, etc. Cause and effect relationships are maintained by these simulators usually through some sort of event-driven scheduling of functional blocks, but detailed timing information is limited to major clock phases. There exist general purpose languages— SIMULA or LISP, for example—which can also be very useful in simulating architectures at this high level. Since these tools are not specific to VLSI, this paper will not discuss them further, but most architectures are simulated at this high level before being committed to silicon, and information from these simulations is often used to verify subsequent lower-level simulations.

A final word of warning: all simulators are based on models of actual behavior. As with any model, discrepancies are likely to exist between the model predictions and the actual behavior of a circuit. The tools described here attempt to be conservative, but this cannot be guaranteed. Thus, it is important that the designer become acquainted with the inner workings of the models and their shortcomings. The tools perform a calculation one could do by hand, only faster and with greater accuracy and consistency—they should *not* be treated as black boxes.

The following sections focus on each of the modeling approaches. The discussion provides an introduction to the various topics, many of which are major disciplines in themselves. References are provided at the end of the paper for those who wish to pursue a particular topic in more depth. In particular, [SV80] and [Newton80] are excellent detailed introductions to circuit-level simulation; [Vlach83] is a good reference for the potential implementor.

2. Circuit-level Simulation

The goal of circuit-level simulation is to provide detailed electrical information about the operation of a circuit. As mentioned above, this level of detail is a two-edged sword: such detail is necessary to successfully design some components, but it is so expensive to generate that only selected pieces of a circuit can be simulated at this level. Fortunately there is some opportunity to trade speed for accuracy; some of these techniques are outlined below.

Circuit-level simulators all use the same basic recipe:

(i) Choose the *state variables* of the circuit, *e.g.*, capacitor voltages and inductor currents. The values of these variables will tell us all we need to know about the past behavior of the circuit in order to predict future behavior.

(ii) Construct a set of circuit equations which embody constraints on the values of the state variables and are derived from physical laws, *e.g.*, Kirchoff's voltage and current laws, or descriptions of a component's operation, *e.g.*, Ohm's law.

(iii) Solve the circuit equations given initial conditions. The presence of nonlinearities and differentials usually dictates the use of numerical solution methods. The discussion below is oriented towards *transient analysis* where we wish to compute the values of the state variables over some time interval, say, $0 \leq t \leq T$.

One can see from the above that circuit-level simulators actually embody solutions to two separate problems: first, how to model the circuit with component models and circuit equations, and second, how to solve the circuit equations arising from application of the models. In principle, one can address the problems separately—this is the approach taken below—but in practice one would not choose, say, highly accurate component models and then adopt a quick and dirty solution technique. Often simulators leave the final choice to the designer: SPICE, for example, has a repertoire of three different MOSFET models and allows one of two integration methods to be used during the solution phase.

Some *mixed-mode* simulators allow one to mix and match models and solution techniques within a single simulation run, using simpler, more efficient approaches for some pieces of the circuit, saving the more accurate (and expensive) analysis for "critical" subcircuits. A subcircuit may be "critical" because it lies along some path of particular interest to the designer, or because it is not modeled correctly by simpler techniques. It is this latter possibility which gives rise to an important caveat when using mixed-mode simulators: in the interests of efficiency, one may erroneously assume a subcircuit can be modeled as a noncritical component, leading to incorrect predictions without any indication that the simulation has gone awry.

Circuit simulators are prized for their "accuracy," in particular, SPICE is often used as the metric against which other simulation techniques are judged. However, it is important to keep in mind that circuit simulators are *not* infallible oracles concerning circuit performance; what they offer is accurate solutions for systems of equations. Unfortunately, the equations themselves are often not nearly as accurate as their carefully derived solutions. Over-simplified component models, missing parasitics, poorly chosen input waveforms, etc., all contribute to erroneous predictions. Subtler effects such as the inability to deal with unknown voltages (*e.g.*, from a storage element which has just been powered up) are often overlooked when a printout displays voltages to five decimal places. These points are worth keeping in mind as we explore how circuit simulators do their job.

2.1 Circuit equations and component models

The first step in circuit simulation is the building of a set of circuit equations from a designer-supplied description of the circuit. The description is usually in terms of *nodes* and *branches*, where a branch is formed whenever a component connects two nodes. If a circuit contains n nodes and b branches, the straightforward formulation results in $2b + n$ equations [Hachtel71]:

- Kirchoff's current law provides n equations involving branch currents,

- Kirchoff's voltage law provides b equations relating branch voltages to node voltages, and

- models for individual components provide b equations describing the relationship between branch parameters and other parameters of the circuit,

involving a total of $2b+n$ unknowns, *i.e.*, n node voltages, b branch voltages and b branch currents.

The behavior of a component, which gives rise to the last set of b equations mentioned above, can in general be expressed in terms of an interconnection of *ideal elements*. Only a small repertoire of ideal elements needs to be supported by the simulator:

resistive elements, characterized by algebraic equations relating the branch currents to the branch voltages. This category includes *two-terminal elements* (*e.g.*, resistors and independent voltage and current sources) where the behavior of a branch is described in terms of the branch current or voltage, and *four-terminal controlled sources* where the behavior of a branch is described in terms of the voltage or current across a second pair of control terminals.

energy storage elements, characterized by algebraic equations relating the state of the storage elements (charge for a capacitor, flux for an inductor) to one of the branch variables (voltage for a capacitor, current for an inductor). These equations in turn lead to differential equations relating the branch currents to the change in branch voltages (capacitors), or vice versa (inductors).

If the algebraic equation that describes the operation of an ideal element can be graphed as a straight line passing through the origin, the element is said to be *linear*, otherwise the element is deemed to be *nonlinear*.

There are well known techniques for solving the sets of linear equations arising from circuits containing only linear resistive elements; however, most VLSI circuits also contain nonlinear components (*e.g.*, MOSFETs) and storage elements (*e.g.*, node capacitances). The next section describes techniques for solving the resulting system of mixed nonlinear algebraic and differential circuit equations. Before embarking on that discussion, we briefly turn our attention to the component models themselves.

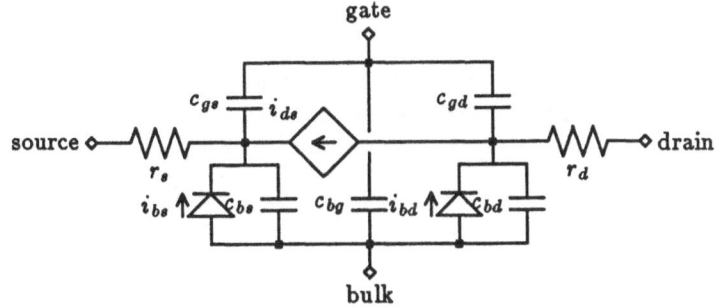

Figure 2.1. *Equivalent network for an n-channel MOSFET*

Figure 2.1 shows how one might model an n-channel MOSFET as an interconnection of ideal elements [pg. 315, Vlach83]. A wide variation in accuracy and computational overhead is possible, depending on how the various parameters of the model are determined.

Simplified models treat many of the parameters as constants computed from user-supplied information about the device, *e.g.*, its geometry. Often, the relatively inexpensive Shichman-Hodges model is used to approximate the current conducted by the device:

$$i_{ds} = \begin{cases} 0 & v_{gs} - v_{th} < 0 & \text{"off"} \\ \frac{\kappa}{2}(v_{gs} - v_{th})^2 & 0 \le v_{gs} - v_{th} \le v_{ds} & \text{"saturated"} \\ \kappa(v_{gs} - v_{th} - \frac{v_{ds}}{2})v_{ds} & v_{gs} - v_{th} > v_{ds} & \text{"linear"} \end{cases}$$

where v_{th} is the threshold voltage of the MOSFET and

$$\kappa = \frac{w}{l}\mu C_{ox} \approx \frac{w}{l}(25\frac{\text{microamps}}{\text{volt}^2})$$

is a constant that depends on the width w and length l of the particular MOSFET under consideration. The Level 1 model of SPICE implements a simplified model.

Analytic models—*e.g.*, the Level 2 model of SPICE—strive to be very accurate and so require detailed information about device geometries, electrical properties of the materials, temperature, etc. These models include deviations from the first order theory used in simpler models, *e.g.*, threshold adjustments, variations in charge distribution along the channel, mobility variations, channel length modulation, short channel effects, subthreshold conduction, charge storage and capacitive effects, etc. Computation of the model parameters is usually quite involved—the Fortran code for the SPICE Level 2 model is seven times as long as that for the Level 1 model. [SV80] reports that for circuits up to 500 nodes, the majority of the simulation time can be spent evaluating the device models. Analytic models are useful for performing simulation at different process corners, since the physical parameters which need to be varied are used directly in the modeling equations.

Empirical models also strive for accuracy, but are based on a curve-fitting approach for deriving the underlying parameters; the model parameters may have no direct physical interpretation. The models are somewhat cheaper computationally, and are very useful when trying to match predictions with actual measured values (since one can simply work backwards from the measured values). The Level 3 model of SPICE is a semi-empirical model.

Table-driven models can be used to avoid the expense of evaluating complicated formulas at simulation time; with some care the computer time spent during model evaluation can be reduced by an order of magnitude. The MOTIS simulator [Chawla75] used two tables in calculating the source/drain current of a MOSFET:

$$i_{ds} = T_d(v_{gs} - T_b(v_{bs}), v_{ds})$$

Other parameters needed for the MOSFET model were also approximated using these tables. To use the tables, voltages must first be quantized; [Fan77] reports that reasonably accurate results are possible if one provides about 100 different entries for each dimension. No interpolation is required unless the solution technique attempts to iterate to convergence (see below), in which case a quadratic interpolation scheme should be used. The contents of the tables can be derived from analytic formulas or taken directly from device measurements.

Finally, *macromodels* can be used to characterize the terminal behavior of larger functional blocks such as op amps, logic gates, modulators, etc. with considerable savings in the number of modeling elements that have to be managed by the simulator. This, of course, translates into reduced simulation time. Macromodels are particularly useful for those portions of the circuit which are not of direct interest to the designer.

2.2 Circuit-level simulation techniques

Numerical solution techniques for systems of mixed nonlinear algebraic and differential circuit equations have received a lot of attention. This section provides a quick tour of the standard approaches; more detail and good bibliographies can be found in [SV80] and [Vlach83].

We can write our system of circuit equations as

$$F(\dot{\mathbf{x}}, \mathbf{x}, t) = 0$$

where $\mathbf{x}$ is the vector of state variables with $2b + n$ elements. In general, no closed form solution exists, so instead we develop the solution incrementally for a series of time steps $t_0, t_1, ..., t_n$ using a *linear multistep method* which computes $\mathbf{x}_{n+1}$ from values of $\mathbf{x}$ and $\dot{\mathbf{x}}$ at earlier steps, subject to the initial conditions specified by the designer. Many simulators use single step methods which involve only the information from the previous time step.

Explicit methods use equations incorporating only information from earlier steps

$$\mathbf{x}_{n+1} = f(\mathbf{x}_n, \mathbf{x}_{n-1}, ...)$$

and so are quite efficient since one already knows all the parameters needed to compute new values for the state variables. The Forward Euler method is particularly inexpensive:

$$\mathbf{x}_{n+1} = \mathbf{x}_n + (\Delta t)\dot{\mathbf{x}}_n.$$

where Δt is the step size. Unfortunately, explicit methods suffer from numerical stability problems and are not suitable for high accuracy simulators. Nevertheless, explicit methods have been used successfully as the basis for EMU, an inexpensive timing analysis program [Ackland81] tailored for digital MOS circuits. The authors of EMU argue that the errors introduced during simulation do not get out of hand due to the high-gain and voltage-clamping properties of digital LSI circuitry. Use of these techniques is rewarded by a speed improvement of several orders of magnitude.

Implicit methods build a set of equations

$$\mathbf{x}_{n+1} = f(\mathbf{x}_{n+1}, \mathbf{x}_n, \mathbf{x}_{n-1}, ...)$$

which improve the stability of the solution technique, but at the cost of having to solve sets of simultaneous equations in order to determine values of the state variables for the next time step. The accuracy of the solution is affected by the size of the time step; in general, if the method is chosen with care, the error can be driven to zero as the size of the time step is decreased. Circuit simulators choose the time step so as to bound the accumulated error while maximizing the progress through time.†

Returning to the problem of solving the circuit equations, using an implicit multistep method we can reduce the original problem to one of solving sets of equations of the form

$$F(\dot{x}_{n+1}, x_{n+1}, t_{n+1}) = 0.$$

Since we don't know $\dot{x}_{n+1}$, we use an *integration rule* to eliminate the differentials from the set of equations. For example, using the Trapezoidal rule:

$$\dot{x}_{n+1} = 2\frac{x_{n+1} - x_n}{\Delta t} - \dot{x}_n$$

Substituting, we are left with a set of algebraic nonlinear equations

$$F(x_{n+1}, t_{n+1}) = 0.$$

These equations can linearized by applying a multidimensional version of Newton's method called Newton-Raphson (NR), and then solved using Gaussian elimination or LU decomposition. Several NR iterations may be required to achieve a sufficiently accurate solution.

To summarize, the steps used by traditional circuit analysis programs for computing values of the state variables at a new time step are

   a) update values of independent sources at t_{n+1}

   b) apply integration formulae to capacitors and inductors

   c) linearize nonlinear elements using using NR

   d) assemble and solve linear circuit equations

† The successful analysis of "stiff" circuits—those which contain both quickly responding and slowly responding components—requires the use variable sized time steps; also, the choice of which linear multistep method to use has a large effect on the amount of computation required.

e) check for convergence in NR method; if not achieved, return to step (c)

f) check error estimates, and modify time step if necessary.

The search for new circuit analysis algorithms is still underway. This is hardly surprising, given the important role played by circuit analysis in the design of integrated circuits. For example, [Newton83] describes two promising new approaches: *iterated timing analysis* which applies relaxation techniques at the nonlinear equation level, and *waveform relaxation* which applies the same techniques at the differential equation level. Both approaches offer the potential for dramatic improvement in simulator performance and seem particularly suitable for implementation on multiprocessors.

3. A linear model for MOS networks

This section discusses RSIM, a logic-level simulator built with the goal of being able to simulate entire VLSI circuits with acceptable accuracy. Rather than perform a detailed simulation of each transistor's operation, RSIM uses the linear model to directly predict the logic state of each node and to estimate transition times if the nodes change state. The net effect is to trade some accuracy in the predictions for an increase in simulation speed. When the linear model is conservatively calibrated, its predictions can be used to identify potential problem circuits in need of more accurate analysis. A large portion of most circuits pass the scrutiny of RSIM and so the expense associated with detailed simulation of the whole circuit is avoided.

The transistor model in RSIM can be quite simple since it is only used to predict the final logic state of a node and the length of time each state transition takes. As an example of how the model works, consider a simple inverter: one can think of the effective resistance of its component devices at any moment as

$$R_{eff:pullup} = \frac{v_{ds:pullup}}{i_{ds:pullup}} \qquad R_{eff:pulldown} = \frac{v_{ds:pulldown}}{i_{ds:pulldown}}.$$

Although the effective resistances of the transistors change as their terminal voltages vary, it might be possible to use "average channel resistances" to characterize the transistors' behavior. The other salient feature of a transistor's operation is its switch-like behavior: with certain voltages on a transistor's terminal nodes it makes no connection at all between its source and drain terminals—the transistor is "off". As the relative terminal voltages change, the transistor turns "on", conducting current between its source and drain terminals. Of course, the transistor is more "on" at some times than others, but distinctions between different "on" states might be ignored for simplicity.

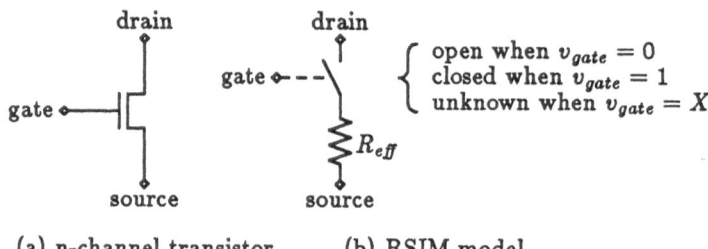

(a) n-channel transistor   (b) RSIM model

Figure 3.1. RSIM *model for an n-channel MOSFET*

One can build on the observations made above to construct a linear n-channel transistor model for the simulator, shown in figure 3.1. It is easy to tabulate the sort of connection that exists between the source and drain terminals as a function of the gate voltage:

$$R_{ds} = \begin{cases} R_{eff} & \text{switch closed} & (v_{gate} = 1) \\ \infty & \text{switch open} & (v_{gate} = 0) \\ [R_{eff}, \infty] & \text{switch unknown} & (v_{gate} = X) \end{cases}$$

Note that uncertainty about the state of the switch leads naturally to an interval describing the resistance of the source-drain connection. In fact, all the network calculations use interval arithmetic, and the bounds of the resulting intervals are used when converting voltages to logic states, etc.; no other mechanisms are needed to deal successfully with X states in the network. Models for other types of transistors differ in the way the position of the switch is determined from v_{gate}.

The effective resistance, R_{eff}, is determined separately for each transistor and depends on type of simulation one wishes to perform. In the linear model, R_{eff} depends on

width, length; dimensions of the active transistor area. Non-linear effects make R_{eff} a more complicated function of the transistor geometry than simply length divided by width.

type. Most MOS circuits contain more than one type of transistor. The different types are distinguished by, among other things, different values for their threshold voltage. Since the current conducted by a transistor is a function of its threshold voltage, the modeling resistance naturally depends on the transistor type.

context. Accuracy in choosing the effective resistance can be improved by distinguishing several contexts in which a transistor may appear: for example, an enhancement transistor can be used as a pulldown or source-follower in addition to the more general pass gate configuration. Surprisingly few contexts need to be recognized to encompass a large portion of digital MOS designs.

The determination of R_{eff} is made once for each transistor and does not depend on any dynamic properties of the circuit to be simulated. During simulation using the linear model the only device information used about a transistor is its effective resistance.

Voltages in this model are quantized into one of three values; this corresponds to our intuition for *digital* logic and greatly simplifies the simulation calculations. If all node voltages are normalized to fall in the range $[0, 1]$, then the possible quantized values are

0   logic low—voltages in the range $[0, v_{low}]$;

1   logic high—voltages in the range $[v_{high}, 1]$;

X   intermediate or unknown voltages.

where v_{low} and v_{high} are the predetermined logic thresholds.

How is the value of a node determined? RSIM characterizes the effect of the network on a particular node by the Thevenin circuit equivalent for all pieces of the network that directly influence the value of the given node (see figure 3.2).

V_{thev} a voltage interval $[V_{thev-}, V_{thev+}]$ in the range $[0, 1]$ specifying the possible voltages the output node may have.

R_{drive} a resistance interval in the range $[0, \infty]$.

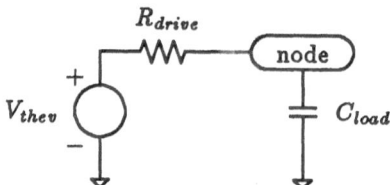

Figure 3.2. *Equivalent circuit for a network node*

V_{thev} and R_{drive} are, in general, intervals since the equivalent transistor resistances from which they are derived might themselves lie in an interval. A node's final value is determined by comparing V_{thev} with the low and high logic thresholds and choosing the appropriate logic state. If the calculation of a node's final value yields a result different from the node's current value, a transition has been discovered and the simulator must predict how long it will take for the node to cross some predefined switching threshold v_{thresh}. Given the model shown in figure 3.2, an obvious choice for the transition time is $R_{drive}C_{load}$. With suitable definitions of R_{drive} and C_{load}, this is the approach adopted by RSIM.

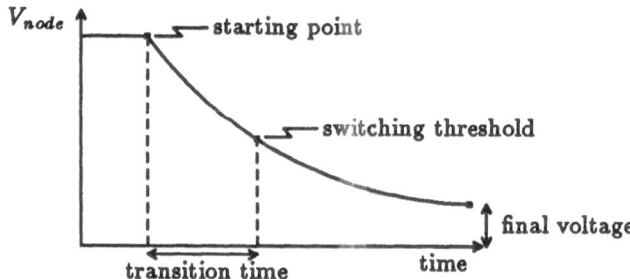

Figure 3.3. *R_{eff} used to predict (1) transition time and (2) final voltage*

Actually RSIM uses not one, but three effective resistances for each transistor. To see why, recall that RSIM is trying to predict the transition time and final voltage, as shown in figure 3.3. One would like to calibrate the model to give accurate predictions for both parameters, but that is impossible with a single set of resistances. To solve this problem RSIM uses three resistances for each transistor:

R_{static} used when calculating V_{thev}, the node's final voltage.

R_{dynlow} used when calculating $R_{low-drive}$ for high-to-low transitions.

$R_{dynhigh}$ used when calculating $R_{high-drive}$ for low-to-high transitions.

Two "dynamic" resistances are used so that the asymmetric behavior of "pass" devices can be accurately predicted. All three values (V_{thev} and two versions of R_{drive}) are calculated simultaneously, so the extra overhead introduced by multiple modeling resistances is not very large.

3.1 The RSIM simulation algorithm

Basic to the operation of RSIM is the notion of an *event*. An event specifies (i) a node in the network, (ii) a new logic state, and (iii) a time at which the node's value is to be changed to the new logic state. RSIM maintains a list of events, sorted by time, that tells what processing remains to be done. Whenever the user changes an input, an event is added to the list; when the list is empty, the network has "settled" and RSIM waits for further input.

When started on an initial event list, RSIM sequentially processes events from the list, stopping (1) when the list is empty, (2) when a node the user is tracing changes value, or (3) when the specified amount of simulated time has elapsed. Processing an event entails

(a) removing the event from the event list.

(b) changing the node's state to reflect its new value, generating the appropriate reports.

(c) calculating any consequences, *i.e.*, new events, resulting from the node's new value. First all nodes that might be affected by the change are found and marked—this requires a tree-walk of the network starting at the source and drain nodes of transistors for which the changing node is the gate. The tree-walk follows source and drain connections, stopping at input nodes or non-conducting transistors. For each marked node two calculations are made: (1) a charge-sharing calculation that models changes of state due to charging/discharging of the node capacitances and (2) a final-value calculation that determines the node's ultimate state.

As seen in step (c) the network is naturally partitioned into stages, each stage consisting of nodes "shorted" together by source-drain connections. The values for all nodes in a stage are recalculated whenever nodes are added or removed from the stage because of a transistor turning on or off. Since nodes are only added to the event list when their values change, portions of the circuit unaffected by the current set of changes are not re-evaluated—the algorithm is event driven.

A node can have two events pending:

(1) a charge-sharing event describing an immediate change in the node's state due to the redistribution of charge among the capacitors for nodes in the current stage. This type of event is generated only when a node is added to a stage (*i.e.*, when a transistor turns on).

(2) a final-value event describing what the final, driven state of the node will be. This type of event is generated when $R_{drive} < \infty$.

The charge-sharing calculation models what happens when two or more charged nodes in different logic states are connected. In this case, all the connected nodes will reach the same logic state; this state is determined by the relative capacitances and initial logic states of the nodes in the stage. For example, if a large (high capacitance) node such as a data bus were connected by a pass transistor to a small node such as the input to a register cell, then the small node would "share" the charge of the large node as its final value regardless of the charge it had initially. In such cases, the charge-sharing value is determined from two capacitance intervals

$$C_{high} = [C_{high-}, C_{high+}] \quad \text{and} \quad C_{low} = [C_{low-}, C_{low+}]$$

computed during the tree walk of the surrounding network. C_{high} (C_{low}) reflects the total amount of capacitance in the stage which is currently charged high (low); this value might be an interval due to neighboring nodes with an X value or connections through transistors with a gate node at X. For example, during the tree walk, the capacitance of an X node is added to both C_{high+} and C_{low+} (but not C_{high-} and C_{low-}); capacitance of a 0 node is added to both C_{low-} and C_{low+}; and so on. Similarly, capacitance information about subcircuits on the other side of transistors with a gate node at X affects only C_{high+} and C_{low+}. The capacitance intervals are used to determine the charge-sharing value of the node:

$$V_{share-} = \frac{C_{high+}}{C_{low-} + C_{high+}} \qquad V_{share+} = \frac{C_{high-}}{C_{low+} + C_{high-}}$$

Since nodes at logic state X contribute an undetermined amount of charge to the result, V_{share} is an interval whose bounds represent conservative assumptions about the actual values of X nodes, *i.e.*, we want to make V_{share-}

as large as possible and V_{share+} as small as possible. This interval is compared with the logic thresholds when calculating the charge-sharing value:

$$\text{Charge-sharing value} = \begin{cases} 0 & V_{share+} \leq v_{low} \\ 1 & V_{share-} \geq v_{high} \\ X & \text{otherwise} \end{cases}$$

If $R_{drive} < \infty$, and the node is not an input, the final state of a driven node is calculated from the V_{thev} interval $[V_{thev-}, V_{thev+}]$:

$$\text{Final value} = \begin{cases} 0 & V_{thev+} \leq v_{low} \\ 1 & V_{thev-} \geq v_{high} \\ X & \text{otherwise} \end{cases}$$

If this value differs from the charge-sharing value then the appropriate event is scheduled $R_{drive}C_{load} + \Delta_{input}$ seconds in the future where

$$R_{drive} = \begin{cases} R_{high-drive} & \text{final value} = 1 \\ R_{low-drive} & \text{final value} = 0 \\ \min(R_{high-drive}, R_{low-drive}) & \text{final value} = X \end{cases}$$

$$C_{load} = \begin{cases} C_{low+} & \text{final value} = 1 \\ C_{high+} & \text{final value} = 0 \\ C_{low-} + C_{high-} & \text{final value} = X \end{cases}$$

The lumped capacitance represented by C_{load} is overly conservative; fortunately, more accurate models are available [Penfield81,Horowitz83] for future incorporation.

The analysis in [Terman85] of the propagation delay of logic gates indicated that an RC time constant is a very good estimate for the delay of a gate when the input waveform is a voltage step. However, the analysis concludes that a simple RC time constant underestimates the actual propagation delay if the input waveform is other than a step, *e.g.*, a voltage ramp with a rise/fall time of δ. It was shown there that a correction factor, Δ_{input}, can be added to produce a conservative estimate of the propagation delay:

$$\Delta_{input} = \begin{cases} \frac{\delta}{2}(v_{thresh} - v_{tc}) & \text{final value} = 1 \\ \frac{\delta}{2}(1 - v_{thresh}) & \text{final value} = 0 \\ 0 & \text{final value} = X \end{cases}$$

where the correction factor depends only on parameters of the input waveform. Since RSIM does not calculate δ directly, we'll need the following expression:

$$\delta = \begin{cases} \frac{\tau_{in}}{v_{thresh}} & \text{rising input} \\ \frac{\tau_{in}}{1 - v_{thresh}} & \text{falling input} \end{cases}$$

where τ_{in} is the transition time calculated by RSIM for the input waveform. Combining these equations with estimates of the parameters for a typical 5μ nMOS process yields

$$t_{plh} \leq \begin{cases} R_{drive:high}C_{low+} + (0.34)\tau_{in} & \text{rising input} \\ R_{drive:high}C_{low+} + (0.27)\tau_{in} & \text{falling input} \end{cases}$$

$$t_{phl} \leq \begin{cases} R_{drive:low}C_{high+} + (0.64)\tau_{in} & \text{rising input} \\ R_{drive:low}C_{high+} + (0.50)\tau_{in} & \text{falling input} \end{cases}$$

$$t_{px} \geq \min(R_{drive:low}, R_{drive:high})(C_{low-} + C_{high-})$$

as our final equations for estimating a node's transition time. Note that to be conservative, RSIM strives to overestimate transition times to 0 and 1, and underestimate transition times to X.

3.2 Experience with RSIM

RSIM has been in use at both university and industrial environments for several years. During that time it has simulated several hundred designs, ranging in size from very small to approximately 50,000 transistors. Because RSIM was fast enough to simulate whole circuits, it often uncovered circuit flaws that had fallen between the cracks during simulation of the individual components. The trend has been to think of RSIM as a companion to circuit analysis, using it for all logic-level verification and preliminary timing analysis, and then resorting to circuit analysis for those paths identified as critical by RSIM.

The simulation algorithm is embedded in a Lisp-like command language that has been used to write quite elaborate programs to drive the simulation and process the results. Since programs to prepare simulation input are much less tedious to construct than the input itself, designers have been able to conduct more exhaustive tests than they were able to do using earlier simulators. For example, it is a simple matter to take a set of test vectors used to drive an RTL simulation, use those vectors as input for an RSIM run, and compare the predicted outcomes, all under program control.

With careful calibration, RSIM's predictions for combinational logic are within 30% of those of SPICE. For circuits relying on analog behavior (sense amps, bootstrapped nodes, etc.) or chains of "pass" devices, the predictions are less accurate. To compensate, several "escape" mechanisms exist: it is possible to specify the logic thresholds and transition times for individual nodes so that one can incorporate the results of more detailed simulation into RSIM. Usually this mechanism need be invoked for only a few critical

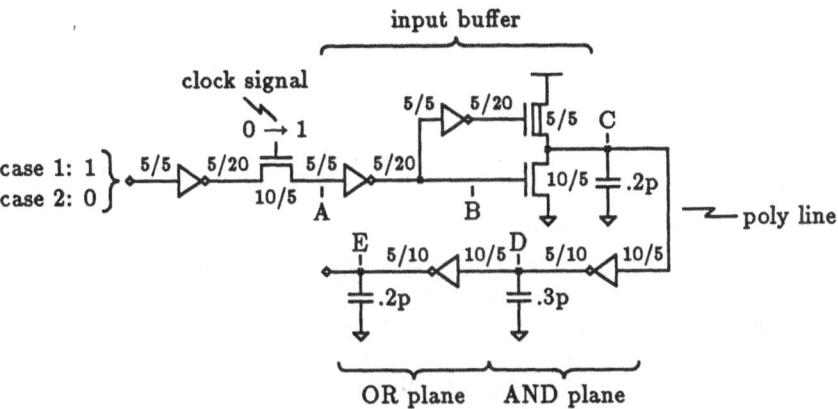

Figure 3.4. *Sample circuit showing path through PLA*

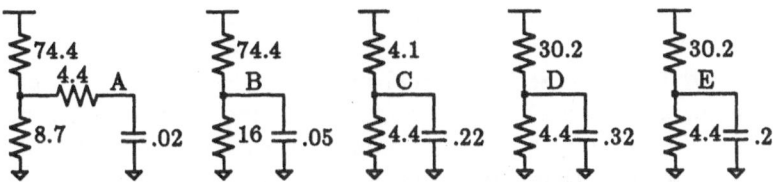

Figure 3.5. *Equivalent RC network for PLA example*

	node	transition	τ	$\sum \tau$	SPICE
	A	↓	0.2	0.2	0.8
	B	↑	3.8	4.0	3.5
Case 1	C	↓	2.9	6.9	6.8
	D	↑	10.4	17.3	15.5
	E	↓	7.4	24.7	20.7
	A	↑	1.6	1.6	0.6
	B	↓	1.7	3.3	1.9
Case 2	C	↑	1.4	4.7	3.3
	D	↓	2.1	6.8	6.4
	E	↑	6.6	13.4	12.1

Table 3.1. *Simulation results for PLA*

nodes (*e.g.*, clock driver outputs). Another approach is to identify problem subcircuits and replace them with logically equivalent circuits that can be simulated easily; a network preprocessor that performs subcircuit matching and replacement is available and has been used to good effect.

To illustrate RSIM's use, the transition times for signals in the sample circuit of figure 3.4 are analyzed below (see also [Terman85]). Transistor sizes are given in microns as width/length. When the clock signal goes high, the input signal (buffered by the inverter on the left) propagates through the input buffer and the two PLA planes. Figure 3.5 shows the equivalent resistor/capacitor network; resistances are given in KΩ and capacitances in pF. Note that the pullup for node C is recognized as a depletion source-follower without considering the actual voltage on its gate. Since depletion devices are always on, the inverter which leads from node B to the gate of the pullup is ignored by the simulator, and the timing for node C is always controlled by node B. Also note that the resistance chosen for the pulldown for node B reflects the threshold drop of node A.

When calculating the transition time using the linear model, one must identify which resistors are actually participating in the network at the moment, and then use series-parallel reduction to compute the effective resistance for the network. For example, a rising transition for node A takes $(74.4 + 4.4)(0.02) = 1.6\,\text{ns}$; a falling transition takes $((74.4 \| 8.7) + 4.4)(0.02) = 0.2\,\text{ns}$. Using this approach, table 3.1 shows the results of propagating two different data values through the PLA. The time of each node's transition is shown in nanoseconds, as predicted by the linear model and SPICE. As one can see, the linear model overestimates the transition times with reasonable consistency. (One expects overestimates because of the inequalities in the equations above.) The estimate for Case 1 is 19% greater than the SPICE prediction; for Case 2, 11% greater.

4. A switch model for MOS networks

If a designer is only interested in the logical properties of a circuit, *i.e.*, those properties independent of performance issues, it is possible to simplify the linear model of the previous section even further by modeling each transistor as an on/off switch whose state is determined by the type of transistor and the state of its gate node. While it would be possible to use the formulas presented in the previous section (suitably modified), it is more profitable to rethink our approach and develop a simpler, more efficient computation that takes advantage of the simpler model. Before presenting the switch model in more detail, a small digression on the representation of node values is in order.

4.1 Representing node values

The success or failure of a logic-level simulator often hinges on the choice of the set of possible node values. If the set is too small, the actual node value may not be precisely described by any one of the available values and the simulator must choose an approximation. Usually the approximation involves some variant of the X (unknown) value which may carry logical implications beyond what the network itself imposes—such a choice is termed either "conservative" or "pessimistic" depending on one's point of view. If the set is large, it becomes difficult to establish whether the simulator's calculations are correct in all cases. Relying on the accumulated evidence of many simulation runs when arguing correctness lacks the rigor that leads to total confidence in the algorithm. This section develops criteria for evaluating a set of node values.

There are three major influences on the choice of the node-value set:

(1) the need to report node values to the user;

(2) the need to determine the state of each network component from the values of its terminal nodes; and

(3) the need to represent intermediate values during an incremental simulation calculation.

If only the first two influences are considered, a three-value set—0, 1, and X—will suffice for logic-level simulation.† Users and component models cannot reasonably expect more information than provided by this set, since most logic-level algorithms cannot support more detailed deductions from arbitrary MOS networks with any degree of accuracy. It is the third influence that leads to all the complication.

Almost all logic simulators analyze a network piece by piece, modifying their estimates for node values as the effect of each piece of the network is

---

† It might be useful to distinguish X', an unknown, but legitimate logic value (*e.g.*, the output of a pair of cross-coupled inverters) from other types of X values. X' values are well behaved in logic operations, for example, $B + \neg B = 1$ if the value of B is X', but equals X if the value of B is X. Such distinctions can be important during initialization.

determined. Until the new-value computation is completed, the intermediate node values serve as accumulators that store all the information the
simulator has about the effects of network pieces already examined. Thus,
distinct values are needed for all qualitatively different intermediate states;
e.g., a node currently at logic high might have that value because examination of the network to date revealed that it was (i) storing charge, (ii)
connected to a depletion pullup, or (iii) being precharged by an enhancement device. The simulator must distinguish among these possibilities,
since the final value of node may be different in each case if, for example,
further network processing discovers a pulldown for the node. The exact
number of values needed depends on the details of the simulation computation; most simulators fall into one of the two categories discussed below.
As will be seen, the two categories are distinguished by their approach to
X values.

Cross-product value sets

One intuitively appealing approach to choosing a set of node values is to
think of each value as having several distinct attributes chosen from independent categories. Thus, for example, one might characterize a node's logic
state and the "strength" of the value separately. The logic state is usually
one of 0, 1, or X; sometimes a high-impedance state, Z, is included to represent the output of tri-state logic gates. [Flake80, Holt81]. The strength
indicates what sort of network connection exists between the source of the
value and the current node:

   input. Node is a designated input (*e.g.*, VDD or GND). The value of
   an input node can only be changed by explicit simulator commands—
   the assumption is that inputs supply enough current to be unaffected
   by connections (possibly shorts to other inputs) made by transistor
   switches.

   driven. Node is connected by closed switches to inputs or other driven
   nodes. Driven nodes can affect the value of weak or charged nodes
   without being affected themselves, but may be forced to an X state if
   shorted to an input or driven node that has a different logic level.

   weak. Node is connected to an input node by a depletion-mode transistor. Weak nodes can affect charged nodes without being affected
   themselves, but are forced to a driven state when connected to another driven or input node. A weak node returns to the appropriate
   weak state when completely disconnected from driven or input nodes
   (*i.e.*, a weak node can never enter the charged state).

charged. Node is connected, if at all, only to other charged nodes. Until reconnected to some other part of the network, charged nodes maintain their current logic state indefinitely (charge storage with no decay). This is the default state of all non-weak nodes.

Other strengths can be included to model the effects of differently sized transistors, node capacitors, etc.

The plethora of 9-, 12-, and 16-state logic simulators (see [Newton80]) use values chosen from the set formed by the cross product of the various value attributes. For example, a 9-state simulator might use the values shown in table 4.1. Note that in this formulation, X is treated as sort of a third logic value on a par with 0 and 1; presumably X's are generated by the simulator to model invalid combinations of 0's and 1's. The implication is that one can determine if a value should be X without any consideration of strengths. (Remember that the main motivation of forming the cross product is that the various attributes are independent.) This can lead to pessimistic predictions, as is shown in an example below.

		logic state		
		0	1	X
	driven	DL	DH	DX
strength	weak	WL	WH	WX
	charged	CL	CH	CX

Table 4.1. *Typical cross-product value set*

It is useful to order the possible signal values according to their relative strengths. Intuitively, value A is stronger than value B, written $A > B$, if value A predominates when both signals are shorted together. Of course there are situations where neither value emerges unscathed—for example, when two signals of the same strength but opposite logic states are shorted—in which case neither signal is said to be stronger than the other. The notion of strength can be formalized using a lattice of node values, as shown in figure 4.1. The node value λ is used to represent the null signal, *i.e.*, no signal at all.

Referring to the lattice, given two values A and B, $A > B$ if A is not equal to B and there is an upward path through the lattice that starts at B and

Figure 4.1. *Lattice of node values for a 9-state simulator*

reaches A. For example

DX is greater than all other signals,

DH is greater than WL, but

WL is not greater than WH.

The least upper bound (l.u.b.) of two values A and B, written $A \cap B$, is defined to be the value C such that

(i) $C \geq A$

(ii) $C \geq B$

(iii) for every value D, if $D \geq A$ and $D \geq B$, then $D \geq C$.

Examining the lattice above, it is easy to see that the l.u.b. always exists for any two node values. Note that if $A > B$, $A \cap B = A$; the l.u.b. captures our intuition about what should happen when two signals of different strengths are shorted together. With the appropriate placement of X values in the lattice, the l.u.b. can be used to predict the outcome when any two signals are shorted.

The interpretation of X values captured by the lattice above is quite appropriate for describing the logic state of nodes involved in a short circuit (see figure 4.2). Assuming the two transistors are the same size, the middle node's value is the result of merging two equal strength signal values. According to our lattice, this merger yields an X value. Short circuits are the mechanism by which X's are introduced into a network previously containing only 0's and 1's.

However, the situation is not as straightforward when one considers connections formed by transistors with a gate signal of X. The resulting values cannot be computed directly using the $\cap$ operation on the source and drain

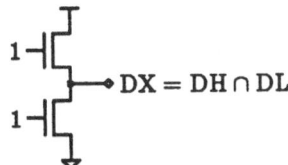

Figure 4.2. *A short circuit leading to an X value*

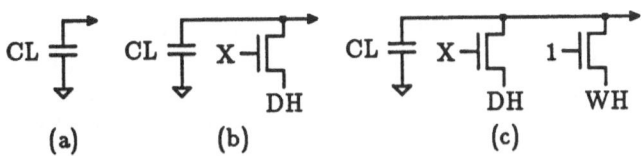

Figure 4.3. *Incremental analysis of a simple network*

signals, and once that hurdle has been surmounted, there is some difficulty in choosing which value to use from the cross-product value set. Consider the following analysis of a node with stored charge and connection to two transistors. Before any connections to the node have been discovered (figure 4.3(a)), the node maintains the charge of its last driven value, say, logic low; the simulator would assign the node a value of CL. After the first transistor is discovered (figure 4.3(b)), the facts change:

(i) Because of the X on the gate of the transistor, one cannot be certain what type of connection exists between the node in question and the DH on the other side of the transistor. Thus, the new logic state of the node should be X.

(ii) The strength of the new value is uncertain, but clearly "weak" or "charged" would be inappropriate since they understate the strength in the case where the unknown gate value was actually a 1.

Since a weak or charged value could be overridden by an enhancement pulldown discovered later on, mistakenly leading to DL value, the simulator has no choice but to select a driven value. The conclusion: DX is the only state available that handles all eventualities in a conservative fashion. Of course, with knowledge of what the rest of the network contains, the simulator could make a more intelligent choice, but this is beyond the ken of an incremental algorithm.

By the time a connection to a depletion pullup is discovered (figure 4.3(c)), the die has been cast: the previously chosen DX value overrides any contribution by the pullup (DX ∩ anything yields DX). While this answer is not wrong, it is more conservative than required; at this point the logic state of the node should be 1. The pullup guarantees a logic 1 with the unknown connection to DH, only leaving doubts about the strength of the value (somewhere between weak and driven).

Proponents of cross-product value sets might point out that the analysis would have generated a different answer if the transistors had been discovered in a different order. The somewhat embarrassing ability to produce two different answers for the same network, both correct, is caused by the fact that the merge operation is not associative when connections are made through transistors with X gates. In fact, most incremental simulators that use cross-product value sets perform the incremental analysis in an order that yields a reasonable answer on the example above. Unfortunately, it is usually possible to confound them with more complex circuits containing X's; while such circuits are not commonplace, they often crop up during network initialization when all nodes start off at X.†

In conclusion, it is possible to build effective simulators using cross-product value sets; however, they can make conservative predictions on circuits that contain X's. In practice, this leads to difficulty in initializing some circuits and to occasional over-propagation of X values.

Interval value sets

The difficulties with the cross-product value set arise because of its separation of the notion of strength and logic state. Once a node value is set to an X value at some strength, it cannot return to a normal logic state unless overpowered by a stronger signal; if a node is set to the strongest X value, it stays at that value for the rest of the computation. As in the example above, this leads to conservative predictions when the strongest X value is chosen because of the lack of suitable alternatives. Specifically the difficulty came about because the simulator had to pick the highest strength to be on the safe side; there was no value available that would indicate that the logic low signal which contributed to the intermediate X value was of very low strength and hence might be overridden by later network components.

† [Bryant81] suggests using an incremental calculation only for subnetworks of nodes connected by non-X transistors. Once these values have been computed, a separate computation merges subnets connected by X transistors. Since this computation has global knowledge of the network, it can avoid the problems mentioned here.

This suggests a different approach to constructing the set of possible nodes values, one based on intervals. First one starts with a set of node values with a range of strengths and 0/1 logic states, for example, the six non-X states used above: {DH, DL, WH, WL, CH, CL}. Then additional values are introduced by forming intervals from two of the basic values; if there are six basic values, then there are $\binom{6}{2} = 15$ such intervals, leading to a total of 21 node values altogether.

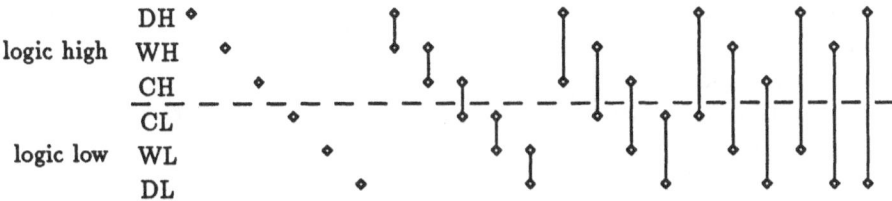

Figure 4.4. *The 21 node values of the interval value set*

Intervals represent a range of possible values for a node. The size of the range is related to the strength of its end points. If we arrange the six basic values in a spectrum ranging from the strongest 1 (DH) to the strongest 0 (DL), the possible node values can be shown graphically, as illustrated in figure 4.4. Intervals that do not cross the center line correspond to a valid logic state: intervals above the line represent logic high values, and those below the line, logic low. Intervals that cross the center line represent X values. (The X values of the previous section correspond to intervals with equal strength end points: DX = [DL,DH], WX = [WL,WH], and CX = [CL,CH].) Thus, X values result from ambiguity about which of the base values best represents the true node value. As will be seen below, this is more satisfactory than thinking of X as a third, independent logic state.

When the simulator merges two node values, it chooses the smallest interval that covers all the possible node states. However, unlike the cross-product value set, the interval set can represent X values without loosing track of the strengths of the signals that lead to the X values. Consider the problems raised by figure 4.3(b). Using an interval value set, the resulting node value is naturally represented by [CL,DH], an interval that corresponds to an X logic state. When the pullup is discovered (figure 4.3(c)), the simulator can narrow this interval to [WH,DH] since the pullup overpowers the weaker CL value. This corresponds to a logic high signal—a sensible answer.

An algebra for calculating the result of merging two interval node values is developed in [Flake83]. With an interval value set, the merge operation is commutative and associative, and the network can be processed in any order without affecting the final node values. The extra 12 values introduced by the interval value set are needed to carry sufficient information about how the current value was determined, to ensure that the final answer is independent of the processing order.

The examples above suggest the following conjecture about the correct size of a node value set. Assuming that one has s different signal strengths and two logic levels (0 and 1), then $2s + \binom{2s}{2}$ values are needed to ensure that the signal algebra is well-formed. In simulators with too few states, some states take on multiple meanings; for example, the DX value in the cross-product value set is used to describe nodes that fall into 5 separate values in the interval value set:

$$[DL,DH] \quad [WL,DH] \quad [CL,DH] \quad [WH,DL] \quad [CH,DL]$$

This lack of expressive power on the part of cross-product value sets is what leads to pessimistic predictions for node values in certain networks.

4.2 Developing the switch model

Switch models of MOS circuits are of interest since a switch is the simplest component that meets the criteria outlined in Section 1: switches are inherently bidirectional and the logic operations they implement can be computed with acceptable efficiency in large networks.

Randy Bryant [Bryant79], one of the first to apply switch-level simulation to MOS transistor networks, viewed the network as divided into equivalence classes. Two nodes are equivalent if they are connected by a path of closed switches. Nodes in the same equivalence class as VDD are assigned a logic high state; those equivalent to GND, a logic low state. A pullup (a depletion-mode transistor which is always on in the switch model) gives the node to which it is attached a special property: if an equivalence class of nodes does not contain either VDD or GND, but does contain a pulled-up node, all the nodes in the class are assigned a logic high state. Finally, if an equivalence class contains neither an input nor a pulled-up node, it is "storing charge" and maintains whatever logic state it had last.

The simulator based on this switch model iteratively calculates the equivalence classes for all the nodes in the network until two successive calculations return the same result (*i.e.*, no nodes change state). Unfortunately this pure switch model has some deficiencies:

(i) Switches in indeterminate states (those with gate nodes of X) make the equivalence calculation somewhat more difficult. The desired computation is inefficient since it involves a combinatorial search; all combinations of on/off assignments to switches in the X state need to be investigated to determine whether a switch's state makes a difference. If the network is unaffected by a switch's state, the switch can be ignored; otherwise all affected nodes are assigned the X state.

(ii) The equivalence calculation is much more time consuming than necessary since it deals with the whole circuit rather than focusing only on the parts which are changing.

(iii) In certain circuits transistor size is important, and the notion of size cannot be expressed in the pure switch model. A pullup is a trivial example: viewed as a switch it was always on, but more "weakly" than the "strong" switches in the pulldown. The size of transistors also determines the "strength" of various driver circuits; for example, it is common for the write amplifier of a static memory to force a value into a memory cell by simply overpowering the weaker gate in the cell itself.

The remainder of this section investigates an approach to solving the first two problems outlined above. The third problem is addressed with some success by the linear model which uses size information not only to calculate node values but to provide timing information as well.† The simulator adopts a model where each node value is computed via a *global* examination of the network. The result is a calculation very similar to that implemented by the linear model, except that abstract "logical" resistances ($R_{eff} = 0, 1,$ and ∞) are substituted for the "real" resistances used in the linear model.

4.3 The global switch model

The global simulator calculates a node's value by computing the effect of each input on the node of interest. The simulation is global in that each node value is based directly on the values of the inputs to which it is connected. Thus, the values of non-input nodes do not enter into the computation. This means that 0, 1, and X will suffice as final node values; a

† Bryant [Bryant81] proposes extending the switch model to include a hierarchy of switch sizes, a generalization of the *ad hoc* solution for pullups. His thesis develops an algebra, in the spirit of Boolean algebra, for dealing formally with such networks.

node state need only capture the logic state of the node and no strength information is necessary.

Node values in the global switch model

Each transistor switch in the network is assigned a state determined from the transistor's type and the current value of its gate node. This state models the switch-like qualities of the source-drain connection without trying to capture any more detailed information about the connection—a simplification of the linear model.

The state of a transistor switch summarizes the type of connection that exists between its source and drain nodes. For MOS circuits, the possible switch states are:

open   no connection, the state of a non-conducting n-channel (gate = 0) or p-channel (gate = 1) transistor.

closed   source and drain shorted, the state of a conducting n-channel (gate = 1) or p-channel (gate = 0) transistor.

unknown   uncertain connection between source and drain, the state of an n- or p-channel transistor whose gate is X.

weak   the state of a depletion transistor. Depletion devices are always assigned this state, regardless of the state of their gate nodes.

The relationship between a switch's state, its types, and its gate value is summarized in the figure 4.5.

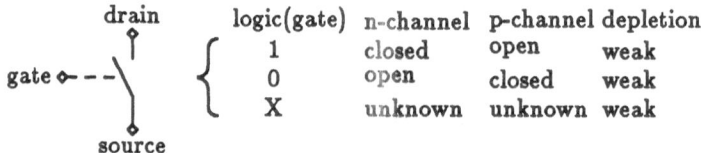

drain	logic(gate)	n-channel	p-channel	depletion
	1	closed	open	weak
gate	0	open	closed	weak
	X	unknown	unknown	weak

source

Figure 4.5. *Switch state as a function of transistor type and gate voltage*

In the global simulator, the value of a node is determined by the inputs to which it is connected and the states of the intervening switches. During the calculation of a node's value, the simulator uses the interval node-value set presented in figure 4.4. When the calculation is complete, the resulting interval is used to determine the final logic state of the node, using the following formula.

$$\text{final logic state} = \begin{cases} 0 & \text{CL [CL,WL] [CL,DL] WL [WL,DL] DL} \\ 1 & \text{DH [DH,WH] [DH,CH] WH [WH,CH] CH} \\ X & \text{all other values} \end{cases}$$

The calculation of a node's value begins by discovering all the inputs which can be reached from the node by paths of closed, weak, and unknown switches. If no inputs can be reached, the final logic state of the node is determined by a charge sharing calculation described in the next section. If one or more inputs can be reached, their contribution to the node's value is determined by an incremental calculation which starts at the inputs and works its way back toward the node.

The value of a logic low input is DL; the value of a logic high input is DH. As the calculation works back toward the node of interest, it computes an effective value that indicates the effects of intervening switches on the original input value. The effect of a switch on a value it transmits is specified by the *switch* function, shown in figure 4.6. The effect of a switch on a value is a function of the value and the switch's state; the relationship is tabulated in table 4.2. A new value, λ, is introduced to describe the value transmitted by an open (non-conducting) switch, *i.e.*, no value at all. The value λ is weaker than CH or CL, and corresponds to a logic state of X.

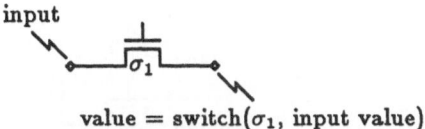

input

value = switch(σ_1, input value)

Figure 4.6. *Effective value of an input after passing through a switch*

When two paths merge, their effective value is determined using the ∩ operation introduced earlier, as shown in figure 4.7. The ∩ operation is defined using the lattice shown in the figure 4.8. Following the procedure outlined in figure 4.7, the contributions of all inputs connected to the node of interest can be reduced to a single interval. This interval is merged (using ∩) with the contribution from the node's current logic state

$$\text{contribution of current state} = \begin{cases} \text{CL} & \text{if current logic state} = 0 \\ \text{CH} & \text{if current logic state} = 1 \\ \text{[CH,CL]} & \text{if current logic state} = X \end{cases}$$

to give the final interval characterizing the node's new logic state.

value	open	closed	switch state weak	unknown
DH	λ	DH	WH	[DH,λ]
[DH,WH]	λ	[DH,WH]	WH	[DH,λ]
[DH,CH]	λ	[DH,CH]	[WH,CH]	[DH,λ]
[DH,CL]	λ	[DH,CL]	[WH,CL]	[DH,CL]
[DH,WL]	λ	[DH,WL]	[WH,WL]	[DH,WL]
[DH,DL]	λ	[DH,DL]	[WH,WL]	[DH,DL]
WH	λ	WH	WH	[WH,λ]
[WH,CH]	λ	[WH,CH]	[WH,CH]	[WH,λ]
[WH,CL]	λ	[WH,CL]	[WH,CL]	[WH,CL]
[WH,WL]	λ	[WH,WL]	[WH,WL]	[WH,WL]
[WH,DL]	λ	[WH,DL]	[WH,WL]	[WH,DL]
CH	λ	CH	CH	[CH,λ]
[CH,CL]	λ	[CH,CL]	[CH,CL]	[CH,CL]
[CH,WL]	λ	[CH,WL]	[CH,WL]	[CH,WL]
[CH,DL]	λ	[CH,DL]	[CH,WL]	[CH,DL]
CL	λ	CL	CL	[λ,CL]
[CL,WL]	λ	[CL,WL]	[CL,WL]	[λ,WL]
[CL,DL]	λ	[CL,DL]	[CL,WL]	[λ,DL]
WL	λ	WL	WL	[λ,WL]
[WL,DL]	λ	[WL,DL]	WL	[λ,DL]
DL	λ	DL	WL	[λ,DL]

Table 4.2. *switch(σ,value) as a function of σ and value*

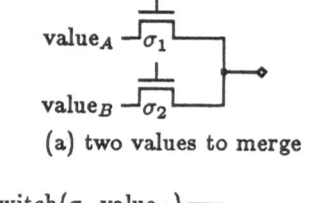

(a) two values to merge

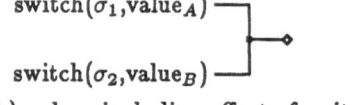

(b) values including effect of switches

switch(σ_1,value$_A$) ∩
switch(σ_2,value$_B$)
(c) merged value

Figure 4.7. *Merging the values for two paths which join*

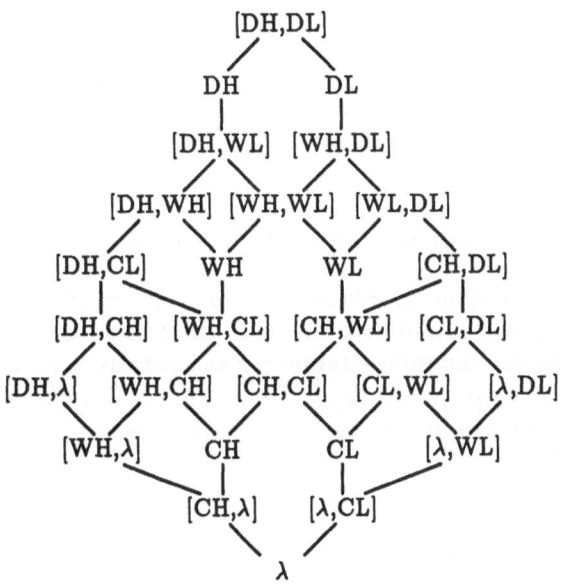

Figure 4.8. *Lattice for interval-node value set*

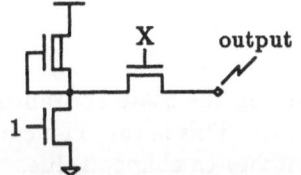

Figure 4.9. *Example circuit*

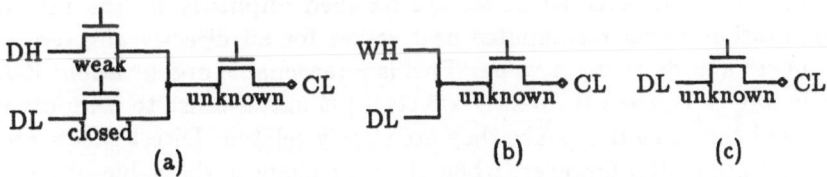

Figure 4.10. *New-value calculation for circuit in figure 4.9*

As an example of how the new-value calculation works, consider the circuit shown in figure 4.9. Assume that the current logic state of the output is 0. The new-value calculation for this circuit is shown in figure 4.10. The final

interval for the output node is CL ∩ [λ,DL] = [CL,DL] which corresponds
to a logic low state. This makes sense; the previous state of the output
node was logic low, so the uncertain connection to the inverter does not
affect its logic state, just the strength with which its driven. Note that it
is important to merge the values of paths that join before continuing with
the calculation since

$$switch(\sigma, \alpha \cap \beta) \neq switch(\sigma, \alpha) \cap switch(\sigma, \beta)$$

when using this particular value set and *switch* function. For example, if
the WH and DL values had been merged *after* transmission by the switch in
the unknown state, the final interval for the output node would have been
[DH,WL], which corresponds to an X logic state. The calculation described
here performs all possible merges *before* transmitting the result through
the appropriate switch.

The global simulation algorithm

This section outlines the basic steps for propagating new information about
the inputs to the rest of the network, recalculating node values (where
necessary) using the global value calculation in the previous section.

When a node changes value, it can affect the network in one of two ways:

(i) directly, through source/drain connections of conducting transistors.

(ii) indirectly, by affecting the state of transistor switches controlled
by the changing node. This is turn can cause the source and drain
nodes of those switches to change value.

The global simulator accounts for these two effects using to different mechanisms. Directly affected nodes are handled implicitly by the new-value
computation which recomputes new values for all directly affected nodes
whenever a node changes value. This is a reasonable organization: if A directly affects B, then B directly affects A; it makes sense to compute both
values at the same time since they are closely related. Direct effects are not
handled implicitly, however, when the user changes the value of an input
node. In this case, the simulator invokes the new-value computation on the
input, not to recompute the input's value (which is set by the user), but to
recompute the values of all directly affected nodes.

The indirect effects of a value change are managed by an *event list* that
identifies all transistor switches that have changed state. Actually, the event

list keeps track of the nodes that have changed, but this is equivalent since the network data base maintains a list of transistors controlled by each node. The simulator operates by removing the first node from the event list, and then performing a new-value computation for the sources and drains of all transistors controlled by that node. The new-value computation accounts for all the direct effects of the new transistor state and adds events to the event list if indirect effects are present. This process continues until the event list is empty, at which point the network has "settled" and the simulator waits for further input.

```
while event list not empty {
    n := node associated with first event on event list
    remove first event from event list
    update logic state of n to new value
    for each transistor with n as gate node
        set COMPUTE flag for source and drain
    for each transistor with n as gate node {
        if COMPUTE still set for source, compute new value for source [fig. 4.14]
        if COMPUTE still set for drain, compute new value for drain
    }
}
```

Figure 4.11. *Main loop of global simulation algorithm*

Finding nodes affected by an event is straightforward; recomputation of values is needed for the sources and drains of all transistors with the changing node as gate. For example, if the node marked (*) in figure 4.12 changes, nodes B and C need recomputation. Of course, node D also needs to be recomputed, as will be discovered during the processing of B and C (see below).

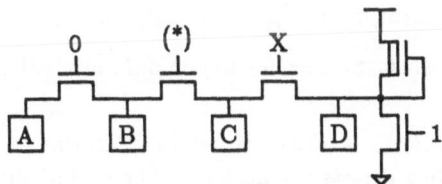

Figure 4.12. *Event for node (*) involves nodes B and C*

To recompute the value of a given node, the simulator first makes a *connection list* containing all nodes connected to the first node by a path of conducting transistors. The idea is to start with a node known to be affected by an event, and then find that node's electrical neighbors, and so on, halting whenever an input is reached. In the example above, if the (*) node's value is 1, the connection list for node B contains nodes B, C, and D. If the (*) node's value is 0, the connection list for node B contains only node B. Node A is not included in the list in either case because it is not connected to node B by a path of conducting transistors. In the code in figure 4.13, which computes the connection list for a given node, the terms "source" and "drain" are used to distinguish one terminal node of a transistor from the other, and do not imply anything about the terminals' relative potential. The connection list drives the new-value computation, shown in figure 4.14.

```
initialize list to have starting node as only element
set pointer to beginning of list
while pointer not at end of list {
    n := node currently pointed at
    for each "on" transistor with source connected to n
        if drain is not an input and is not on list
            add drain to end of list
        advance pointer to next list element
}
```

Figure 4.13. *Non-recursive routine to build connection list*

```
make connection list starting with given node [fig. 4.13]
for each node on connection list {
    compute interval value for node [fig 4.15]
    determine new logic state
    if different from old logic state
        enqueue new event
}
reset COMPUTE flag for each node on connection list
```

Figure 4.14. *Subroutine to compute new value for node*

The value of each node is determined in accordance with the procedure described above. New events are added to the end of the event list whenever a node changes value. If a changing node is already on the event list, nothing happens (the node is not moved to the end of the list).

For efficiency, each affected node's value is only computed once while processing a given event. The connection list ensures that all affected nodes are recomputed; the COMPUTE flag ensures that once a node has appeared on some connection list, it will not be resubmitted for processing during the current event.

The computation of a node's value is easily described by a recursive procedure which analyzes the surrounding network (see figure 4.15). The variable LOCAL-IV is a stack-allocated local variable of the subroutine. Returning to the example in figure 4.12, assuming that the (*) node's value is 1, and that the old values for B, C, and D are $B = 1$, $C = 0$, and $D = 0$, figure 4.16 shows the calls which are made when computing the new value for node C:

```
if node is an input
    return DL, DH, or DX, as appropriate
else {
    LOCAL-IV := contribution of current logic state
    set VISITED flag for current node
    for each "on" transistor, t, with source connected to current node
        if drain does not have VISITED flag set {
            recursively determine interval value for drain node
            LOCAL-IV := LOCAL-IV ∩ switch(σₜ, drain's interval value)
        }
    reset VISITED flag for current node
    return LOCAL-IV
}
```

Figure 4.15. *Subroutine to compute interval value for node*

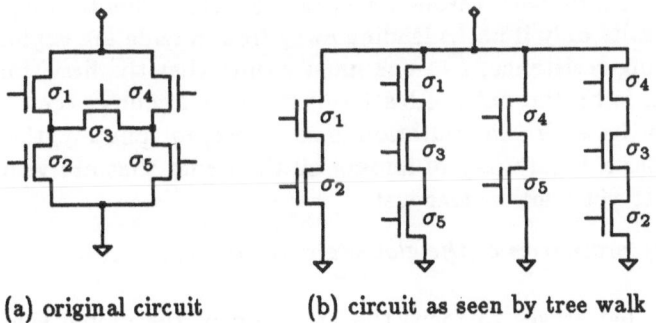

(a) original circuit   (b) circuit as seen by tree walk

Figure 4.17. *The tree walk traces out all possible paths*

Marking each visited node (by setting its VISITED flag) avoids cycles; this keeps the tree walk expanding outward from the starting node. The VISITED

```
compute-params(C)
  LOCAL-IV = CL
  compute-params(D)
    LOCAL-IV = CL
    compute-params(VDD)
      return DH
    LOCAL-IV = CL ∩ WH = WH
    compute-params(GND)
      return DL
    LOCAL-IV = WH ∩ DL = DL
    return DL
  LOCAL-IV = CL ∩ [λ,DL] = [CL,DL]
  compute-params(B)
    LOCAL-IV = CH
    return CH
  LOCAL-IV = [CL,DL] ∩ CH = [CH,DL]
  return [CH,DL]
```

Figure 4.16. *Trace of interval value computation for example in figure 4.12*

flags are reset as the routine backs out of the tree walk, so all possible paths through the network are eventually analyzed. If the network contains cycles (see, *e.g.*, figure 4.17), the tree walk might lead to more computation than a series/parallel analysis; this is a problem for circuits containing many potential cycles (such as barrel shifters), especially during initialization when many of the paths are conducting because control nodes are X. To speed up the calculation, a node's VISITED flag can be left set, restricting the search to a single path through a cyclic network. This technique produces correct results only if paths leading away from a node are explored in order of increasing resistance, *i.e.*, one must ensure that the first time a node is reached, it is by the path of least resistance. Of course, the flags must be reset once the entire computation is complete; fortunately, the connection list provides a handy way of finding all the nodes that are visited without resorting to yet another tree walk.

Interesting properties of the global algorithm

The event list serves to focus the attention of the global simulator; new values are computed only for nodes which appear on the event list or which are electrically connected to event-list nodes. Portions of the network that are quiescent are not examined by the simulator. Algorithms that have this property are said to be *selective-trace* or *event-driven* algorithms and generally run much faster than algorithms which are not event driven [Szygenda75].†

An interesting implication of selective trace is that special care must be taken to ensure that "constant" nodes, such as the output of an inverter with its input tied to GND, are processed at least once (otherwise they will have the wrong values). One technique is to treat VDD and GND as ordinary inputs when first starting a simulation run—sort of a power-up sequence as VDD and GND change from X to 1 and 0 respectively. Computing both the direct and indirect consequences of changes in VDD and GND might involve a tremendous amount of computation since the whole circuit is affected; often only computing the indirect consequences is a sufficient and less costly alternative.

Although there is no explicit mention of time in the global simulator, the first-in, first-out (FIFO) processing of events imposes some ordering on the changes of node values. This ordering is similar to, but not the same as, the unit-delay ordering used by many gate-level simulators. In an event-driven unit-delay algorithm, the output of each gate that had an input change is recomputed using the current values of the input nodes. The new output values are saved and imposed on the network only after processing all gates. The net effect is that each computation cycle (representing a unit of time) propagates information through one level of gate, *i.e.*, each gate has unit delay. Because changes in node values are imposed all at once, values change simultaneously, which can lead to problems in circuits containing feedback paths.

The global simulator implements a pseudo unit-delay algorithm. New events are added to the end of the event list, so the oldest changes are processed before any consequences of those changes are processed. Thus, FIFO event management leads to the same sequence of gate evaluations as a unit-delay algorithm. However, because the global algorithm changes values in the network incrementally rather than all at once, it is possible to find circuits that behave differently under the two simulators, *e.g.*, the circuit shown in figure 4.18. A 0-1 transition on the input causes a unit-delay algorithm to loop forever. The global algorithm predicts only one

† Exceptions to this rule are some hardware-based simulation algorithms, such as programs run on the Yorktown Simulation Engine [Pfister82]. The builders of the YSE point out that simulations might well run slower because the extra communication and branching needed to implement selective trace would compromise the parallelism and pipelining used to great advantage in the YSE. However, if sufficiently large portions of the circuits could be ignored, the overhead of selective trace is probably worth the investment.

transition—the output of whichever gate it processes first. Neither answer is
completely correct; the actual circuit enters a meta-stable state on a 0-1 input transition, eventually settling to a particular configuration determined
by subtle differences in the gains of the two gates. It will not remain in
the meta-stable state forever, so an infinite oscillation is a poor prediction.
On the other hand, the final configuration chosen by the global simulator
depends on the order of some list in the network data base. The predicted
outcome is the same each time, not necessarily the best prediction.† The
global simulator does not offer a general solution to the oscillation problem;
both simulators will oscillate on circuit shown in figure 4.19.

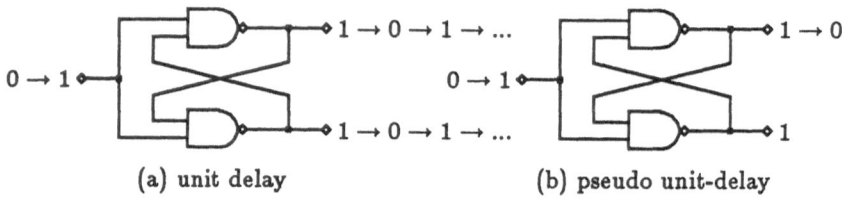

(a) unit delay   (b) pseudo unit-delay

Figure 4.18. *Circuit that distinguishes unit-delay from pseudo unit-delay*

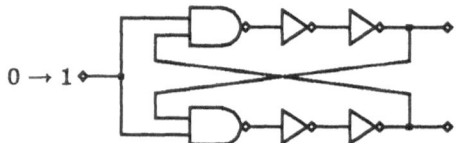

Figure 4.19. *Circuit which causes both simulators to oscillate*

Along the same lines, the global simulator predicts that the output of the
circuit in figure 4.20 will oscillate when the input changes from 1 to 0. The
actual output quickly rises to the balance point of the pullup/pulldown
combination. In a logic-level simulation, this corresponds to finding a solution to the equation $\alpha = \neg\alpha$ which has the solution $\alpha = X$ (a reasonable
logic-level representation for the balance point). This example is drawn
from a larger class of circuits where a node is both an input and output of
the circuit. Since the new-value computation uses current transistor states

---

† [Bryant81] suggests that the oscillation can be detected and the offending node
values replaced by X, but the technique for determining the number of oscillations
to allow yields answers so large for circuits of any substantial size that this is not a
very practical alternative.

(determined by current node values) to predict the new values, it is impossible to predict the value of a node that depends on its own value. This limitation has not proven to be a problem in practical circuits.

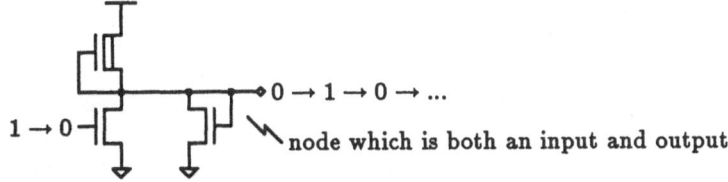

Figure 4.20. *Circuit with a node that is both an input and output*

In conclusion, use of the switch model as a basis for simulation provides a good compromise between the need for accuracy and the need for speed. Switch-level simulators are emerging as the ideal companion for circuit-level simulators—together the simulators provide for most of the simulation needs of current-generation VLSI design.

5. Gate-level Simulation

Gate level simulation is similar in many ways to the switch level techniques presented in the previous section. The major difference is that the basic modeling element is a unidirectional *gate* rather than a bidirectional switch. Each gate has one or more inputs and a single output; if several gate outputs are shorted together in the actual circuit—*e.g.*, on a tristate bus—it is customary for the circuit preprocessor to add additional gates to accomplish the wired logic function explicitly. The net result is that the value of each node is determined by exactly one gate. The *fanin* of a gate is a list of nodes which are the inputs of the gate; the *fanout* of a gate is a list of other gates which have the first gate's output node as an input. Whenever a node in the fanin list of a gate changes value, the gate should be resimulated; whenever a gate's output changes value, all the gates on the fanout list should be resimulated. These lists provide a natural database for event-driven simulation.

Inside the simulator, each type of gate is modeled by a subroutine (or simple table) which computes the new value of the output from the current values of the inputs. Gate types may range from simple Boolean functions (NOT, AND, NOR, etc.) to complex modules (arithmetic logic units, UARTs, etc.). More complex modules may have internal state in addition to that provided by the fanin and fanout nodes; this state is maintained in a per gate database constructed when the network was read into the simulator. The digital logic implemented by gates is well matched to the operations implemented by a computer, so it is not surprising that gate-level simulators offer the best performance of all the simulators discussed in this paper.

The technology behind most gate-level simulators is similar to that for switch-level simulation presented in the previous section. Node values are typically drawn from a cross-product value set with the number of strengths determined by the circuit technology: a single strength suffices for bipolar gate arrays (often with a fourth high impedance logic state added); three or four strengths are commonly used for MOS logic. The HILO gate-level simulator uses the interval value set presented in section 4.1. Recently there has been some escalation in the size of the value sets as various vendors strive for product differentiation; however, even these simulators follow the basic strategy outlined in section 4.1.

Event-driven simulators are the rule, although some gate-level simulators specialized for synchronous logic do away with events altogether and evaluate each gate once per clock cycle in an order determined by the gate's distance from the circuit's inputs. Events are scheduled with either unit delay or a variable delay which depends on the particular gate and direction of the transition. Some simulators provide for a *min* and *max* delay and augment the possible node values to include R (rising) and F (falling) states. Many simulators implement an *inertial delay* model where the output of a gate is constrained not to change more rapidly than its intrinsic delay. Output transitions disallowed under this model are reported as spikes to the user, and, in some simulators, the output node is set to an error state.

5.1 Fault simulation

Gate-level simulators are most useful when a node-by-node modeling of a circuit is desired, but a one-to-one mapping between circuit components and model components is not required. One application in which gate-level simulation excels is *fault simulation* where the designer wishes to determine which potential circuit faults are detected by a set of *test vectors*. The most common fault models, called *stuck-at faults*, fix the values of certain nodes in the circuit, *i.e.*, individual nodes are stuck at 1 or stuck at 0. The designer specifies which nodes are to be faulted, and then has the simulator determine what percentage of the faults are detected by the test vectors. Since the faults are on a per node basis, gate-level simulation is well suited to this sort of application—the more detailed component models provided by switch-level and circuit-level simulation are not needed.

Since the number of potential circuit faults is proportional to the number of nodes, a large number of test vectors is required to diagnose a circuit of any reasonable size. This implies that large amounts of simulation time will be required to develop the test vectors and verify their coverage. One alternative available to the designer is *concurrent fault simulation* [Ulrich73]. A concurrent simulator is a gate-level simulator which has been modified to run many simulations of the same circuit simultaneously. The simulator maintains a single copy of the circuit database and as many copies as required of node values and pending events. In concurrent fault simulation, all the simulations use the same test data, but differ in which nodes have been faulted. In the worst case, for *n* simultaneous simulations, the node value and event storage might be *n* times as large as for a single simulation, and no improvement in runtime will be seen over that for *n* sequential single-machine simulations. However, it is usually the case that the states of many of the faulty machines *converge* after only a small amount of simulation and the databases for those machines can be collapsed into a single database. If the fault is detected early on, the database for the corresponding machine can be merged with that of the "good" machine simulation. In either case, the total amount of work required to complete *n* simulations is usually far less than that required for *n* separate simulations. A carefully implemented concurrent fault simulator can achieve a performance improvement of several orders of magnitude.

5.2 Hardware accelerators

A second approach to speeding up gate-level simulation is to provide special purpose hardware tailored for that purpose. Some vendors have simply provided special microcode for their processor as a way of tailoring the machine's operations to the requirements of simulation. In such cases, speed improvements are usually limited to at most a factor of ten—similar to what one can achieve by moving to a faster computer.

Others have turned to multiprocessing architectures in order to achieve more impressive gains. The IBM Yorktown Simulation Engine (YSE) [Pfister82] uses up to 256 processing elements interconnected by a high bandwidth 256x256 crossbar. Each processing element is a pipelined table-lookup unit which can evaluate a five-input logic equation using a four-valued logic once every 80ns. The inputs to the equation are read from a local node memory; the output can be stored in the local memory or transmitted over the crossbar to other processing elements. In any given cycle, the processing element either evaluates an equation or reads and stores a value from the crossbar. The total capacity of the machine is four million equations, one million nodes, and an ability to process over three billion events per second! The machine is not event-driven; each processing element evaluates its equations in order. Often the equations can be ordered and distributed among the processing elements so that only a single pass is needed to simulate the actions of a large digital circuit. Using a straightforward compilation strategy, feedback in the circuit may necessitate several passes through the equations before the simulation is complete. More sophisticated compilers can eliminate this extra overhead for most circuits.

A different strategy is pursued by Zycad [Zycad83] and, more recently, Silicon Solutions: they have chosen to build a slower, event-driven engine. The basic operation of these machines is an equation evaluation by table-lookup similar to that of the YSE; however, changes in node value are remembered on an event list which is used to determine which equations need re-evaluation. The machines have multiple processor elements interconnected by a medium speed message bus used to transmit event information. Each processing element can process approximately 500,000 events per second, about twenty times slower than the YSE. Both companies have recently announced a concurrent fault simulation capability.

The jury is still out as to which architecture provides the best performance. In many digital designs, most of the circuit is active every clock cycle, suggesting that event-driven techniques provide only a small advantage over the strategy of complete re-evaluation. Thus, for single-machine simulations, the YSE architecture may be the better performer. However, the event list is an important part of concurrent simulation—it provides the part of the database needed to detect when two machines can be merged for the remained of the simulation. Thus, for concurrent simulation, event-driven architectures may be the best choice. Stay tuned for further developments.

To summarize: gate-level simulators have much in common with switch-level simulators, differing mainly in the basic modeling element. Gate-level simulation really shines in applications, such as fault simulation, where the lack of detail at the component level doesn't really matter to the designer. The ease with which hardware accelerators can be built for gate-level operations probably means that gate-level simulation will remain an important tool in the design of VLSI circuits for the foreseeable future.

6. References

6.1 Circuit-level simulators

B. Ackland and N. Weste, "Functional Verification in an Interactive IC Design

Environment," *Proceedings of the Second Caltech Conference on Very Large Scale Integration*, January 1981.

V. Agrawal, *et al*, "A Mixed-mode Simulator," *Proceedings of 17th Design Automation Conference*, June 1980.

B. Chawla, H. Gummel, and P. Kozak, "MOTIS—An MOS Timing Simulator", *IEEE Transactions on Circuits and Systems*, Vol. CAS-22, No. 13, December 1975.

S. Fan, M. Y. Hseuh, A. Newton, and D. Pederson, "MOTIS-C: A New Circuit Simulator for MOS LSI Circuits," *Proceedings IEEE International Symposium on Circuits and Systems*, April 1977.

G. D. Hachtel, *et al*, "The Sparse Tableau Approach to Network Analysis and Design," *IEEE Transactions on Circuit Theory*, Vol. CT-18, January 1971.

E. Lelarasmee, A. Ruehli, and A. Sangiovanni-Vincentelli, *The Waveform Relaxation Method for Time Domain Analysis of Large Scale Integrated Circuits*, Memorandum No. UCB/ERL M81/75, Electronics Research Laboratory, University of California, Berkeley, June 1981.

L. Nagel, *SPICE2: A Computer Program to Simulate Semiconductor Circuits*, ERL Memo No. ERL-M520, University of California, Berkeley, May 1975.

H. Nham and A. Bose, "A Multiple Delay Simulator for MOS LSI Circuits", *Proceedings of 17th Design Automation Conference*, June 1980.

A. Newton, "Timing, Logic and Mixed-mode Simulation for Large MOS Integrated Circuits", *Computer Design Aids for VLSI Circuits*, P. Antognetti, D.O. Pederson, H. De Man eds., Sijthoff & Noordhoff, Rockville, Maryland, 1981.

A. Newton and A. Sangiovanni-Vincentelli, *Relaxation-based Electrical Simulation*, University of California, Berkeley, 1983.

K. Okasaki, T. Moriya, and T. Yahara, "A Multiple Media Delay Simulator for MOS LSI Circuits," *Proceedings of 20th Design Automation Conference*, June 1983.

A. Sangiovanni-Vincentelli, "Circuit Simulation", *Computer Design Aids for VLSI Circuits*, P. Antognetti, D.O. Pederson, H. De Man eds., Sijthoff & Noordhoff, Rockville, Maryland, 1981.

J. Vlach and K. Singhal, *Computer Methods for Circuit Analysis and Design*, Van Nostrand Reinhold, New York, New York, 1983.

W. Weeks, *et al*, "Algorithms for ASTAP—A Network Analysis Program," *IEEE Transactions on Circuit Theory*, Vol. CT-20, November 1973.

6.2 Switch-level simulators and hybrid models

R. Bryant, *Logic Simulation of MOS LSI*, M.I.T. Laboratory for Computer Science TR-259, 1981.

P. Flake, P. Moorby, and G. Musgrave, "An Algebra for Logic Strength Manipulation," *Proceedings of 20th Design Automation Conference*, June 1983.

M. Horowitz, "Timing Models for MOS Pass Networks," *Proceedings of the IEEE International Symposium on Circuits and Systems*, 1983.

N. Jouppi, "TV: An nMOS Timing Analyzer," *Proceedings of the Third Caltech VLSI Conference*, 1983.

J. Ousterhout, "Crystal: A Timing Analyzer for nMOS VLSI Circuits," *Proceedings of the Third Caltech VLSI Conference*, 1983.

P. Penfield and J. Rubinstein, *Signal Delay in RC Tree Networks*, M.I.T. VLSI Memo No. 81-40, January 1981.

D. Pilling and H. Sun, "Computer-Aided Prediction of Delays in LSI Logic Systems," *Proceedings of 10th Design Automation Workshop*, June 1973.

C. Terman, "Timing Simulation for Large Digital MOS Circuits," *Advances in Computer-Aided Engineering Design, Volume 1*, JAI Press Inc., 1985.

6.3 Gate-level simulators

Z. Barzilai, *et al*, "Simulating Pass Transistor Circuits using Logic Simulation Machines," *Proceedings of 20th Design Automation Conference*, June 1983.

G. Case, "SALOGS-IV—A Program to Perform Logic Simulation and Fault Diagnosis," *Proceedings of 15th Design Automation Conference*, June 1978.

M. Denneau, "The Yorktown Simulation Engine," *Proceedings of 19th Design Automation Conference*, June 1982.

P. Flake, P. Moorby, and G. Musgrave, "Logic Simulation of Bi-directional Tri-state Gates," *Proceedings of IEEE International Conference on Circuits and Computers*, October 1980.

D. Holt and D. Hutchings, "A MOS/LSI Oriented Logic Simulator," *Proceedings of 18th Design Automation Conference*, June 1981.

R. McDermott, "Transmission Gate Modeling in an Existing Three-value Simulator," *Proceedings of 19th Design Automation Conference*, June 1982.

G. Pfister, "The Yorktown Simulation Engine: Introduction," *Proceedings of 19th Design Automation Conference*, June 1982.

W. Sherwood, "A MOS Modelling Technique for 4-State True-Value Hierarchical Logic Simulation," *Proceedings of 18th Design Automation Conference*, June 1981.

S. Szygenda, "TEGAS2—Anatomy of a General Purpose Test Generation and Simulation System for Digital Logic," *Proceedings of 9th ACM Design Automation Workshop*, June 1972.

S. Szygenda and E. Thompson, "Digital Logic Simulation in a Time-Based, Table-Driven Environment," *IEEE Computer*, Vol. 8, March 1975.

E. Thompson, *et al*, "Timing Analysis for Digital Fault Simulation Using Assignable Delays," *Proceedings of 11th Design Automation Conference*, June 1974.

E. Ulrich and T. Baker, "The Concurrent Simulation of Nearly Identical Digital Networks," *Proceedings of 10th Design Automation Workshop*, June 1973.

E. Ulrich, "Non-integral Event Timing for Digital Logic Simulation," *Proceedings of 13th Design Automation Conference*, June 1976.

LE-1000 Series Logic Evaluator Intermediate Form Specification, Release 1.0, Zycad Corporation, Roseville, MN, 1983.

4

Aspects of Computational Circuit Analysis

W. M. Coughran, Jr.   Eric Grosse
*AT&T Bell Laboratories**   *AT&T Bell Laboratories**

Donald J. Rose[†]
Duke University[‡]

Abstract

A hierarchical formulation of the differential-algebraic systems describing circuit behavior is presented. A number of algorithms that have proven effective are reviewed. These include multidimensional splines that preserve monotonicity, sparse direct and iterative methods for the linear equations, damped-Newton and Newton-iterative techniques for the nonlinear equations, continuation methods, and low-order time-integration formulae. Some aspects of time macromodeling are described.

1 Introduction

Circuit simulation has been of interest to the engineering community for a number of years. Although a variety of timing and logic simulators are now available, analog simulation of crucial subcircuits or even significant portions of an entire integrated circuit is often done. Simulation continues to be more cost effective than repeated fabrication of integrated circuits.

The emphasis here is on general purpose numerical algorithms that are applicable in circuit analysis rather than on the details of a particular implementation. It is our belief that modern numerical analysis can play an important role in the construction of an effective simulator.

This paper is organized as follows. The formulation of the circuit equations is described in the next section (§ 2), with emphasis on voltage-controlled circuit elements. The remaining sections discuss table representations for device constitutive functions (§ 3), linear-algebra techniques for linear DC operating-point analysis (§ 4), Newton-like methods for nonlinear DC operating-point computations (§ 6), continuation algorithms for transfer analysis (§ 5), low-order time-integration

---

*Murray Hill, New Jersey 07974, USA.

[†]This work was partially supported by AT&T Bell Laboratories, the Microelectronics Center of North Carolina, and the Office of Naval Research under Contract N00014-85-K-0487.

[‡]Department of Computer Science, Durham, North Carolina 27706, USA.

schemes (§ 7), and circuit macromodeling (§ 8). Note that much of the material in § 3–7 can be found in [9,2,1,10].

2 Formulation of the circuit equations

The overall behavior of a circuit is governed by the individual devices. (We will concentrate on voltage-controlled devices to simplify our discussions.) For example, a nonlinear resistor could be represented mathematically as

$$i_2 = f(u_2 - u_1) \tag{1}$$

where u_1 and u_2 are the node voltages at the terminals and i_2 is the current associated with the second terminal. Conservation of current implies that $i_1 = -i_2$. As a further example, a nonlinear capacitor obeys the following relation

$$i_2 = \frac{d}{dt} q(u_2 - u_1), \tag{2}$$

where $q(v)$ represents the charge.

Nonlinear resistors and capacitors are simple devices whose currents are governed by differences of node voltages at the terminals. More complicated elements are easily constructed. Consider, for example, a nonlinear resistor and capacitor in series (see Fig. 1). The current is given by

$$i_2 = f(w - u_1) = \frac{d}{dt} q(u_2 - w) \tag{3}$$

where u_1 and u_2 are the terminal voltages and w is the internal node voltage between the resistor and capacitor. Once again conservation of current guarantees $i_1 = -i_2$. Clearly, the internal voltage state w must be solved for given $u^0 = u(t_0)$.

Figure 1: Series RC.

The most important device for MOS integrated circuits is the transistor. A transistor is four-terminal device whose terminal currents obey the following relation

$$\begin{pmatrix} i_s \\ i_g \\ i_d \end{pmatrix} = f(v_{ds}, v_{gs}, v_{bs}) + \frac{d}{dt} q(v_{ds}, v_{gs}, v_{bs}) \tag{4}$$

where $v_{\alpha\beta} \equiv u_\alpha - u_\beta$ and s, g, d, and b correspond to the source, gate, drain, and bulk terminals, respectively. Kirchhoff's current law implies $i_b = -(i_s + i_g + i_d)$.

Note that this form assumes the so-called *quasi-static* approximation. In nMOS technology, $i_d \approx -i_s$ becomes appreciable when $v_{ds} \neq 0$ and v_{gs} exceeds the threshold voltage v_T, which is positive for enhancement transistors; depletion ($v_T < 0$) transistors are often used as nonlinear resistors (loads). Unless otherwise stated, assume nMOS technology is being discussed below.

The idea of a *macroelement* was informally introduced above with the description of a series RC circuit, where the internal voltage state w was not of particular interest. Another simple 2-terminal macroelement is the inverter (see Fig. 2), which obeys Kirchhoff equations of the form

$$i_{out} = i_d^{enh} + i_s^{dep} \tag{5}$$

$$i_{in} = i_g^{enh}. \tag{6}$$

The inverter has no internal voltage states. On the other hand, the two-input NAND (see Fig. 2) has an internal voltage state w, and w is usually of little interest to the circuit designer.

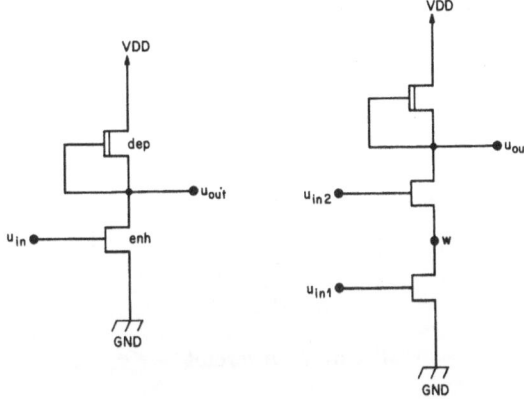

Figure 2: Inverter and two-input NAND.

We will say a node is a boundary node if its voltage with respect to ground is known (connected to a grounded voltage source). A node voltage associated with a nonboundary node is an unknown and must be computed. (Similar descriptions can be used with currents.) A recursive definition of a circuit is then given by:

Definition 1 *A circuit is a set of k-terminal (k variable) subcircuits and nodes* $N = \{n_j\}$ *such that:*

1. *Each terminal τ of a subcircuit has an associated node n_τ, a node voltage u_τ, and a current i_τ. Moreover, i_τ is determined by the terminal voltages $\{u_\alpha\}$ and state voltages w_s at (state) nodes $\{n_s\}$, where $\{n_s\} \cap N = \emptyset$.*

2. *Kirchhoff's current law holds, that is, the sum of currents at each nonboundary node $n_j \in N$ is zero.*

For each k-terminal subcircuit, indexed by l, with terminal nodes $U_l \subseteq N$ and state nodes W_l, we have the $|W_l|$ "internal" Kirchhoff equations

$$F_l^I(u_l, w_l) + \frac{d}{dt} Q_l^I(u_l, w_l) = 0 \tag{7}$$

and the $|U_l|$ output equations

$$i_l = F_l^E(u_l, w_l) + \frac{d}{dt} Q_l^E(u_l, w_l). \tag{8}$$

At each nonboundary node n_j

$$\sum_{l,\tau\, :\, n_\tau \in U_l,\, n_\tau \equiv n_j} i_{l,\tau} = 0. \tag{9}$$

Note that we suppress the explicit dependence of these (and later) equations on t, which arises from boundary nodes.

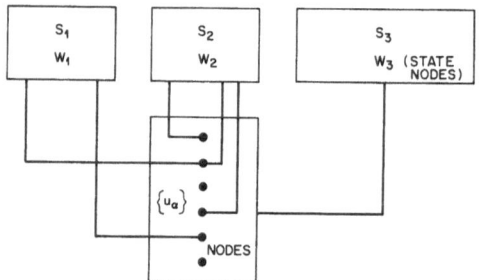

Figure 3: A circuit.

The global circuit equations can then be summarized as a coupled differential-algebraic system for the internal (w) and external (u) voltages

$$h^I(u, w) \;\equiv\; f^I(u, w) + \frac{d}{dt} q^I(u, w) = 0 \tag{10}$$

$$h^E(u, w) \;\equiv\; f^E(u, w) + \frac{d}{dt} q^E(u, w) = 0; \tag{11}$$

the first equation follows from Eq. (7) and the second from Eqs. (8)–(9). Traditional circuit simulation makes no distinction between the u's and the w's (see [21,37,22,26,9]). We will concentrate on this traditional view until we take up macromodeling in a later section so the equation of interest becomes

$$h(u) \equiv f(u) + \frac{d}{dt} q(u) = 0. \tag{12}$$

This form is a generalization of the usual (two-terminal) equations based on Kirchhoff's current and voltage laws

$$Ai \;=\; 0 \in \mathbb{R}^n \tag{13}$$

$$v \;=\; A^T u \in \mathbb{R}^m \tag{14}$$

where $A \in \mathbb{R}^{n \times m}$, u, v, and i are the reduced-incidence matrix, node voltages (which includes unknowns and boundary values), branch voltages, and branch currents (which includes unknowns and boundary values), respectively, and global constitutive relations

$$K(i,v) \equiv i - \left(\frac{d}{dt} q(v) + f(v) \right) = 0 \in \mathbb{R}^m. \tag{15}$$

Eqs. (1), (2), and (4) are typical components of the latter.

If Eq. (15) is linearized about a point, the resulting equation is

$$\Delta i + K_v \Delta v = r \tag{16}$$

where K_v represents the partial derivatives of K with respect to v. The linearized circuit equations can then be written in terms of the reduced-tableau matrix [21,9]

$$\begin{bmatrix} I & K_v A^T \\ A & 0 \end{bmatrix} \begin{pmatrix} \Delta i \\ \Delta u \end{pmatrix} = s. \tag{17}$$

Let us consider the special case of assembling that part of the linearized circuit equations associated with a transistor T_1. Suppose the terminals of T_1 are attached to four nodes $n_1, \ldots, n_4$, not necessarily distinct. Assume n_1 is connected to the source terminal so it is the reference node. The piece of the reduced-incidence matrix associated with T_1 is

$$\begin{bmatrix} 1 & 0 & 0 \\ 0 & 1 & 0 \\ 0 & 0 & 1 \\ -1 & -1 & -1 \end{bmatrix}. \tag{18}$$

We can write

$$K_v = \begin{bmatrix} K_{v1} & 0 \\ 0 & K_{vR} \end{bmatrix} \tag{19}$$

where $K_{v1} \in \mathbb{R}^{3 \times 3}$ corresponds to the constitutive relation for T_1 (Eq. (4)) while K_{vR} represents the remainder of the circuit. Let us partition A, i, and u similarly as $A = [A_1 \; A_R]$, $i^T = (i_1^T, i_R^T)$, and $u^T = (u_1^T, u_R^T)$, where $i_1 \in \mathbb{R}^3$ and $u_1 \in \mathbb{R}^4$. We can then write the reduced-tableau matrix as

$$\begin{bmatrix} I & 0 & K_{v1} A_1^T \\ 0 & I & K_{vR} A_R^T \\ A_1 & A_R & 0 \end{bmatrix} = \begin{bmatrix} I & 0 & 0 \\ 0 & I & 0 \\ A_1 & 0 & I \end{bmatrix} \begin{bmatrix} I & 0 & K_{v1} A_1^T \\ 0 & I & K_{vR} A_R^T \\ 0 & A_R & -A_1 K_{v1} A_1^T \end{bmatrix}. \tag{20}$$

Eq. (20) indicates how T_1 "assembles" into the lower right block of nodal equations; this can also be done using Eqs. (7)–(8). This process can be iterated to obtain the nodal equations for voltage-controlled elements and can be extended to non-voltage-controlled elements [9]. Thus, the assembly of circuit equations resembles finite-element assembly [35]; moreover, the global incidence matrix is never formed since the assembly is done device by device.

3   Table representations of devices

The functions f and q in Eq. (4) have classically been approximated in circuit simulation by polynomials and exponentials. These are chosen to follow physical properties such as "current increases with voltage." Different expressions are needed for various operating regions of the transistor. Ensuring continuity of these expressions at the interfaces, fitting various unknown parameters, and extending the models for new device behavior is labor intensive.

3.1   Variation-diminishing splines

First consider the univariate problem. We wish to approximate a smooth monotone function f, given data at uniformly spaced sample points $t_j^* = (j-1)h$ on the interval $[0,1]$ where $h = 1/(n-1)$ and $1 \leq j \leq n$. Since in our application only a C^1 approximation is needed, we elect to use quadratic splines. Take knots $t_j = (j-2.5)h$ midway between the sample points t_{j-2}^* and t_{j-1}^*. (These are chosen so that $t_j^* = (t_{j+1} + t_{j-1})/2$, as is required for variation-diminishing splines.) Using the data as B-spline coefficients gives the variation-diminishing spline [33]

$$S(x) = \sum f(t_j^*) B_j(x). \tag{21}$$

(See [5] for the definition of $\{B_j\}$, which are written as $\{B_{j,3,t}\}$ in that reference.) Because of the local support of the B-splines, if $t_j \leq x \leq t_{j+1}$, then $S(x)$ depends only on f at t_{j-2}^*, t_{j-1}^*, and t_j^*. In the trivial case $n = 1$, take $S(x) = f(t_1^*)$.

Note that we do not use the customary multiple knots at the endpoints. We thereby obtain B-splines that are all identical up to translation

$$B_j(x) = B_0(x - jh) \tag{22}$$

and avoid introducing an irregular sample point near the boundary. But the definition of $S(x)$ for $x < h/2$ refers to $f(t_0^*)$, an imaginary sample outside $[0,1]$ indicated by the dotted circle in Fig. 4. We implicitly estimate this by linear extrapolation from $f(t_1^*)$ and $f(t_2^*)$. This implies that for $x < h/2$ the spline reduces to a linear function. Here and in the following, we only discuss the left boundary and implicitly treat the right boundary symmetrically. This technique is used in computer graphics under the name of "phantom vertices" [4]. This nonstandard definition retains the properties of variation-diminishing quadratic splines defined with the usual multiple knots [10]. In particular, *if f'' is Lipschitz continuous, then S has the following properties: $S \in C^1$; if f is linear, then $S = f$; $S(0) = f(0)$, $S(1) = f(1)$; if f is monotone or convex, then so is S; if f is quadratic, then $S' = f'$; $\|f - S\|_{L_\infty[0,1]} = O(h^2)$; $\|f' - S'\|_{L_\infty]0,1[} = O(h^2)$; and $|f' - S'| = O(h)$ at 0 and 1.*

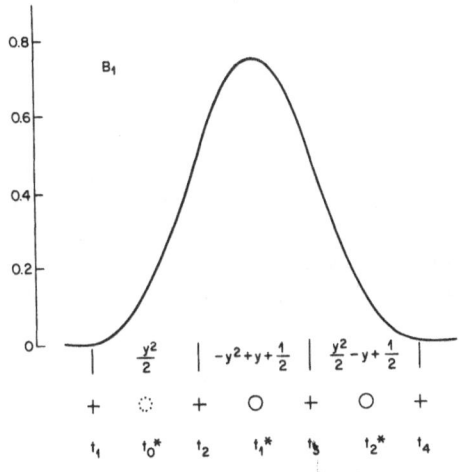

Figure 4: B_1 with knots indicated by pluses. Function values are sampled at the points indicated by o's.

The restriction to a uniform mesh saves a factor of eight in execution time over efficient general spline codes [5] with $k = 3$ and leads to improved convergence of the derivative. Note that higher order splines would not give higher order convergence to f, though they of course would give more continuous derivatives. A more complicated code could handle multiple endpoint knots without much loss of efficiency. A minor advantage of variation-diminishing splines is that no preprocessing of the data is required.

3.2 Tensor products

Any linear univariate approximation process can be extended to several variables through the use of tensor products [20]. For the variation-diminishing spline, a two-dimensional tensor variant with the same knots in each variable is given by

$$S(x,y) = \sum_{jk} f(t_j^*, t_k^*) B_j(x) B_k(y). \tag{23}$$

(In practice we use different numbers of knots in different variables; we have simplified here to avoid the otherwise bewildering indices.) *If the second derivatives of f* are continuous, then $\|f - S\|_{L_\infty[0,1]} = O(h^2)$, where *h* is the larger of the sample spacings in *x* and *y* [10].

Define a bivariate function *f* to be monotone increasing if $x \leq x^\star$ and $y \leq y^\star$ implies $f(x,y) \leq f(x^\star, y^\star)$. If *f* is monotone, then so is *S* [10].

The extension to three variables is immediate:

$$S(x,y,z) = \sum_{jkl} f(t_j^\star, t_k^\star, t_l^\star) B_j(x) B_k(y) B_l(z). \tag{24}$$

The computational costs of a tensor spline in *p* dimensions is $O(\prod n_j)$ space given n_j sample points for the *j*th variable and $O(3^p)$ time per evaluation [20].

Our FORTRAN implementation takes 25 microseconds on a Cray XMP using the CFT 1.15 compiler, including subroutine call overhead, for a single evaluation of *S* and its partial derivatives with $p = 3$. In the transient circuit-simulation application, for each set of applied transistor voltages, we need the steady-state current and three charges, assuming no bulk leakage currents. By running these four evaluations together, the cost per evaluation drops to 13 microseconds, which is comparable to the cost of compact analytical models. (A version of the code has also been used in a timing simulator [36].)

Since *S* is linear near the endpoints, there is a natural C^1 linear extension to $\mathbb{R}$. This is often an excellent approximation in transistor modeling and allows the Newton iteration in the circuit simulator to temporarily step outside the physically realizable region (§ 5).

If data can be sampled on general grids, coordinate transformations such as square root in the v_{gs} variable would reduce somewhat the number of coefficients needed. (We would still use uniform knots, but in the transformed variable.)

We have assumed implicitly a transistor of specific length and width; for a typical circuit simulation perhaps a dozen tables would be required. Sometimes the width can be treated as simply a scale parameter so that fewer tables are needed.

More recently, we have experimented with linear B-splines. On coarse grids these lead to significant errors in the Jacobians, but the Newton method (§ 5) usually seems to be robust enough to converge anyway. It is not yet clear whether the cheaper evaluations save enough to overcome the increase in iterations.

4   Linear-algebra techniques

After device constitutive relations have been represented (§ 3), the next problem of interest is solving for *x* in

$$Ax = b \in \mathbb{R}^n \tag{25}$$

where $A \in \mathbb{R}^{n \times n}$ is large and sparse (few nonzero entries per row). This problem arises during the Newton iterations that are employed in the DC operating-point and transient analyses (§ 5).

Only the nonzeros of A and appropriate pointers need to be stored. We use either the standard IA,JA,A format or a variant that assumes the matrix is structurally symmetric, that is, $a_{ij} \neq 0$ implies $a_{ji} \neq 0$. In our implementation, the matrices are effectively structurally symmetric.

Sparse direct methods compute a factorization

$$PAP^T = LDU \qquad (26)$$

where P is a permutation matrix, chosen by the minimum degree algorithm (once) in an attempt to minimize fill-in [12,19], D is diagonal, and L and U are unit lower and upper triangular, respectively. (Such a factorization is well defined when $A + A^T$ is positive definite.) Eq. (25) is then solved by a forward and backward substitution. If $n \gg 1$, the cost of computing the factorization is usually much larger than the forward and backward substitutions, which can be exploited in nonlinear equation algorithms (§ 5).

The matrix formats and codes are documented in the literature [14,13,12,1] and, due to space constraints here, we will defer to these earlier papers.

Iterative methods, such as conjugate gradients, seem attractive because the cost per iteration is much lower than the cost of computing an LDU factorization. Unfortunately, the nonsymmetry of A make the convergence of iterative methods problematical. It is possible, however, to enhance the performance of a Newton-Richardson algorithm (§ 5) by using Orthomin as discussed in [1].

5 Newton-like methods

The underlying circuit equations for the DC operating point or a transient time step are of the form

$$h(z) = 0 \in \mathbb{R}^n \qquad (27)$$

where $z \in \mathbb{R}^n$ is the vector of unknown node voltages (and possibly currents). There are a number of approaches to solving such nonlinear systems but we will concentrate on Newton-like methods here.

Newton's method produces a series of iterates z_j and is motivated by the following Taylor expansion (requiring a sufficiently smooth h)

$$h_{j+1} = h_j + h'_j(z_{j+1} - z_j) + \int_0^1 \{h'(z_j + s(z_{j+1} - z_j)) - h'_j\}(z_{j+1} - z_j)\, ds \qquad (28)$$

where $h_j \equiv h(z_j)$ and $h'_j \equiv (\partial h/\partial z)(z_j)$. If we assume $h_{j+1} = 0$ and neglect the integral remainder term above, we obtain the usual Newton correction equation

$$h'_j x_j = -h_j, \qquad (29)$$

which is well defined if h'_j is nonsingular. Note that Eq. (28) implies $-(h'_j)^{-1}h_j$ is a descent direction, that is, $z_j - \epsilon(h'_j)^{-1}h_j$ results in a smaller value of $\|h\|$ for small enough ϵ.

The basic Newton procedure is as follows:

Algorithm 1 Let z_0 be an initial guess to a solution z^* for which $h(z^*) = 0$.

1. Set $j = 0$;

2. Do

    a. Solve $h'_j x_j = -h_j$ for x_j;

    b. Set $z_{j+1} = z_j + x_j$;

    c. Replace $j \leftarrow j + 1$;

    d. If $j \geq m$, then report failure to converge;

  until $\|x_{j-1}\| \leq \epsilon_1 \|z_j\|$ and $\|h_j\| \leq \epsilon_2$.

Step 2 is one possible stopping criterion for this iterative procedure; the two ϵ parameters can be adjusted to make both the change x_j and the function value small. The m parameter limits the total number of iterations allowed. (Note that the 2-norm, $\|x_j\|^2 = x_j^T x_j$, is used in the remainder of this paper.)

It is well known that Newton's method exhibits quadratic convergence, that is, $\|z_{j+1} - z^*\| = O(\|z_j - z^*\|^2)$, in a neighborhood of z^* [27]. There is no guarantee, however, that $\lim_{j \to \infty} z_j \to z^*$ for an arbitrary z_0.

The basic Newton procedure can be made more robust in a variety of ways. One possibility is to use damping in an attempt to force $\|h_j\| \to 0$ in a monotone way, which is motivated by the fact that $-(h'_j)^{-1} h_j$ is a descent direction. A damped-Newton algorithm is given by:

Algorithm 2

1. Set $j = 0$, $l = 0$, $s_{-1} = 1$, and $s_0 = 1$;

2. Do

    a. Solve $h'_j x_j = -h_j$ for x_j;

    b. Set $z_{j+1} = z_j + s_j x_j$;

    c. If $1 - \|h_{j+1}\|/\|h_j\| < \epsilon_M s_j$, then

        i. Replace $l \leftarrow l + 1$;

        ii. If $l > L$, report failure to reduce the norm;

        iii. Replace $s_j \leftarrow s_{j-1} (\epsilon_M \|z_j\|/\|x_j\|)^{(l^2/L^2)}$;

        iv. go to step 2b to redefine z_{j+1};

    d. Set $s_{j+1} = s_j/(s_j + 0.2(1 - s_j)\|h_{j+1}\|/\|h_j\|)$;

    e. Reset $l \leftarrow 0$;

    f. Replace $j \leftarrow j + 1$;

    g. If $j \geq m$, then report failure to converge;

  until $\|x_{j-1}\| \leq \epsilon_1 \|z_j\|$ and $\|h_j\| \leq \epsilon_2$.

Here ϵ_M is the machine epsilon and L is usually set to 9. Step 2c enforces the so-called sufficient-decrease condition described in [2]. Substep iii decreases s_j more and more rapidly on consecutive failures to reduce $\|h_{j+1}\|$ while step 2d geometrically increases $s_{j+1} \to 1$ after a successful step.

With appropriate hypotheses, this damped-Newton method is globally convergent and is quadratically convergent in a neighborhood of the solution [2]. One of the necessary hypotheses is not always satisfied for circuit simulation problems. In particular, the theory requires that $\|(h'_j)^{-1}\| \leq C$ for all j but h' may be (nearly) singular when a subcircuit is (nearly) decoupled from the rest of the circuit. The simplest example of this difficulty arises in the DC operating-point problem for the NAND (see Fig. 2) with two zero inputs. The "don't care" node between the two enhancement transistors does not have a well-determined value, which is reflected in the condition number of h'.

There are various schemes for dealing with the singularity of h'. One possibility is to use so-called two-parameter damping where Eq. (29) is replaced by

$$(h'_j + \lambda_j \|h_j\| I) x_j = -h_j; \qquad (30)$$

the diagonal shift often produces a nonsingular equation but $\lambda_j > 0$ cannot be too large if the results of [2] are to apply. Another possibility is to recognize disconnected subcircuits and solve each of them independently by arbitrarily setting one node voltage (deflation). Other approaches manipulate λ_j without damping ($s_j = 1$) [3] or treat the problem via nonlinear least squares [11]. Finally, homotopy methods may be applicable [15]. Our present implementation uses damped Newton (Alg. 2) with two-parameter damping, trivial deflation [9], or a variant of the method described in [3] as options.

A number of codes employ the more general update equation

$$z_{j+1} = z_j + D_j x_j \qquad (31)$$

where D_j is a diagonal matrix. This scheme allows component-wise chopping of values so an individual component of z_{j+1} can be constrained to a physically reasonable value. There is no guarantee that $D_j x_j$ is a descent direction since x_j and $D_j x_j$ are not co-linear in general (unless, for example, $D_j = s_j I$). Moreover, it is possible to evaluate $h(z)$ for arbitrary values of z using our tensor-product splines (§ 3) so we have no need for component-wise chopping.

Step 2a of Alg. 2 is often costly, particularly if sparse direct methods are used to compute an LDU factorization of h'_j followed by a forward and backward substitution to compute x_j (§ 4). One scheme to reduce the overall expense, the so-called Newton-Richardson iteration, is to reuse the sparse LDU factorization of an old Jacobian in an inner iteration. The Newton-Richardson concept starts with a uniformly convergent splitting of h'_j given by

$$h'_j = M_j - N_j \qquad (32)$$

where $\|M_j^{-1}N_j\| = \|I - M_j^{-1}h_j'\| \leq \rho < 1$ for all j. Typically, M_j represents the sparse LDU factors of an old Jacobian h_{j_0}'. The actual Newton-Richardson algorithm is the same as Alg. 2 except step 2a is replaced by the following iterative procedure:

Algorithm 3

1. Set $k = 0$ and $y_{j0} = 0$;
2. Do

    a. Solve $M_j(y_{j,k+1} - y_{jk}) = -(h_k'y_{jk} + h_k)$ for $y_{j,k+1} - y_{jk}$;
    b. Replace $k \leftarrow k + 1$;

   until $\|h_j'y_{jk} + h_j\| \leq \alpha\|h_j\|^2/\|h_0\|$;

3. Set $x_j = y_{jk}$.

Here $0 < \alpha < 1$ is an experimentally determined parameter. The Newton-Richardson algorithm with the stopping criterion given by step 2 above results in quadratic convergence [2]. This is intuitively appealing since the stopping criterion requires more inner iterations as $\|h_j\|$ becomes smaller.

6   Continuation methods

Computing transfer curves is an obvious application for predictor-corrector continuation methods [23,15] since most of these problems amount to solving for $z(\lambda)$ in $h(z, \lambda) = 0$ with the voltage λ ranging over $[\underline{\lambda}, \overline{\lambda}]$. The predictor-corrector approach is motivated by parameterizing the equation by arc-length, s,

$$h(z(s), \lambda(s)) = 0 \qquad (33)$$

and then differentiating with respect to s (denoted by dots) to obtain

$$h'\dot{z} + h_\lambda\dot{\lambda} = 0 \qquad (34)$$
$$\|\dot{z}\|^2 + |\dot{\lambda}|^2 = 1, \qquad (35)$$

where h_λ represents the partial derivatives of h with respect to λ.

If the solution (z_j, λ_j) and its unit tangent $(\dot{z}_j, \dot{\lambda}_j)$ are known for some s_j, then we can advance to s_{j+1} by predicting a new set of values with forward Euler

$$\begin{pmatrix} z_{j+1} \\ \lambda_{j+1} \end{pmatrix} = \begin{pmatrix} z_j \\ \lambda_j \end{pmatrix} + (s_{j+1} - s_j)\begin{pmatrix} \dot{z}_j \\ \dot{\lambda}_j \end{pmatrix} \qquad (36)$$

and then applying a Newton-like method to the corrector equations

$$h(z_{j+1}, \lambda_{j+1}) = 0 \qquad (37)$$
$$N(z_{j+1}, \lambda_{j+1}) = 0 \in \mathbb{R}. \qquad (38)$$

The augmenting equation, $N = 0$, is often taken to be the forward-Euler *pseudo-arc-length* equation

$$N(z_{j+1}, \lambda_{j+1}) \equiv \dot{z}_j^T(z_{j+1} - z_j) + \dot{\lambda}_j(\lambda_{j+1} - \lambda_j) - (s_{j+1} - s_j) = 0 \qquad (39)$$

which, under appropriate hypotheses, ensures the nonsingularity of the Jacobian of Eqs. (37)–(38) [23], even at simple limit points where $h' \in \mathbb{R}^{n \times n}$ has rank $n - 1$ and $h_\lambda \notin \text{Range}(h')$. Other forms for N have been proposed but Eq. (39) suffices for curve tracing. Note that once (z_{j+1}, λ_{j+1}) has been determined then Eq. (34) implies $\dot{z}_{j+1} = -\dot{\lambda}_{j+1}(h'_{j+1})^{-1}h_{\lambda_{j+1}}$ and Eq. (35) normalizes $(\dot{z}_{j+1}, \dot{\lambda}_{j+1})$ so $(\dot{z}_{j+1}, \dot{\lambda}_{j+1})$ can be determined up to a choice of sign; moreover, the solution of the appropriate linear system is needed for block Gaussian elimination (see below) so little additional work is required to compute the tangent.

If there are no limit points, then it suffices to predict a value using Eq. (36), fix λ_{j+1}, and solve Eq. (37) for z_{j+1}. This approach reacts to the local curvature of the solution and provides a natural step size in λ unlike methods that take a fixed step in λ. Of course, this leaves the problem of regulating $s_{j+1} - s_j$.

Selecting the step size $s_{j+1} - s_j$ can be a complicated matter. Since we are interested in following the curve fairly closely, we could monitor the truncation error of the forward-Euler predictor (Eq. (36)) using divided differences. Our current implementation uses a modification of Algorithm III in [31], which regulates the angle between the tangent $(\dot{z}_j, \dot{\lambda}_j)$ and the secant $(z_j - z_{j-1}, \lambda_j - \lambda_{j-1})/(s_j - s_{j-1})$. It is sometimes necessary to repeat a step if $\|z_{j+1} - z_j\|$ (or $|\lambda_{j+1} - \lambda_j|$) turns out to be too large. This step-size selection procedure could be improved.

The augmented linear system for the Newton correction equation associated with Eqs. (37)–(38) is of the form

$$\begin{bmatrix} A_{11} & a_{12} \\ a_{21}^T & a_{22} \end{bmatrix} \begin{pmatrix} x_1 \\ x_2 \end{pmatrix} = \begin{pmatrix} b_1 \\ b_2 \end{pmatrix}, \qquad (40)$$

where $A_{11} \in \mathbb{R}^{n \times n}$, $a_{12}, a_{21}, x_1, b_1 \in \mathbb{R}^n$, and $a_{22}, x_2, b_2 \in \mathbb{R}$. In our application, $A_{11} = h'$, $a_{12} = h_\lambda$, $a_{21} = \dot{z}$, and $a_{22} = \dot{\lambda}$.

Block elimination is often used to solve Eq. (40). This approach is motivated by the block factorization

$$\begin{bmatrix} A_{11} & a_{12} \\ a_{21}^T & a_{22} \end{bmatrix} = \begin{bmatrix} I & 0 \\ a_{21}^T A_{11}^{-1} & I \end{bmatrix} \begin{bmatrix} A_{11} & a_{12} \\ 0 & a_{22} - a_{21}^T A_{11}^{-1} a_{12} \end{bmatrix}. \qquad (41)$$

Then the basic block-elimination algorithm is given by the following:

Algorithm 4

1. Solve $A_{11}w = a_{12}$ and $A_{11}y = b_1$ for w and y, respectively;
2. Set $x_2 = (b_2 - a_{21}^T y)/(a_{22} - a_{21}^T w)$;
3. Set $x_1 = y - x_2 w$.

However, it is known that the Jacobian, represented here by A_{11}, becomes singular at limit points so that block elimination can be ill-conditioned. One possibility is to use the deflated-block-elimination algorithm due to Chan [7]:

Algorithm 5 Let ψ_0 (say $(1, 1, \ldots, 1)^T$) be a guess for the left-singular vector (associated with the smallest singular value) of A_{11}.

1. For $j = 1, \ldots, m$ (say $m = 2$), do

   a. Solve $A_{11}\tilde{\phi}_j = \psi_{j-1}$ for $\tilde{\phi}_j$;

   b. Set $\phi_j = \tilde{\phi}_j / \|\tilde{\phi}_j\|$;

   c. Solve $A_{11}^T \tilde{\psi}_j = \phi_j$ for $\tilde{\psi}_j$;

   d. Set $\sigma_j = 1/\|\tilde{\psi}_j\|$;

   e. Set $\psi_j = \sigma_j \tilde{\psi}_j$.

2. Set $\alpha = \psi_m^T a_{12}$ and $\beta = \psi_m^T b_1$;

3. Solve $A_{11}w = a_{12} - \alpha\psi_m$ and $A_{11}y = b_1 - \beta\psi_m$ for w and y, respectively;

4. Set $\gamma_1 = b_2 - a_{21}^T y$, $\gamma_2 = a_{22} - a_{21}^T w$, and $\gamma_3 = a_{21}^T \phi_m$;

5. Set $\delta_1 = \alpha\gamma_1 - \beta\gamma_2$, $\delta_2 = \beta\gamma_3 - \sigma_m\gamma_1$, and $\delta_3 = 1/(\beta\gamma_3 - \sigma_m\gamma_2)$;

6. Set $x_2 = \delta_2\delta_3$ and $x_1 = y + \delta_3(\delta_1\phi_m - \delta_2 w)$.

Step 1 of the algorithm is just inverse iteration to obtain approximations to the smallest singular value and the corresponding left- and right-singular vectors [34,7]; it often suffices to do two or even one-and-a-half iterations ($1/\|\tilde{\phi}_j\|$ also approximates the smallest singular value).

This algorithm is effective when A_{11} has at least rank $n - 1$ (also see [8]). Obviously, there is the cost of doing a few extra forward and backward substitutions to approximate the singular value and vectors but this only need be done near limit points and can often be reused for several continuation steps. There can be difficulties near a limit point if the circuit has (nearly) disconnected subcircuits and, hence, more complicated singularities, as mentioned earlier.

7 Time-integration schemes

Circuit equations (Eq. (12)) are not ODEs but are differential-algebraic systems (DASs), that is, q' may be singular. Nevertheless, there is a history of applying backward-differentiation formulae to such problems [17,6]. (It has recently been made clear that DASs can be degenerate [29,18,32], but some appropriate mathematical software exists [28].) Here we will emphasize a simple second-order scheme that is well suited for use with a Newton-Richardson algorithm (§ 5); second-order methods appear to be reasonably efficient for the circuit simulations we have performed, which agrees with the previous experience [37,26].

We make use of a trapezoidal-rule/backward-differentiation-formula (TR-BDF2) composite method. Consider integrating Eq. (12) from $t = t_n$ to $t_{n+1} \equiv t_n + \Delta t_n$. We apply TR to go from $t = t_n$ to $t_n + \gamma \Delta t_n$

$$2q_{n+\gamma} + \gamma \Delta t_n f_{n+\gamma} = 2q_n - \gamma \Delta t_n f_n. \tag{42}$$

This implicit scheme has a Jacobian of the form $2q'_{n+\gamma} + \gamma \Delta t_n f'_{n+\gamma}$. We then apply the second-order backward-differentiation formula (BDF2) to go from $t = t_n + \gamma \Delta t_n$ to t_{n+1}

$$(2 - \gamma)q_{n+1} + (1 - \gamma)\Delta t_n f_{n+1} = \gamma^{-1}q_{n+\gamma} - \gamma^{-1}(1 - \gamma)^2 q_n. \tag{43}$$

This implicit scheme has a Jacobian of the form $(2 - \gamma)q'_{n+1} + (1 - \gamma)\Delta t_n f'_{n+1}$.

The TR and BDF2 Jacobians have the same form if

$$\frac{2}{\gamma} = \frac{2 - \gamma}{1 - \gamma} \tag{44}$$

which implies $\gamma = 2 - \sqrt{2} \approx 0.59$. Assume γ takes this value for the remainder of this paper. With a Newton-Richardson algorithm (§ 5), it is often possible to reuse Jacobian factorizations and still retain rapid convergence to the solution of Eq. (42) or Eq. (43) [1].

Consider applying a one-step method $y_{n+1} = A(\lambda \Delta t)y_n$ to the usual scalar test problem $\dot{y} = \lambda y$ with $\Re \lambda < 0$, where $\Re \lambda$ denotes the real part of λ. Recall that the one-step method is said to be A-stable if $|A(\lambda \Delta t)| < 1$ for all $\lambda \Delta t$ with $\Re \lambda \Delta t < 0$; the method is said to be L-stable if it is A-stable and $|A(\lambda \Delta t)| \to 0$ as $|\lambda \Delta t| \to \infty$ [24]. However, A-stability alone may not be strong enough for extremely stiff problems. For example, if the A-stable TR method is used when $\Re \lambda \ll 0$, then $y_{n+1} \approx -y_n$ unless $|\lambda \Delta t| = O(1)$ so, without proper error control, "ringing" may occur, which can be exacerbated by a nonlinear problem. But restrictions on Δt are anathema for stiff problems. On the other hand, BDF2 is known to be L-stable so there are no restrictions on Δt. BDF2 has a higher truncation error than TR and is trickier to implement since it is not a one-step method, however.

The principal truncation term for a step of TR-BDF2 (Eqs. (42)–(43)) is

$$C(\Delta t_n)^3 q^{(3)}(\xi) \tag{45}$$

where

$$C = \frac{-3\gamma^2 + 4\gamma - 2}{12(2 - \gamma)} \approx -0.04. \tag{46}$$

Note that $|C(\gamma)|$ is minimized when $0 < \gamma = 2 - \sqrt{2} \le 1$.

The composite TR-BDF2 procedure is an easily restarted, second-order, one-step, composite-multistep algorithm, which is nearly as simple to implement as TR. It is compatible with the Newton-Richardson algorithm. Finally, the scheme is suitable for stiff problems requiring moderate accuracy since *the TR-BDF2 method with $\gamma = 2 - \sqrt{2}$ is L-stable* [1].

In order to regulate the step size Δt_n, we need to estimate the local truncation error (LTE). (Note that error estimation for DASs can be much more difficult than for simple ODEs [29,28,18].) We have found that the divided-difference estimator

$$\tau_{n+1} = 2C\Delta t_n \left[\gamma^{-1}f_n - \gamma^{-1}(1-\gamma)^{-1}f_{n+\gamma} + (1-\gamma)^{-1}f_{n+1}\right] \approx C(\Delta t_n)^3 q^{(3)} \quad (47)$$

approximates the LTE in terms of q reasonably well and is inexpensive to compute. (There are a variety of alternatives for estimating the LTE [17,29,28,1].)

Given a per component LTE estimate, we can predict a new candidate step size $\Delta t^\star$, expected to satisfy a specified error tolerance, by

$$\Delta t^\star = \Delta t_n / \sqrt[3]{r} \quad (48)$$

where

$$r = \frac{\|\tau_{n+1}\|}{\|e_{n+1}\|} \quad (49)$$

$$e_{n+1,i} = \epsilon_1 |q_{n+1,i}| + \epsilon_2. \quad (50)$$

Here the second subscript on q represents the component and ϵ_1 and ϵ_2 are absolute- and relative-error parameters, respectively, and the cube root reflects the second-order nature of the scheme. The error measure represented by r can be insensitive to large relative changes in small values so more conservative schemes may be appropriate at times [1].

If $r \le 2$, the step is accepted; otherwise the step is repeated with $\Delta t_n \leftarrow 0.9\Delta t^\star$, where the 0.9 is a "paranoia" factor. If for some reason the nonlinear equations cannot be solved in a small fixed number of iterations, the step is repeated with $\Delta t_n \leftarrow \Delta t_n/2$. If the step is accepted, the next step size is taken as $\Delta t_{n+1} = \min(0.9\Delta t^\star, 2\Delta t_n)$ subject to minor adjustments as mentioned below. This last rule restricts the rate of increase in order to avoid step-size oscillations.

Since input wave forms often have natural breakpoints that should be sampled exactly at the corners for graphical reasons, we have employed the device described in [16], which further limits the step size. In particular, if the integration is to stop at t_s, we take

$$\Delta t_{n+1} \leftarrow \frac{t_s - t}{\lceil (t_s - t)/\Delta t_{n+1} - \epsilon \rceil} \quad (51)$$

where ϵ is a small multiple of the machine epsilon.

Some circuit-analysis packages try to get away with much simpler step-size control schemes. One choice is to cut back the step size when the previous time step took more than a certain number of Newton iterations and to increase the step size when the previous step took less than another certain number of Newtons. This approach tries to maintain a "reasonable" number of Newtons per time step. If this scheme is applied to a linear RC network, the Newton procedure will always converge in one iteration resulting in continual step-size increases and, thereby, arbitrarily bad truncation errors.

8 Macromodeling of circuits

Recall that Definition 1 implies a hierarchy of circuit variables in the sense that we write the global circuit equations in terms of internal w and external u voltages

$$h^I(u, w) \equiv f^I(u, w) + \frac{d}{dt} q^I(u, w) = 0 \tag{52}$$

$$h^E(u, w) \equiv f^E(u, w) + \frac{d}{dt} q^E(u, w) = 0. \tag{53}$$

The goal of macromodeling is to decouple the computation of the u's, which are of interest, from the w's, which are of lesser concern. Ideally, we would like to "eliminate" the w variables and assemble equations that model the macroelements. This is not completely possible in the transient simulation context since initial conditions on the w variables play a role. However, momentarily suppressing the dependence on initial conditions, we will see that macromodeling in a general sense can be viewed as nonlinear elimination.

8.1 Macromodeling as nonlinear elimination

Assume that given u, $h^I(u, w) = 0$ determines $w(u)$; that is, the internal equation is used to solve for, and hence eliminate, w given u. Note that h^I is a block-diagonal function of w for fixed u since the internal voltages of one macroelement do not interact directly with the internal voltages of another macroelement. Suppose we then solve the remaining equation in u, $h^E(u, w(u)) = 0$, by a damped-Newton scheme

$$\left[h_1^E + h_2^E \frac{dw}{du} \right] \Delta u = -h^E(u) \tag{54}$$

$$u \leftarrow u + s\Delta u, \tag{55}$$

where h_1^E represents the partial derivatives of h^E with respect to its first variable. We need the quantity dw/du. From $h^I(u, w(u)) = 0$, we obtain

$$h_1^I + h_2^I \frac{dw}{du} = 0. \tag{56}$$

This can be rewritten as

$$\frac{dw}{du} = -(h_2^I)^{-1} h_1^I. \tag{57}$$

The Newton correction equation (54) then becomes

$$\left[h_1^E - h_2^E (h_2^I)^{-1} h_1^I \right] \Delta u = -h^E(u). \tag{58}$$

Eqs. (54)–(58) can be interpreted in the context of block Gaussian elimination.

The Jacobian of the coupled system for the internal and external voltages (Eqs. (52)–(53)) is

$$h' = \begin{bmatrix} h_2^I & h_1^I \\ h_2^E & h_1^E \end{bmatrix}. \tag{59}$$

Block Gaussian elimination gives the Schur complement in the lower right (2,2) position

$$h_1^E - h_2^E (h_2^I)^{-1} h_1^I, \tag{60}$$

which is what we had before. Eqs. (54)–(58) can be seen as solving

$$\begin{bmatrix} h_2^I & h_1^I \\ h_2^E & h_1^E \end{bmatrix} \begin{pmatrix} \Delta w \\ \Delta u \end{pmatrix} = - \begin{pmatrix} 0 \\ h^E(u) \end{pmatrix} \tag{61}$$

by block elimination as indicated by Eqs. (59)–(60). Note that the first block equation of Eqs. (61) and (56) imply the notationally obvious

$$\Delta w = \frac{dw}{du} \Delta u. \tag{62}$$

Indeed Eq. (62) and the zero in the right-hand side of Eq. (61) are consequences of $h^I(u, w) = 0$; a usual Newton-type iteration on the complete system would, of course, replace the zero in Eq. (61) by h^I (which would, in general, be nonzero) and the relation given by Eq. (62) would not be valid. Even so, the interpretation represented by Eq. (61) indicates that a *macromodel iteration* (Eqs. (54)–(58)) requires the same information as a complete Newton iteration. The consequence of this interpretation is the observation that macromodeling will pay off if getting w from $h^I(u, w) = 0$ is inexpensive and enforcing $h^I = 0$ requires fewer overall Newton iterations than would be necessary to solve the complete system (also see [30]). In table-oriented macromodeling for the DC case, we will see that we also get a bonus of obtaining the Schur complement essentially free.

8.2   Macromodeling using tables

The static (or DC) macromodeling problem is simpler since the charges q are identically zero. Thus, the equations are nonlinear algebraic instead of nonlinear operator equations.

For the DC operating-point problem, let us consider a particular macroelement and suppose its terminal voltages u_l are given. Then the internal state voltages $w_l(u_l)$ come from solving

$$F_l^I(u_l, w_l) = 0. \tag{63}$$

We approximate (fit) the related terminal current function

$$i_l = F_l^E(u_l, w_l(u_l)) \equiv G_l(u_l). \tag{64}$$

With our tensor-product variation-diminishing spline approach, most of the Schur complement is returned as $G_l'(u_l)$ along with i_l [9]. In practice, we see quadratic convergence of $h^E(u, w(u)) = 0$ to a u^* correct to within the tolerance provided by the tables.

In combinatorial logic circuits, the circuit equations can become acyclic leading to the possibility of "numerical logic simulation."

In [9] we summarized the behavior of a damped-Newton method used to compute a DC operating point for a one-bit full adder made up of nine two-input NANDs with the NANDs represented by DC macromodels (Eq. (64)). (The NAND macromodel (Eq. (64)) was constructed by simulating a NAND (see Fig. 2) for various prescribed values of u_{in1}, u_{in2}, and u_{out}.) The advantages of such DC macromodeling include reducing the size of the nonlinear system (since the internal voltage w is not needed) and alleviating some of the inherent singular behavior of DC operating-point problems. With our current Newton schemes and models, the same one-bit adder simulation with DC macromodels and zero inputs as described in [9] saves at least a factor of three in execution time (due to reduced Newton iterations and smaller nonlinear systems) for moderate accuracy requests. Additional savings are seen if the Newton stopping criteria are tightened. Moreover, the NAND's internal node can be ill-determined (§ 5) and this is reflected by an increased need for damping and linear Newton convergence, unless the DC macromodel is employed.

This static table approach can be extended to the transient case. The internal equation, $h^I(u, w) = 0$, discretized by a one-step method, say backward Euler, with fixed time step Δt is

$$\frac{q^I(u, w)}{\Delta t} + f^I(u, w) = \frac{q^I(u_o, w_o)}{\Delta t} \equiv r_o. \qquad (65)$$

Given u_o, w_o, and u, we can obtain $w(u; u_o, w_o)$. As noted before, the above equation can be viewed as that of an individual macroelement since the global h^I is block diagonal. The external equation, $h^E(u, w(u)) = 0$, can then be discretized similarly and u can be be computed by a Newton-like method using the $w(u; u_o, w_o)$ from above.

The $r_o(u_o, w_o; \Delta t)$ above is a linear combination of q's (and, in general, f's), which arises naturally during the assembly of the global equations. Splines can be used to fit the internal equation solutions

$$w = g(u, r_o; \Delta t), \qquad (66)$$

which gives $w(u)$ and dw/du. Derivatives in the r_o variable are not needed.

This approach is the natural extension of the static macromodeling case. However, r_o and Δt are new parameters. It may be possible to create a few tables for several Δt values and interpolate to vary the time step.

The scheme works for the two-input NAND where

$$w = g(u_{in_1}, u_{in_2}, u_{out}, r_o; \Delta t) \qquad (67)$$

but the four-dimensional tables are near the practical limit.

8.3   Operator-based macromodeling

The table approach is limited to macroelements with few internal states and external connections. We will now study operator-based methods that have the potential of being used on arbitrary macroelements.

Consider solving the block-diagonal system $h^I = 0$ with some scheme suitable for differential-algebraic systems (DASs) and then iterating on h^E with a Newton-like method. This requires taking Fréchet derivatives. The outer Newton correction equation is

$$\frac{d}{dt}\left[q_2^E \frac{dw}{du}\Delta u + q_1^E \Delta u\right] + f_2^E \frac{dw}{du}\Delta u + f_1^E \Delta u = -h^E(u, w(u)). \qquad (68)$$

dw/du is still needed but is no longer easily obtained. The action of dw/du on a waveform Δu can be derived by taking

$$\left[\frac{d}{du}h^I(u, w)\right]\Delta u = 0. \qquad (69)$$

Let

$$\Delta w \equiv \frac{dw}{du}\Delta u. \qquad (70)$$

From Eqs. (52), (69), and (70), we obtain

$$\frac{d}{dt}\left[q_2^I \Delta w + q_1^I \Delta u\right] + f_2^I \Delta w + f_1^I \Delta u = 0. \qquad (71)$$

Now we can solve the coupled DAS represented by Eqs. (68) and (71). Obviously, a general waveform method could be cumbersome. (Variants of nonlinear operator Jacobi and Gauss-Seidel [27] have proven effective for circuits with limited feedback [25,38].)

This operator method can be substantially simplified if we restrict $u(t)$ waveforms to be, say, piecewise linear, that is, $u(t) = u_0 + \alpha(t - t_0)$ on an interval $[t_0, t_1]$. The algorithm to advance from $t = t_0$ given u_0 and w_0 then becomes:

Algorithm 6 Let α_0 be the initial guess for the slope.

1. Set $j = 0$;

2. Do

    a. Solve the block-diagonal DAS $h^I = 0$ for $w(t)$;

    b. Compute $[(dw/d\alpha)(\alpha_j)](t)$ by solving small decoupled linear ODE systems;

    c. Using the previous step, solve $[(dh^E/d\alpha)(\alpha_j)]\Delta\alpha = -h^E(\alpha_j)$ for $\Delta\alpha$;

    d. Set $\alpha_{j+1} = \alpha_j + s\Delta\alpha$;

    e. Replace $j \leftarrow j + 1$;

until converged.

We are currently investigating a couple of variants of these operator-based schemes. Aspects of error and time-step control and how to exploit the hierarchical circuit structure require additional study.

Acknowledgement

We thank Don Erdman for his critical evaluation of earlier manuscripts.

References

[1] R. E. Bank, W. M. Coughran, Jr., W. Fichtner, E. H. Grosse, D. J. Rose, and R. K. Smith. Transient simulation of silicon devices and circuits. *IEEE Trans. CAD IC Sys.*, CAD-4:436–451, 1985.

[2] R. E. Bank and D. J. Rose. Global approximate Newton methods. *Numer. Math.*, 37:279–295, 1981.

[3] R. E. Bank and D. J. Rose. Parameter selection for Newton-like methods applicable to nonlinear partial differential equations. *SIAM J. Numer. Anal.*, 17:806–822, 1980.

[4] R. H. Bartels, J. C. Beatty, and B. A. Barsky. *An introduction to the use of splines in computer graphics.* Technical Report CS-83-09, Univ. Waterloo, 1985.

[5] C. de Boor. *A Practical Guide to Splines.* Springer-Verlag, 1978.

[6] R. K. Brayton, F. G. Gustavson, and G. D. Hachtel. A new efficient algorithm for solving differential-algebraic systems using implicit backward differentiation formulas. *Proc. IEEE*, 60:98–108, 1972.

[7] T. F. Chan. Deflation techniques and block-elimination algorithms for solving bordered singular systems. *SIAM J. Sci. Stat. Comp.*, 5:121–134, 1984.

[8] T. F. Chan and D. C. Resasco. *Generalized Deflated Block-Elimination.* Technical Report RR-337, Yale Univ. Comp. Sci. Dept., 1985.

[9] W. M. Coughran, Jr., E. H. Grosse, and D. J. Rose. CAzM: a circuit analyzer with macromodeling. *IEEE Trans. Elec. Dev.*, ED-30:1207–1213, 1983.

[10] W. M. Coughran, Jr., E. H. Grosse, and D. J. Rose. Variation diminishing splines in simulation. *SIAM J. Sci. Stat. Comp.*, 7:696–705, 1986.

[11] J. E. Dennis, D. M. Gay, and R. E. Welsch. An adaptive nonlinear least-squares algorithm. *ACM Trans. Math. Software*, 7:348–368, 1981.

[12] S. C. Eisenstat, M. C. Gursky, M. H. Schultz, and A. H. Sherman. Yale Sparse Matrix Package I: the symmetric codes. *Int. J. Numer. Meth. Engin.*, 18:1145–1151, 1982.

[13] S. C. Eisenstat, M. C. Gursky, M. H. Schultz, and A. H. Sherman. *Yale Sparse Matrix Package II: The Nonsymmetric Codes.* Technical Report 114, Yale Univ. Comp. Sci. Dept., 1977.

[14] S. C. Eisenstat, M. H. Schultz, and A. H. Sherman. Considerations in the design of software for sparse Gaussian elmination. In J. R. Bunch and D. J. Rose, editors, *Sparse Matrix Computations*, pages 263–274, Academic Press, 1976.

[15] C. B. Garcia and W. I. Zangwill. *Pathways to Solutions, Fixed Points, and Equilibria.* Prentice-Hall, 1981.

[16] C. W. Gear. *Efficient Stepsize Control for Output and Discontinuities.* Technical Report UIUCDCS-R-82-1111, Univ. Ill. Comp. Sci. Dept., 1982.

[17] C. W. Gear. The simultaneous numerical solution of differential-algebraic systems. *IEEE Trans. Circ. Th.*, CT-18:89–95, 1971.

[18] C. W. Gear and L. R. Petzold. ODE methods for the solution of differential/algebraic systems. *SIAM J. Numer. Anal.*, 21:716–728, 1984.

[19] A. George and J. W.-H. Liu. *Computer Solutions of Large Sparse Positive Definite Systems.* Prentice-Hall, 1981.

[20] E. H. Grosse. Tensor spline approximation. *Lin. Alg. Applics.*, 34:29–41, 1980.

[21] G. D. Hachtel, R. K. Brayton, and F. G. Gustavson. The sparse tableau approach to network analysis and design. *IEEE Trans. Circ. Th.*, CT-18:101–113, 1971.

[22] C. W. Ho, A. E. Ruehli, and P. A. Brennan. The modified nodal approach to network analysis. *IEEE Trans. Circ. Sys.*, CAS-22:504–509, 1975.

[23] H. B. Keller. Numerical solution of bifurcation and nonlinear eigenvalue problems. In P. Rabinowitz, editor, *Applications of Bifurcation Theory*, pages 359–384, Academic Press, 1977.

[24] J. Lambert. *Computational Methods in Ordinary Differential Equations.* Wiley, 1973.

[25] E. Lelarasmee, A. E. Ruehli, and A. L. Sangiovanni-Vincentelli. The waveform relaxation method for time domain analysis of large scale ICs. *IEEE Trans. CAD IC Sys.*, CAD-1:131–145, 1982.

[26] L. W. Nagel. *SPICE2: A Computer Program to Simulate Semiconductor Circuits.* PhD thesis, Univ. of Calif., Berkeley, 1975.

[27] J. M. Ortega and W. C. Rheinboldt. *Iterative Solution of Nonlinear Equations in Several Variables.* Academic Press, 1970.

[28] L. R. Petzold. A description of DASSL: a differential/algebraic system solver. In *Scientific Computing*, pages 65–68, North-Holland, 1983.

[29] L. R. Petzold. Differential/algebraic equations are not ODE's. *SIAM J. Sci. Stat. Comp.*, 3:367–384, 1982.

[30] N. B. Rabbat, A. L. Sangiovanni-Vincentelli, and H. Y. Hsieh. A multilevel Newton algorithm with macromodeling and latency for the analysis of large-scale nonlinear circuits in the time domain. *IEEE Trans. Circ. Sys.*, CAS-26:733–741, 1979.

[31] C. den Heijer and W. C. Rheinboldt. On steplength algorithms for a class of continuation methods. *SIAM J. Numer. Anal.*, 18:925–948, 1981.

[32] S. Sastry, C. Desoer, and P. Varaiya. Jump behavior of circuits and systems. In *Proc. IEEE Int. Symp. Circ. Sys.*, pages 451–454, 1981.

[33] I. J. Schoenberg. On spline functions. In O. Shisha, editor, *Inequalities*, pages 255–291, Academic Press, 1967.

[34] G. W. Stewart. On the implicit deflation of nearly singular systems of linear equations. *SIAM J. Sci. Stat. Comp.*, 2:136–140, 1981.

[35] G. Strang and G. Fix. *An Analysis of the Finite Element Method.* Prentice-Hall, 1973.

[36] P. Subramaniam. Table models for timing simulation. In *Proc. Custom IC Conf.*, pages 310–314, Rochester, New York, 1984.

[37] W. T. Weeks, A. J. Jimenez, G. W. Mahoney, D. Mehta, H. Qassemzadeh, and T. R. Scott. Algorithms for ASTAP — a network-analysis program. *IEEE Trans. Circ. Th.*, CT-20:628–634, 1973.

[38] J. White, R. Saleh, A. L. Sangiovanni-Vincentelli, and A. R. Newton. Accelerating relaxation algorithms for circuit simulation using waveform Newton, iterative step size refinement, and parallel techniques. In *Proc. IEEE Int. Conf. CAD*, pages 5–7, Santa Clara, Calif., 1985.

5

VLSI CIRCUIT ANALYSIS, TIMING VERIFICATION AND OPTIMIZATION

Albert E. Ruehli and Daniel L. Ostapko
IBM T. J. Watson Research Center
P. O. Box 218
Yorktown Heights, NY 10598

ABSTRACT

In this paper, we give an overview of the state-of-the-art in Circuit Analysis, Timing Verification, and Optimization. Emphasis is given to circuit analysis, timing verification and optimization since simulation is covered by C. Terman in this book. Also, the optimization of large circuits is receiving new attention due to the need for timing performance improvement in silicon compilation.

1. INTRODUCTION

In this paper we give an overview of the state-of-the-art in VLSI circuit analysis, timing verification and optimization. Simulation is covered in detail in this book by C. Terman [24]. Hence, we will only cover aspects of the topic which are relevant for the other section in this paper. Circuit analysis is a subset of circuit simulation. Mainly, in circuit analysis, we employ numerical analysis type algorithms, and aim at accurate solutions.

Time simulation is faster and requires less storage than circuit analysis with a commensurate decrease in waveform accuracy. The difference in the waveform representation for time simulation and circuit analysis was discussed in a recent paper [25]. Only very few data points are used to represent the waveforms for simulation. As a consequence, a speedup results for simulation techniques and time waveforms can be computed for a large number of logical gates. However, the waveform accuracy may not be sufficient for high performance VLSI circuits.

In contrast, circuit analysis aims at waveforms with an accuracy in the order of 1 percent. However, the actual accuracy of the actual waveforms may be limited by the transistor models employed. Mathematically consistent numerical analysis algorithms are employed. This obviously comes at the cost of an increase in compute time as compared to simulation. To counteract this, new algorithms and techniques have been invented. One of these new approaches is the waveform relaxation (WR) technique [25] - [28]. This technique has resulted in an increasing number of logic circuits which can be analyzed simultaneously. The WR approach has been shown to have the potential for the analysis of circuits with 10,000 to 20,000 transistors if the parasitic and interconnect circuits do not contribute excessively to the number of nodes.

Another area which promises to have an impact on the future of circuit analysis is parallel processing. The WR approach is well suited for parallel processing [31-33], [35] and we expect that the gain will be even larger for VLSI circuits with 20000 or more transistors. Parallelism of the order of 10, if obtained, could have a considerable impact on circuit analysis on both mainframes and workstations. The most profound effect may be for workstations where a factor of 10 may make the analysis of reasonably sized circuits possible.

Logic circuits delay time optimization has been attempted for more than two decades and the literature is surprisingly rich [4] - [23]. Both bipolar e.g. [18] and MOSFET [4]-[17], [19-23] circuits have been considered with an ever increasing complexity. Earlier, several authors attempted to increase the performance of single logic gate circuits for both bipolar [1] and MOSFET [3] circuits. The design variables for bipolar circuits are usually a set of resistors and less frequently bipolar device parameters. The main design parameters for MOSFET devices is the size of the FET gates.

2. CIRCUIT ANALYSIS

A problem of some circuit simulators and even some analyzers is the lack of flexibility. In fact, this problem has limited the utility of several interesting approaches like macromodeling. An approach which is based on specialized modeling of circuit configurations will always have limitations. Each new circuit configuration has to be modeled before it can be used. Methods which can solve a subclass of circuits like arbitrary configurations of MOSFET circuits have a far greater utility. It is obvious today that the generality of circuit analysis like ASTAP [29] or SPICE [30] is one of the reasons for their success.

General purpose analysis programs like SPICE and ASTAP will always find numerous other applications in time domain analysis besides transistor circuit analysis. Further, they represent a standard against which all new analyzers and simulators are measured not only in terms of compute time requirements but also accuracy. In fact, accuracy is one of the most important aspects of these programs since many simulators have problems with complicated pass transistor circuits. We cannot expect that the speed of general purpose analyzers will increase drastically in the future since they are based on widely known techniques which have been improved for more than a decade. Thus, we should expect to see the development of special purpose analyzers to address classes of important problems.

Special purpose programs can be divided into two classes. In the 1970's macromodel analyzers were devised in parallel with approximate simulators. However, they could only compute waveforms for circuits for which macromodels were constructed. This usually involved a time consuming process. Most of these programs failed to find wide use.

A newer generation of programs is finding wider use since they can treat a larger class of circuits. However, they are special purpose tools since they are restricted to *dc* and *transient* analysis only. Also, many of them are limited to one technology such as MOSFET transistors. At present, the waveform relaxation analyzers Relax [27] and Toggle [28] are two examples. These programs have the potential of finding a large user community if they are able to handle arbitrary topologies.

The Waveform Relaxation (WR) approach has been applied to a variety of circuits [26] - [28]. However, most practical implementations of WR are limited in generality. Mainly, the partitioning problem has to be solved for the general class of MOSFET circuits. The majority of the work has been done with MOSFET circuits. The special purpose WR programs like Relax [27] and Toggle [28] have to compete effectively with general purpose circuits analyzers like ASTAP and SPICE. Hence, they provide stiff competition to any special purpose program at least for small to medium size circuits. Also, they can handle a mixture of MOSFET and Bipolar circuits. Combined analog and digital circuits can easily be mixed in both approaches. Mainly, WR programs use a small SPICE like program as the analysis "engine". Hence, analog circuits are accommodated by using this engine and properly partitioning the analog part into a single partition.

The largest circuit analyzed to date with Toggle is an ALU circuit with 9000 transistors which took 75 minutes of IBM 3090 computer time which is about a factor 100 faster than a SPICE type program.

The improvements in compute time for the WR are due to many factors. A few are listed below:

1. The circuit matrix (Jacobian) is subdivided into smaller matrices of size N_s. The matrix solution time for a subsystem is N_s^p where $1.2 < p < 2$ for sparse matrix code. Hence, the sum of all subsystem matrices can be solved faster than the sparse solution of the full system Jacobian which takes time proportional to N_A^p where N_A is the size of the Jacobian of the entire system.

2. The decoupled subsystems are integrated at their own rate. Hence, this multirate decoupling of the subsystems prevents unnecessary computations. Specifically, the most time consuming task is the evaluation of the transistor models and many unnecessary time steps are taken in the subsystems in a SPICE like program. Unfortunately, computations of this type cannot be avoided in a conventional *incremental* program.

3. A small change in a circuit can easily be updated in a WR program by utilizing previously computed waveforms. In a conventional program like SPICE, the entire solution must be recomputed.

Circuit timing simulation uses up a large portion of the CPU compute time of the entire VLSI design budget. Hence, special purpose hardware can often be justified. Both specially designed hardware and more general parallel processing configurations provide viable approaches to decreasing the computation time. The two approaches can, in fact, be used simultaneously to effectively use two levels of parallelism [32, 33].

The lower level of parallelism would typically exploit parallelism within direct methods of circuit analysis while the higher level of parallelism might use the parallelism inherent in WR. The parallelism in the direct method comes from exploiting the parallelism in several of the steps of the direct method algorithm namely, prediction and integration, forming linear equations, solving linear equations, and checking errors and convergence. The parallelism at the higher level, which has larger granularity, could employ the Gauss-Jacobi or time-point pipelining version of WR. However, a mixed scheme that orders the computations so that subcircuits in parallel "chains" are computed in parallel is a good compromise [35]. This approach results from the observation that digital circuits tend to be "wide" in that gates fan out to more than one subcircuit. It should be possible to find a parallelism of 10 for circuits of moderate size.

The gain obtained for parallel processing from the WR approach is expected to be even more significant for VLSI circuits with 20,000 or more transistors. If the parallel approach selected does not rely on specific hardware accelerators, then the same parallel approach may be used on both parallel mainframes and parallel workstations. These parallel structures are being investigated and built at several universities and industrial firms [36, 37]. The most profound effect may be for workstations where a factor of 10 may make analysis of reasonably sized circuits possible. Although improvements in processing time have been demonstrated, the amount of data storage and movement must also be carefully managed if the full potential of the available parallelism is to be realized.

3. TIMING VERIFICATION

A good timing verification model based on relatively simple ideas is given in this section. It is important for both timing verification as well as optimization as will be apparent from the next section. Here we will present a model for timing verification for combinational circuits which finds wide use in IBM [34].

The key advantage of verification over simulation is in the compute time. Simulation by path tracing is an exponential process while this algorithm is linear in the number of gates.

In this model, all the logic circuits are described as inverting or non-inverting. Both the rising and falling delays are defined by delay equations independent of the function of the logical gates as shown in Fig. 1. This total delay must be considered for completeness for combinational logic. The short delay check is needed when there is clock overlap. This insures that signals do not arrive too early so that the latches are not disturbed. The long delay check is the one which is usually considered so that the timing constraints are met. We can define the following signal times and delays:

a_r, b_r = rising waveform arrival times for signals a and b
a_f, b_f = falling waveform arrival times for signal a and b
d_r, d_f = delay of output rising or falling respectively
c_r, c_f = rising and falling output arrival times

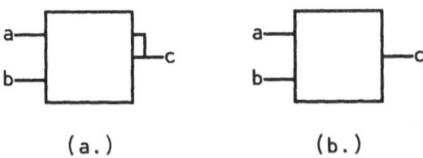

(a.)   (b.)

Fig. 1(a). Noninverting circuit, (b) Inverting Circuit

With these definitions, the long delay equations are for the inverting circuit in Fig. 1(a).

$$c_r = \max\ (a_f, b_f) + d_r$$
$$c_f = \max\ (a_r, b_r) + d_f \tag{1}$$

while the short delay equations are

$$c_r = \min\ (a_f, b_f) + d_r$$
$$c_f = \min\ (a_r, b_r) + d_f \tag{2}$$

For the non-inverting circuit in Fig. 1(b), the long delay equations are:

$$c_r = \max \ (a_r, b_r) + d_r$$
$$c_f = \max \ (a_f, b_f) + d_f$$

$$(3)$$

while the short delay equations are

$$c_r = \min \ (a_r, b_r) + d_r$$
$$c_f = \min \ (a_f, b_f) + d_f$$

$$(4)$$

Using this model signal arrival times can be calculated in order to verify that they arrive neither too early nor too late. This is done by levelizing the circuits and by adding up the arrival times at each node without considering the function of the gates. This model has been used also for timing optimization as will be apparent from Section 4.

4. CIRCUIT OPTIMIZATION

New techniques have been invented for the optimization of large circuits on VLSI chips. Mainly, only a few design parameters should be used for each logical circuit since this may severely limit the number of logical circuits which can be optimized. Ideally, the timing model has only one global parameter per logic circuit. An example of such a model is a MOSFET circuit where all the FET gates are adjusted in proportion to a single design parameter W.

VLSI circuit optimization has gained much more importance with the advent of silicon compilation. In fact, high performance silicon compilers must take timing into account. For example, it has been shown that minimizing the total wire length does not guarantee that the delay due to the wires is a minimum [7]. Hence, new techniques have been deviced to optimize the electrical performance of VLSI circuits at the large scale levels.

Probably the most common way to adjust the timing of MOSFET integrated circuits is the adjustment of the FET gates [4] - [11], [23]. Two fundamentally different algorithmic techniques are employed. Analytical nonlinear optimization techniques are used in several of the approaches [4] - [11], [13].

In [4] - [7] a Newton optimization technique is employed. Circuits with more than 1000 parameters were optimized and no limitations were found. Both the compute time and storage requirements are moderate if the Hessian matrix problem

$$\Delta t = H^{-1}g. \tag{5}$$

is solved in sparse form. In Eq. (5), H is the Hessian matrix and g the gradient. Table I gives a comparison of some of the results obtained with different optimization methods for a 138 gate circuit with 68 design variables. The circuit was optimized until the same minimum power was obtained for all the methods tested.

TABLE I: COMPARISON OF OPTIMIZATION METHODS

Method	No. of Function Calls	Compute Time
Davidson Fletcher Powell	1405	33.98
Conjugate Gradient	1709	32.59
Newton	409	10.77

These results show that the Newton method is very economical for this type of circuit. Further, it was found that methods based on an approximation of the Hessian matrix like the popular Davidon Fletcher Powell method could not be employed with more than 100 variables. The main problem is the iterative computation of the Hessian.

The Newton optimization scheme requires that the macromodel have analytical derivatives. In [4]-[7] a simple model is employed of the form

$$W = \frac{A}{d - B} \tag{6}$$

where d is the delay and A,B are functions of the device capacitances while W is the FET gate size. For NMOS circuits the power is proportional to the gate size, or

$$P = kW \tag{7}$$

Hence, the total power for a circuit is given by

$$P_T = \sum_i \frac{A_i}{d_i - B_i} \tag{8}$$

Assuming a long delay constraint given by Eqs. (1) and (3), the timing equations can be written in the form

$$d_i = t_o - \max(t_1, t_2, ...) \tag{9}$$

where d_i is the delay to the output and $t_1, t_2, ...$ are the arrival times at the inputs. To obtain a function with analytical derivatives, the max function in Eq. (9) is replaced by the smooth max function.

$$\text{smax}(t_1, t_2, ...) = \alpha^{-1} \ln(e^{-\alpha t_1} + e^{-\alpha t_2} + ...) \tag{10}$$

where α is adjusted during the iterations. Finally, the transition times at the different nodes are found from

$$t_i = t_{i-1} - \beta \Delta t \tag{11}$$

where Δt is given by Eq. (5) and β is found by a scalar minimization of P_t with respect to β.

This approach has successfully been applied to the optimization of several circuits [7]. However, the most appropriate analytic formulation depends on several factors. The type of technology has a profound impact on the approach. It is necessary to distinguish between NMOS, CMOS and dynamic circuits. Also, the objective function which is minimized can be power, area, sum of gate size, the placement of the circuits, or combinations of these functions. Further, different constraints can be imposed. An example in the above formulation is imposing a timing constraint by fixing the times at the outputs of the circuits. We could also keep the power constraint while minimizing the timing.

A number of other techniques have been employed besides the analytical techniques presented above. A major problem exists in the methods based on path delays, rather than the timing verification model presented above, since the path problem is of exponential complexity.

The circuit placement is another design parameter which has been explored by several authors [5], [7], [12], [17] - 18]. Mainly, the critical circuits can be sped up by placing them closer together, a step which results in a decrease in the capacitances. An example is the delay weighted force directed pairwise relaxation [5], [7] where the position of a circuit connected to other circuits is weighted by 1/delay.

Finally, a recent addition to the VLSI optimization techniques is the redesign of the logic at hand [14], [21], [22], [23]. For example, the circuits which participate in a critical path can be redesigned to decrease the length of the critical path by reducing the number of logical stages. This is done at the cost of delay in the non-critical stages. This does not represent a problem since all paths are equally critical in an optimized circuit.

All the above mentioned optimization techniques can be combined in a silicon compiler by applying them sequentially. For example, in [7] the following algorithmic steps were used:

1. Placement for distance

2. Global routing

3. Gate size optimization

4. Placement to minimize power

5. Return to 2. if not converged.

One can easily include a logic redesign optimization step in this algorithm. We should perform steps 1. and 2. before the logic optimization step since it is desirable to have a good estimate of the capacitances for this step.

It is clear that this brief overview does not cover this new area of research completely. Many new results will be obtained in the near future in spite of the fact that VLSI timing optimization already has an amazingly rich history.

REFERENCES

Timing Optimization

[1]   G. D. Hachtel and R. A. Rohrer, "Techniques for Optimial Design and Synthesis of Switching Circuits," Proc. IEEE, Vol. 55, No. 11, pp. 1864-1877, Nov. 1967.

[2]   D. R. Friedman, M. P. Patel, "Performance Simulation with Circuit Level Models," IEEE Int. Solid-State Circuit Conf.Dig., pp. 40-41, 1974.

[3]   R. K. Brayton, S. W. Director, "Computation of Time Delay Sensitivities for Switching Circuit Optimization, IEEE Trans. Circ. and Systems, Vol. CAS-22, pp. 910-920, Dec. 1975.

[4]   A. E. Ruehli, P. K. Wolff, Sr. and G. Goertzel, "Power and Timing Optimization of Large Digital Systems," IEEE Proc. Int. Symp. on Circ. and Sys., ISCAS, Munich, pp. 402-405, April 1976.

[5]   B. J. Agule, J. D. Lesser, A. E. Ruehli, P. K. Wolff, Sr., "An Experimental System for Power/Timing Optimization of LSI Chips," Proc. 14th Des. Autom. Conf., New Orleans, LA, June 1977.

[6]   A. E. Ruehli, P. K. Wolff, Sr., G. Goerzel, "Analytical Power/Timing Optimization Techniques for Digital Circuits," Proc. 14th Des. Autom. Conf., New Orleans, LA, June 1977.

[7]   P. K. Wolff, A. E. Ruehli, B. J. Agule, J. D. Lesser, G. Goertzel, "Power/Timing: Optimization and Layout Techniques for LSI Circuit," Design Automation and Fault Tolerant Computing, pp. 145-164, 1978.

[8]   A. M. Mohsen, C. A. Mead, "Delay-Time Optimization for Driving and Sensing of Signals on High-Capacitance Paths of VLSI Systems," IEEE Trans. on Elect. Device, Vol ED-26, No. 4, pp. 540-548, April 1979.

[9]   B. Vergnieres, "Macro Generation Algorithm for LSI Custom Chip Design," IBM J. Res. Develop., Vol. 24, No. 5, pp. 612-621, Sept 1980.

[10]   C. M. Lee, H. Soukup, "An Algorithm for CMOS Timing and Area Optimization," IEEE Journal of Solid-State Circ. Vol. CS-19, No. 5, pp. 781-787, Oct. 1984.

[11]   M. D. Matson, "Optimization of Digital MOS VLSI Circuits," Chapel Hill Conference, Computer Science Press, Rockville, MD, H. Fuchs ed., 1985.

[12]   M. Burstein, M. N. Youssef, "Timing Influenced Layout Design," Proc. 22nd Des. Autom. Conf., pp. 124-130, June 1985.

[13]   K. S. Hedlunds, "Electrical Optimization of PLA's," Proc. 22nd Des. Autom. Conf., pp. 681-687, June 1985.

[14]   A. J. deGeus and W. Cohen, "A Rule-Based System for Optimizing Combinational Logic," IEEE Design and Test, pp. 22-32, August 1985.

[15]   J. Fishburn, A. Dunlop, "Tilos: A Posynominal Programming
         Approach to Transistor Sizing," IEEE Int. Conf. Comp. Aid. Des.
         ICCAD, Santa Clara, pp. 326-328, Nov. 1985.

[16]   S. Trimberg, "Automated Performance Optimization of Custom
         Integrated Circuits," in Advances in Computer-Aided Design, A.
         Sangiovanni-Vincentelli, ed., Vol. 1, Jai Press Inc. Greenwich, CT,
         pp. 225-283, 1985.

[17]   S. Teig, R. L. Smith, J. Seaton, "Timing-Driven Layout of
         Cell-Based IC's," VLSI System Design, pp. 63-73, May 1986.

[18]   Y. Ogawa, T. Ishii, T. Shiraishi, H. Terai, T. Kozawa, K. Yuyama, K.
         Chiba, "Efficient Placement Algorithms Optimizing Delay for
         High-speed ECL Masterslice LSI's," Proc. 23rd Des. Autom. Conf.,
         pp. 404-410, July 1986.

[19]   J. D. Pincus, A. M. Despain, "Delay Reduction Using Simulated
         Annealing," Proc. 23rd Des. Autom. Conf., pp. 690-694, July 1986.

[20]   M. Glesner, J. Schuck, R. B. Stede, "A New Statistical Timing
         Verifier in a Silicon Compiler System," Proc. 23rd Des. Autom.
         Conf., pp. 220-226, July 1986.

[21]   D. Gregory, K. Barlett, A. de Geus and G. Hachtel, "SOCRATES:
         A System for Automatically Synthesizing and Optimizing
         Combinational Logic," Proc. 23rd Des. Autom. Conf., pp. 29-85,
         July 1986.

[22]   W. H. Joyner, L. H. Trevillyan, D. Brand, T. A. Nix, S. C.
         Gundersen, "Technology Adaptation Logic Synthesis," Proc. 23rd
         Des. Autom. Conf., pp. 94-100, July 1986

[23]   G. DeMicheli, "Performance-oriented Synthesis in the
         Yorktown-Silicon Compiler," IEEE Int. Conf. Computer Aided Des.
         ICCAD, Santa Clara, CA, Nov. 1986.

Other References

[24]   C. J. Terman, "Simulation Tools for VLSI," in this book.

[25]   A. Ruehli, G. Ditlow, "Circuit Analysis, Logic Simulation, and
         Design Verification for VLSI," Proc. IEEE, Vol. 71, No. 1, pp.
         34-42, January 1983.

[26]   E. Lelarasmee, A. E. Ruehli, A. L. Sangiovanni-Vincentelli, "The
         Waveform Relaxation Method for the Domain Analysis of Large
         Scale Integrated Circuits," IEEE Trans. CAD, Vol 1, No. 3, pp.
         131-145, July 1982.

[27]   J. White and A. L. Sangiovanni-Vincentelli, "Relax 2.1-A Waveform
         Relaxation Based Circuit Simulation Program," Proc. 1984 Int.
         Custom Integrated Circ. Conf., Rochester, New York, June 1984.

[28]   P. Debefve, J. Beetem, W. Donath, H. Y. Hsieh, F. Odeh, A. E.
         Ruehli, P. Wolff, Sr. and J. White, "A Large-Scale MOSFET Circuit
         Analyzer Based on Waveform Relaxation," IEEE int. Conf. on
         Comp. Design, Rye, New York, October 1984.

[29]   W. T. Weeks, A. J. Jimenez, G. W. Mahoney, D. Mehta, H.
         Quassemsadeh and T. R. Scott, "Algorithm for ASTEP - A Network
         Analysis Program," IEEE Trans. on Cir. Theory, Vol. CT-20, pp.
         618-634, November 1973.

[30]   L. W. Nagel, "SPICE2: A Comptuer Program to Simulate Semiconductor Circuits," Electronic Research Laboratory, Rep. No. ERL-M520, University of California, Berkeley, May 1975.

[31]   J. White, R. Saleh, A. R. Newton, A. Sangiovanni-Vincentelli, "Accelerating Waveform Relaxation Algorithms for Circuit Simulation Using Waveform Newton, Iterative Step Refinement and Parallel Techniques," IEEE Int. Conf. Comp. Aided Des., ICCAD, Santa Clara, November 1985.

[32]   J. T. Deutsch, T. D. Lovett, M. L. Squires, "Parallel Computing for VLSI Circuit Simulation," VLSI System Design, pp. 46-52, July 1986.

[33]   J. White, N. Weiner, "Parallelizing Circuit Simulation - A Combined Algorithmic and Specialized Hardware Approach," IEEE Int. Conf. on Comp. Design, ICCD '86, Rye Brook, New York, October 1986.

[34]   R. B. Hitchcock, "Timing Verification and the Timing Analysis Program," Proc. IEEE/ACM 19th Design Automation Conf., pp. 594-604, June 1982.

[35]   A. E. Ruehli, J. K. White and D. L. Ostapko, "Time Solution of Large Scale VLSI Circuits," Proc. Scientific Applications and Algorithm Design for High Speed Computing Workshop, Urbana, IL, April 1986.

[36]   C. L. Seitz, "The Cosmic Cube," Comm. ACM, Vol. 28, January 1985, pp. 22-33.

[37]   G. F. Pfister, et. al., "The IBM Research Parallel Processor Prototypes (RP3): Introduction and Architecture," Proc. of ICPP, St. Charles, IL, August 1985, pp. 772-781.

6
CAD Tools
for
Mask Generation

Jonathan B. Rosenberg

Microelectronics Center of North Carolina

ABSTRACT

CAD Tools for Mask Generation is a general title that refers to Computer-Aided Design software that produces as its output mask-level descriptions of integrated circuits. The current state-of-the-art for general mask generation in a custom design environment is that of symbolic design. So this chapter is about symbolic design and how it generates a mask. Included are discussions of design capture, circuit description languages, compaction and technology encapsulation all in the context of the symbolic design environment. General overviews are not given in most cases. Instead, a feel for this technology is given through specific examples of a system most familiar to the author.

INTRODUCTION

Outline of chapter

Advantages of symbolic design discusses the rationale and philosophy behind this method of designing integrated circuits. The goal in the end, of course, is an economic one: decreasing time, increasing productivity are ways to decrease cost. But it may also allow certain designs to be done that were totally infeasible otherwise and it may allow individuals to do a design that would not have done designs otherwise.

Mechanisms for symbolic design outlines the steps in the design process under this methodology. Many of the steps are the same as when the design is directly laid out using a "rectangle pusher" but at each step there are differences that are significant.

ABCD — a symbolic-level hardware description language is a description that gets into some detail of a particular hardware description language that is used for symbolic design. It has some very important features that recommend it as a hardware description language and it is no small part of the overall improvement in the design process afforded by such a symbolic design system.

Design capture in a symbolic design environment is another important part of the symbolic design process. The symbiosis between design capture and the hardware description language (and its circuit database role) are keys to efficient symbolic design. The user interface is fundamentally different from that used in a "rectangle pusher." The description of this tools function gets quite detailed in order to fully explain the user interface.

Compaction – the key to symbolic design recognizes that without effective compaction the rest is quite academic and the entire system becomes nothing more than a toy. Some comparisons of different compaction strategies is made since this is still a very hot area of research and many schemes have been proposed in the literature.

Technology encapsulation is the separating out from the application tools the information about the technology and process that will be used to fabricate the design under consideration. It is important to make this separation so that software remains immune to changes in technology. But it is equally important so that one can really take advantage of the inherent "delayed binding" that takes place in a symbolic design environment. I refer to the fact that a design need not specify the exact rules and models of the target process until the moment the design is generated at the mask level via compaction. And the same design can be re-compacted (and re-fabricated) on different fabrication lines at will.

Software engineering of a production CAD system discusses interesting insights into software engineering issues that come from the development of a production CAD system. The purpose of making these observations is to benefit others attempting the same thing.

Why is better CAD important?

Its all a matter of economics in the final analysis. Increasing productivity is important because design time is a critical issue in the production of all products and can make or break a product and a company. Expert designers are a rare commodity and they are expensive so if they can get more design done in less time that is money saved. Symbolic design increases productivity because, 1) a designer works with an abstraction of a physical layout and so manipulates one symbol that represents many he would have manipulated in the strictly mask layout environment, and 2) compaction generating design-rule error free layouts means no worry about the difficult and complex details of design rules during the layout phase.

All well and good but why do layout by hand at all, symbolic or mask? Why not just do standard cell layout or even better, just do gate array designs? Certainly these are very important design strategies where appropriate but there are times when they aren't appropriate. Even without symbolic layout many design groups have favored the performance-area-flexibility versus design-time tradeoff in favor of custom layout. As alluded to, their reasons are usually due to higher performance, higher silicon density, architectural flexibility or combinations of all three. With symbolic layout, the down side of this tradeoff, design-time, is lessened so it is easier to justify going with full custom. Furthermore, even in standard cell environments, symbolic design is making inroads as people see the enormous benefits of symbolic standard cell libraries compacted to mask cells at each technology change. The economic issue with full custom design is simply that success of a design may require performance out of the range of any other method; or, area is an all important factor since it gets very expensive to have more boards in a system; or, the project requires some technology or architectural feature not possible any other way.

Performance, silicon area, and architectural constraint are also intimately related to changes made in the fundamental technology of implementation as well. Delayed binding, that is, an independence from complete specification of technology and process parameters until late in the design process, allows a design to fully exploit any technology advances. And even once a design is "completed", since the actual compaction to mask happens very late in the design process, it can be re-implemented very easily as technology of fabrication advances. The economic advantages inherent in this have already been shown.

Why is it hard?

Why is it hard? is a question about which much could be said but I will suffice it by only making a few points as way of an answer. First, Computer-Aided Design software, since it is solving difficult algorithmic problems while continuously interacting with the designer who must perform a wide variety of design tasks, is large and complex. It is difficult to design the best possible user interface for such a wide variety of tasks. The users of such a system are very demanding – they have used software before and they expect to get a significant improvement out of a tool they take the time to learn. They are sophisticated users and expect the software to be easy to use, well documented, perform perfectly and do so efficiently. But, most of all, they expect this new tool to be a substational improvement over their last tool.

Second, consistency is important within this system and between this system and those with which it must integrate. For it is clearly unwise to consider ones own system to be the only one a design team will use. This is not just consistency of user interface but also of compatible circuit database formats, command inputs, and the like.

Third, the basic VLSI algorithms are difficult and much more work is needed in placement, routing and compaction among other algorithms.

Finally, in spite of the very strong motivation to provide higher-level design support, the goal of providing tools to do so remains quite elusive. This is partly due to the fact that the design process itself is still not well understood.

Where do we stand?

With respect to symbolic CAD, where do we stand? A great deal of experience has been gained in the CAD field in general with interactive graphics for design capture. While the issue of improved user interfaces is certainly not closed each new interactive CAD tool seems to be very similar to existing ones which may indicate a maturing in this area.

In the area of compaction, key to symbolic design as already stated, several different algorithms have been proposed and implemented. The single most important requirement for a compactor is that it make area efficient layouts. It is assumed that once the compactor is debugged it does not produce design-rule errors. The difficulty the implementors of compactors have had is in keeping the implementation as clean as the basic algorithms because these programs have a real tendency to get filled with special cases. Until compactors regularly get very close to hand design silicon area (currently the best ones get to within 20% on average if no pathologies exist) there will be a lot of motivation to improve them. Hierarchical compaction is also being done but not very satisfactorily.

A stated goal of every CAD system should be to have it smoothly integrate with existing CAD tools. But this is far from the case and proves to be very difficult. As is the case in many areas, lack of standardization is a serious impediment (e.g. home video cassette formats). The lack of standardization in CAD means lots of translators must be used to get two systems to communicate.

The design process is not a pure and simple oneway path – it has skips, gaps and loops all during the process. This is difficult to deal with and leads one to want effective common representations for circuits and to allow mixed representations. Unfortunately it is not well-understood how to do this and this remains a goal of most CAD development efforts.

Finally, it should be pointed out that we are just taking baby steps. We would like to follow the analogy of software high-level language compilers and provide such effective high-level silicon compilers that design at the mask level would no more be done than programming directly in assembly or machine language.

ADVANTAGES

The symbolic virtual-grid design methodology presents significant advantages over other methods for custom VLSI design. This methodology has facilitated development of a single integrated system which provides designers several important features: higher productivity; technology independence for a wide range of MOS processes (CMOS, nMOS, SOI); scale-independent circuit designs; an open architecture that simplifies integration with existing tools and creation of new tools; fast simulation to layout loop; and fully automatic mask-generation and chip assembly.

Higher productivity is offered by this approach due to a designers manipulating abstractions of circuit elements as opposed to the mask component rectangles. This is referred to as object-oriented editing and it allows the designer to think in terms of the circuit being designed instead of doing mental synthesis of circuit into rectangles continuously. Not having to worry about design rules enhances the power of this abstraction and so only relative positioning of circuit objects concerns the designer. It has been shown in other domains such as software, that a powerful abstraction such as this allows designers to rip-up and re-do erroneous circuit fragments as opposed to the dangerous practice of patching. Thus not only is productivity is enhanced but quality as well since designers now can take a more global view of their design.

Technology independence is a most important advantage. It means the same tool works over a broad range of MOS technologies and, within a specific MOS technology, a given design tracks changes in the design (or ground) rules without redesign. Tools are configured at run-time to a technology and a specific process within that technology so the full capabilities, but no more than the available capability, are made available to the designer in a natural way.

Whereas directly laid out mask designs do not scale (wire widths scale at a different rate than contact cuts, for example) symbolic virtual-grid designs scale perfectly. In an era of "planned obsolescence" this approach provides delayed obsolescence of designs. It also provides another economic benefit: second sourcing even on dis-similar process lines. Correctness-by-construction is promoted by eliminating the mask layout step. The compactor generates the mask layout, therefore, designers do not have to be aware of the design rules in order to create an error-free layout. Technologies with many layers get prohibitively complex to design at the mask level (even CMOS is significantly more complex than nMOS) but they are equivalent in the symbolic virtual-grid environment. Process lines frequently must simplify their ground rules at the cost of less than optimal layouts and/or performance for the sake of the designers working at the mask level — this is no longer necessary.

The symbolic virtual-grid approach to CAD has facilitated an open architecture and very natural interfaces. Clearly defined modules have well-specified inputs and outputs so integration with other systems is greatly simplified. Once a base-line system is established, many tools can be added on, taking advantage of this methodology. For example, structure generators, special-purpose routers, or silicon compilers have hidden from them the details of design rules so their developers can concentrate only on the hard algorithms of the problem. Research and development of modules can continue and individual modules can easily be replaced without disrupting the rest of the system.

Quick estimated timing simulation of the symbolic layout provides a very fast simulation to layout loop. Designers are then afforded the ability to explore the function and performance of a circuit while remaining at the symbolic level. This greatly improves a designers productivity. Once compaction is done the details of the circuit are known to allow detailed timing simulation such as SPICE.

Mask layouts are generated automatically by a compactor. The layouts produced are guaranteed to be design-rule error free. Rules for compaction are extracted from a table which is easily modified as design rules change. Hierarchical compaction preserves the advantages of hierarchical designs by only compacting members of arrays of common structures once. Chips are composed of large blocks of compacted structures routed together and routed to pads.

MECHANISMS

Symbolic, virtual-grid layout

Symbolic, virtual-grid layout can be viewed as an evolutionary refinement of mask layout. In mask layout, the designer specifies the circuit by drawing a set of polygons that indicate how to create a mask for each layer in the fabrication process. At the mask level, the basic elements of circuit design (such as transistors or contact cuts) are composite structures. Each transistor or contact cut is composed of polygons on several layers that are sized and positioned according to the design rules of the target fabrication process. In mask layout, each time one of these composite structures is needed, it is re-created from the component polygons. Symbolic layout provides a solution that eliminates this tedious and error-prone task.

With symbolic layout, symbols are provided to represent the most common structures. The designer organizes the symbols into a layout and the computer translates them into the proper mask representation. In its simplest form, the translation is done by replacing the symbol with a fixed collection of polygons that implement the desired structure. (Many mask layout systems provide translation with a "macro" feature.) A more flexible approach to symbol translation is to associate parameters with the symbols and to have a program use the parameters for generating a broad range of structures. For example, the symbol for a transistor might be accompanied by two parameters that specify the width and length of the gate region. The transistor generation program would then use the parameters to size the transistor when constructing the mask layout.

Like symbolic design, virtual-grid layout is an extension of mask design. In mask design, the layout is usually created on a grid. The spacing of the grid represents some "real" spacing (for example, 3μ) and the designer uses the grid as an aid to establish correct spacing between objects. The function of the virtual grid is the same as for a "real" grid except that the spacing between grid lines does not represent a fixed physical spacing. A symbol's placement captures only the relative geometry of the circuit. (For example, transistor A is above and to the right of transistor B.) The actual spacing between two adjacent grid lines is determined by the compactor program. The compactor examines the objects on adjacent grid lines and, based on the design rules, determines the correct spacing between the grid lines. This approach results in an appropriate division of labor — the designer makes the global decisions about the circuit topology and the computer performs the detailed geometric construction.

Layout verification

A symbolic design system must provide tools for verifying symbolic virtual-grid layouts. Two such tools are: a symbolic level circuit extractor and an interactive circuit simulator. Design verification directly from the symbolic, virtual-grid layout rather than from the mask layout offers the advantage of fast response. This quick response allows the designer to perform extensive circuit debugging early in the layout process.

The symbolic level circuit extraction is performed by a static semantic circuit checker. This tool references the technology database to calculate the electrical parameters associated with each circuit element. The calculated values are, by necessity, estimates since the mask generation has not been performed. However, these estimates are relatively accurate for all of the primitives except wires, which are directly dependent on the final size of the layout. Reasonable estimates of wire length can be obtained by assuming that the spacing between the virtual grid lines will average out over the design. This average grid spacing parameter is coded in the technology database and can be tuned by the designer according to the technology being used and the performance of the compactor. The extraction process is relatively fast since the extractor does not have to go through the costly process of inferring the circuit structure from the mask geometry.

Circuit simulation is performed by a circuit-level timing simulator designed to work from symbolic circuit descriptions. The simulator has been designed for MOS simulations and can be used with circuits as large as several thousand devices. The speed of this simulator results from its selection of models and internal structure. Only MOSFET models are used and it precalculates tables of simulation values before beginning a simulation. Because of its simpler modeling and use of symbolic, virtual-grid extraction, such a simulator does not provide the accuracy of a full network analysis program. However, it fills a gap between such programs and logic level simulators. It is faster than a detailed circuit simulator but still accurate enough to provide the waveform information necessary for debugging the analog behavior of a circuit.

Mask generation

The creation of a mask description from a symbolic, virtual-grid layout is accomplished by a hierarchical compactor. The compactor reads symbolic circuit design descriptions and generates a rectangle-based mask description of the circuit. The compactor operates in two distinct steps: leaf cell compaction and hierarchical compaction.

Leaf cell compaction

Leaf cells contain only circuit primitives: wires, devices, contacts and pins arrayed on the virtual grid. The compactor translates the symbols into their mask representation and then spaces them according to the design rules. Much of the difficulty of compaction arises because these two steps are not independent. In particular, the placement of wires and contacts on rigid structures (such as transistors) is dependent on the location of adjacent circuit elements. The compactor solves this problem by augmenting rigid structures with flexible wires that the compactor may extend when making connections. These wires "decouple" the rigid structures from the rest of the layout and allow the compactor to treat the mask-spacing problem in a uniform manner.

Mask spacing is determined during two passes (one vertical and one horizontal) across the cell. During each pass the compactor positions the grid lines relative to a frontier that represents the previously compacted portion of the cell. The compactor partitions each grid line into groups of grid points such that all points in a group are connected by devices or wires. It then determines the minimum spacing possible between each group and the frontier. Once the spacing has been established for each group, the grid line is positioned and offsets from the grid line are assigned for each group. These offsets in effect "break" grid lines and allow the groups to be positioned with minimum spacing from the frontier. Without sacrificing the speed and predictability of virtual-grid compaction, this strategy provides a significant area improvement over previous virtual-grid compactors.

There are two additional constraints imposed on the compaction process that enable the hierarchical portion of the compactor to automatically pitch-match cells and that improve the compactor's predictability when used for leaf-cell generation. The first is that a grid line is not allowed to move past another grid line during compaction. This constraint prevents the compactor from interlocking adjacent grid lines which would prevent cells from stretching during pitch-matching. The second is that the beginning and ending group on a grid line are positioned at the same location. This constraint allows the designer to lock the position of signal lines together and insure that a cell will pitch match to itself. Although this constraint is not important if the design is entirely symbolic, it is valuable if the system is being used to generate leaf cells for use with mask-layout tools.

Hierarchical compaction

The hierarchical portion of the compactor compresses the hierarchy into a set of cells that completely cover the circuit layout. For each distinct leaf cell, the compactor analyzes all of the environments in which the cell occurs and generates a "worst-case" version of the cell. The compactor then performs a leaf-cell compaction and obtains a mask cell that can be placed in any of the original environments without causing design-rule errors. Once all of the leaf cells have been compacted, the mask cells are assembled according to the original layout.

The main problem in the assembly phase is pitch matching the adjacent cells. Since each leaf cell is compacted separately, there may be wires on abutting cell boundaries that matched on the virtual grid but are offset now that the physical spacing has been established. The compactor, however, retains information about the original virtual-grid layout so it can stretch the cells by the appropriate amount to insure a match. The cells cannot be compressed to achieve a match because the leaf-cell compactor has already produced the smallest layout possible for each cell. The final output of the compactor is a mask description of the circuit in terms of rectangles and layers.

Mask Layout

The generation of a mask description by the compactor is the last step in the symbolic portion of the system and is the first step in the mask portion. A symbolic design system is intended to function as a front end to a mask-layout system.

In many cases it is possible to design an entire chip with the system and remain independent of mask-layout tools; however, the addition of I/O pads and final routing must be done at the mask level. A chip assembler (or cell composition system) is used to combine mask-level blocks and I/O pads into a complete chip.

Technology database overview

Underlying the entire symbolic design system is a technology encapsulation and database referred to as the Master Technology File (MTF). The symbolic design system self-configures to a particular process technology and environment by consulting the MTF. The MTF contains all of the system's knowledge of circuit primitives. It controls their representation by defining the names of device types and process layers. Also, it controls the appearance of the circuit display by defining the symbol shapes, colors, and stipple patterns. It controls the rendering of circuit primitives in mask-layout form by providing symbol-to-mask translation rules. Finally, it contains information about the primitives' electrical properties, capacitances, sizing, and "best," "worst," and "average" case transistor models.

ABCD

This section describes in some detail the ABCD language. ABCD is the symbolic-level circuit description language used in the VIVID System — the CAD system developed at the Microelectronics Center of North Carolina. The VIVID System is the fruits of research into symbolic, virtual-grid CAD systems and is becoming widely available as a foundation for both design work and further software work as well.

Introduction

Taxonomy of languages

Hardware description languages range from very high-level functional and behavioral descriptions like ISPS down to the very low-level physical mask descriptions like CIF and Calma Stream. ABCD is near the lower end of this scale being a layout language. Sticks is a similar language but it is lower still since it is just an abstraction of mask rectangles. ABCD is more than that since it contains information about circuit primitives like transistors and wires. Additionally, ABCD has a lot in common with higher languages that describe circuits at the net-work level since ABCD contains information about nets.

Interaction with rest of system

ABCD is the interchange language and the database format for all the symbolic-level tools. It is created by generators, the interactive design capture tools, or by simple text editors. It is read by the interactiave design capture tools, the static circuit semantic checker, the place and route tools, and the compactor. Figure 1 shows the relationship between ABCD and the rest of the symbolic VLSI design system known as VIVID. ABCD as implemented in VIVID is both a language and a library of support routines to manipulate descriptions of circuits in the language. It is a parser for the language, it contains routines to generate legal statements in the language, it defines the data structure to be used internally by all tools as well as access routines to interact with those data structures (i.e. it is a database). It is integrated into all symbolic-level tools in the system so all such tools share the same code and thereby avoid any inconsistencies.

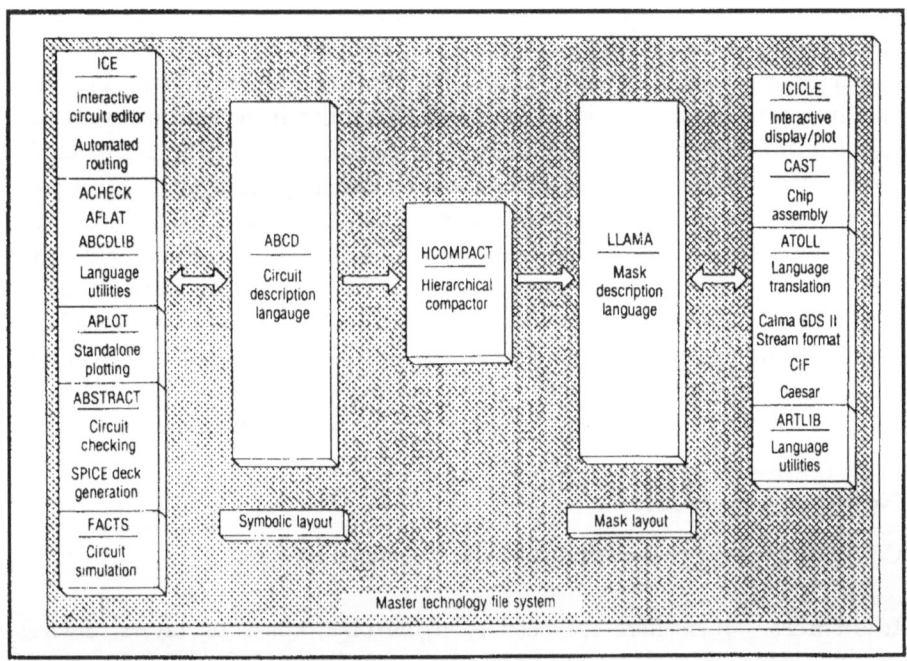

Figure 1: Block diagram showing ABCD in a symbolic CAD system

Major features

ABCD, besides being the basis for all symbolic-level circuit descriptions, has a number of other noteworthy features. Each primitive language statement provides for extensions which allow the language to adapt to changes in CAD requirements over time. These extensions are not handled directly by any language routines but are passed along into and out of the data structure intact so that tools that need information contained in these extensions can use it and tools that don't can ignore it. The language parser that is part of the ABCD support library provides for incremental or hierarchical parsing which allow a tool to parse in only parts of a hierarchical design. This is a vast improvement over strictly depth-first parsing which required every last leaf cell in a design be parsed before any manipulation of the design could begin. For the first time, a symbolic-level language is providing support for analog circuit primitives: the resistor and capacitor. These objects can be freely intermixed in symbolic designs allowing both pure analog designs and hybrid designs to be done symbolically. Primitives that have electrical significance are considered to be part of a net and the information about nets is maintained in the ABCD data structures allowing tools to maintain net-lists. This is a great aid to correctness-by-construction since it helps a designer know of electrical net errors early in the design process.

Notational conventions

Conventions used to describe the ABCD language follow.

Identifiers. Identifiers are used to name pins, contacts, layers, devices and instances. Identifiers are composed with the alphabetic and numeric characters, the underscore sign (_) and the hyphen (−) and are restricted in length to 32 characters. An identifier must begin with an alphabetic character. Identifiers are case-sensitive.

Comments. A comment is any string that begins with a number sign (#) and ends with a carriage return.

Keywords. A set of words is reserved explicitly for use in the ABCD language and cannot be used as identifiers. Keywords are composed of the lowercase alphabetic characters. They include:

    begin
    end
    device
    wire
    contact
    pin
    label
    group
    instance
    capacitor
    resistor

Continuation. Most statements in the ABCD language begin and end on the same line. A continuation character (\) allows long statements to be continued on succeeding lines.

Backus-Naur form. Conventions used to describe ABCD syntax in Backus-Naur form are listed in the following table.

BNF Notation	
Convention	Example
Reserved keywords are shown in bold lowercase characters. Keywords must be written exactly as shown in the BNF description	wire device instance
Variables are represented in italicized lowercase characters.	cname units width
Values are represented in bold lowercase characters, which cannot be expanded.	n-type poly xoff
[] Brackets enclose an optional item.	[*num*] [*pname* :]
| A vertical bar separates options, one of which must be selected.	**auto** | **vss** | **vdd**
() Parentheses group choices	(*num* , *str*)
{} Braces indicate zero or more repetitions of the enclosed expression	{ (*width*|*length*|*orien*) }

Text and graphics

An important design constraint on the ABCD language was that it be compatible with interactive graphic design capture techniques. This leads us to the new phrase *What you see is what we have.* There is a one-to-one correspondence between primitives in the language and graphical objects as seen

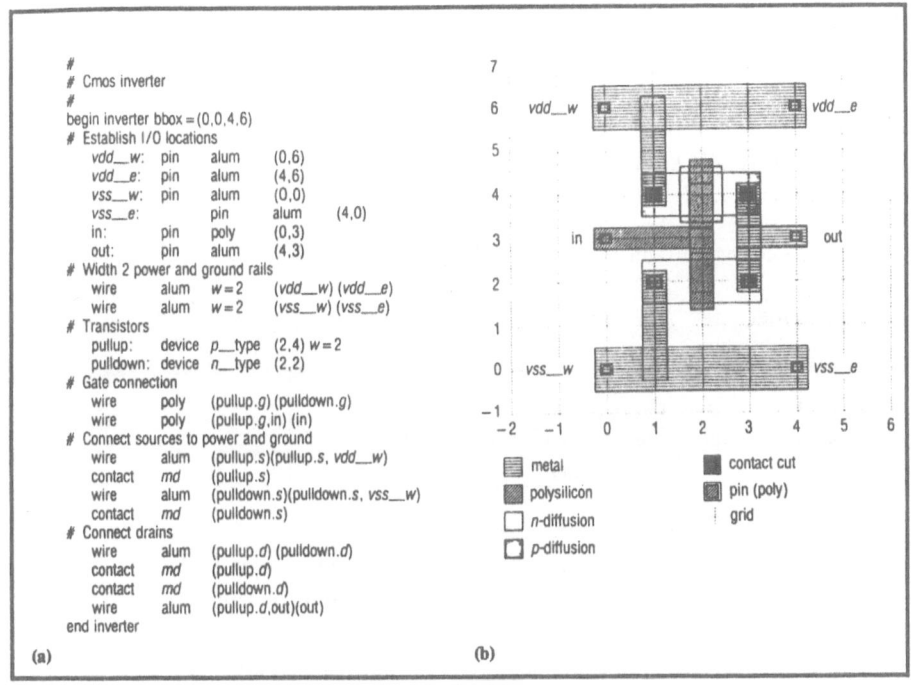

Figure 2: ABCD description of CMOS invertor (a); its symbolic layout (b)

through the symbolic editor. However, the language needed to be a good target for generators and silicon compilers so the language is also very flexible when represented textually. Primitives and locations can be named and other primitives can be placed relative to these making for self-documenting descriptions. Design modification can be done with the interactive graphics tools or standard text editors as suits the designer. Figure 2 shows the text and graphics for a small design cell.

Language Elements

Extensions. Each primitive statement in the language can contain user defined fields of the form $x = y$ and programs can be written to interrogate, modify, create, and delete these extensions in the internal (parsed) ABCD structures. The ABCD parser places the information for all unfamiliar fields of this form into the extension data structure attached to that

ABCD primitive.

BEGIN and END declarations.   Each ABCD cell description, not including comments, must begin with a declaration statement. This statement is identified by the keyword **begin**. The final statement in a cell description marks the end with the **end** keyword.

```
begin cellname tech elements depth [exten] bbox

cellname   ::=   identifier
tech       ::=   tech=(nmos | cmos)
elements   ::=   elements= num
depth      ::=   depth= num
exten      ::=   identifier=identifier [exten]
bbox       ::=   bbox=(x1,y1,x2,y2)
```

cellname   is the name of this cell

tech   is a parameter specifying the technology to be used for the cell. All cells within a design must use the same technology.

elements   tells how many primitives are in this cell.

depth   indicated how many levels of hierarchy are below this cell. -1 is used to indicate unknown and 0 means this cell is a leaf cell.

bbox   is a parameter indicating the size of the cell's bounding box. A bounding box is specified by the coordinates of diagonally opposite corners (usually lower-left and upper-right) of the area covered by this cell.

Here is an example begin line of a leaf cell:

```
begin reg1 tech=cmos elements=16 depth=0 bbox(0,0,10,8)
```

Transistors

ABCD transistors help to highlight the differences between mask descriptions and symbolic, virtual-grid descriptions. Although the transistor is graphically represented as a rectangular box with fixed connection points, in reality, this box represents a more complex physical and structural entity. If, for example, the design is fabricated in the CMOS technology, the box of diffusion material is surrounded by tubs and wells, which are also connected to other elements. ABCD models your knowledge of transistors by storing technology-dependent details separately from the ABCD file and using them only when generating actual layout.

```
[devicename:] device dtype [exten] location dmods

devicename   ::=   identifier
dtype        ::=   Any MTF supported device type
                   enh|dep|load or similar in nMOS
                   n-type|p-type or similar in CMOS
location     ::=   See page 32
exten        ::=   identifier=identifier [exten]
devicenet    ::=   { snet | gnet | dnet }
snet         ::=   snet= identifier
gnet         ::=   { gnet= identifier | net= identifier }
dnet         ::=   dnet= identifier
dmods        ::=   { width | length | orien }
width        ::=   w== ( num [ . 0|5])
length       ::=   l== ( num [ . 0|5])
orien        ::=   n | s | e | w
                   ne|nw|se|sw|en|es|wn|ws
```

dname   is the name by which the device is symbolically referenced.

dmods   represents any number (including zero); they can be specified in any order. Those not specified assume the default values, which are given in the master technology file database.

width   specifies the width of the transistor. If an integer is specified, it is multiplied by a predefined default width. If a fixed-point number is given, it specifies the device width expressed as a factor by which to multiply the default-size transistor.

length   specifies the length of the transistor. If an integer is specified, it is multiplied by a predefined default length. If a fixed-point number is given, it specifies the device length expressed as a factor by which to multiply the default-size transistor.

orien   specifies the orientation with respect to the drain of the transistor as either north, south, east, or west. For transistors that are not symmetrical, all eight orientations are available.

As device width increases, more connection points to the source and drain are needed for reduction of parasitic resistance. When points are connected to a gate, the connection is always made to the center of the device.

For example, an ABCD description of a double-width device is given:

```
device n-type (x,y) w=2
```

The resulting double-width device, as compared with a single-width and triple-width device, is shown as follows:

Single, Double, and Triple-Width Devices

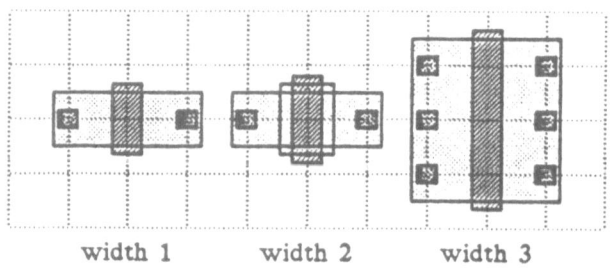

width 1   width 2   width 3

In this illustration, the gate connections are shown as rectangles filled with a stipple pattern, and the drain and source connections are shown as black squares. Connections to a device can be specified symbolically using a dot notation signifying connections to the gate, source, or drain of the device. In MOS, the drain and source nodes of a transistor depend on the relative voltages at each end of the device. The ABCD language uses these terms to identify locations of device parts rather than to specify electrical behavior. For example, if a device is named *d1*, then *d1.g* denotes the gate, *d1.s* is the source, and *d1.d* is the drain. Furthermore, for non-unit-width devices having more than one connection point, a number can be appended to the name, such as *d1.d[1]*. This number, which can be positive or negative, is interpreted as an offset from the default connection point in the direction perpendicular to the source-drain axis.

The number of device connection points depends on the device's width; the number of gate connections, however, is fixed at the value of 1. If the width is odd, the number of source and drain connections is equal to the width of the device; if the width is even, the number of source and drain connections is equal to a value one less than the width of the device. The length of a device determines the location of the source and drain connections relative to the gate. Normally, these locations are 1 virtual-grid unit apart but this distance becomes greater for longer devices.

You do not have to specify device-sizing information; defaults are provided. Defaults, which are technology-dependent, are read from a master technology file when the cell is processed. Device sizes should be dependent on the application and topology for a specific design.

Wires

The **wire** keyword is used to specify connections of circuit elements within cells and it also performs the composition function needed for cell interconnection. When wires are specified, a layer, a width, and a list of points are needed.

```
wire layer width network [exten] loclist

layer    ::=  any MTF supported layer
              e.g. metal m| poly | ndiff | pdiff
width    ::=  w=(num [.0| 5]
network  ::=  net=identifier
exten    ::=  identifier=identifier [exten]
loclist  ::=  location { location }
```

layer   is the fabrication material specification for the wire. There is no default layer for wires. The layer must be explicitly named in each wire statement.

width   is the wire width specification. The width remains constant for the entire length of the wire. There is a default width associated with each technology and layer. The expression, w=2, means that the default width is doubled. If a number containing a decimal point is used, the default width is multiplied by this factor.

network is the electrical net specification for this wire. Electrical connectivity is maintained in ABCD descriptions.

loclist is the list of points through which the wire travels.

An example of a **wire** statement follows:

```
wire metal (0,4) (12,4)
```

This **wire** statement defines a default-width wire in the metal layer between virtual-grid points (0,4) and (12,4). Additional points can be specified by adding them to the end of the list. The width can be changed by adding a width parameter. For example, the wire can be extended and widened using this statement.

```
wire metal w=2 (0,4) (12,4) (12,0)
```

Only perpendicular wire placements are supported in the ABCD language.

Contacts

If circuit elements are located on different layers, the **contact** statement is used to ensure electrical connection.

[cname:] **contact** contype [exten] location cmods		
cname	::=	identifier
contype	::=	Any MTF supported contact type
		auto|**via**|**vss**|**vdd** or similar in CMOS
		auto|**buried** or similar in nMOS
location	::=	See page 32
exten	::=	identifier=identifier [exten]
dmods	::=	{ orien | xoff | yoff }
orien	::=	**n** | **s** | **e** | **w**
		ne|**nw**|**se**|**sw**|**en**|**es**|**wn**|**ws**
xoff	::=	**xoff**= num
yoff	::=	**yoff**= num

    cname   is the name by which a contact is symbolically referenced.

    contype   specifies the contact type. The possible contact types include **auto** and body contacts, **vss** and **vdd**, in the CMOS technology. Legal contacts in the nMOS technology include **auto** and **buried**. The body contacts connect power and ground to their respective substrates.

    cmods   represents a set of contact modifiers.

    orien   represents a set of orientations for non-square contacts.

    xoff   is an offset is the amount of distance (in mask units) to leave between the location specified and the true center of the contact on the x coordinate. Offsets are useful for mask generation and they become effective during compaction.

    yoff   is the same as *xoff*, except it is used on the y coordinate.

A simple example of a metal wire connected to the diffusion of a device with a metal-diffusion contact follows:

$$x: \text{device n-type } (12,4) \text{ or=w}$$
$$\text{cxscontactauto x.s}$$
$$\text{wire}\quad \text{metal cxs } (11,0)$$

The device named x is numerically positioned at grid location (12,4) with its drain pointing west (or left). A contact *cxs* is symbolically referenced to the source of device x. A metal wire is connected between contact *cxs* and numeric coordinate (11,12). If the transistor is moved, the contact and wire also move to track the transistor.

Pins

The **pin** keyword is used for several functions. It is the entry point to a cell since it is legal to connect a cell only through its pins. Pins also provide you with useful feedback when a cell is displayed since the names of pins usually reflect their functions. The pins, in essence, form the link between a conventional circuit diagram and its ABCD description.

```
[pname:] pin layer [exten] location [network]

pname    ::=   identifier
layer    ::=   any MTF supported layer
               e.g. metal m| poly | ndiff | pdiff
location ::=   See page 32
exten    ::=   identifier=identifier [exten]
network  ::=   net=identifier
```

pname is the name by which a pin is symbolically referenced.

layer is the layer to which this pin should be considered
         electrically connected. There is no default layer for
         pins. The layer must be explicitly named in each **pin**
         statement.

Pin names are composed of two parts that are separated by an underscore (_). These two-part names are used when physically distinct nodes are
treated as the same electrical node. The **pin** statement specifies layers because pins are always connected to one layer at their position. An example
follows:

$$\text{vdd\_w:} \quad \text{pin metal (0,12)}$$
$$\text{vdd\_e:} \quad \text{pin metal (20,12)}$$
$$\text{vss\_w:} \quad \text{pin metal (0,0)}$$
$$\text{vss\_e:} \quad \text{pin metal (20,0)}$$

Four pins on the metal layer identify two distinct electrical nodes for
power and ground connections. Pins are important in symbolic bonding
since initial connection descriptions are specified only in terms of transistor
nodes and pins.

Groups

A group type name can be assigned to a circuit description, which precedes
a **begingroup** and **endgroup** grouping, to facilitate the automated wire
router in its routing and unrouting activities. Especially in cases where wires
overlap, the group name can identify for the router only the specific wires
to be affected by a route or unroute action. Furthermore, group names are
helpful in identifying instances to be expanded or unexpanded when they
overlap in the grid work area. (Refer to the next section for further details.)

```
begingroup [exten] gname

... any number of ABCD primitives
exten    ::=    identifier=identifier [exten]
gname    ::=    identifier
```

All devices specified in a circuit description are presumed to be in a group. Instances, however, are excluded from such a group.

```
begingroup  type=route wr123
        wire   routed  (7,3) (12,4)
        wire   metal   (7,3) (12,4)
        wire   poly    (12,3) (12,4)
     contactauto       (12,3)
endgroup    wr123
```

The **group** keywords surround the statements that describe a pseudo wire and its subsequent conversion to metal and polysilicon wires and a contact.

Labels

The **label** keyword allows you to assign a symbolic name to a location on the virtual grid to facilitate relative placement of other devices on the grid. Arbitrary text can be associated with this point to label a design as well.

```
[lname:] label ltype [exten] lleft [uright] qstring

lname    ::=    identifier
ltype    ::=    type= ( error | point | info )
exten    ::=    identifier=identifier [exten]
lleft    ::=    location
uleft    ::=    location
qstring  ::=    " any printable characters "
```

lname   is the name by which the labeled point is symbolically referenced.

An empty cell could be given a desired area, such as in floorplanning, by specifying labeled points at diagonally opposite corners with the label construct. An example follows:

```
begin blankcell
    ll: label type=point (0,0)
    ur:label type=point (12,20)
end blankcell
```

This example produces a cell with an area of 12 by 20 virtual-grid units.

Instances

An instance is an explicit arrangement of a cell at a specific virtual-grid location and orientation, possibly replicated into an array. The instance primitive gives ABCD its hierarchical structure.

A typical design includes a set of leaf cells, which do not contain instances, and a set of composition cells, which contain instances. Composition cells are used to "glue" together parts of the design. Composition cells can instantiate other composition cells and, consequently, the hierarchy can contain multiple levels. To replicate a cell into an array, you specify the repetition factor and the spacing between array elements. The array elements can be positioned into one of eight orientations as follows:

↕SE	↕SW	↠EN	↞WN
↟NE	↟NW	↠ES	↞WS

The eight possible orientations are derived from the arrangement of two arrows within a bounding box. The longer arrow is the major axis; the shorter arrow intersecting the major axis the minor axis. The group of orientations is divided in half; the selections on the left side are more frequently used than those on the right. In the preceding figure, the default orientation, **NE** (major axis points north, minor axis points east), is in the lower-left position. The **NW** orientation is produced from mirroring the minor axis of the default. **SW** mirrors the major axis of **NW** and **SE** mirrors the minor axis of **SW**.

In the second group, **ES**, is produced by rotating the default in a 90-degree clockwise direction. The remaining orientations in the second group are formed in an identical manner to the first group.

```
[iname:] instance cname iplace imods [connect]

iname        ::=   identifier
cname        ::=   identifier
iplace       ::=   [corner = ] location
corner       ::=   ll | ul | lr | ur
exten        ::=   identifier=identifier [exten]
imods        ::=   {reps | dir | space | orien}
reps         ::=   n= num
dir          ::=   dir= ( h | v )
space        ::=   space= num
orien        ::=   ne | nw | se | sw | en | es | wn | ws
connect      ::=   connect ( connections )
connections  ::=   { identifier : identifier [,connections] }
```

iname   is the name by which the instance is symbolically referenced.

cname   is the name of the cell being instantiated.

iplace   specifies the placement of the instance.

index   allows the specification of a particular repetition of a cell within an instance that has a specified repetition factor.

corner   specifies one of the four corners of the bounding box of the instance to be the reference point for placement on the grid.

imods   represents the instance modifiers; those not specified presume the default values provided in *rep*, *or*, *dir*, or *space*.

rep   is a number that specifies the repetition factor; the default is 1.

dir   is the direction in which the cell is positioned when successive cells are arranged; its value is either horizontal or vertical. The default is horizontal.

space   specifies the number of virtual-grid units of space to leave between individual cells within an instance. Use of this field is discouraged in favor of separate instance calls. The default is zero (0).

orien   specifies which of eight possible orientations is to be applied to the instance. The orientation is applied to each cell individually before replication. The default is **ne**.

connect specifies cell-to-cell connections to be made between the cell being instantiated and another that is abutted to this one.

A leaf cell contained in file *inv.ab* has its lower-left corner at the virtual-grid point (0,0) and its upper-right corner at (4,8). Thus, the bounding box has these coordinates: (0,0,4,8).

A composition cell that instantiates *inv.ab* uses this statement:

```
cinv1:  inst cinv 11=(0,0) or=ne
connect(in:neta, out:netb, vdd:Vdd, vss:Groundnet)
```

The composition cell *inv1*, which instantiates the leaf cell *inv*, positions the lower-left corner of *inv* at coordinates (0,0) in the virtual grid.

At this point, the lower-right corner of *cinv1* is at location (4,0). The pins with names "in", "out", "vdd", and "vss" are said to be connected to the parent cells nets or pins "neta, netb, Vdd, Groundnet".

A statement to place a second copy of cell *inv* into the composition cell *inv1* so that the left edge of *inv2* is shared with the right edge of *inv1*. Note that the network which was specified as being connected to inv1's output is now said to connect to inv2's input. Because these cells abut, this connection should be redundant. The connectivity verification programs will report any inconsistencies between this designer intention and the actual layout connectivity. In top down design, the inverter cell may not have been designed and the inconsistency may arise long after the connect statment was created. Inconsistency may also arise if an existing consistent inverter design is modified.

cinv2: **inst** cinv ll=cinv1[lr] or=ne
connect(in:netb, out:netc, vdd:Vdd, vss:Groundnet)

The preceding two statements can be replaced by one statement:

inst cinv ll=(0,0) n=2 dir=h or=ne
connect(in:neta, out:netc, vdd:Vdd, vss:Groundnet)

Because the connection information is optional, this could also be:

inst cinv ll=(0,0) n=2 dir=h or=ne

A lower-left corner point (**ll**) refers to the corner of the entire block of instantiated cells. The repetition factor specifies the number of repetitions of the cell in the instance (**n = 2**), and a horizontal direction (**dir=h**) is specified.

The orientation specification in an instance statement applies to each cell to be instantiated. There is no concept of rotation or mirroring about an arbitrary point or axis; individual cells are oriented as specified and then placed at a position according to the corner and direction declarations.

In places other than the connection specification, one may wish to refer to pins inside an instance for symbolic placement purposes. In this case, the following form is used:

instance_name [*instance_index*] **.pin_name**

In a repeated instance, the index must be specified in the range 1 to n for n cells. In an instance of a single cell, the index can be omitted.

A sample ABCD cell description — a CMOS shift register that instantiates two cells, *inv* and *tgate1* — follows:

```
begin reg1 tech=cmos elements=16 depth=0 bbox(0,0,10,8)
# a shift register composed of an inverter
# whose outputs go through a transmission gate
# controlled by phi1.
#
# instances:
#
    inverter:  instance inv ll=(0,0) or=ne
    tgate:     instance tgate1 ll=inverter[lr] or=ne

# the register's pins:
#
    vdd_w:   pin      metal inverter.vdd_w
    f2br_w:  pin      metal inverter.f2br_w
    f1br_w:  pin      metal inverter.f1br_w
    in:      pin      metal inverter.in
    f2_w:    pin      metal inverter.f2_w
    f1_w:    pin      metal inverter.f1_w
    vss_w:   pin      metal inverter.vss_w

    vdd_e:   pin      metal tgate.vdd_e
    f2br_e:  pin      metal tgate.f2br_e
    f1br_e:  pin      metal tgate.f1br_e
    out:     pin      metal tgate.out
    f2_e:    pin      metal tgate.f2_e
    f1_e:    pin      metal tgate.f1_e
    vss_e:   pin      metal tgate.vss_e

end reg1
```

This notation facilitates relative placement of instances, which illustrates another application of symbolic bonding and the textual design of circuit descriptions. It is also a powerful mechanism when used in a floorplanner. When any cell is modified in a way that changes its area, symbolic bonding automatically implements the change throughout the circuit by shifting all symbolically placed instances to the right and up.

In the above example, the **bbox** may or may not be correct since it depends upon the size of the two instantiated cells. In general, these numbers are used to facilitate top-down design but are recomputed whenever a full design is parsed and analyzed.

Symbolic Referencing

Each primitive in the ABCD language requires placement information to determine its location in the virtual grid. The location parameter contains both symbolic and numeric specifications. Symbolic specification refers to the placement of a primitive on the virtual grid, relative to another primitive. You can place two primitives at the same location or on the same horizontal or vertical grid line. Any primitive that is referenced symbolically must be defined somewhere in the current cell; however, a primitive definition does not have to precede a reference to the primitive.

location	::=	(xval , yval)
	|	symbol
xval,yval	::=	num
	|	symbol
	|	[instname] edge
symbol	::=	instname [index] pin_name
	|	instname [index] [corner]
	|	pinname
	|	labelname
	|	devicename [.(s | g | d)][*offset*]
	|	contactname
instname	::=	identifier
pinname	::=	identifier
labelname	::=	identifier
devicename	::=	identifier
contactname	::=	identifier
edge	::=	.n | .s | .e | .w
		|.ne | .nw | .se | .sw
offset	::=	[num]

num,num   is an ordered pair of values that specify a numeric reference.

symbol   is a symbolic reference to a point, contact, or pin. It is a name given to a primitive in the current cell. It is a local symbol that must be unique only within the current cell.

edge   refers to the edge (for a single coordinate) or a corner (for an entire point) of the current cell. This will be used primarily for pins and wire ends which should always stay on the edge of the cell. The actual grid coordinates of this symbolic points are not computed until all other primitives have been located and the bounding box has been computed.

For this reason, only point primitives (eg. NOT devices and instances) can be symbolically bonded to an edge and no other primitives should be bonded to a primitive that is bonded to an edge.

exten   allows you to specify a connection to the drain, source, or gate (the default) of a device. Additionally, on non-default width devices having more than one drain and source connection, an offset can be specified relative to the center of the device. For example, $t1.s[-1]$ specifies one grid space down from the default source connection point.

An example of symbolic referencing follows:

```
p1:   pin      metal    (43,0)        # (numeric,numeric) form
d1:   device   p-type   w=3 (43,56)   # (numeric,numeric) form
      wire     metal    (p1) (d1)     # lsymbol form & devref form
      contact auto      (d1.d)        # devref (drain) form
      contact auto      (d1.d[-1])    # devref (offset drain) form
```

A metal pin named "p1" is positioned at numeric grid location (43,0). The drain of a P-type CMOS transistor, "d1", is positioned at location (43,56). The third line represents a metal wire connecting pin "p1" and the gate (by default) of device "d1". The contact statement specifies that a contact cut is symbolically connected to the center of drain "d1" using the specification "d1.d". Finally, the last contact cut is symbolically connected to the drain of "d1", offset one grid line down from the middle drain location using the specification "d1.d[-1]".

Parsing Strategies

Parsing is the process of reading a clear text file in some language and translating it into an internal form suitable for further processing by a computer program. Most circuit description languages are hierarchical meaning that complex objects are made by including one or more less complex objects in them. We use the terms "leaf cell" and "composition cell" to differentiate between cells that have nothing but circuit-level primitives in them (i.e. "leaf cells") and those that contain calls to other cells (i.e. "composition cells"). When parsing a hierarchical design (one where the starting cell is a composition cell) several methods of parsing can be used. The most important issue to consider when comparing methods is user wait time.

User wait-time. We must be concerned at all times with user wait-time because excessive wait-time will frustrate users and decrease their willingness to use the CAD tool. Users have usually got a preconceived notion of which tasks are difficult and which are not so they will wait some "reasonable" amount of time. If they are planning to do some small manipulation with an interactive tool on a very small, localized piece of the design then they do not view it as reasonable to have to wait for the parser to process every cell in the entire design.

Recursive depth-first parsing. The most common strategy for parsing is recursive depth-first parsing. It is most intuitive and easiest to implement. The parser process statements it reads from the input file in order and when a call to a subcell is found it process that statment by calling the parser recursively. The result is a depth-first expansion of the tree representing the hierarchical design. It results, however, in worst-case user wait-time.

Hierarchical parsing. Another approach to parsing available with the ABCD parser is parsing of one level in the hierarchy at a time. This minimizes user-wait time (if the user wants to only modify a small piece of the design — if the user wants the entire design on hand the time to parse is identical to recursive depth-first parsing) and spreads the time of parsing out.

DESIGN CAPTURE

Introduction

This section is about interactive design capture for creating nd viewing
ABCD designs. The particular tool we will explore is called ICE and like
ABCD it is part of the VIVID Symbolic VLSI CAD System.

Interactive design capture is the most convenient way to create and edit
ABCD. The one-to-one mapping between ABCD primitives and circuit objects that are represented as single graphical entities makes this straight
forward. ABCD is a shared library used by ICE as it is by other parts of
VIVID. Being a shared library makes it possible to implement shared memory when two programs occupy the same address space and communicate
via common ABCD access routines.

ABCD is the circuit database format used throughout the system. ABCD
primitives are indivisible objects that are translated into graphical objects at
plotting time. ICE provides atomic editing functions on the circuit objects
not the constituent rectangles. ICE is fully hierarchical so it used the ABCD
instance construct to build-up arbitrarily deeply nested composition cells.

The editing paradigm, that is the model used to manipulate objects, is
one of: set up the attributes of an object and then do something to that
object. This paradigm works better in an object oriented setting than the
paradigm of "painting" used frequently in mask-level rectangle pushers.

The virtual-grid is explicit in ICE — it is a visual grid on the graphics
screen to simplify object placement during editing. The designer just used
this grid to assist in correct placement of objects but to the user, the grid
remains evenly spaced while using ICE even though compaction will make
spacing between grid lines uneven.

ICE is a technology independent and process independent tool. At run
time a choice of MTF database configures ICE to the users choice of technology and process. This includes color maps, names of process layers, types
of transistors and contacts, etc.

The normal method of use for ICE is to interactively create, edit and
modify a design cell by cell. Cells already in libraries can be brought into
a design. ICE can be used as a viewing tool to get hard copy of a design.
In many cases, with semi-automatic generators such as routers and placers,
ICE can be used to view and/or modify intermediate results.

ICE performance is critical as it is for any highly interactive design capture tool. This is especially true for parsing (already discussed in reference

to ABCD) and plotting. But users make their overall evaluation of design capture performance more on the "snappiness" of the buttons during normal editing — how long after a button is hit can another be hit. This requires some very special data structures and access methods that will be discussed later.

Structure

VIVID is logically divided into two major parts separated by the compactor: symbolic virtual grid on one side and mask-level on the other. ICE, as the principal means of manipulating ABCD, is the central figure in the symbolic virtual-grid domain.

ICE uses almost every subroutine library contained in VIVID; ABCD is its library circuit data-base access routines, Z is the graphics library for device independent interactive graphics, MTF is the technology database library of access routines, and WOMBAT is the library of general-purpose (i.e. miscellaneous) routines.

User Interface

Screen usage is a critical issue for design capture tools. It needs to be clean and simple so it is pleasant to use but all the information a user requires must be there. Even 25 inch screens with over 1 million pixels represent a resource that must be managed carefully. In ICE, the majority of the usable screen space is dedicated to active editing — it is the work area. Below that is status information and keyboard echo. To the right of that is a facsimile of what is on the work surface with a "you-are-here" indicator arrow. When the work area is zoomed in for high magnification the facsimile still shows where that is relative to the un-magnified view. The entire right-hand strip of the screen is dedicated to menus. A representation of this screen layout is shown in figure 3.

Input is managed via a mouse or data tablet primarily. Hits on menus change the state of the editor, hits on the work surface cause editing functions to be invoked. The keyboard is used for controlling editor attributes and for *scripts*. Scripts are short hand commands for a series of longer commands like menu hits. Usually, users would define single keys to correspond to commonly needed more complex operations — 'd' to mean *delete*, 'p1' to mean *width 1 p-type CMOS transistor*, etc. Additionally, strokes also can map to these scripts. Strokes are user defined symbols "drawn" using the data tablet puck or mouse. The purpose of a stroke is to allow the user's

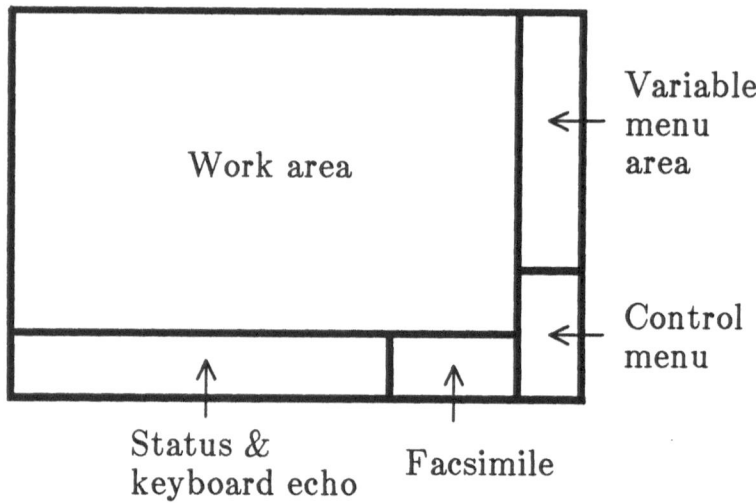

Figure 3: Screen layout for symbolic design capture

eyes to remain on the worksurface concentrating on the important information — the design. A stroke can be done anywhere on the work area without losing ones concentration. Finally, voice input is also an important offerring; again, it maps to a script. The voice recognizer matches a spoken word or short phrase to one already known about and if it makes the match correctly it sends the appropriate command sequence to ICE.

Menus are an important aspect to the user interface of a design capture tool. They should be simple and clear but functional enough so that users are not always hitting menu items to get their function performed. Pull-down menus may be a poor choice in a CAD design capture tool as they are in many applications. Instead, the ICE menus use colors, groupings and meaningful icons to aid in quick comprehension while providing the necessary functions to minimize menu switching.

The user's use of files is primarily those ABCD files containing the design description on which s/he is working. Because VLSI designs do not nicely map onto hierarchical file structures provided by operating systems like UNIX, a flexible mapping scheme is needed that performs a more natural mapping. A scheme was developed in VIVID to allow a user to specify mappings by file, by module, or by file type in an arbitrary way.

Flow of control — how you get from point 'a' to point 'b' — in a design capture tool also has great impact on the user interface. Frequently traveled paths should be short while less frequently traveled paths can be longer. The most frequent complaint from users of such tools is about how clumsy it is to do certain functions. In ICE the functions to change the state of the editor (e.g. grid on or off) are always accessible via keyboard hits. Common editing functions are always available like delete, modify and insert. To get from inserting one primitive to any other primitive is only one menu hit. Contacts and wires are available from the same menu since they are usually placed simultaneously. There is not separation between leaf cell editing and composition-cell editing but warnings are issued if hierarchy is mixed with circuit primitives in the same cell.

Function

This section describes how this symbolic virtual-grid design capture tool known as ICE functions.

Editor attribute setting. An alphanumeric menu containing the list of all settable attributes and their current values can pop-up over the work area any time the user requests it. But without it being visible, the user can still at any time change these attributes. The list of attributes is shown in the following table.

CMD	Attribute	Value type	Description
af	audio feedback	on / off	on=click when button hit
ap	active primitives	list of prims	list of prims that will be plotted
aw	ABCD warnings	on / off	on=display ABCD warnings
b	background	on / off	on=grey background; off=black
cc	command char	char	: is normal character to preface "long" commands such as changing these attributes
cd	current dir	string (legal directory)	the current project directory
db	debugging	list of debugging flags	an encoded debugging flag
fm	fill mode	solid / outline	solid=objects will be drawn using filled rectangles
gs	grid spacing	1..n	grid lines drawn on every virtual grid line
gt	grid type	off / ticks / dots / full	three types of grids can be drawn or none at all
ib	inst bonding	on / off	use symbolic referencing when floorplanning so that the floorplan can expand
ii	inst identify	cell / inst name	cell=use cell names when displaying instances
ir	inst resolve	on / off	on=resolve the symbolic referencing between instances as each editing operation is made
is	inst snap	on / off	on=snap instances to their nearest neighbor for abutment
ld	level display	1..n	display the entire design at level x in the hierarchy
lt	label text	on / off	on=show the label text
pb	preserve bonds	on / off	on=preserve the symbolic references when editing
pf	pin file	string (legal file name)	the name of the file where all the signal names exist
rc	RCS usage	on / off	on=invoke RCS source code control when writing cells
sh	shell name	string (legal shell name)	usually /bin/sh or /bin/csh
wi	wire ins mode	norm / orth	rubber-band wires or strictly orthogonal
wl	wire layers	mtf wire layer list	displayable wire layers from MTF list
z	zoom factor	2..n	the multiplication factor to use when zooming in and out

Top-level functions. A series of commands are also available at all times via keyboard hits. They are listed in the following table.

CMD	Action	Explanation
A	Save attributes	Save the current state of all the attributes (shown above) so that the next time the program is run it will begin with these values
D	Delete & reparse	Delete the current cell from memory and reparse it from its unchanged disk file to restore memory to that cell before any editing was done
F	Figure subsystem	Enter the figure subsystem to produce pictures to be placed in documents like design specifications or reports
G	Get new active cell	Yet another way to bring a new cell into the system
H	Hard copy subsystem	Get into a subsystem that allows the setting of a special set of display attributes for hard copy
L	List cells in memory	Get a list of all the cells in the system with some statistics about them
W	Write cell to disk	Yet another way to write the current cell to disk
!	Escape to shell	

CONTROL menu.

The top six items are icons for primitive menus: **device** for transistors, **pin**, **wire/cut**, **label**, **analog** (resistors and capacitors) and **instance** (for composition cells). When one of these items is selected the large menu resting on top of this one changes to be the correct primitive menu.

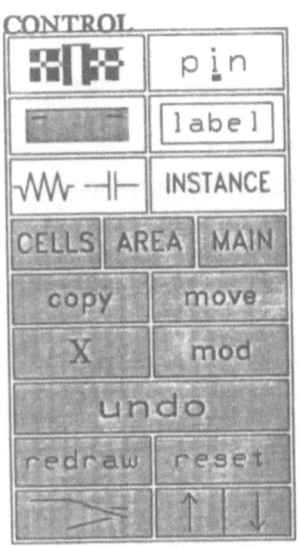

Beneath that are three items that also cause a change in the large menu above: **CELLS**, **AREA**, and **MAIN**. **CELLS** is a menu to control the ABCD cells resident in memory — parsing, writing and selecting which cell is currently being edited. **AREA** is a menu allowing global area operations such as area delete, area move, area copy and area modify as well as grid operations such as adding or deleting a grid line. This later operation is very useful if the editing area gets too crowed for the complex circuitry necessary. This is an operation that is trivial in the symbolic virtual grid environment and is all but impossible in the mask environment. **MAIN** is the menu just described previously.

Beneath that row of three items is a group of five items relating to editing: **copy**, **move**, **X** (delete), **mod**, and **undo**. **copy** allows one primitive to be copied to a new location preserving the attributes of the copied object. **move** moves one primitive keeping all attributes except its location constant. **X** is for deleting primitives — it changes the mode of the editor. **mod** is for modifying one or more attributes of a primitive (e.g. changing a p-type transistor to n-type). **undo** simply un-does the last editing change that was made whether it was an area delete of 800 transistors or the changing of a width 1 wire to a width 2.

Finally, there are four items for changing the display. **redraw** erases the graphics display and refreshes it so the picture is correct — necessary because of occasional hardware and software glitches that cause lint on the screen. **reset** changes the magnification and centering so the design under scrutiny is back to its default view — centered and full size on the display. **zoom** zooms up or down depending on which part of the icon is hit — the magnification of the display on the design changes usually by factors of 2. **up/down** select different details of view when looking at a hierarchical design. Each 'down' selected shows one more level of detail until the internals of the lowest leaf cells has been displayed.

MAIN menu. The top two items are related to each other. **edit-in-place** allows one to edit a cell in context no matter what its orientation — this is critical since one usually cares how a cell communicates with its neighbors. **copy & edit** is similar except one may not want to edit every instance of a cell as with **edit-in-place** so this makes a new cell out of the desired cell in its current orientation and allows one to edit it.

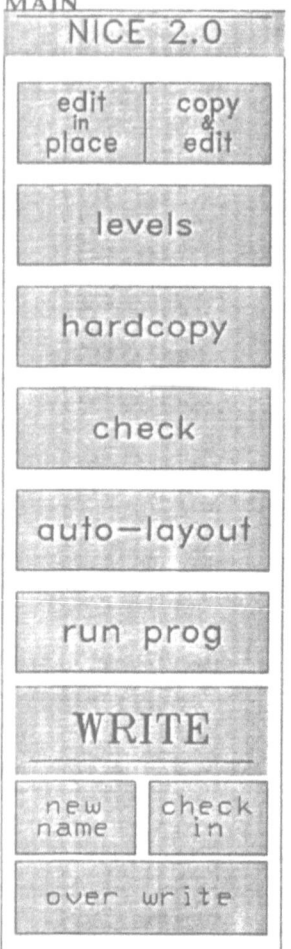

levels controls selective level display on an instance by instance basis allowing detailed views of some cells while others remain only in outline form.

hardcopy makes a plot on the current hard copy output device of what is currently displayed.

check runs a static semantic circuit check on the current cell to identify electrical problems such as shorts and opens.

auto-layout is an interface to an experimental technique for translating a net-list to symbolic layout.

run prog is an interface to external tools; when selected the current cell is written out, the specified program is run with that cell as an argument, then the cell is read back in and normal operation continues.

WRITE writes the current cell out to disk; it has associated with it some special items to manage this process.

Primitive menus.

TRANSISTOR Includes items for setting transistor length and width, controlling the user extension fields, transistor orientation, and the technology specific transistor types (shown here is a CMOS version with p-type and n-type only).

WIRE / CUT Includes items for the technology specific layer, wire width, extensions, routing and jog insertion on the wire portion; it includes only technology specific cut type and feedback of x-offset, y-offset and orientation on the contact cut portion.

LABEL Includes items for label type, text formatting options and extension information.

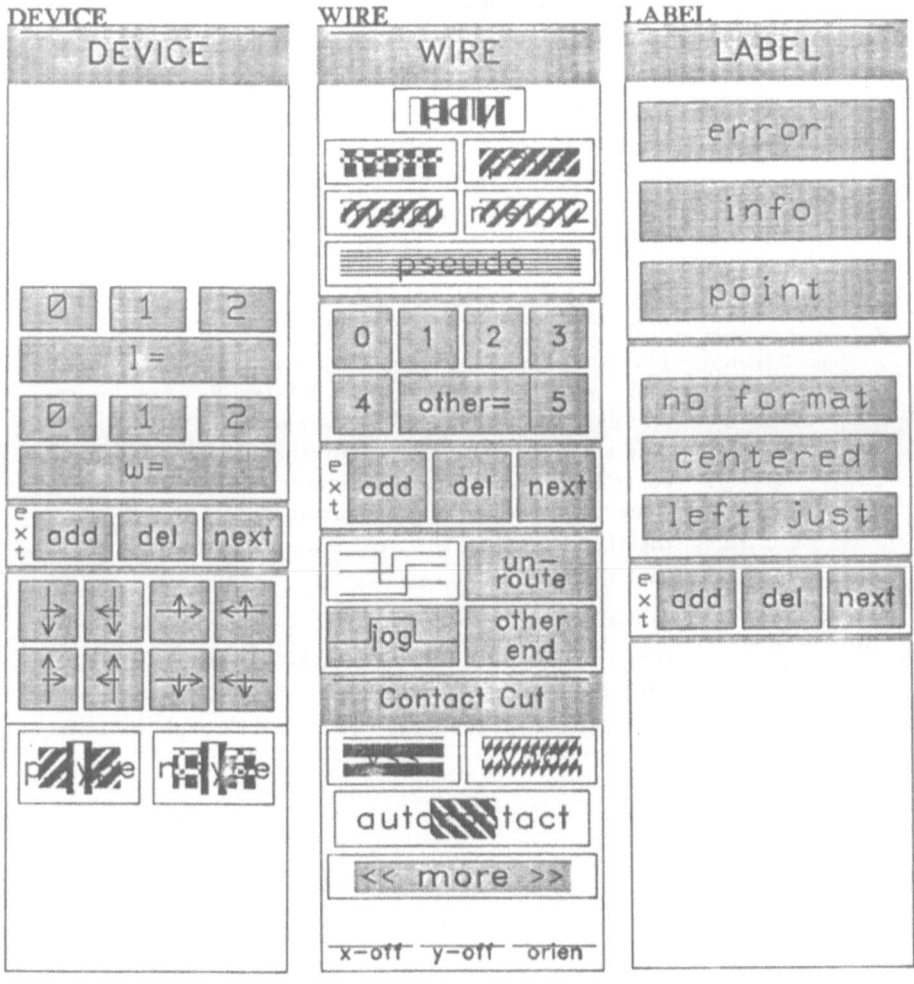

Hierarchy. This menu is the interface to editing composition cells — sometimes referred to as floor-planning.

cell name is a scroll region containing all the cells known to the program with items allowing perusal of this list as well as adding new cells to this list.

INSTANCE

INSTANCE

cell name

rd | wr | inc

inst name

ll lr ul ur

2-D

ext add del next

edit in place   copy & edit

llama cell   erase pins   show pins

route ✓ connections ✓ force vectors ✓ congestion

inst name is also a scroll region of instance names which uniquely identify each instance of a cell.

corner is a group of items allowing specification of which corner of an instance is to be placed on the grid.

direction is a group of two items to specify a horizontal or vertical array of cells to form an instance.

repetitions is an item to specify how many cells are part of this arrayed instance.

2-D is an interface to a method for placing 2-dimensional arrays of cells as would be used in memory arrays.

exten is the standard group of items allowing changes to be made to the extension field of the ABCD instance statement currently under consideration.

edit-in-place and **copy & edit** are duplicated here from the MAIN menu as a convenience.

pins pins can sometimes be an annoyance when floorplanning so they can be selectively turned on and off.

placement is an interface to a general cell placement system is provided here that allows different views of the current design that indicate how the placer will perform. The different views offerred are: show the connections that will control how communication between cells will be affected; show the force vectors that are pulling certain cells closer together; and, show the areas where congestion is a factor that will affect placement.

global routing is an interface to an AI-based global router. The route can be executed, it can be un-done, what the router will do can be shown before it runs, and the results of the route can be modified through this interface.

Circuit net-list. This menu relates to pins — these are non-physical entities that form the backbone of the electrical specification of a circuit.

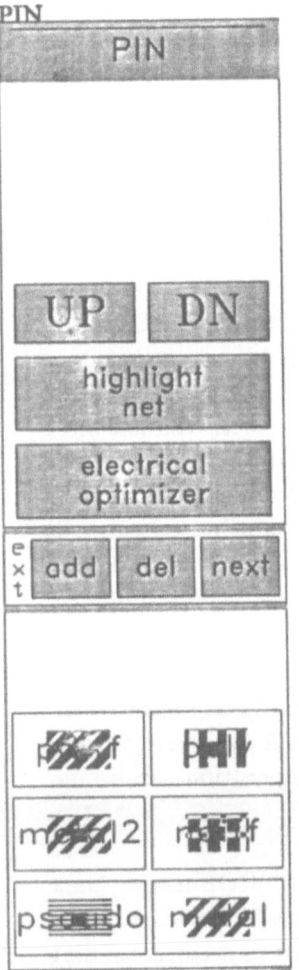

signal the open area at the top of the menu contains a scroll region of all signal names known to the program with the ability to accept new names. One name in this list will be considered the current signal name of the pin being edited.

UP/DN pulls pins in subcells up to the current level or, alternatively, it pushes pins in the current level down into subcells of the current cell. This aids bottom-up design or top-down design respectively.

highlight net displays all the primitives on the current net identified by signal name.

exten is the standard interface to the pin extension information (the most likely one to be used extensively).

layer a group of technology-specific items for specifying which layer the pin is associated with — this is the link between the physical and the electrical realms.

electrical optimizer is an experimental interface to a tool that aids in finding critical paths and, having found such a path, it helps the designer make the speed/power tradeoff by sizing transistors appropriately.

Editing operations.

INSERT is the normal mode of the ICE design capture tool. During this mode, the primitive menu currently displayed controls the type of object that will be inserted in the current cell when appropriate buttons are hit. With the cursor over the work area, if the middle button is hit the current type of primitive is inserted into the cell at that point. During this mode the right-most button is an *eraser* for the current type of primitive. This means that objects of the same type being inserted can be deleted just by hitting the right button when the cursor is over them.

DELETE is an editing mode of the ICE design capture tool. Selection of the **X** menu item in the CONTROL menu puts ICE in this mode. Once in this mode, objects can be deleted with hits of mouse or data tablet puck buttons in an appropriate manner. Items are selected with the left-most button. With the displayed cursor over an item or items this button is selected to *pick* an object. The first object picked will be of the same type as the current primitive menu displayed (if any) if such a primitive exists at this spot. Successive hits while at the same location will *cycle* through all the objects that intersect this location. The object selected will be hilighted. The top (or middle) button will delete the selected object.

UNDO is available to un-do some editing operation at any time by selecting the **UNDO** item on the CONTROL menu. A stack of previous operations is maintained so more than just the last operation can be un-done.

MODIFY is an editor mode selected from the CONTROL menu. Once in this mode the object to be modified can be selected which will cause the appropriate primitive menu to appear with the attributes set corresponding to the selected primitive. Selection of different attributes is immediately reflected by changes in the selected primitive.

COPY is a one-time function selected from the CONTROL menu. In this function one can make copies of primitive objects which have identical attributes except for location.

AREA is a mode that is selected from the CONTROL menu. Selection of this item causes a new menu to appear in the upper menu region. Area operations are considered very important by most users. It is because they are such a time saver: large blocks of objects are manipulated as a group. In any order desirable the user chooses what operation is to be performed (move, cut, copy, modify, or paste), what objects are to be affected, what orientation to apply (in the case of move, copy or paste), and what area is to be affected.

operation gives choices of **move, cut, copy, modify, paste**, and **show buffer**. **move** takes all selected objects, deletes them from their current locations and moves them, maintaining their relationship to one another, to a new set of locations after applying any orientation change. **cut** deletes all selected objects and saves them in a buffer for a later paste operation. **copy** takes all selected objects and makes a new copy of them, without destroying the originals, at a new set of locations after applying any orientation change. **modify** applys some attribute change to all selected objects at the same time. **paste** puts copies of all objects in the buffer at a new set of locations after applying any orientation change. **show buffer** shows what is currently in the buffer without placing it anywhere in the circuit.

object selection can be done as a group or individually. This set of items allows all objects in the selected area to be selected or de-selected (whichever reverses their current state). Individual objects can be toggled by selecting them with the cursor.

orientation allows one of the standard eight orientations to be used when doing a move, copy or paste operation.

grid change provides four items to add or subtrace vertical or horizontal grid lines from the current cell.

Performance

Three areas were considered most crucial in making the performance of this design capture tool acceptable: very fast pick operations, optimization of graphics display functions, and appropriate use of the ABCD parsing strategies.

Pick is the operation used when finding all the objects that intersect a given point in the virtual grid. It must be fast since it is used constantly during editing and can greatly impact the feeling of "snappiness" the user perceives. Most interactive design capture tools maintain the objects being edited in a simple linear linked list since this is a good structure for insertion and deletion and is simple to implement and maintain. However, if 1000 objects are maintained in such a list the constant searching of the entire list to find all the objects at a given point slows down the system prohibitively. A very fast data structure for allowing a fast pick is simply a 2-d grid of pointers to linked lists of objects. Each cell in this 2-d array represents a grid location and the list at that cell is the list of objects that intersect that grid location. Pick is then a constant time operation. This 2-d grid could get quite large so it is restricted to be no larger than 50 by 50 grid units. This window is as large a window as designers edit in. The window slides around as the designer moves around but there is a lot of locality in editing so it doesn't have to move very often. The grid is not used during editing of composition cells because there a relatively small number of objects covers a relatively large area and the linked list is adequate.

Display optimization looked for ways to minimize the amount of processing required to draw a circuit description on the graphics display. This is a procedure that occurred frequently as the designer moves around on the design getting different views to work from. One aspect of this optimization was to examine the way clipping was done. Clipping is the process of comparing the outside dimensions of an object to be displayed with the current viewing window — it the object is outside don't draw it at all, if it is inside then draw it and if it intersects the window draw only part of it. But clipping can be done on a cell by cell basis and eliminate the examination of the contents of entire cells as well. Transformations was another aspect of this optimization. In VLSI CAD, unlike three dimensional modeling, there are only a few transformation matrices ever used and these are sparse with lots of 1 and 0 entries. Therefore, the matrix multiplication by these types of matrices is much simpler than the general case. For each of the small number of possible transformation matrices ever used in VLSI CAD a very fast procedure that does the minimal number of operations was built to optomize the transformation, and therefore the display, speed.

Implementation

The symbolic virtual-grid design capture system described here is a second implementation of such a tool. The first implementation was developed in an exploratory environment as the issues and ideas of symbolic virtual-grid layout were being explored. This meant that the tool was designed without any prior specifications. And it meant that as time went on more and more was added on to the system making the code horrendous to understand and maintain. It became clear that this was the one to throw away and try again on a second system.

Second systems have some pitfalls that must be avoided if the second system is to indeed be an improvement over the first. Most important of these is a tendency to put in "too much". Once you have seen how the first system performs, and, more importantly, what it doesn't do, there is a temptation to put in every feature that can be dreamed up. But this can be countered just by being aware of the temptation. One is provided an opportunity to develop complete and accurate specifications for the first time. This helps keep the system coherent and well-integrated. A cleaner implementation is possible both from the maintainers point of view but from the user's as well. This implementation of ICE was done initially under Berkeley UNIX 4.3 in C. It, like all of VIVID, is ported to SUN and MicroVAX II workstations. VIVID is made available to universities for a small distribution fee and in special cases is also made available to commercial firms.

COMPACTION

Introduction

What is compaction?   In the symbolic virtual-grid environment, designs
are captured in a symbolic-level circuit description language like ABCD;
generation of mask-level information and correct spacing on design objects
are completed by the compactor. Compaction is process independent by
virtue of the fact that design rules and complex object synthesis details are
table driven from the technology database. The output of the compactor is
mask-level rectangles suitable for further editing if necessary by a rectangle
pusher.

Hierarchical context.   Virtual-grid compaction as opposed to constraint-
based compaction, is used because it allows the cells produced to be pitch-
matched to abutting cells. This is essential if hierarchical compaction is to
function and this, in turn, is essential if large designs are to be attempted.

Modes of use.   Experience has shown that it is very difficult to get compactors to give consistently good results irregardless of design style — some
knowledge on the part of the designer about how it works is needed for designers to get consistently good compaction results. Therefore, on critical
cells, designers usually get into a loop of symbolic layout — compaction —
examine mask output — repeat until the results are satisfactory. Once the
basic cells are designed, designers usually wait to run the full hierarchical
compactor until nearing completion. Another common mode of operation
exists in the standard cell environment. Here, the goal is freedom from
design rule and process changes necessary to preserve the integrity of very
large cell libraries. Cell libraries are carefully designed at the symbolic level
and maintained that way. As processes change and/or design rules change,
the entire cell library is run through the compactor giving a new mask-level
cell library using the full features of the target process line.

Virtual-grid leaf cell compaction

The basic algorithm in virtual-grid leaf cell compaction is quite simple. Grid
lines are kept intact and spacing between grid lines is established by determining the maximum design rule spacing between each object on one grid
line and its neighboring grid line. First the process is run horizontally, then
vertically producing a spacing between each pair of grid lines. It is clear

that this makes pitch matching between abutting cells trivial because only the inter-grid spacing needs to be examined and made larger if necessary to make connections meet. But, it is also clear that one gets a worst case spacing determined in many cases by one pair of objects at one end of a cell that makes a much larger than necessary space on the other end of the cell. And this gets progressively worse as the cells get larger and larger. This is why severe criticism has been leveled at compactors using this simple algorithm.

In most other compactors a constraint-graph is used. Nodes in this graph are rigid structures that cannot be stretched or broken. Edges in the graph are spacing constraints. The result is that the output cell is fixed — it cannot be pitch matched. This precludes the use of this type of compactor in a fully hierarchical compactor and this in turn means that it could not be used in large designs. Hope remains for the virtual-grid approach if careful modifications of the basic algorithm and data structures are made.

Improvements

Two basis strategies present themselves: 1) position symbols, then translate; or, 2) translate and then position the resulting rectangles. Most compactors use technique number 1 — compact symbols first. This results in technology dependent code. The internal data structures are, of necessity, non-uniform and this results in excessively complex and unreliable code. A better approach is to translate the symbols into a flexible mask representation first and then do compaction on this representation. In this way all technologies can be represented since any mask-level description can be represented as rectangles. There is only one single atomic data type — the rectangle. The code is simpler and inherently technology independent.

The VMR array. The data structure so devised is called the virtual-mask rectangle (VMR) array. A leaf cell becomes a (two-dimensional) array of grid points. At each point is a list of rectangles. Some rectangles have missing edges and represent wires. Figure 4 shows an example of the VMR array at a contact cut with some fixed rectangles and some extended rectangles.

A compactor based on this new data structure has been implemented. Using a very careful symbolic layout style resulted in mask-level output cells as compact as hand designed mask cells. However, many problems were discovered with this simple approach which led to several evolutionary enhancements.

Several problems found are worth further explanation. There are cases

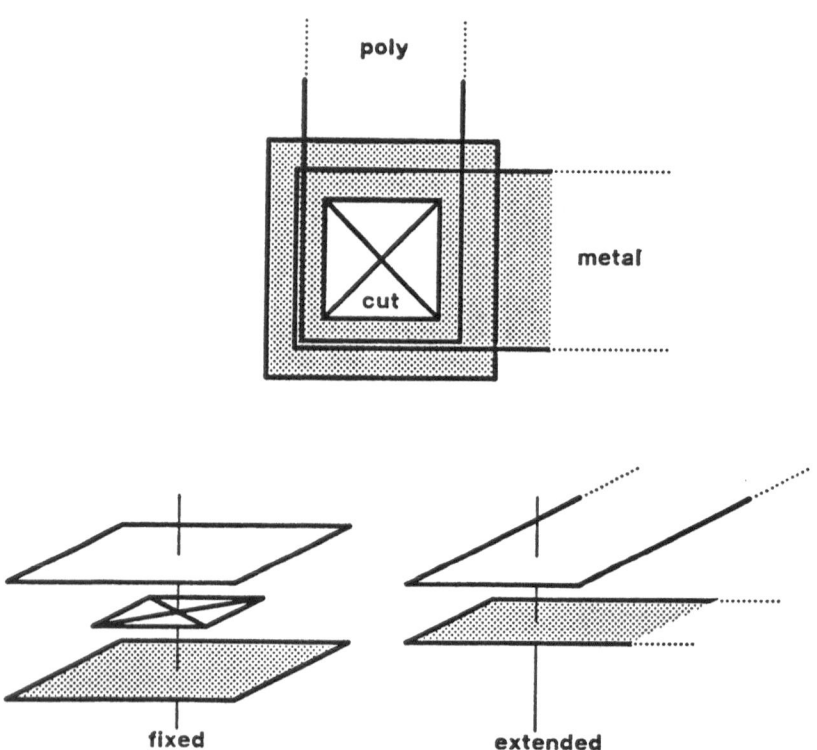

Figure 4: VMR array at a contact cut

where ostensibly hidden edges were not really hidden. For example, the diffusion rectangles of a transistor came in two parts because the two electrical nodes of source and drain caused each to be represented as a separate rectangle. When spacing was being done, the interior edges of these two rectangles caused undesirable (i.e. too much) spacing that really need not have occurred. This compactor had to have a very elaborate rule set and this set was very difficult to debug. Interactions between rules were common and fixing a bug in one rule usually caused some other rule to cause excessive spacing since the rules interacted so much. An additional problem is that critical information is missing. The polygons that are still to be placed can not be viewed as a simple leading edge of material as in this approach — material further inside the polyginal region affect placement at the frontier.

Edge-corner compaction. The examination of these problems led to refinements that resulted in a new compactor called the edge-corner compactor. Here, the data is explicitly represented as rectilinear polygons. The resulting data structure is more elaborate but the rules for how to do spacing are much simpler. Important is that technology independence is preserved.

The internal representation for polygons is just to maintain the significant edges and corners. The bottom and top edge are significant (compaction is always done top to bottom — to do left to right the cell is just rotated 90 degrees). Convex corners are significant. Wires are represented as holes with concave corners. The diagram in figure 5 can help illustrate these points.

Groups. The next important step was the addition of the *group* concept to this compactor. The idea is to cluster grid points into groups based on connected grid points (usually connected by a wire). The first and last point of a grid line are in the same group so pitch matching will work correctly. Within the cell interior, groups are given offsets relative to a grid line (i.e. "breaking" the grid line). The position of each group is determined independently.

The results of these enhancements were impressive. The density of compacted cells is comparable to those done using a constraint-graph compactor; yet, the cells were still pitch-matchable. The groups used in this approach are larger than the similar concept in the constraint-graph approach and constraint-graph compactors have more small-scale flexibility since a rather simple model was used to determine what objects are in what group.

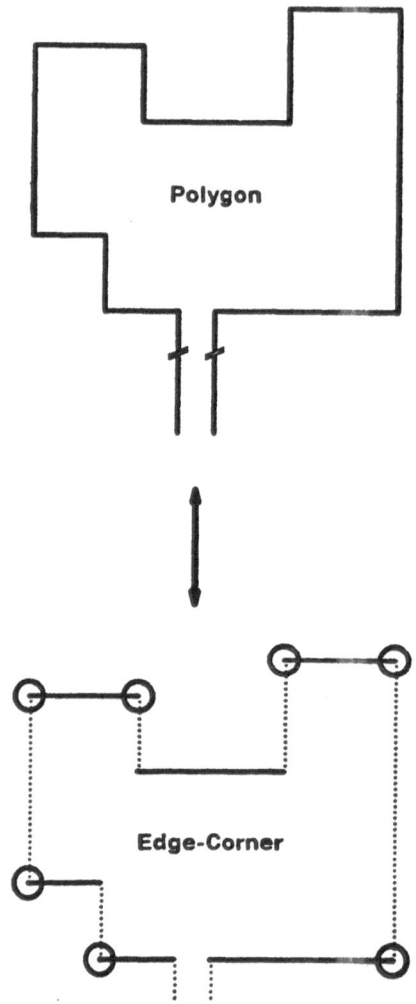

Figure 5: Polygon representation

Hierarchical compaction

A hierarchical compactor takes advantage of the fact that a typical hierarchical design uses the same cell in several different places. Each cell is first examined in all of its contexts. The leaf-cell compactor next compacts each leaf-cell to its minimum possible size in its worst-case environment. The last step is to abut instances of the cells according to the original symbolic layout specification. During this pitch-matching step, instances of some cells may increase the spacing between grid lines in order to properly match the grid spacing in neighboring instances. The algorithm used in this process is outlined in the following steps.

Remove intermediate-level cells. Hierarchy impedes rapid communication between neighboring leaf cells during compaction. It may be necessary to ascend the hierarchy to the top-level cell and then descend just to find something out about a neighbor cell. The speed of this type of inquiry can be speeded up by temporarily eliminating intermediate levels of the hierarchy. To do this first all mixed cells are "smashed" to contain only primitives. Second, all intermediate-level cells are replaced with instances until the result is just a two-level hierarchy.

Reorient cells. Inter-cell communication speed is so important it is also worth reorienting all cells to a single default orientation. So any cells not already oriented in this way are processed once so that when looking at primitives later, not transformations are necessary.

Net-list extraction. The electrical connectivity is extracted from the symbolic design next. This information is important when deciding whether or not to merge rectangles on physically separate grid lines. Some spacing rules need not be applied between rectangles on the same electrical node. For example, a U-shaped wire need not have the spacing rules applied inside the U.

Instance placement. Instance locations of the top level of the two level hierarchy are stored in a corner-stitched data structure. This data structure is very fast at neighbor operations and the approach used to hierarchical compaction described here does a great deal of neighbor operations.

Worst-case environment extraction. Because we wish to compact each unique leaf cell only once, the hierarchical compactor looks in each cell's total environment (i.e. at all of the cell's neighbors) and finds the largest environing features at each point on the cell bounding box. By spacing the cell's internal features far enough away from these worst case features, the compactor guarantees that all of the cell's instances can be placed without danger of inadvertent contact between the cell's internal features and the edges of other instances.

When abutting cells have an empty grid point in common, both cells have to decide how far to space their internal elements from their edges. Because they do not know what is in their neighbor's interior, each is allowed to come only so close to the edge. Under this "mutual non-aggression pact," both cells agree to keep each feature one half the maximum distance for that feature from the edge. This agreement, along with worst-case environment extraction, allows leaf cell grid spacing to take place without having to communicate with the interiors of neighboring cells.

Leaf cell compaction. Each leaf-cell is now compacted by the techniques described earlier.

Pitch matching. Instances that represent cells that were compacted separately may not, when placed adjacent to each other, communicate correctly. Their pitch is different and lines that communicated when in the symbolic are now broken. Communication is restored by expanding the grid spacing of some instances until the broken line is repaired. Broken grid lines that do not affect communication are not repaired.

Anti-feature elimination and output. The leaf-cell grid spacing algorithm leaves certain rectangles too close together creating small spaces or gaps called "antifeatures." The leaf cell compactor is allowed to violate spacing rules in this way whenever the material on both sides of a space is electrically connected.

The antifeature eliminator finds these spacing violations and fills them in. It sorts the non-grid-based rectangles and then expands them by one half of the spacing rule. All of the rectangles are then merged into polygons which are then shrunk by one half of the spacing rule distance and then fractured back into rectangles. This is a standard technique known as *bloating and shrinking.*

Future directions

There is a lot still to be done. Some of those things are listed here.

Contact offsetting and wire jogging. This requires some decoupling of the groups. The two issues are related and the results could be dramatic. Since contacts are usually larger than minimum wires offsetting them to one side of a wire or the other can make a dramatic effect on material being held away by a contact spacing rule. Likewise, even simple wire jogging can make a dramatic difference in the size of the final mask cells.

Handling pre-compacted chunks. In many cases, a designer wants to incorporate some hand designed cells (especially memory) in a symbolic design. The compactor will have to treat this as an unstretchable blob. The really hard part is how to communicate with this blob.

Producing (smaller) hierarchical output. The output of the hierarchical compactor is very large. This is because the uniformity of instances of the same cell is broken. Each instance of a cell has a different environment and so each ends up a different size; so, for all intents and purposes they are, in the output, different cells. Either under a designer's control or through some heuristics it should be possible to insist that like cells come out the same in the final output unless doing so really enlarges the design.

Produce exact circuit parameters. The compactor know the exact sizes of wires and transistors and it has access to the technology database so it could produce the output one normally gets from a detailed mask-level circuit extractor. One needs this information for a detailed circuit-level timing analysis such as with SPICE. This circuit extraction should be produced directly by the hierarchical compactor.

TECHNOLOGY ENCAPSULATION

Introduction

The diverse group of tools in a system like VIVID, relies on a single component, the Master Technology File (MTF) system for technology encapsulation. The information in the MTF includes technology characteristics and process parameters as well as information that governs behavior of graphics devices used by design tools. Because all design tools use a single source of technology information, consistent operation is guaranteed.

Technology database

The MTF System allows designers who operate VIVID System tools (the users) to specify completely the characteristics of a target fabrication process on both the symbolic and mask levels. User-defined process characteristics recognized by the MTF System include these categories:

- **Process Layers** The user can create both symbolic and mask-level process layers. In addition he can create special layers that are used to display menus and other bookkeeping information.

- **Transistors and Contacts** The symbolic-level transistors and contacts existing in a process are declared by the user by specifying the process layers from which they are constructed.

- **Graphical Attributes** The colors and stipple patterns used to display each layer can be defined.

- **Graphical Shapes** The appearance of circuit primitives supported by the ABCD (A Better Circuit Description) language can be described by the user. Graphical shapes can be tailored according to a circuit primitive's type, size, and orientation.

- **Virtual Shapes** The virtual-grid points occupied by symbolic-level transistors can be specified.

- **Symbolic to Mask-level Translation** The MTF System allows the VIVID System compaction program to be technology-independent by including two types of information. One master technology file contains instructions describing the translation of symbolic-level circuit primitives into constituent mask-level rectangles. Another file includes a list of mask layers and the physical distances that must separate each pair of layers.

- **Electrical and Physical Constants** The MTF System provides a mechanism for specifying the physical and electrical constants of the process layers, wires, and transistors existing in the target process.

Features

Two key strategies are responsible for the design of the MTF System:

- **Compilation**

- **System-wide technology independence**

The decision to compile technology specification files rather than to interpret them or deal with them as a data base has resulted in a system with three separate parts, a procedural interface, a group of compilers, and the master technology files themselves. The second principle demanded that the MTF System give technology-independence to an entire spectrum of design tools, not to individual tools in a fragmented way. This section will discuss the features that these strategies have given to the MTF System.

Compilation Both users and tool developers benefit from the MTF System's use of compilation. The human-readable master technology files are written in a language that is quickly mastered. The binary representations produced by the compilers are more easily manipulated by machines than the human-readable form. The compilation step not only transforms the files into a format that is more easily handled by machines, but it also arranges the contents so that operations, such as searches, can be performed faster. The overhead imposed on VIVID System tools by the MTF System is minimal. Routines within the procedural interface take advantage of the structure imposed on the master technology files by the compiler, making the routines quick and efficient.

Languages Each master technology file is written in either the C programming language or the MTF language. The MTF language is a simple system of keywords and statements. Since there are only a few syntax rules, the designer can begin using the language quickly. The MTF language itself is defined with the UNIX tools **Lex** and **Yacc**, so that it can be easily expanded to accommodate new design tools. A sample of the MTF language appears below. Keywords appear in boldface type.

```
transistors {

        n_type {
                source = ndiff;
                gate = poly;
                drain = ndiff;
                model_name = en1;
                substrate = node_vss;
        }

        p_type {
                source = pdiff;
                gate = poly;
                drain = pdiff;
                model_name = ep1;
                substrate = node_vdd;
        }
}
```

Master technology files that express more complex design features, such as the graphical appearance of circuit primitives, require the power of a general-purpose programming language. For this reason, the MTF System allows some files to be written with the C programming language. All features of the language can be used. However, the use of external variables or subroutines is forbidden, so that the resulting object code can be loaded at run time and linked to the executing program. Process features expressed with this language include the specification of graphical and virtual shape, the values of physical and electrical constants, and the mask-level representation of symbolic circuit elements.

Consistency Since the MTF System was developed to support an entire range of VIVID System design tools, there is a high degree of coordination among the components of the MTF System. The format and contents of technology rules allow individual design tools to share technology information. The ability to share information provided by the MTF System insures consistency and relieves the user of a considerable programming burden.

Partitioning   The MTF System reduces the memory required by a design tool by partitioning the information contained in the MTF System. For example, graphical editors do not require information about compaction and simulation. Consequently, the MTF System allows graphical editors to obtain only the information they need for editing. Technology rules concerning compaction and simulation are ignored. This feature minimizes the memory required by a design tool, thereby increasing its preformance.

Dynamic Loading   The MTF System allows design tools to load and link additional object code after they have begun execution. This technique is called dynamic loading. Design tools employ dynamic loading in order to obtain technology specifications that have been written using the C programming language and compiled into object code. This object code is linked to the design tool at run time, allowing the user's specifications to be executed without a time-consuming re-compilation of the design tool.

Modular Construction   The structure of the MTF System is not dictated by the VIVID System tools that it supports. Completely new design tools can be designed around the MTF System and its companion VIVID System software libraries.

Benefits of Technology Independence   Finally, the MTF System provides all the benefits associated with technology-independent systems. Technology independence relieves the tool developer of the burden of creating an individual tool for every process. Software maintenance becomes much easier and changes in a process do not force software modification and re-compilation. Technology independence offers the tool user the ability to modify the tool's behavior. Adaptations can be made quickly, without the assistance of a programmer. If a designer works with more than one technology, he can use a single tool for all of them by specifying appropriate sets of rules.

SOFTWARE ENGINEERING

The purpose of developing a CAD system must be to build real VLSI systems with it. For this to happen, it must be given to some designers who have the ability to design real systems. Designers have seen a lot of software come and go and they are quite sophisticated and busy so they will not give new software the time of day if they do not see it as being to their benefit. If for no other reason than this (and there are other reasons) it is worth making a serious CAD program into a programming systems product. A programming systems product is written in generalized fashion, is thoroughly tested and has thorough documentation. It is vastly harder to do this than just producing a CAD program.

For a new idea in CAD like the symbolic virtual-grid methodology to succeed, a CAD programming systems product supporting this methodology must be built. VIVID is such an attempt and it has been largely successful but it also taught all concerned some lessons.

A lot was done well in VIVID. Conceptual integrity is *the* most important consideration in systems design and conceptual integrity is very evident in VIVID. A system is best when one can specify things with simplicity and straightforwardness; in VIVID simplicity has been a high priority from the outset. A programming systems product must be generalized and VIVID is portable, accepts a wide variety of inputs, is very modular and offers an open architecture. VIVID was extensively tested by non-developers (but, it turns out, still got out with numerous bugs). The documentation is voluminous including both tutorials and references.

The team approach used to develop VIVID was very productive. A leader of each project was clearly defined and tis leader had one or more assistants. All groups shared the services of software librarian, quality control technical, secretary, technical writer and legal assistant. Communication was mostly informal until near code freeze dates when it became formal and sometimes confrontational.

The second system effect of "too muchness" was avoided mostly by being worried about it creeping. Extra self discipline was exerted and more complete specifications were expected in the case of second systems.

The work environment was ideal: good computer support, an excellent programming environment, personnel policies of flexibility and informality and no shortage of team spirit and enthusiasm. The typical traps associated with over-zealousness were not avoided however.

Those are the traps that arise out of the classical optimism of programmers — this small team believed it could do it all. This was to become a serious problem when associated with enormous and unreasonable external pressures. Those pressures, not adequately counteracted by program managers, forced software to be released much too early which meant it became unpopular software because it was, as is to be expected, very buggy. Problems became even more severe as the group then attempted to market the software — also, much too early. The enthusiasm of the software group and its technical managers, combined with unwise marketing practices and the continued counterproductive external pressures were bad enough. But these forces were not tempered by good business-like management higher up in this non-profit organization that should have put on the brakes. Politics became a divisive force and the external and marketing pressures were too much for the group to withstand — it destroyed the group in the end.

VIVID has made a real impact by showing the viability of symbolic VLSI design. As follow-ons to VIVID become commercialized by major CAD suppliers, custom VLSI design will reach a wider and wider group of designers and just may cause a mini-CAD revolution.

7

DESIGN AND LAYOUT GENERATION AT THE SYMBOLIC LEVEL

Carlo H. Séquin

Computer Science Division
Electrical Engineering and Computer Sciences
University of California, Berkeley, CA 94720

ABSTRACT

A pipeline of three tools for the construction of high-quality macro modules or library cells is described. TOPOGEN is a synthesis tool that takes a logic description at the gate level and converts it into a symbolic layout of a static CMOS circuit on a virtual (coarse) grid. EDISTIX is an interactive virtual grid editor for the creation or modification of symbolic sticks diagrams. ZORRO is a two-dimensional compactor using the concept of 'Zone refining' to generate the mask geometry from the symbolic layout. The generation of the final layout of a cell is a two-step process using an intermediate symbolic representation on a virtual grid. In this intermediate state, the user can interactively make changes.

1. INTRODUCTION

The creation of high-quality cells and macro modules is a corner stone of automatic and semi-automatic chip synthesis. This is true regardless whether a full custom or a semi-custom standard-cell approach is taken.

The rapid progress of VLSI fabrication technology renders existing standard-cell libraries obsolete rather quickly, so that they must be adapted periodically to a new set of mask layers and new design rules. Because of the repeated usage, density and performance of these cells is important, and thus a lot of effort is normally spent to obtain optimal cell layouts.

Emerging "Silicon Compilers" normally work in hierarchical stages. Hand-designed library cells and procedurally generated modules are assembled at the chip level by powerful placement and routing tools. The latter tools recently have started to outperform human designers for complicated tasks with many blocks. However, automatically generated cells and macro module rarely achieve the performance and density of hand-designs by a good designer. An exception are some special modules such as PLA's, but there the gain stems primarily from logic minimization and from topological folding rather than from the actual layout.

In the last two years we have concentrated some of our efforts on tools that make the production of high-quality cells and macro modules easier and more automatic. The emerging system consists essentially of three parts. TOPOGEN is a synthesis tool that takes a logic description at the gate level and converts it into a symbolic layout of a static CMOS circuit on a coarse virtual grid. EDIS-TIX is an interactive virtual grid editor for the creation or modification of symbolic sticks diagrams. ZORRO is a two-dimensional compactor using the concept of 'Zone refining' to generate the mask geometry from the symbolic layout.

The generation of the final layout of a cell thus becomes a two-step process: conversion of the circuit into a good topology on the virtual grid, and then the fleshing out of the sticks elements and their geometric compaction into a dense layout in accordance with a given set of geometrical design rules. At the intermediate level, the designer has the option to review and possibly improve the topology of the cell with the interactive program EDISTIX. It is also in this intermediate format that the design of the cell should be stored for rapid generation of a new cell when there are small changes in the implementation technology.

2. THE ROLE OF SYMBOLIC REPRESENTATION

The direct conversion of a circuit into a dense layout is too big a step to be taken directly, — this is true for the human designer as well as for a computer program. The concerns of finding a good topology for the layout and of arranging the components to satisfy all design rules are independent enough, so that these two issues can be resolved separately in a two-step process. Between the two steps lies the symbolic representation of the layout in some sticks-like format.

We will briefly discuss the requirements for the representation at this symbolic level and then discuss our chosen representation.

2.1. Requirements for a Symbolic Representation

In the choice of the primitives at the symbolic level, one tries to combine various diverse goals.

The representation should be lean and uncluttered to make it easy for the designer to address the concerns he has at this stage of the design. These are to find an optimal topology for the module under construction that will produce a module of a desirable aspect ratio, place the connections to the external world at the proper sides of the module, and produce direct and minimal internal wiring as well as simple geometry for the well regions.

To be able to make reasonable choices on the topology, the symbolic representation must be expressive enough to render the tricks that are routinely used in the hand-layout of dense library cells. One such trick is to run metal wires over large transistors and to produce, if necessary, a cross-under for a signal that enters the drain/source diffusion on one side of this metal connection and gets picked up on the other (Fig. 1). This construction cannot be represented if the transistor at the symbolic level is viewed as a point device with only four possible connections, one each in the four major directions.

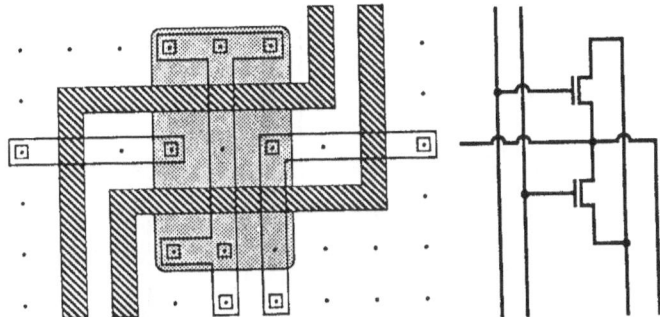

Figure 1. *Cross-under produced by a diffusion region between two transistors. This could not be expressed if the FET were a point device.*

And finally, the symbolic representation should be efficient. It must represent succinctly and unambiguously the geometrical and electrical properties of the circuit, so that the sticks diagram can be checked for functionality and evaluated qualitatively for the area required by the final cell. Of course, it is preferable to keep the size of the file describing the cell at the symbolic level as small as possible.

2.2. Virtual Grid and Raster Components

We have chosen a coarse virtual grid as the basic design space. It allows for a terse representation and makes the geometrical part of the data structure very simple. Further, the determination whether two components are actually connected is straight-forward; this makes the checks for possible illegal interference of components rather simple.

Every component in this representation occupies a number of grid points. The set of basic components selected for our symbolic representation is show in Figure 2. All components can be viewed as linear elements spanning one or more grid points. For wires, this representation is an obvious choice. It also applies for the port, a formal terminal that can serve as a connection to the outside world. If the port extends over more than one grid point, it is still considered an equipotential node. Contacts are also linear equipotential elements, typically represented as rows of contact holes spaced one grid unit apart.

Transistors are slightly more complicated. They occupy three rows of grid points next to one another, one each for the source diffusion, the gate, and the drain diffusion. They still fit the paradigm of a line element, as internally only the "stick" for the gate is represented explicitly, and the adjacent diffusion areas are implied and derived on the fly when needed for some check or for display on the screen. This gives the symbol set a cohesiveness that makes the various data

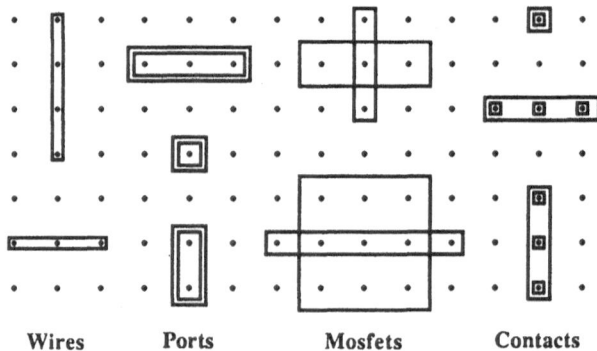

Figure 2. *Virtual grid components in EDISTIX.*

structure manipulations more regular.

In addition there is an auxiliary component called the joint. It is used wherever two or more wires join together. Joints are strict point elements. We first tried an implementation with a data structure that did not need these joints and connected wires directly to one another. The resulting data structure and its manipulation became rather cumbersome. The addition of the extra joints, where needed, simplified things. These joints need not be represented explicitly in the file that describes the circuit symbolically; they are introduced and deleted on demand whenever a wire end is not explicitly connected to a terminal, contact, or transistor.

Every tool described below has its own internal representation of these sticks elements that is most appropriate for the task that the particular tool has to perform. The information is passed between the various tools by means of terse ASCII files. The format of these files is very simple: Every element is represented by a keyword that implies its type and layer and the integer coordinates of its endpoints. In addition, ports and transistors can take names for identification. This decoupling through the use of these intermediate files makes the databases for each tool simpler and more efficient and permits separate tool development.

3. EDISTIX

The virtual grid components described above can give a reasonably accurate description of the layout organization and of the achievable packing of the components. This is necessary to allow the designer or an automated tool to find an optimal module topology. Because of the central role of the symbolic representation, we will first give more details on EDISTIX, rather than present the pipeline of tools in the sequence that an evolving design would see. EDISTIX acts as the glue between the other two tools, and its internal data structures are a good example how one can deal efficiently with the described sticks components.

3.1. The Function

EDISTIX is an interactive virtual grid editor that relieves the designer of many of the chores associated with the modification of symbolic sticks diagrams. The purpose of this tool is to make it easy to enter symbolic designs from scratch or to inspect and modify the ones that come out of a tool like TOPOGEN.

In the first case, the goal is to make sticks entry as fast as sketching on a pad of paper, but with all the potential advantages of having a smart checking program looking over your shoulder and preventing you from making simple mistakes such as tying 'Power' and 'Ground' together. Particular attention was thus given to the user interface, with the goal of minimizing the necessary actions during the entry of circuit elements.

In the second case, the main goal in EDISTIX is to make it easy to change the topology of a layout without changing its connectivity. If a designer wants to improve the layout topology (in cases where TOPOGEN gives less than optimal results) he should be able to spend most of his attention on finding an optimum topological arrangement without having to worry that the interconnections might be changed accidentally in the process. Thus, in this mode, EDISTIX keeps the internal netlist unchanged and tries to reroute all interconnections accordingly when components are moved.

3.2. Data Structures

Considerable effort has been spent to find efficient data structures to represent the geometrical as well as the electrical aspects of a design.

In the *geometrical data structure*, because of the limited size of non-hierarchical macro modules or library cells, and since in good topological arrangements practically all vertical and horizontal grid lines contain at least one component, it is reasonable to represent all the rows and columns of the drawing area explicitly, rather than using sparse matrix techniques. Thus for every row and column we list all the vertical and horizontal line-elements, respectively. In each of the two directions, these elements are grouped into five linked lists sorted by element types (Fig. 3). Thus we store in separate lists: wires and links, contacts, joints, ports, and FET's. This makes it easier to search for a particular element type and to provide the different processing routines necessary for different element types.

Since the elements are either horizontal or vertical sticks, their geometry is fully captured with three numbers: their row/column number and two values for the second coordinates of their endpoints.

In the *electrical data structure*, a distinction is made between equipotential *nodes* such as ports, contacts, FET-terminals, or joints, and binary *connection elements* such as wires and links (Fig. 4). All nodes are connected in a linked list in

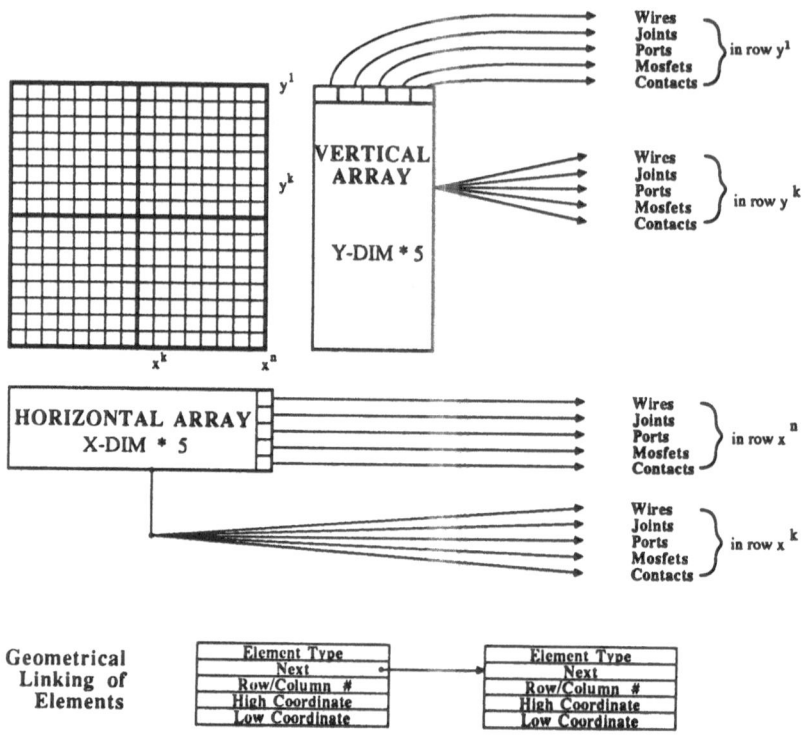

Figure 3. *Geometrical data structures in EDISTIX.*

the order they are created. They may carry an optional name. They also have a pointer that points to the first child, i.e., any attached wire or link, or is nil.

Wire or links are two-ended elements that are attached to two nodes. At each end they have two pointers, one pointing to the node to which they are connected, and the other pointing to any 'siblings', i.e. other wires or links attached to the same 'parent' node.

All the geometrical and the electrical information is contained in the same structure representing both aspects of an element.

3.3. Operations

The many possible operations can be grouped into various classes: edit operations, selection and query commands, clean up operations, rearranging the topology, analysis, and output.

Edit operations are used to build a circuit from scratch or to modify a given circuit. They include the standard operation to add or delete an element, and to

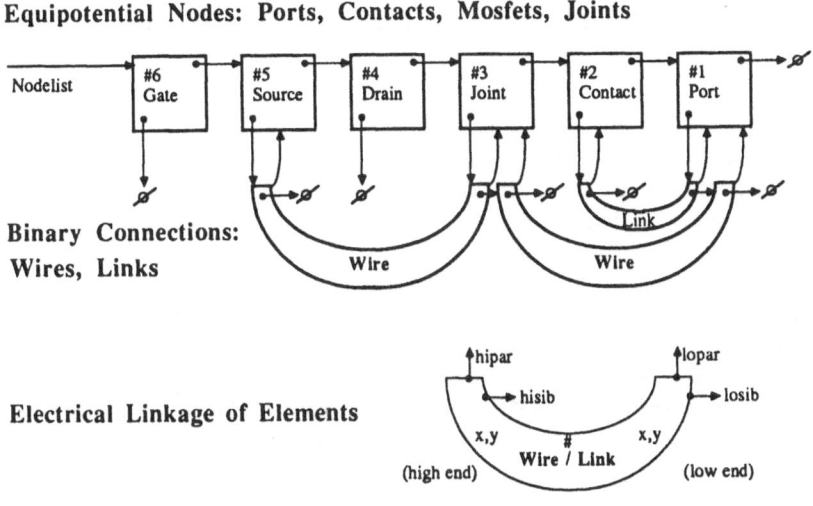

Equipotential Nodes: Ports, Contacts, Mosfets, Joints

Binary Connections: Wires, Links

Electrical Linkage of Elements

Figure 4. *Electrical data structures in EDISTIX.*

select and modify an element. When adding an element, the program watches for illegal constructs such as running a poly wire across a diffusion area, or it warns you of questionable configurations such as level crossings of wires that will lead to an implicit electrical connection. The program maintains up-to-date information about all electrical connections, and it will warn the user when nets with different names are connected or when a loop is formed in a net.

Selection and query commands permit the user to pick one or more elements on the screen and then see a listing of the detailed information on that element as well as a list of other elements it is connected to. Some fields such as the name can be changed.

Clean-up operations remove dangling wires and contacts and merge pairs of collinear wires on the same level. This brings the internal representation into a minimal consistent state.

Rearrangement commands allow the user to change the layout without changing the underlying circuit. These operations are using a generalized block move operation. A group of elements, selected individually or by an 'area select' command, are moved jointly by a given displacement vector. Connections that go beyond the selected area and connect to components that remain fixed have to be recreated. The system does as much rerouting as possible and shows the remaining connections that it cannot handle in contrasting color. It is then up to the designer to find a feasible implementation for these wires, or to make further changes that enable the wiring to be completed.

Analysis commands (not yet implemented) will eventually allow the user to interact directly with a simulator or a timing verifier. In this way it will be easy for the designer to verify functionality or to get a first estimate on performance. For the time being, the designer will have to produce an output file with one of the various drivers for a particular simulation tool, and then run that tool separately on this file.

Output of the stored data on the design can be viewed in many different forms. Elements can be listed in geometrical order, going through the various types of devices on a row by row and column by column basis. This is the default scan mode used when the user wants to write out a file of the database in the ASCII format. Alternatively, the nodes with all the attached children can be listed in the order in which they were generated. Finally the whole network can be traversed in a depth first manner; this is the mode that is used when one wants to create an output file in the format for simulators such as SPICE[1] or ESIM.[2] There is also a possibility to create a file in the format of the OCT data base[3] so that the other tools of the Berkeley Design Environment[4] can be run on the cells generated with EDISTIX.

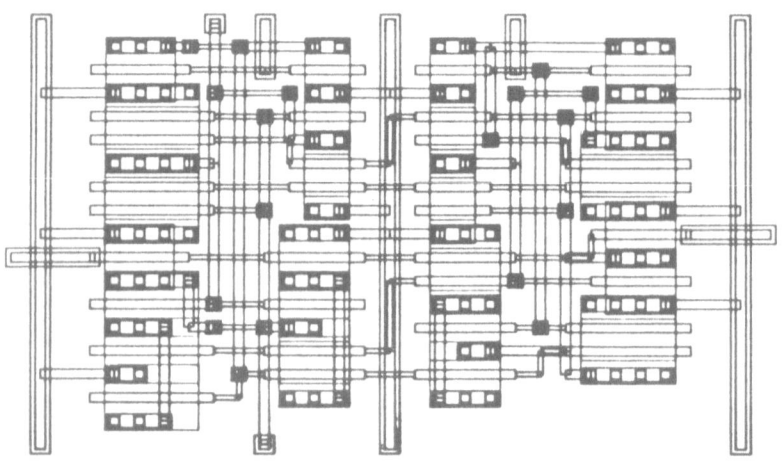

Figure 5. *Virtual grid representation of a flipflop composed with EDISTIX.*

3.4. Results and Discussion

Figure 5 shows the sticks representation of a flipflop as it would appear on the EDISTIX screen. EDISTIX has been under continued development for a couple of years. It has been rewritten from scratch at least four times, first a couple of times in Pascal, more recently in C. The general features discussed in this section have been rather stable over the last few versions, and we are confident that they represent a good solution to capturing a symbolic layout. It gives a rather good idea of what the final layout might look like.

4. TOPOGEN

TOPOGEN is a generator program that takes a functional description at the logic gate level and converts this into a symbolic layout on a virtual grid. The first version of TOPOGEN is aimed at standard cells for a static CMOS family. So far, the layout style is restricted to a single row of transistor pairs with one diffusion strip each for the p-channel and n-channel FETs, respectively, TOPO-GEN is organized in a modular fashion, so that one can experiment with different algorithms for the various steps mentioned below.

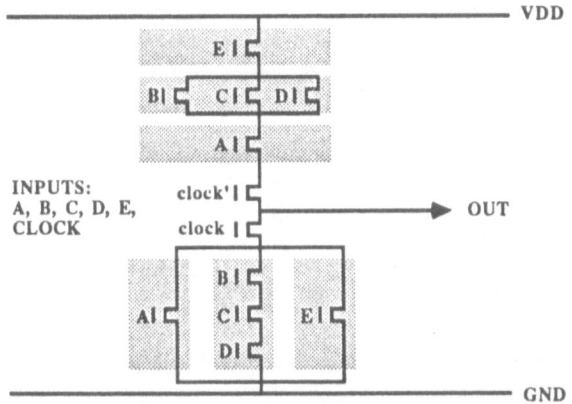

Figure 6. *Circuit generated form the following TOPOGEN input: (evalgate (output OUT) (npt CLOCK) (or A (and B C D) E)).*

4.1. Circuit Generation

The translation of the logic description into a corresponding circuit is straight-forward. TOPOGEN accepts nested AOI expressions that are converted to the corresponding series / parallel networks of transistors. The program looks at every AOI gate in the input stream separately. In the sequence in which the logic inputs appear in the original description, corresponding transistors are placed from left to right and from output rail towards the power/ground lines (Fig. 6). In addition, single or paired clocked switches can be specified. These clock inputs can be placed next to the output rail or next to the supply lines, depending on whether the clock input specification appears before or after the description of the Boolean logic block.

4.2. Gate Optimization

The circuits obtained in the manner outlined above are now arranged as a linear sequence of transistor pairs. In each gate the sequence of the transistor pairs is arranged so that the mutual sharing of the diffused drain/source areas is maximized and thus the length of the rows of transistors is minimized. This amounts to finding corresponding Euler paths through all the transistors of either polarity. We use the method of adding a pseudo input in every series / parallel block with an even number of components[5] since that makes the construction of an Euler path trivial. These pseudo inputs correspond to turned off gates or isolation zones in the final diffusion strips. Their number is minimized by permuting the sequence of the children at every node of the AOI tree (Fig. 7). Multiple adjacent isolation zones can then be collapsed into one.

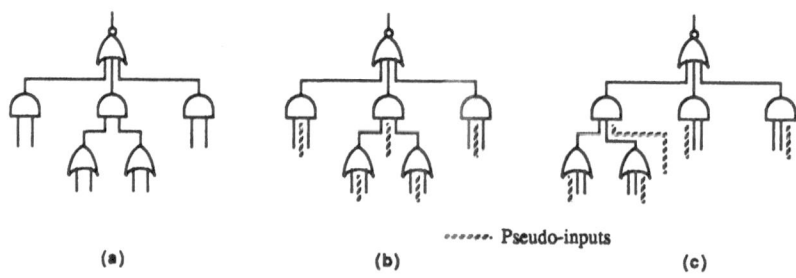

(a)   (b)   ------ Pseudo-inputs   (c)

Figure 7. *Gate rearrangement to produce an Euler path through a circuit.*

4.3. Gate Placement

TOPOGEN subsequently rearranges these individual and-or-invert gates with the goal to minimize the width of the wiring channel between the two diffusion areas. A pairwise interchange algorithm is used to step through all the gate positions once, comparing the potential gains in exchanges with all the gates that lie ahead in the line. The cost function to be minimized is the width of the resulting strip, i.e., the maximum of the sum of the width of both the P and N transistors and the local density in the wiring channel. Since good channel routers can wire a channel without exceeding its density, this evaluation function is quite appropriate.

4.4. Wiring

When a suitable gate arrangement has been determined, all the necessary interconnections in the area between the two diffusion strips are generated. We use the latest channel router available to us. We have had good success with YACR II[6] and we are currently experimenting with others such as CHAMELEON[7] and MIGHTY.[8] TOPOGEN simply writes an ASCII file specifying the routing problem in the particular format that the router needs and subsequently reads the generated file with the wiring description.

In trying to modularize our design environment, we are in the process of defining "standard" format for the description of a routing problem and for the generated solutions. To be general enough, we permit the routing region to be any arbitrary rectagon, signal input pins can lie on this rectagon boundary or inside, and there can also be obstacles inside the routing region on one or more layers. Issues that need to be resolved concern the transformation of the signals from the layers given by the original problem situation to the levels that the router is prepared to handle, and questions whether the router can introduce a level change right at the location of a signal pin.

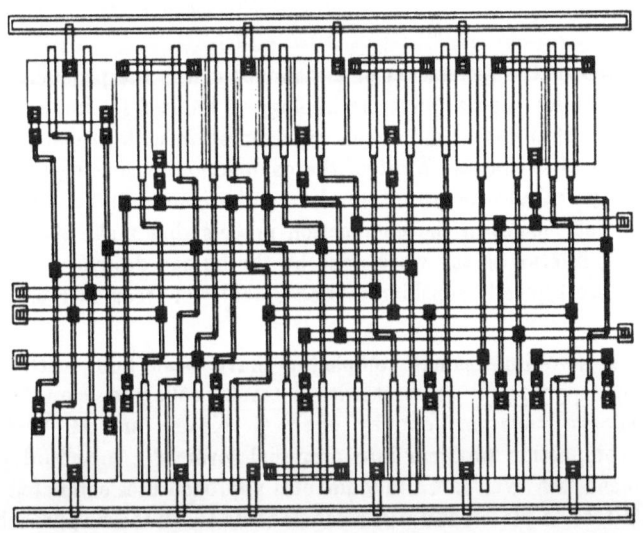

Figure 8. *Sample output generated by TOPOGEN.*

4.5. Output

The final phase is to write an output file in the format understandable by EDISTIX. This is fairly straight-forward since TOPOGEN internally has built up all transistor positions and wirings on the same kind of coarse virtual grid used by EDISTIX. A typical output for a small group of simple gates is shown in Figure 8.

4.6. Results and Discussion

In its current form, TOPOGEN is a useful tool for clusters of gates totaling about a hundred transistors. The layouts are not yet competitive with a hand design. The main reason is that TOPOGEN carries out each phase of the chosen design process without much concern for the other phases and without any iterative feedback loop. We are in the process of reducing this design gap by incorporating more sophisticated routing algorithms that can route over large transistors. In order to handle larger gate clusters, We have started to extend the basic approach to modules with multiple strips of complementary transistor pairs; in this case the gate placement is a harder problem that requires more sophisticated techniques than simple pairwise interchange with the goal to minimize channel density.

5. ZORRO

The third step in the generation of a standard cell is the production of the final mask geometry for the particular technology to be used for implementation, i.e. the compaction of the symbolic circuit representation with proper dimensioning and spacing of all elements. Most of the compaction or spacing programs in practical use today can alter only one coordinate of a component at a time. This leads to certain deficiencies in the compaction process that make the automatic spacing of layouts inferior to the work done by the human designer. The resulting inefficiencies are typically considered unacceptable for frequently used library cells.

Experimental two-dimensional compactors have been built with different approaches. One approach is to start with a totally collapsed layout and then remove the distance violations one by one.[9] G. Kedem and H. Watanabe[10] translated the compaction problem into a special form of a mixed-integer programming problem. An even more fundamental approach uses simulated annealing techniques[11] for the placement of the components.[12] All these approaches typically show non-polynomial growth in runtime for large circuits.

We have taken a less expensive approach to 2-dimensional compaction. Only a small part of the circuit is opened up for two-dimensional motion of the components. This open zone is swept through the precompacted layout in a strong analogy to the *zone refining* technique used in the purification of crystal ingots.

5.1. Zone-Refining

In close analogy to zone refining of crystals (Fig. 9a), we start from a circuit layout that has been "crystallized" by precompaction with a one-dimensional compactor. In our case the "impurities" that we want to sweep out of the "crystal" are the unnecessary voids between circuit components. Starting from the bottom, individual circuit components or small clusters of components are peeled off row by row from the precompacted layout and are reassembled after they have

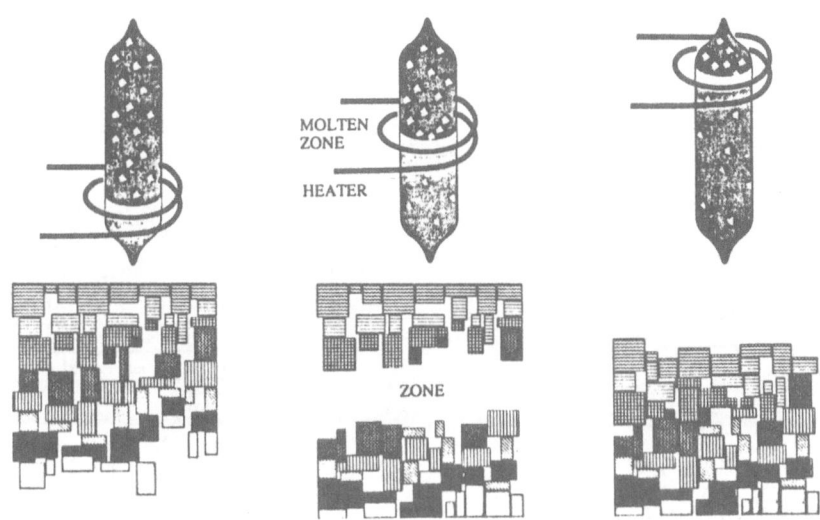

Figure 9. *Zone refining: (top) of crystal ingots and (bottom) of layouts;
the preferred direction of compaction is vertical the direction of the sweep,
but the blocks in the zone can also make lateral movements.*

been moved across an open zone (Fig. 9b). As the components pass this free zone,
they can move laterally to a more advantageous position that will result in a
denser layout. In the process of reassembling the components at the other end
both coordinates of the moved components can be altered and jogs can be introduced in the connecting wires between the circuit components. These additional
degrees of freedom permit a higher packing density in the newly formed part of
the layout than can be achieved with a one dimensional compaction process.

The geometrical design rules are observed by maintaining and using the constraint graphs in both the x- and y-direction.

5.2. Data Structures

The main data structure is the *adjacency graph*, here illustrated on the simple example of a packing problem involving rectangular boxes (Fig. 10a,c,d). The
positions of all blocks are represented in the nodes of the graph. All horizontal
and vertical adjacencies are represented as two types of corresponding arcs
between the nodes (Fig. 10b,e). These arcs are labeled with the minimal allowable
horizontal or vertical separations between the centers of the block; this adjacency
graph can thus be turned into a constraint graph for properly placing the blocks
without overlap. For an actual circuit layout, the constraints attached to these
arcs become more complicated and contain upper as well as lower bounds.

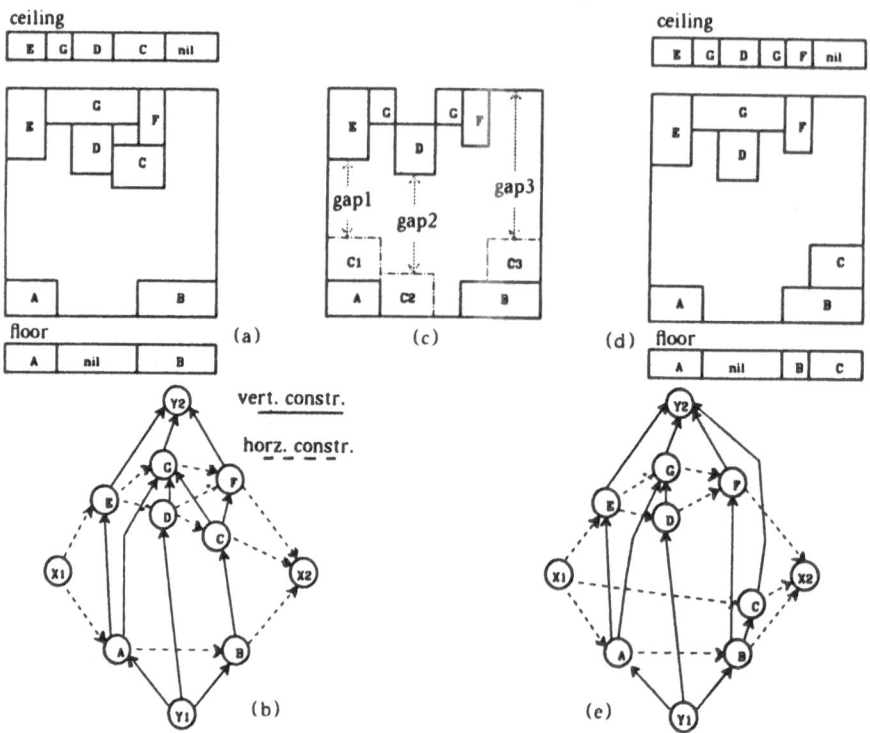

Figure 10. *Example of box packing in progress. (a) Intermediate constellation of boxes and floor and ceiling data structures. (b) Corresponding adjacency graph. (c) Box C has been selected to be moved; three candidate places C1, C2, C3 are evaluated. (d) Box constellation after box C has been placed and new floor and ceiling structures. (e) Updated adjacency graph.*

A second data structure is associated with the moving refinement zone and contains the currently active components that must be referred to frequently in each block move. All elements that form the boundary of the free zone, above and below, are joined together in the 'ceiling' and 'floor' data structures, respectively (Fig. 10a,d). They permit an efficient evaluation of the best position for the elements that are being moved across the zone.

5.3. Zone-Refining Algorithm

Elements are moved from the top part to the bottom part of the circuit across the open zone with the following algorithm. In the ceiling an element is selected that hangs farthest down. For simple box packing, an individual box is selected. For actual circuits, where the components are connected with wires, a whole cluster of components that is connected by horizontal wires without jogs in them must be moved at once. The selected components are removed from the ceiling data structure and from the horizontal adjacency graph. They are now free to float around in the zone.

Now the best location for placing the component on the floor has to be found. We are looking for the position that maximizes the narrowest part of the zone, because then we know that the two halves of the circuit can fit together with minimum total height. In the case of box packing, all grid positions from the left extreme to the right extreme are evaluated. For circuits, the lateral motion is much more restricted. In the first version of ZORRO, components or clusters of circuits are only moved laterally within the freedom allowed by the attached wires. Wires can be moved to the extreme positions of terminals, and horizontal parts of wires can be stretched, but no new jogs are introduced at this point.

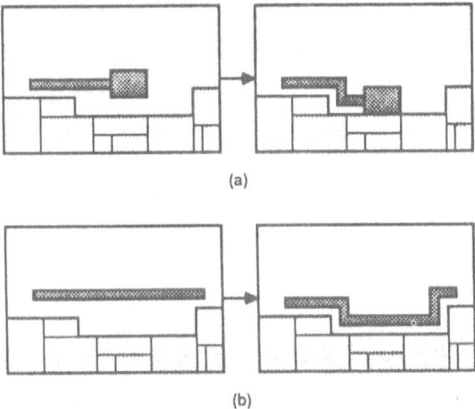

(a)

(b)

Figure 11. *Automatic jog introduction in horizontal wires.*

Once the optimal x-position has been found, the box or the circuit cluster is moved onto the floor and is properly integrated into the floor data structure and into the two adjacency graphs. Updating the horizontal and vertical adjacency graphs is done in an incremental manner. When a component is moved in the vertical direction, its horizontal arcs are removed. Once it is in the new y-position, the new adjacencies are detected by sweeping a scan line across the height of the component and checking what other components get intersected. New horizontal arcs are formed for all discovered adjacencies. Corresponding operations on the vertical adjacency graph are carried out whenever a component is moved horizontally.

For the case of circuit compaction, all attached wires have to be placed properly, once the best place for the moved component cluster has been found. Jogs may have to be introduced in the horizontal wires to permit the component to move all the way to the floor (Fig. 11a). To maximize vertical compression, we will also bend some of the horizontal wires that span over large enough regions of empty space (Fig. 11b).

5.4. Results and Discussion

Figure 12 illustrates the zone refining process on a simple example of box packing. We start from a randomly generated array of rectangles and compact it in the upward direction; the overall height of the array is reduced from 80 units to 63 units. A first zone refining pass, where the boxes move downward across the open zone, reduces the height to 53 units. The second zone refining pass in the opposite direction brings the total height to 47 units. This is the limit; additional zone refining passes do not reduce the height of the constellation any further.

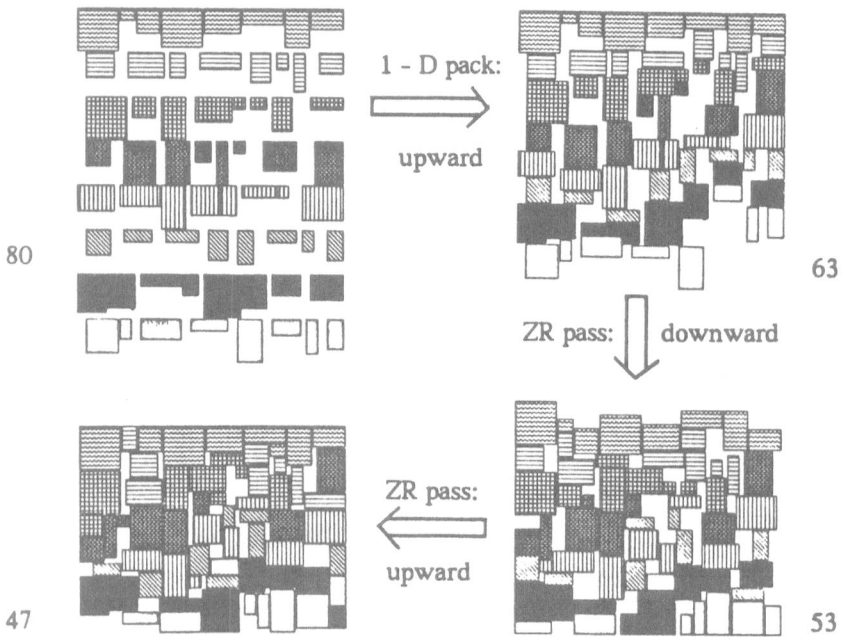

Figure 12. *Example of box packing with zone refining.*

Figure 13 shows various phases in the compaction process of a real circuit. First it shows the precompacted circuit with merged contacts and nets. The next two figures illustrate an intermediate and the final state of the first zone refining pass on this circuit. The last figure shows the result after four more zone refining passes in the vertical direction; these passes also include jog generation in horizontal wires. The obtained reduction in area is 33% compared to the result of simple one-dimensional compaction.

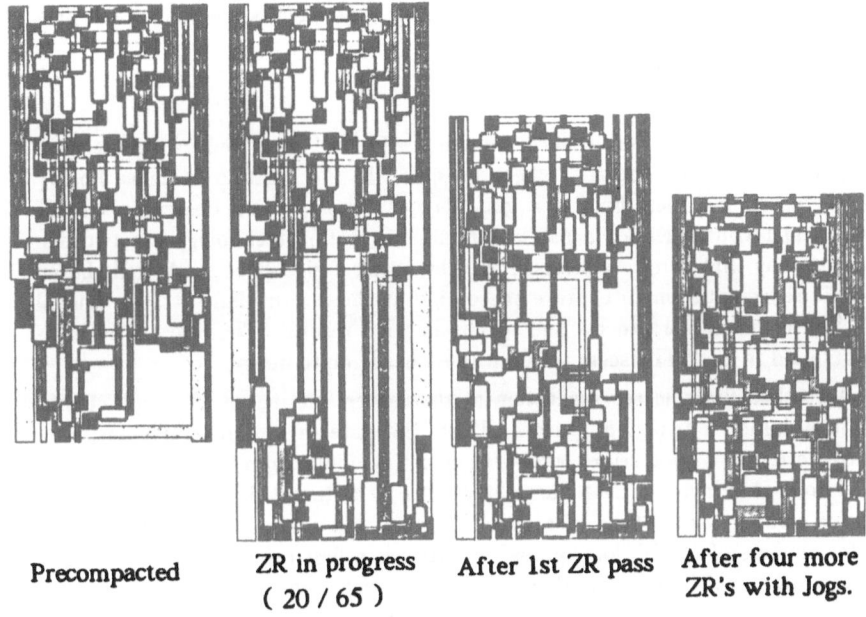

| Precompacted | ZR in progress
(20 / 65) | After 1st ZR pass | After four more
ZR's with Jogs. |

Figure 13. *Example of circuit compaction.*

For box packing problems, zone refining can reduce the area occupied by up to 30% beyond what a one-dimensional compactor can do, at a cost in total run time that is 10 to 30 times longer, depending on the number of passes. For circuits similar improvements have been observed, but because of the complications introduced by the attache wires, and the need for jog generation, total run time can be up to 100 times longer than that required for one-dimensional compaction.

Interconnections play a crucial role in the performance of the circuit, and the given topology of the circuit often has been chosen based on considerations at the micro-architecture level. Thus we do not want the compaction tool to make profound changes to the topology of the circuit; this is the task of a different kind of tool that can take properly into account concerns beyond observation of the geometrical design rules. Thus for the zone refining process we assume that we start from a good topology, given for instance in the form of a symbolic sticks diagram. The given basic ordering is maintained in the compaction process, distinguishing our approach from the more general problem of block placement and routing.

The advantage of the zone refining approach over global two-dimensional placement algorithms is that the number of components that must be considered at any one time is dramatically reduced, and the complexity of the algorithm thus is only of polynomial complexity. In addition, just as in the physical zone refining process, the compaction process can be repeated if the results are not yet satisfactory after the first pass.

6. CONCLUSIONS

"Silicon Compilers" as well as human designers like to reduce design complexity by separating concerns, where possible. In the creation of dense library cells or macro modules, finding a good layout topology and observing all the geometrical layout rules for a particular implementation technology are two distinct concerns that can be addressed in subsequent design phases. A well-chosen symbolic representation to capture the design at the intermediate state is crucial to facilitate the design process and to obtain good results. The coarse-grid components used in EDISTIX seem to fulfill these needs quite nicely.

With this symbolic representation at the center, the design of a high-quality module becomes a two-step process. First, the gate or circuit-level description gets converted to a good sticks layout, then this symbolic representation gets compacted to a real layout. Both these steps can be automated. With TOPOGEN we have created a prototype of a generator that will produce acceptable topologies for clusters of CMOS logic gates. ZORRO is a first prototype of a new class of two-dimensional compactors that can convert sticks-representations to practical layouts. Before long, the process of module generation will be largely done by computers.

7. ACKNOWLEDGEMENTS

Over the last two years several people have worked on and contributed to the tools described in this paper. Special thanks go to the most recent set of developers who have also given me constructive criticism on this paper: Ping-San Tzeng, Glenn Adams, and Hyunchul Shin.

This work is supported in part by the Semiconductor Research Corporation and by the State of California under the MICRO program.

References

1. L.W. Nagel and D.O. Pederson, "Simulation Program with Integrated Circuit Emphasis," *Proc. 16th Midwest Symp. Circ. Theory*, Waterloo, Canada, April 1973.

2. C.M. Baker and C. Terman, "Tools for Verifying Integrated Circuit Designs," *Lambda*, vol. 1, no. 3, pp. 22-30, 4th Q. 1980.

3. D. Harrison, P. Moore, A.R. Newton, A.L. Sangiovanni-Vincentelli, and C.H. Séquin, "Data Management in the Berkeley Design Environment," *submitted to ICCAD-86*, Santa Clara, CA, Nov. 1986.

4. C.H. Séquin, "VLSI Design Strategies," in *Proceedings of the Summer School on VLSI Tools and Applications*, ed. W. Fichtner and M. Morf, Kluwer Acadmic Publishers, 1986.

5. T. Uehara and W.M. VanCleemput, "Optimal Layout of CMOS Functional Arrays," *Trans. Comp.*, vol. C-30, no. 5, pp. 305-312, 1981.

6. A. Sangiovanni-Vincentelli, M. Santomauro, and J. Reed, "A New Gridless Channel Router: YACR II," *IEEE Trans. Comp.-Aided Design*, vol. CAD-4, pp. 208-219, 1984.

7. A. Sangiovanni-Vincentelli, D. Braun, J. Burns, S. Devadas, H.K. Ma, K. Mayaram, and F. Romeo, "CHAMELEON: A New Multi-Layer Channel Router," *Proc. Design Autom. Conf.*, *Paper 28.4*, Las Vegas, July 1986.

8. H. Shin and A. Sangiovanni-Vincentelli, "MIGHTY: A 'Rip-up and Reroute' Detailed Router," *submitted to ICCAD-86*, Santa Clara, CA, Nov. 1986.

9. M. Schlag, Y.Z. Liao, and C.K. Wong, "An Algorithm for Optimal Two-Dimensional Compaction of VLSI Layouts," *Integration* , pp. 179-209, 1983.

10. G. Kedem and H. Watanabe, "Graph-Optimization Techniques for IC Layout and Compaction," *IEEE Trans. CAD of ICAS*, , vol. 3, no. o 1, 1984.

11. S. Kirkpatrick, C. Gelatt, and M. Vecci, "Optimization by Simulated Annealing," *Science*, vol. 220, no. 4598, pp. 671-680, 1983.

12. R. Mosteller, "Simulated Annealing for IC layout," *privat communication*, 1983.

8

Overview of the IDA System:
A Toolset for VLSI Layout Synthesis

Dwight D. Hill
Kurt Keutzer
Wayne Wolf

AT&T Bell Labs
Murray Hill, New Jersey, USA

The Integrated Design Aides (IDA) toolset is a set of VLSI CAD software programs that have been developed to make the most effective use possible of a designer's time. IDA incorporates a number a layout synthesis tools capable of generating both structured circuits, such as ALU's, and random logic. The system centers around a constraint-based, symbolic language called **IMAGES** and a compacter methodology. This paper describes IDA, its capabilities, techniques, and status.

1. Introduction and Background

IDA was developed by researchers at Bell Lab's Murray Hill facility to assist in the design of their own custom MOS chips. In this environment, it is not uncommon for one or two people to develop a non-trivial, full-custom chip in a few weeks, from ideas to mask, for use in their research projects. In fact, for many projects the effort involved in building a custom chip with IDA is comparable to that of building a single TTL breadboard, with the considerable advantage that multiple copies are available with no additional effort or delay.

In order to make this possible, the IDA environment has grown into a range of tools supporting almost all aspects of design, from entry to fabrication, including layout-rule checkers, simulators, routers, *etc.*, all of which are designed to work with each other and present a uniform interface to the designer. These tools embody a set of design principles that have proven to be effective for VLSI CAD. The next section outlines the most important principles; Sections 3,4 and 5 elaborate them. Section 6 summarizes the remaining aspects of IDA and sketches the implementation strategy. Section 7 contains a short summary and concluding remarks.

2. *Key Ideas in IDA*

Experience has indicated a few ideas are important to the success of a VLSI CAD system. Some of these principles relate to the design methodology used to build chips; some relate to the design of the programs themselves. Specifically, the IDA design methodology relies on:

- Layout by geometric constraint. Designers specify the relative positions of components on the chip symbolically, and the tools determine their final numeric positions.

- Symbolical connectivity. When a design is specified by geometric constraints, it is simultaneously constrained electrically. This means that the intended connectivity can be compared with the actual topology, resulting in a higher degree of confidence in design correctness.

- Compaction and assembly. Leaf cells are compacted and assembled into larger symbols which represent subsystems on a chip. This compaction step serves several purposes. In particular, it frees the design from the details of geometric design rules, and makes it easier to parameterize cells both geometrically and electrically.

- Automatic layout synthesis. In order to get maximum productivity from the limited number of skilled designers available, IDA incorporates a number of layout synthesizers, called *generators*. Some of these generate regular, or fixed-floorplan cells, such as counters, while other generators accept arbitrary logic specifications. Practical chips require both

This hardware design methodology is backed up by a software methodology that makes the tools easier to develop and use. The key points here are:

- A common language. The IDA tools communicate in the IMAGES design language. Tools incorporate the IMAGES translator to read IMAGES files, build standard data structures from them, and write IMAGES files. The semantics of IMAGES makes many CAD programs simpler, because the programs are provided with a wide range of common semantics without duplicating functionality.

- Technology independence. The fabrication technology description is read from a technology file at the start of execution of each program. Programs access technology data exclusively through a data structure built from this file. Because the IDA software programs do not "hard-code" technology assumptions, they can be easily ported to new technologies.

- The UNIX^{TM} system. All of the IDA tools work under UNIX and take advantage of its capabilities. This influences both the chip methodology and the tools themselves. For example, not only are the tools recompiled with the "make" facility*, but most non-trivial chips are assembled by a series of operations controlled by "make".

These hardware design and tool-building strategies have evolved over time in light of experience using IDA to build chips. The next three sections discuss most critical aspects of these strategies: the IMAGES language, which is the glue that holds IDA together; the compacter, and how they combine to

3. IMAGES: a Symbolic, Constraint-Based, Generator Design Language

Unlike a layout language that passively captures the mask-level information of an integrated circuit, IMAGES is a language for designing macro-cell generators, and serves as the primary medium for designing integrated circuits in the IDA environment. A language-based design philosophy has been adopted for a number of reasons, including conciseness of description, the availability of a wide variety of tools for building and manipulating languages, and because the effective maintenance of information across levels of hierarchy is often easier in languages than in graphical systems.

Among IMAGES' features are its facilities for modularity, maintenance of electrical connectivity, and user customizability. The IMAGES language supports both virtual and fixed grid representations of a design. In the virtual-grid mode, the user works on a coarse grid, where each grid point exactly fits one wire, one contact, or one connection to a transistor. This speeds up and simplifies editing, since everything automatically snaps into place. Because the user does not have to worry about the spacing of circuit elements or design rules, writing generators in IMAGES is easier than writing generators in "L" [MCB85], HILL [LeMe84], ALLENDE [Mo85] or other languages that do not operate in a virtual-grid design environment. The constraint-based nature of the language also gives it an advantage over alternate languages in virtual-grid environments, such as ICDL [AcWe83] and ABCD [Ro84].

IMAGES is a successor to the "i" language developed by Steve Johnson [Jo82]. IMAGES' features reflect needs and interests of IC design groups as

---

* "Make" is a UNIX utility that controls system software based on the modification times of files. It is normally used to automatically recompile programs when source files are newer than the corresponding executable images.

well as collective experience with the Gate Matrix [Lop80], GRED, (early) IDA [Hi84b] and MULGA [AcWe83] design environments. As a language for writing macro-cell generators IMAGES can be compared to other descendants of "i" including HILL and "L", as well as ALI [LN82] and its successors [LV83] [Mo85].

3.1 Principal constructs

IMAGES programs consist of a list of packages, marked by the keyword PACKAGE, each containing a list of symbols, which are sometimes referred to as cells, and which are marked by the keyword SYM Symbols contain primitive circuit elements including devices such as n and p type MOS transistors (DEVICE), contacts or vias (CONTACT), wires (WIREs), ports, which are sometimes called pins or terminals (PORTs), and primitive pieces of mask geometry such as polygons (BLOB) and rectangles (RECT). In in order to support hierarchy, symbols may also contain instances of previously defined symbols(INST).

Other IMAGES statements exist for manipulating the geometric placement and electrical connectivity of primitive circuit elements. The position constraining statement (BIND) is used to constrain geometric placement of circuit elements. Another statement (PASTE) is used to constrain geometrically and electrically connect instances of symbols. To help specify geometric relations, arbitrary points in the layout may be named (MARK). A simple example of an IMAGES program is given in Figure 1, and the corresponding layout is shown in Figure 2.

```
DIRECTIVE VIRTUAL ;
PACKAGE my_design
SYM inv_v IBEGIN
        DEF_NET out_net;
        DEVICE TP top WIDTH=2 ORIENT=VER ;
        DEVICE TN btm WIDTH=1 ORIENT=VER ;
        PORT  POLY in ;
        WIRE POLY WIDTH=1.2 btm.gn UP 4 TO top.gs;
        WIRE POLY in RIGHT 8 TO btm.gn;
        CONTACT MDP dpout top.de ;
        CONTACT MDP srcpwr top.dw ;
        CONTACT MDN dnout btm.de ;
        CONTACT MDN srcgnd btm.dw ;
        WIRE METAL dpout TO dnout;

        PORT METAL vddleft ;
        MARK METAL vddcenter ;
        PORT METAL vddright ;
        WIRE METAL WIDTH=2 vddleft RIGHT 4
                TO vddcenter RIGHT  8 TO vddright;
        WIRE METAL WIDTH=2 srcpwr UP 8 TO vddcenter;

        PORT METAL gndleft;
        MARK METAL gndcenter;
        PORT METAL gndright;
        WIRE METAL WIDTH=2 gndleft RIGHT 4
                TO gndcenter RIGHT 8 TO gndright;
        WIRE METAL WIDTH=2 srcgnd DOWN 8 TO gndcenter;

        CONTACT MNTUB tubtop vddcenter;
        CONTACT MPTUB tubbtm gndcenter;
        PORT METAL out (dnout,in);
      CONNECT out_net out dnout;
IEND
ENDPACKAGE
```

Figure 1. An IMAGES Leaf Cell

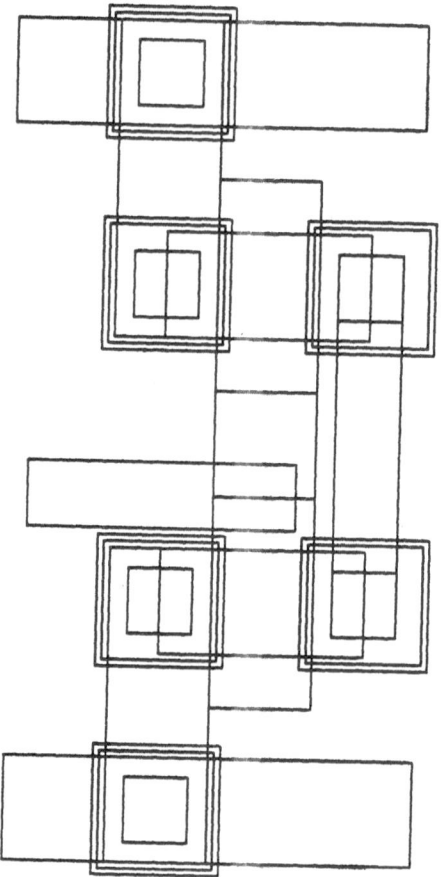

Figure 2. The Resulting Layout

More detail on language features will be given in the sections below.

3.2 IMAGES Use

A typical design path for using the IMAGES language to build a chip might include invoking an *awk*-like or C-like pre-processor whose output would be directed to the IMAGES translator. This would result in a virtual-grid layout which could be viewed graphically. The compacter , and optionally the suite of symbolic routers, would then be invoked to produce a fixed layout. This process is graphically depicted in Figure 3. Once the design is settled, this process may be coordinated by means of UNIX tools, especially the *make* program.

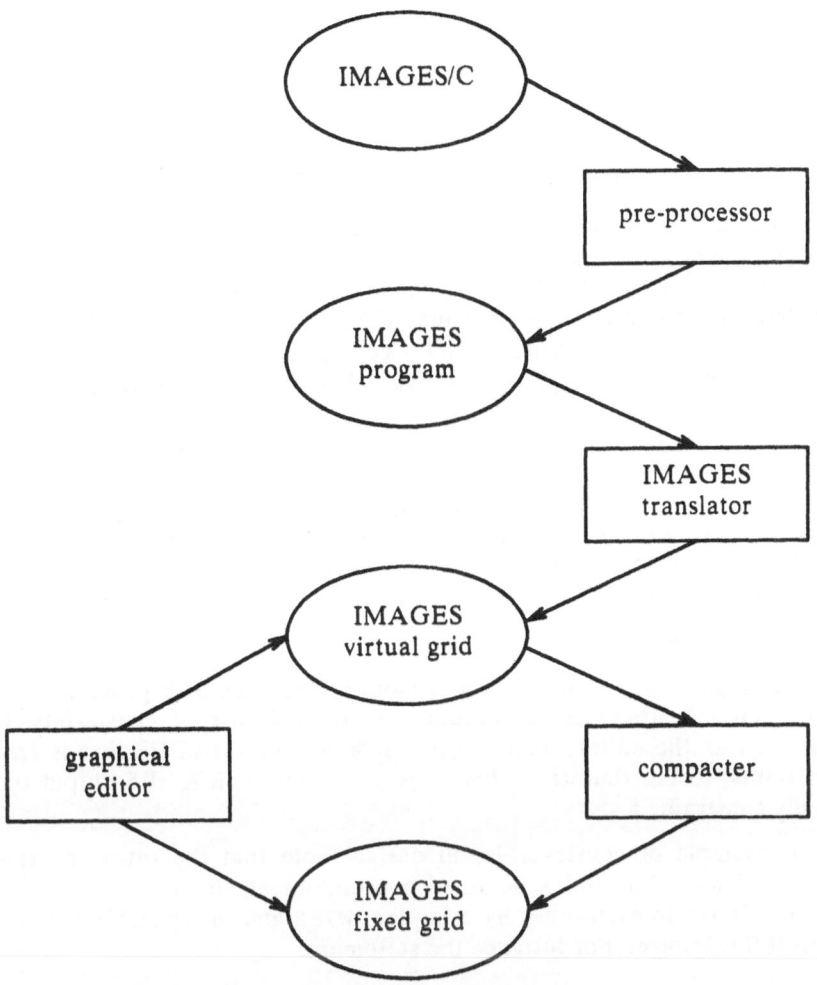

Figure 3. IMAGES Use in the IDA Environment

3.3 Attributes

Each IMAGES primitive circuit element has a number of *attributes*. The IMAGES language provides constructs that allow the user to query the values of attributes of instantiated cells. For instance for an instantiated symbol c with port p the expressions:

```
c.p'X
c.p'LAYER
```

give the value of the *x* coordinate of *p* and its layer respectively. Supported attributes in IMAGES include the geometric coordinates of an object (X and Y), its layer (LAYER) and its net (NET).

3.4 Technology Independence/Technology Accessibility

In order to make the language as technology independent as possible, IMAGES uses a technology file reader and technology database, which will be described in Section 6. The technology file influences the IMAGES translator in two ways: it determines the set of technology words, such as METAL, that will be recognized; and secondly, it provides some quantitative access to the underlying technology. This complements the use of the compacter: many global positioning tasks and wiring need be parameterized by only a few simple constants in order to acheive technology updatability. To accomplish this, a set of technology specific keywords are supported, including TECH'MIN_DESIGN_RULE, TECH'LAMBDA, TECH' METAL_TO_METAL_SPACE and others. These keywords and their values are supplied by the technology database, which means that new categories of information can be added with little effort.

3.5 Constraint-Based Design

Fixed coordinates are rarely used in hand-edited IMAGES programs. Tools which generate IMAGES as output vary in their use of constraints, from tools such as the editor, which writes a dialect of IMAGES that is free of constraints, to the routers, which write generate an IMAGES output that is heavily constrained.

As an example of constraint-based design, note that the inverter example given in Figure 1 contains no fixed coordinates at all. Instead objects are placed relative to each other by using the WIRE and BIND statements in the IMAGES language. For instance the statement:

```
WIRE METAL dpout DOWN TO dnout;
```

constrains, by virtue of the DOWN keyword, the *x* coordinates of dpout and dnout to be the same and the *y* coordinate of dpout to be greater than the *y* coordinate of dnout. The same constraint-based geometric placement of these elements could be accomplished without adding a wire by the following statement:

```
BIND dpout ABOVE dnout;
```

Alternately, dpout could be constrained to be a distance of exactly three units above dnout by any of the following statements:

```
WIRE METAL dpout DOWN  3 TO dnout;
BIND dpout ABOVE dnout BY 3;
```

or, at the time of defining dpout:

```
!place the x coordinate of dpout
!at the x coordinate of dnout
!place the y coordinate of dpout
!at the y coordinate of dnout + 3
CONTACT MP dpout (dnout , dnout + 3);
```

This technique of "design by constraint" is useful for people writing directly in IMAGES, since it eliminates the tedious job of recalculating absolute positions whenever there is a minor layout change. This motivates the use of constraints as a floorplanning tool, which is discussed in a later section.

3.6 Modularity

In a generator design environment a problem of naming conventions often arises when several independently designed generators are integrated into a single layout. To solve this problem in IMAGES a simple analog of the Ada *packages* construct is used. This construct allows designers to use names freely and to "package up" their designs in a way that will allow for easy integration.

Modularity is also supported at the symbol level. IMAGES' symbols are broken down into the external view and the internal view. The external view consists of the ports of a symbol, while the internal view consists of all other circuit elements. This provides for further protection of the global name space.

3.7 Electrical Connectivity

IMAGES provides constructs for specifying the electrical connectivity of a circuit design and uses this information to make inferences about the circuit when processing the design. This facility does not replace the need for circuit extraction but rather complements it. The IMAGES view of electrical connectivity can be checked against the extracted view to ensure that the actual connectivity reflects the designers intention.

Inside IMAGES a *net* is a property associated with a set of IMAGES *connections*. Connections represent electrical connection points in the three dimensional design space. Connections are associated with terminals of devices, and contact cuts, as well as with ports and marks of symbols. Connections differ from simple geometric locations in that they have a layer and a net associated with them.

The IMAGES translator has two rules for deriving electrical connectivity from information supplied by the user:
1)if an IMAGES object that is a member of an net is placed at a connection, and if the layer(s) associated with the connection are consistent with the layer(s) associated with IMAGES object, then a single net results that is the union of the nets associated with the connection and the object.
2)if a wire connects two or more connections, and if the layer(s) associated with the connections are consistent with the layer associated with wire, then a single net results that is the union of the nets associated with each of the connections.
As an example of connectivity derivation, consider the following portion of Figure 1:

```
DEVICE TP top WIDTH=2 ORIENT=VER ;
PORT METAL vddleft   NET=vdd;
MARK METAL vddcenter ;
PORT METAL vddright ;
!place the contact at the diffusion west
!terminal of transistor top
CONTACT MDP srcpwr top.dw ;
WIRE METAL WIDTH=2 vddleft RIGHT 4
     TO vddcenter RIGHT  8 TO vddright;
WIRE METAL WIDTH=2 srcpwr UP 8 TO vddcenter;
```

After the IMAGES translation of this design srcpwr, vddleft, vddcenter, vddright, top.dw, will all share the same user-defined electrical net vdd. The contact srcpwr and the transistor port top.dw will share the same net because srcpwr was placed at top.dw and the first rule comes into play. The connections srcpwr, vddleft, vddcenter, and vddright will all share the same net because they are wired together and the second rule is in effect.

A *set* union-find algorithm [Tar75] is used to dynamically resolve the electrical connectivity of the circuit described by an IMAGES program. The advantage of using a union-find algorithm over simply finding the connected components (*i.e.* nets) of all the nodes (*i.e.* connections) in the circuit is that errors in the user program may be discovered more quickly and more carefully related to the particular element of the user program that is responsible for the error.

3.8 Translator Directives

When a user wants to consider only the electrical connectivity of a circuit, before a real geometric placement has been determined, processing of geometric placement information may be turned off. Alternately, electrical connectivity maintenance may be turned off, and only geometric information represented. As mentioned above, the user may elect only to interpret the external view of a cell and ignore internal cell features. Finally at the statement level a general syntactic construct is provided for passing extra pieces of information to other programs downstream. In other languages this information typically clutters comment fields.

3.9 Summary of the Various Uses of IMAGES

The relationships among the ways that IMAGES is used is summarized in the following table:

	VIRTUAL	FIXED
CONSTRAINT-BASED NUMERIC	output of some generators output of editor	chip assembly output of compacter

3.10 Geometric Constraint Resolution

User-defined constraints in IMAGES are of two types, equality constraints and inequality constraints. The user-defined constraints of an IMAGES program are similar to the design rule constraints that a constraint-based compacter must solve, but several important differences exist. Typically, a compacter, working from a "sticks" or virtual-grid design, has an initial layout and can easily find the topmost or leftmost elements of the layout. Locating these elements is useful for ordering vertices of the design-rule constraint graph associated with the circuit. Since circuit designs in IMAGES have no initial layout, the problem of finding the leftmost or topmost elements of the circuit is equivalent to the problem of finding a feasible layout, which is precisely the problem that the IMAGES constraint resolver is trying to solve. On the other hand, if the initial layout fed to the compacter is design-rule correct, illegal constraints in the design-rule constraint graph should not be present. In a user-defined IMAGES program, however, any number of illegal constraints may have been mistakenly included in the program; moreover, one of the important jobs of the IMAGES constraint resolver is providing intelligent error messages in this situation. A third difference between the problem facing the IMAGES translator and a compacter has to do with the number of fixed objects. In a one-dimensional compaction algorithm the compacter assumes that leftmost and topmost circuit elements are fixed, and compaction is performed with respect to those. In comparison, in an IMAGES circuit design, geometrically fixed elements may appear throughout the design, or alternately, there may be no fixed elements in the design at all. Because of these differences, the IMAGES constraint resolver faces a somewhat more complex problem than the constraint resolver of an ordinary one dimensional constraint-based compacter.

The resolution of constraints in IMAGES is performed using an efficient technique [Ke86] that uses a union-find algorithm for solving equalities [Tar75] [Leng84] and an adaptation of a shortest-path algorithm [John77] [LiWo83] for solving inequalities. Using this algorithm constraint resolution is accomplished in nearly linear time and is always reduced to a small fraction (<10%) of total processing time for an IMAGES program. The bulk of the processing time is inevitably spent in processing the syntax of the

4. Compaction and Assembly

Today almost all IDA designs pass through a compaction stage and an assembly stage. Compaction relieves the designer of responsibility for satisfying the detailed design rules for a technology, while assembly controls the compacter and expedites the task of fitting cells together.

4.1 Compaction

In IDA, compaction is thought of as a technology-binding process, not as a process for making a layout smaller. This is why the IDA compacter is named *ibind*. The compacter accepts a design that was generated without knowledge of the design rules, and a description of the rules for the target technology. It binds the rules to the input symbol, and creates a new symbol that conforms to them. In IDA, the input symbol exists in a different "universe" than the output symbol. The first is on the virtual-grid, the second on the fixed-grid. For that reason, it is impossible to take the output of the compacter and feed it back into the compacter again. However, both are described in IMAGES, and the virtual and fixed-grid representations bear a strong resemblance to each other (though the "uncompacted" one may actually look smaller on the screen).

4.1.1 Combining Cells: Routing and Pitch Matching The compaction-and-assembly process allows designers to choose to make connections using pitch-matching or routing as needed. Pitch-matching is done by selectively de-compacting cells, and then placing them next to each other with a small amount of overlap, so that wires in one subsymbol touch the wires in the next. An alternative technique is to add routing wires between cells. This can be done with or without pitch-matching the cells: If they are pitch-matched, only a tiny section of wire will be needed for each connection. If they are not pitch-matched, more wires, perhaps on several layers, may be required. The IDA system supports both methodologies: the first is most useful for regular arrays such as RAM cells, the second for chip assembly.

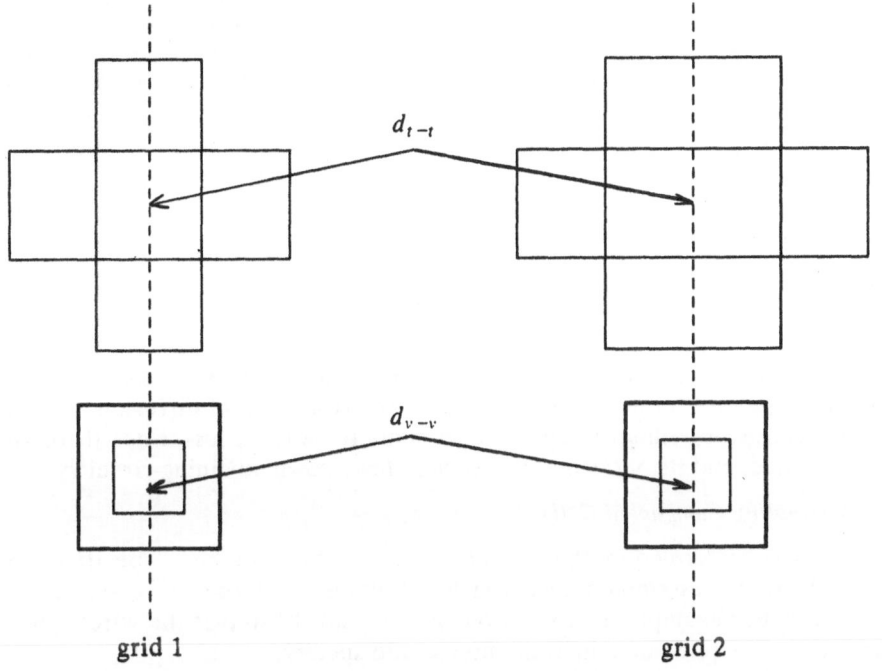

grid 1   grid 2

Figure 4. Spacing the Virtual Grid

At present, the ibind program uses the *virtual-grid algorithm*. Virtual-grid compaction is more straightforward to implement than some other compaction techniques because there are fewer degrees of freedom. In particular, all the objects than begin compaction on a single virtual-grid line, either horizontal or vertical, are normally still lined up after compaction [West81]. The process of compaction is therefore reduced to one of finding a spacing between these lines. In most cases, the distance required between adjacent grid lines is the maximum spacing required between any two components that face each other on the grid. For example, in Figure 4, because d_{t-t} is larger than d_{v-v} it will be used as the distance from grid 1 to grid 2, and there will be some extra unused space between the contacts.

4.1.2 Displaying a Virtual-Grid Symbol - Points Around Transistors A virtual-grid layout is usually a reasonable representation of the actual layout. Transistors are not represented in their exact dimensions: they are approximated by a three-grid by three-grid region. Its center represents the center of the active area, one grid away on each side are the source and drain connections, and one grid away in the orthogonal dimension are the connection points to the gate. These symbolic layout connections to transistors are preserved as transistor sizes are changed. During compaction, the actual size of the transistors is determined by multiplying the default size for each type of device (*e.g.* TP or TN) by the individual size of the device. A typical example might be a pMOS device whose width is 2.5 times the default width for pMOS devices (3 microns), or 7.5 microns. By deferring this multiplication until *ibind* is run, the design can more easily be ported to new technologies with different transistor properties. The compacter also moves the transistor connection points rigidly with the center of the transistor, which may result in a gap between the transistor and the wires connecting to it. The IDA compacter automatically inserts wires into the cell as necessary to fill this gap.

4.1.3 Strapping Wide Transistors When contacts to metal wires are attached to the source or drain of a transistor, the compacter automatically introduces a metal wire and a row of contacts. This process is called *strapping*, and is critical in high performance circuits. Since the resistance of diffusion is high, a large transistor without strapping may actually be a net loss, since its drive is diminished, and its parasitic capacitance slows down adjoining circuitry.

4.2 Assembling Compacted Cells

Assembly is performed by the *pasteup* program. Pasteup takes a specification the cells to be assembled and which terminals on those cells are to be connected. For example, to abut two cells, "a" and "b" so that the wires "gnd" and "vdd" were pitch-matched, the user would specify:

```
PASTE a.gndright TO b.gndleft;
PASTE a.vddright TO b.vddleft;
```

In contrast, the MULGA[Wes81] assembler requires that ports to be connected must be on the same virtual-grid line. That is, if the two virtual-grid cells are plotted next to each other, the connected ports must be coincident. This may require the designer to add virtual-grid lines to cells to achieve connectivity. Once this task is done, it has the advantage the the module is easier to view in the virtual grid, since all the wires line up neatly. However, it is tedious, and may make the individual subcells cells harder to reuse in other designs, since they must to be modified to fit their context.

IDA's pasteup program and finds legal spacings between the PORT's on subsymbol using a graph traversing form of linear programming. It then writes the PORT spacings out to file, which is used by the *ibind* program to stretch the leaf cells. The stretched cells are then assembled with the IMAGES language.

4.3 Beyond Virtual-Grid Compaction

Experience with virtual-grid compaction has led to an appreciation of its virtues and limitations. Although leaf cells compacted by the virtual-grid algorithm are not as small as they could be [Wolf85], in many practical chips it is cell assembly, not leaf-cell compaction, that is the process most in need of improvement. The area inefficiencies of the virtual-grid compaction algorithm are magnified when it is used for cell assembly. The pitch-matching algorithm does not easily allow a cell to contain both subcells and primitive elements, so that random wires, vias, and transistors used to connect cells must be put in specially-created cells. In addition, IDA's facilities for finding the required spacing between cells are, at present, not fully automatic.

To correct these problems, an effort is underway exploring constraint-graph hierarchical compaction as a more efficient means of assembling large chips. Constraint-graph compaction has been used in a variety of systems, such as CABBAGE [Hsueh79]. In constraint-graph compaction, unlike virtual-grid compaction, each component is assigned its position independently, giving the layout elements greater freedom of movement. The design-rule constraints can be written as linear inequalities; these inequalities can themselves be represented as a graph, where nodes reflect components or wires. Weighted, directed edges represent the value and direction of the inequality. Connections between components can be represented by pairs of constraints that specify the upper and lower bounds on the distance between two components. The graph can be solved using a critical-path algorithm to find positions for the components (represented by values of the nodes) that satisfy all the design rules and make the layout as small as possible. The constraint-graph representation can be extended to composite cells. For example, Lava [Ull84] assembles hierarchical layouts using the constraint-graph technique. To reduce the complexity of the compaction problem, an abstraction of the cell is built that represents, in simplified form, the cell's stretchability. The only components represented in the abstraction are the ports. Constraints between the ports define how the cell can change shape during compaction. The port constraints are determined from the design-rule constraints for the cell; the abstraction therefore behaves exactly as would the full cell during compaction. The hierarchical compaction step finds the positions of the ports that match the cell to its environment. The complete layout for the stretched cell is found by recompacting the original cell with added constraints that force the ports to their new positions.

One limitation of some hierarchical compacters is their inability to determine the minimum spacing between cells. The spacing required from a component to a cell can be found only by looking inside the cell to find the objects near the boundary that affect spacing. Some hierarchical compacters use the maximum design rule as the spacing from a cell to any other object. The worst-case spacing is rarely necessary, and the waste can be significant in large arrays of cells. The difference between the worst-case rule and the actual required spacing can be particularly costly in CMOS technologies, where the tub-tub spacing is typically much larger other than spacings.

The *donut* abstraction [Rei86] extends the notion of a cell abstraction with enough information to allow the compacter to determine the spacing between cells or to overlap cells where feasible. The donut abstraction for a cell includes both the cells' external ports and components near the boundary that can affect cell spacing. The abstracted constraint graph for the cell describes how ports and components in the donut stretch during compaction. The compacter can then look at the positions of components within the donut to determine the separation between the cell and other objects.

4.4 Cooperation Between Compaction and IMAGES Translation

Nothing in the IMAGES language or the IDA methodology requires that the virtual-grid compaction algorithm be used. Indeed, because both the IMAGES translation process and many modern compaction algorithms center around resolving a graph of constraints, it seems likely that there may be some advantage to combining the two steps. Research into this possibility is underway.

5. Layout Synthesis

The IMAGES language, combined with the compacter system, can greatly accelerate the design process of full-custom chips. However, to get maximum leverage a higher level of automation is required, in which the primitives are not transistors or contacts, but subsystems more closely tied to the intended architecture. This is the motivation behind the IDA layout synthesis tools. These tools come in two varieties: fixed-floorplan generators, and random-logic generators.

5.1 Fixed-floorplan tools

Even with the assistance of all the general-purpose tools, developing a major subsystem of a chip still requires two things: expertise and effort. In order to get the most from both, the notion of generators was introduced to the IDA system. These generators are software units that contain a carefully worked-out design and allow it to be parameterized so that it can be applied in a number of different contexts, without the effort of redesigning from scratch. This technique captures both circuit and layout expertise in machine-readable form, so that the generator user need not be knowledgeable about the internals of the circuit being created. In addition, generators can be used within generators, further multiplying their utility.

5.1.1 The C-IMAGES and Awk-I Languages
The special purpose generators provided under IDA are written in either the C-IMAGES or Awk-I languages. The first is a mixture of IMAGES + the C language, the second a mixture of IMAGES + awk (Awk is a string processing language under UNIX.) These languages were developed specifically for this purpose. Each line of the source file belongs to either IMAGES or the host language. A preprocessor determines which, and maps IMAGES statements into print statements. In order to do this the preprocessor must understand the technology words such as METAL and MPTUB, so it starts by reading the IDA technology file. The result is then compiled with the ordinary compiler, and the executable file is stored in a library as a generator. Each time it is executed, the user can specify a set of parameters, such as the number of bits in an N-bit counter generator. During execution, the print statements introduced by the preprocessor are executed and generate an IMAGES file. Exactly which print statements are executed, and how many times, is controlled by the host language (C or awk), and this determines the nature of the design being generated.

5.2 Fixed-Floorplan Generators: Parameterized Layouts

In the IMAGES language section, an inverter was used as an illustration. This inverter could be compacted and assembled as a generator, the resulting code shown here.

```
DIRECTIVE FIXED;
gen_row(name, stages) char *name; int stages;
        SYM %sname IBEGIN
        INST inv_1 inv[1..stages];
                FOR j in [1..stages] LOOP
                        PASTEALL inv[j].gndright TO inv[j+1].gndleft;
                        WIRE AUTO inv[j].out RIGHT TO inv[j+1].in;
                        ENDLOOP;
        PORT AUTO in inv[1].in;
        PORT AUTO out inv[stages].out;
        PORT AUTO gnd inv[1].gndleft;
        PORT AUTO vdd inv[stages].vddleft;
        IEND


main (argc, argv) char **argv; int argc;
        gen_row(argv[1], atoi(argv[2]));
        ;
```

This generator produces a symbol with a user-specifiable number of inverters
connecting its input to its output. It accepts the name of the output symbol
and the number of stages as parameters from the UNIX shell. (In practice, a
production-quality generator normally checks its inputs and prints a line
explaining its proper use if the parameters are unreasonable. Very complete
parameter checking and default services are provided in the GENASYS
system, which is a subsystem of IDA developed at Allentown, PA.) This
generator was run with a parameter of 3, the resulting layout is shown below:

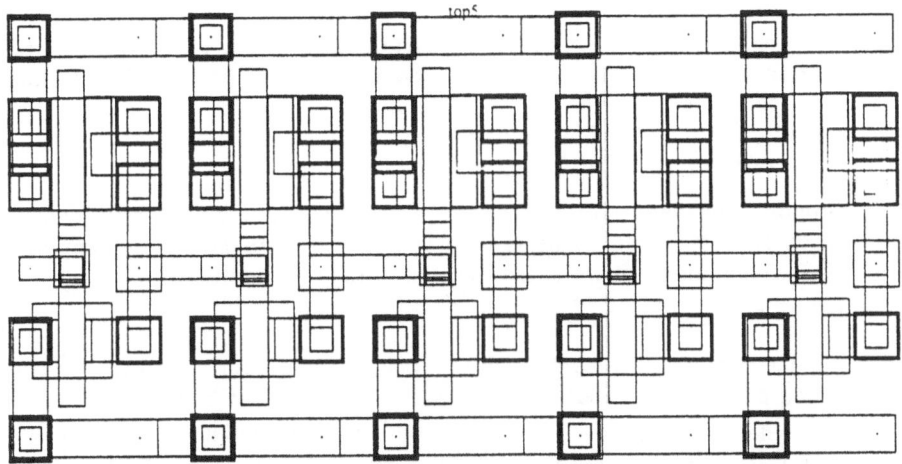

In this case the generator is basically just a programmable repetition of lower-level symbols, with little "value-added." However, the same technique can be applied to far more complex subsystems. For example, one generator in IDA lays out cyclic redundancy code (CRC) counters. An N-bit counter of this type has a period of 2^N, but has only a fraction as much circuitry as binary counter. In order to do this, the generator has to contain a table of irreducible polynomials of degree 2 through 16 (the maximum value of N supported). When it is invoked, it looks up the appropriate polynomial, and then allocates and wires together a series of shift-register cells and XOR gates. One of these counters was used in a dynamic RAM chip to supply row addresses during automatic refresh.

5.3 Technology-Updatable Generators

While developing a custom symbol represents a good deal of effort, developing a generator represents an even greater investment. In order for such an investment to be justified, the generator should be useful for the longest time possible, preferably longer than the lifetime of any one technology. Therefore, the recent work in generators has combined the C-IMAGES language with the compacter, in order to produce layouts that are independent of the detailed design rules. Because cells need to be combined with each other in a variety of ways, the IDA methodology includes three techniques for using compaction in generators:

1. The subsymbols can be compacted once and put in a library. Each time the generator is used, they are constrained together using the IMAGES language. This is convenient when some portion of the cells are not

built with the compacter, as with RAM designs.

2. The cells can be created in the virtual grid once, perhaps with the editor, in the virtual grid, and then compacted and assembled each time the generator is run. This automatically provides two degrees of freedom: pitch matching and transistor sizing. This method is used by most of the IDA fixed-floorplan generators today.

3. Finally, the generator can create a new, custom, virtual-grid symbol each time it is run. This is then compacted, and the job is complete. One such generator, SC2, is discussed in the next section.

5.4 SC2: A Custom-Logic Layout Tool

When a designer needs a medium-sized block of custom logic, and there is no such block pre-developed, he or she may consider using SC2 as an alternative to a developing a symbol for it from "scratch." The choice between using SC2 or a full-custom symbol is normally based on the degree of regularity in the logic, and the speed and size requirements. For small to medium-size, highly random symbols, SC2 can represent an attractive alternative.

The input to SC2 basically consists of a transistor connectivity list. This list can be created in a number of ways:

1. By hand, textually.

2. By graphic schematic capture, using "icon," the IDA editor.

3. By extracting it from a preexisting layout.

4. By synthesis. The PROLOG program "itrans" is available in IDA to convert arbitrary boolean equations into transistors. It understands simple logic transformations, such as DeMorgan's law, and the principles of complementary, domino and zipper logic.

The list of transistors is enhanced with geometric specifications. For example, the user may specify on which side of the circuit module each input or output needs to be located. SC2 parses the input, converts it into CMOS transistor connectivity, orders and orients the transistors, and wires them together in the gate-matrix style. The steps inside SC2 are as follows:

1. Very wide transistors are split into several smaller ones, wired in parallel.

2. pMOS and nMOS devices are grouped into pairs.

3. The pairs are arranged in order from left to right. The algorithm to do this is a min-cut technique invented by Kernigham and Lin [Ker70].

4. The source and drains of the devices are flipped as necessary to maximize the number of abutting diffusion regions, and secondarily, to minimize routing requirements.

5. Routing takes place using a variety of techniques, ranging from the cheapest (in terms of area) to the most costly. A "greedy" channel router is used last.

6. Extraneous contacts are eliminated, and wherever possible polysilicon wires are replaced with metal wires.

7. An IMAGES language file is written.

An example of an input file to SC2, using "itrans," is shown below:

```
symbol(mult2_v). #this is a one bit multiplier
generate :-
        mb := (xi0 * t) + (xi * o),
        b := ( mb * nc ) + ( ( ^mb) * n ),
        cxb := ( ci * (^b) ) + ( (^ci) * b),
        so := ( cxb * (^si) ) + ( (^cxb) * si),
        co := ( (si + b) * ci ) + ( si * b ).
access( xi1 , [left, bottom, right]).
access( xi0, [top]).
access( ci, [left]).
access( si, [right]).
access( o , [top, bottom]).
```

The output is shown here:

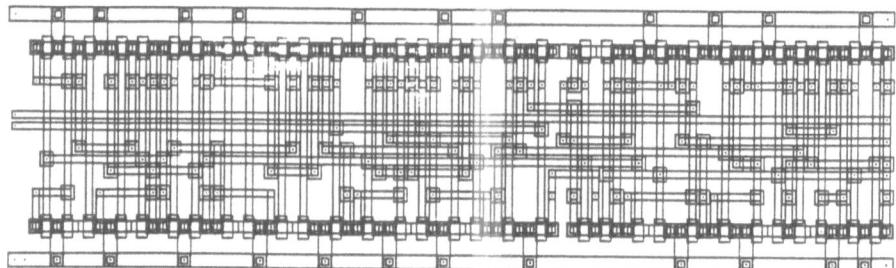

5.5 Routers

Symbols created by icon, SC2, or other means need to be interconnected in order to do useful work. For highly regular layouts, the easist and most efficient way is usually to generate the interconnection wires using a loop in IMAGES. For example, if a and b are subcells, with b below and to the left of a, then the loop

```
FOR i in [1..10] LOOP
        WIRE METAL b.out[i] RIGHT UP TO a.in[i];
        ENDLOOP;
```

would connect the outputs of b to the input of a.

For more randomly-connected layouts more complex routing is needed. To support this, IDA provides a set of routers that take advantage of the IMAGES language. The key idea here is to insert the routing wires in the same cell as the symbols to be routed, not in a rigid subcell. The wires and contacts can then have their positions constrained symbolically to one of the objects being routed, and the wires are allowed to stretch to the other. This way, the objects are still not fixed rigidly even after routing, and may be moved further apart to meet other requirements. Moreover, this method does not require that the designer provide any prediction of the space required for the route.

6. Other Tools and Features of IDA

In addition to the items discussed above, the IDA toolset incorporates several additional facilities to support and extend it. These are summarized here

6.1 Technology Description File

The IDA tools are parameterized by a technology description file [Chu83] that describes the set of layers, transistors, contacts, and other facilities provided by each technology. Each time an IDA program begins execution, it starts by reading a technology description from a file into a global data structure known as the technology database. Programs get their information about a technology through the technology database—no constants are hard-coded into the programs. This makes is easier to write CAD programs that handle designs involving a number of technologies. The technology database has been successfully used to describe a number of different nMOS, CMOS, and wafer-scale technologies.

The first step in describing a technology is to define the layers available. Each layer is defined by a name and a mask level. Some of the symbolic levels, such as NDIFF, may actually translate into several mask layers (such as THINOX and DIFFUSION). Others, such as the ANNOT (annotation) layer, do not translate into any mask, but are just used to "comment" the design. Each layer has a minimum width (the size of the smallest allowed feature), and a default width associated with it. Once the layers have been defined, design rules can be specified between them. A *RULE* statement specifies the spacing between two layers under normal circumstances; a field in the rule statement allows specification of different spacings for elements on the same electrical node and those on different nodes. A *FLAG* statement describes exceptions to those rules: it includes two layers, an integer code identifying the type of exception, and the value of the rule. (The meaning of the flag is defined by the program using the flag information.) For example, in one of the older, 2.5 CMOS technology file, some layers and rules are:

```
#   LEVEL name, minimum-width, real-level, flag, MASK
LEVEL METAL 2.5 1 1 METAL -1 0 -1 0 N70
LEVEL NDIFF 2 1 1 NDIFF NPLUS 2.5 -1 0 N31
LEVEL POLY 2.5 1 1 POLY -1 0 -1 0 N40


#   RULE from, to, net-equivalence, distance
RULE NDIFF POLY both 1.25


#   FLAG from, to, net-equivalence, flag
#    flag=0x02 == NOTCH; flag=0x01 == TRANSISTOR
FLAG NDIFF NDIFF other 1
```

Layers are also used in the definition of transistors. The technology database understands transistor and contact primitives, and allows any number of component types of each primitive to be defined. The IMAGES language defines transistors as stretchable objects: the exact geometry depends on the transistor type information combined with that particular transistor's channel length and width. Examples of transistor specifications are shown below:

```
# Transistor:
# TRAN name substrate min-width min-length default-width
#     default-length effective-chan-length V-sub-t mobility
TRAN TN PTUB 2 3 6 3 1.3 700 640

# The rectangles that make up the transistor:
#   TRLAYER type-name logical-level layer
# left-extension bottom-extension right-ext, top-extension
TRLAYER TN BOUNDBOX BB DEF -6 -9 6 9
TRLAYER TN DIFFUSION NDIFF DEF 0 -3 0 3
TRLAYER TN GATE POLY DEF -3 0 3 0
```

Other parameters define global properties of the database. Available properties include a flag to indicate whether the technology is nMOS or CMOS, the oxide thickness, the name of the transistor type to be used as a pullup, (required for displaying interactive simulation), and the contact to use for power and ground connections to the tubs in a CMOS technology. While this decree of parameterization has complicated the software of IDA to some degree, the effort has been rewarded many times as IDA has been ported to new technologies with minimal effort.

6.2 Quick Access to Large Designs

In the case where all the IMAGES symbols have been produced graphically, all coordinates are fixed and there is no need to parse a design "bottom up." The IDA tools can take advantage of this situation with a "pseudo-database" that provides quick, random access to the individual cells. In this format, each symbol normally resides in its own file. A design can be spread out over any number of UNIX directories, a feature that is heavily used when libraries of standard symbols are needed or when several people cooperate on a design. The text in these cells is a dialect of IMAGES, formated in a way that allows the external information about a cell to be accessed independently from its internal details. Because the structure of these text segments limits the amount of information that must be scanned to enter the editor, editor start-up time is independent of the overall size of the chip.

Because the basic IMAGES medium is textual, it is possible to use a wide range of conventional text manipulation tools on it. For example, under the UNIX operating system a command called "grep" is available that searches for a pattern occurring in any specified set of files. This turned out to be useful for identifying various features when it was necessary to adapt a design to new design rules. For example, to find all nMOS pullup transistors with channel length equal to 2 lambda, the command

```
grep "TI .* WIDTH=\<2\>" *.im
```

was used.

Because IDA can accept a mixture of hand-made IMAGES symbols and machine-edited symbols, it retains flexibility for those parts of the design undergoing current revision and needing IMAGES design-by-constraint. But because non-essential low-level details need not be read in until needed, the session start-up time can be dramatically reduced.

6.3 The Icon Editor

"Icon," the editor supplied with IDA, is capable of supporting both schematics and layout editing [Hill84a]. In schematics mode, the logic designer can use icon to design with logic cells from one or more libraries, and see them on the screen in logic diagram format. In layout mode the designer sees the layout with accurate dimensions. Because the internal structures icon uses for schematics are the same as those used for layouts, it is possible to intermingle the two. This gets around the problem of interfacing schematics with automatically generated layouts, and assists in simulation and documentation.

The icon editor's command structure is based on a "reverse-polish" model. In this model, one first specifies the operands, then the operation. The operands are denoted by a subset of objects on the screen known as the "chosen" group. Icon provides many ways of selecting a portion of a symbol to be part of the chosen group, *e.g.* everying right of the cursor, all the transistors, everything within a certain box, and so forth. Icon then provides a wide range of operations that can be performed on it, *e.g.* deleting them, duplicating them, creating a subsymbol out of them, *etc.* While editing, icon also keeps track of the electrical net associated with each object. This can be tied into the editing operations, For example, one can choose every metal wire associated with the Vdd net and change its width to two times the minimum. Almost every operation in icon can be undone with the "undo" command, including the "undo" command itself.

6.4 SOISIM simulator

In order to help verify the correctness of the logic design, IDA includes an interface to a logic simulator called SOISIM which models MOS logic at the *switch-level*[Szy82]. The model understands the notions of pull-up, pull-down and pass transistors, resistance ratios, and stored charge, and is similar in nature to MOSSIM [Bry80]. Internally, the simulator understands and optimizes the functions of common MOS structures, such as pass transistors and AOI gates.

One unusual feature of the IDA toolset is the ability to combine simulation with interactive graphics. While inside icon, the user can invoke the SOISIM simulator. This forks off a simulator process interacts with it via data streams, and displays the results on the graphic screen. The user can enter values by pointing to nodes on the screen and setting their values. This form of simulation can be performed in schematics, virtual-grid, and fixed-grid modes. In schematic mode, it is relatively easy to make logic changes quickly. After committing a logic design and beginning its layout, the interactive simulation is normally used only to track down bugs, not to modify the logic.

For large designs, interactive simulation gets tedious, so designers write C language programs that drive the SOISIM simulation and print out the results. These two modes of use are complimentary. Experience has shown that the non-graphical interface is invaluable for testing out complete systems when they are nearly correct, but that when there are errors users tend to go back to the graphical mode to puzzle them out.

6.5 Design-Rule Checking

IDA includes a layout-design-rule checker that is parameterized by the technology database. This checker works hierarchically, examining only the boundary of large symbols where they may interact with other adjoining circuitry. Because electrical connectivity and transistor placement are specified textually and explicitly in IMAGES, the checkout tends to be thorough and meaningful. The performance is fast enough to allow a circuit to be checked in time that is perhaps double that required to parse it, which is on the order of tens of minutes for a 30,000 transistor chip on a DEC VAX 11/780, depending on its structure and hierarchy.

6.6 Transistor Sizing

In order to operate at a specified speed, it is mandatory that individual transistors be sized appropriately to the load that they must drive. TILOS is a tool that examines the timing requirements of a circuit, and automatically determines the minimum transistor sizes that will achieve the required speed[Fish85]. TILOS and SC2 can work with each other, and with SOISIM, so that a design with a performance requirement can be simulated and then fabricated automatically.

6.7 Tub Inserter

CMOS chips require that transistors be surrounded by tubs of the opposite polarity. Ida includes an automatic tub inserter to expedite this. The designer or generator is required to provide a reasonable number of tub contacts, the tub inserter works by finding the closest contact to each object requiring a tub, and then surrounding both by a rectangle of tub material. Extraneous or redundant rectangles are then eliminated.

6.8 Circuit Simulation

In order to evaluate analogue circuit performance, IDA can interface with a circuit capacitance extractor called GOALIE [Szy83]. This produces a circuit description file that is read in by a circuit simulator based on SPICE. Because of the effectiveness of SOISIM and TILOS, today only small, critical circuits, such as clocks or carry propagation lines, are normally simulated with ADVICE.

6.9 The IDA Software Environment

The IDA design environment consists of the major tools described above plus many utility programs. The code of the IDA design environment amounts to over 200,000 lines of the C language. At the core of the IDA environment are the IDA common data structures for representing circuit design, as well as the IDA technology data structure, which embodies the information in the technology file. To build and manipulate these data structures IDA employs a library of common code including:

1. a technology file reader;

2. memory allocation and reclamation routines;

3. routines to build the IDA common data structures associated with circuit elements;

4. routines to manipulate the IDA common data structures associated with circuit elements;

5. routines for string and symbol table manipulation;

6. routines for writing the circuit represented by the IDA data structures as an IMAGES program.

In addition, considerable attention has been paid to making the code as portable as possible. Two facets of IDA illustrate this: the facilities for working with multiple host machines from a single, networked set of source files; and the set of extensible graphic terminal drivers that are bound into the code with a "jump-table" at run-time.

7. Summary

The IDA tools have helped to test a number of ideas and demonstrate their feasibility. Specifically, they have clearly demonstrated the utility of:

1. the "design-by-constraint" techniques of IMAGES;

2. interactive graphical simulation;

3. tools based on a technology database;

4. C-IMAGES and AWK-I language generators;

5. the combination of machine and hand-compacted layouts;

and other ideas.

To be sure, there are gaps in the IDA toolset, most noticeably in the area of test generation, where work is just beginning. However, the overall experience in using IDA has been very positive. More than 60 different chips have been designed and fabricated, in technologies ranging from 3.5 micron nMOS, to 1-micron, high-speed nMOS technology, to three different CMOS technologies, and even to a new wafer-scale technology. The most encouraging statistic is that the majority of designs created with IDA have been fabricated with no logic errors. This is due in large part to the accuracy and speed of the SOISIM simulator, which allows a thorough testing of the whole chip down to the transistor level. Equally important has been the accurate (and perhaps conservative) design-rule checker combined with the net specifications of IMAGES, which has made sure that designs do not fail because of unintentional shorts or missing power connections. But there is one other factor which has probably had the most important effect of all: all the CAD programmers are also chip designers. This gives them, first hand, the experience and understanding of which issues are important to building real chips. As a consequence, almost every chip built has had a side benefit of polishing or improving some portion of the IDA toolset, making each new design effort at least a little easier than the one preceding it.

REFERENCES

[AD85] Ackland, B. , A. Dunlop, J. Fishburn, D. Hill, K. Keutzer, H. Moscovitz, J. Tauke, "The IMAGES Symbolic Layout Language: Version 0.0," *AT&T Bell Labs internal memorandum*, (1985).

[AcWe83] Ackland, B., N. Weste, "An Automatic Assembly Tool for Virtual-Grid Symbolic Layout," *Proceedings of VLSI 83*, 457-466, (1983).

[Bry80] Bryant, R. E., *MOSSIM: A Logic-Level Simulator for MOS LSI, User's Manual*. Integrated Circuit Memo 80-21, M.I.T. Department of EECS. 1980.

[Chu83] Chu, K-C., J. P. Fishburn, P. Honeyman, Y. E. Lien, "Vdd - A VLSI Design Database System," *Proceedings of the 1983 Annual Meeting -- Database Week*, IEEE Computer Society Press, 1983.

[Hill83] Hill, D. D., "Edisim -- A Graphical Simulator Interface for LSI Design," *IEEE Transactions on Computer Aided Design of Integrated Circuits*, April 1983.

[Fish85] Fishburn, J. P, and A. Dunlop, TILOS: A Posynomial Approach to Transistor Sizing, *Proceedings of IEEE International Conference on Computer-Aided Design-85*, Santa Clara, Ca. 1985.

[Hill84a] Hill, D., "ICON: A Tool for Design at Schematic, Virtual-Grid and Layout Levels," *IEEE Design and Test*, Vol. 1, 4, 53-61 (1984).

[Hill84b] Hill, D., "CAD Systems for VLSI Design" *Proceedings of the National Communications Forum* Rosemont, September 84, 673-691.

[Hill85] Hill, D., "SC2: A Hybrid Automatic Layout Program," *Proceedings International Conference on Computer Aided Design*, Santa Clara, November 85, 172-174.

[Hsueh79] Hsueh, Min-Yu. *Symbolic Layout and Compaction of Integrated Circuits*. PhD thesis. University of California at Berkeley. December, 1979.

[HFL85] Hill, D., J. Fishburn, M. Leland. "Effective Use of Virtual-Grid Compaction in Macro-Module Generators." *Proceedings 22nd Design Automation Conference*. Las Vegas, June 85. p. 172-174.

[John77] Johnson, D., "Efficient Algorithms for Shortest Paths in Sparse Networks," *Journal of the ACM*, Vol. 24. 1, 1977.

[Jo82] Johnson. S., "Hierarchical Design Validation Based on Rectangles," *Proceedings Conference on Advanced Research in VLSI*. M.I.T., p. 97-100, January 1982.

[Ker71] Kernighan, B.W and S. Lin, "An Efficient Heuristic Procedure for Partitioning Graphs." *Bell Sys. Tech. Journal*, Vol. 49 (2), pp. 291-308, 1970. *Solving Equalities, Inequalities and Shortest Paths*, in preparation. 1986.

[Ke86] Keutzer, K., *Solving Equalities, Inequalities and Shortest Paths*, in preparation. 1986.

[Leng84] Lengauer. T., "On the Solution of Inequality Systems Relevant to IC-Layout." *Journal of Algorithms*, 5, p. 408-421. 1984.

[LeMe84] Lengauer, T., K. Mellhorn, "The HILL System: A Design Environment for the Hierarchical Specification, Compaction, and Simulation of Integrated Circuit Layouts." *Proceedings 1984 Conference on Advanced Research in VLSI*, M.I.T., p. 139-147. 1983.

[LN82] Lipton, R., S. North, R. Sedgewick. J. Valdes, G. Vijayan, "ALI: A Procedural Language to Describe VLSI Circuits." *Proceedings 19th Design Automation Conference*. p. 467-474. 1983.

[LV83] R. Lipton, J. Valdes, G. Vijayan, S. North, R. Sedgewick, "VLSI Layout as Programming," *ACM Transactions on Programming Languages and Systems*, Vol 5, no. 3, 1983.

[LiWo83] Liao. Y-Z., C. Wong, "An Algorithm to Compact a VLSI Symbolic Layout with Mixed Constraints." *IEEE Trans. on Computer Aided Design of Integrated Circuits and Systems*, CAD-2, 2, p.62-69, 1983.

[Lop80] Lopez, A.D., and H-F. Law, "A Dense Gate-Matrix Layout Method for MOS VLSI," *IEEE Transactions of Electron Devices*, Vol. ED-27, No 8, Aug 1980.

[MCB85] Matheson, T., C. Christensen, M. Buric, "A Software Environment for Building Core-Microcomputer Compilers," *Proceedings International Conference on Computer Design*, p. 221-224, 1985.

[Mo85] Montiero da Mata, J., "ALLENDE: A Procedural Language for the Hierarchical Specification of VLSI Layout," *Proceedings 22nd Design Automation Conference*. p. 183-189, 1985.

[Nag80] Nagel, L. W., "ADVICE of Circuit Simulation," *IEEE Symp. on Computers and Systems*, 1980, Houston, TX.

[Nag75] Nagel, L. W., *SPICE2 - A Computer Program to Simulate Semiconductor Circuits*, University of California, Berkeley, ERL Memorandum Number ERL-M520. May. 1975.

[Rei86] Reichelt. M., *Improved Abstractions for Hierarchical Constraint-Graph Compaction*, Master's thesis. M.I.T.. September, 1986.

[Ro84] Rosenberg. J.. "Chip Assembly Techniques for Custom IC Design in a Symbolic Virtual-Grid Environment." *Proceedings Conference on Advanced Research in VLSI*. M.I.T.. p. 213-217. January, 1984.

[Szy82] Szymanski. T. . *unpublished memorandum*, 1982.

[Szy83] Szymanski. T.. C. J Van Wyk. "Space Efficient Algorithms for VLSI Artwork Analysis."

Proceedings of the 20th Design Automation Conference, Miami, pp. 734-739. 1983.

[Tarj75] Tarjan. R. , "On the Efficiency of a Good but Non-Linear Set Union Algorithm," *Journal of the ACM*. Vol. 22, 2, (1975).

[Ull84] Ullman. J. D.. *Computational Aspects of VLSI*, Computer Science Press. Rockville MD, 1984.

[West81] Weste, N. , "Virtual Grid Symbolic Layout," *Proceedings of the 18th Design Automation Conference*, Nashville, Tenn.. 1981.

[Wolf85] Wolf, W., "An Experimental Comparison of 1-D Compaction Algorithms." *Chapel Hill Conference on VLSI*, Computer Science Press, Rockville MD, Henry Fuchs. May. 1985.

9

CAD Programming in an

Object Oriented Programming Environment

James Cherry

Symbolics Inc.
Cambridge, Massachusetts

1 Introduction

NS is an integrated design system which unifies a broad spectrum of different IC design tools. NS currently contains facilities for schematic capture, electrical level simulation, switch level simulation, virtual grid symbolic layout with compaction and pitch-matching, automatic standard cell layout generation, floorplanning, and network comparison between the layout and the schematic representations of a design. Designs may be entered either through a graphical editor or via procedural generation. All facilities of the system are accessible through a single graphical editor; they are all driven from a single data base and they may be manipulated through a single procedural interface.

The technique NS uses for integrating large systems is radically different from that employed by more conventional (UNIX-based) CAD systems. The entire NS system has been implemented in an object-oriented extension of LISP (called Flavors) on the Symbolics LISP Machine. NS relies on the use of a large virtual address space in which all procedures and data are available from the time of their creation until they become garbage or the machine is rebooted. The various facilities in the NS system do not communicate by character-stream oriented techniques (that is, files or pipes); rather, procedures communicate by passing objects as arguments. Our central concern in this paper is to explain how such a system facilitates the construction of a highly integrated VLSI design system supporting a broad range of diverse tools.

Integrated is probably the most abused buzz-word in the lexicon of VLSI. The typical integrated CAD system consists of a collection of separate tools, each with its separate data structures and procedures which communicate by generating and parsing a plethora of file formats. In the best of cases, a "user-friendly" front end isolates the user from this by performing the various file conversions behind the user's back. Even in these cases, however, the level of integration achieved is fairly low. A collection of tools that communicate in this manner require a large collection of conversion routines. It is also difficult to force all of the complicated relationships involved in a sophisticated design to flow through too narrow a pipe. Finally, it is virtually impossible for the end users of such systems to extend or customize the system to their own needs.

This paper describes a design system, called NS, which is in fact highly integrated. There is a single graphical editor for manipulating all graphical aspects of a design (for example, schematic, and layouts), a single set of data representations, and a single programming language for manipulating these representations. These uniform interfaces provide a broad spectrum of facilities. The NS graphics editor facilitates the entry of both schematics and layouts. The editor's data structures represent the fact that at various levels of the design hierarchy, a particular layout is meant to be a faithful implementation of a particular schematic. This correspondence is checked by one of the NS verification tools. Simulators can simulate the behavior of both the layout and the schematic at both the electrical level (SPICE) and the switch level (RSIM), displaying the results through the NS graphical interface. Switch level simulation can be merged with and checked against a high-level functional simulation written in the host LISP language. At no time does a user of NS need to know about or be aware of file formats; the user only thinks about the objects in the design.

The ability to achieve this integration comes from the use of a radically different programming style. NS runs on a Symbolics LISP Machine and makes use of the *Flavor* object-oriented extension of LISP [1]. There are four features that make such an environment a much better vehicle for integrating a large, diverse collection of tools into a single, uniform CAD system.

- Object-oriented programming, which allows a diverse collection of objects to exhibit a generic behavior.

- The use of a large and uniform virtual address space, which holds all procedures and data structures.

- Procedures communicate by passing data structures instead of by character streams or files.

- Programs and data within such an environment are long-lived; they remain in the environment as long as they are needed and are garbage collected automatically by the system when they cease to be useful.

Another form of integration that NS exhibits which distinguishes it from more conventional CAD systems is its integration with the surrounding programming environment. All of NS is implemented in a single language (LISP). LISP is directly accessible from NS, since the top level of NS is an extension of the LISP interpreter. End users of the system can employ all of the program development facilities of the LISP programming environment to create procedures which manipulate the NS data base. In most conventional CAD systems there are typically at least two programming languages with which the end user must be familiar: the command language of the host system and the language of the design system. Often, there is yet a third language which is the programming language of the host system. For example, in the MULGA system [2,3] these languages were the UNIX shell language, ICDL, and C, respectively. In contrast, NS provides an extension of LISP embedded in the LISP environment. There is only one language that a user need learn. As with most graphics editors, one can do quite a bit without learning any programming language at all.

Finally, NS integrates its procedural and graphical elements. The graphics editor can, of course, display designs which were generated procedurally. In addition, the graphics editor can be used to create parameterized cell designs which are represented by generator procedures. This idea was employed earlier in the DAEDALUS [4,5] design system.

This paper has two goals: to present the NS system, and to convey an understanding of how NS is implemented. The following section explains the programming methodology used to construct NS and illustrates how this helped to achieve the high level of system integration. The next section outlines the NS core data structures. The following two sections describe the tools used for functional design and physical design in NS.

The intellectual roots of NS lie in several systems: MULGA suggested the techniques of virtual grid symbolic layout, compaction and the pitch-matching style of layout assembly; DAEDALUS provided a style of user-interface and the representations used to support procedural generation; finally SUDS [6] and SCALD [7] greatly influenced its schematic capture and network extraction techniques.

2 The Programming Environment

The features of the surrounding programming environment which make it possible to build a tightly integrated system as extensive as NS are expanded upon in the following sections.

2.1 Flavors

All objects in NS are implemented in *Flavors*, an object-oriented extension to the LISP language. A flavor corresponds roughly to the idea of a Class in Smalltalk or Simula. Each flavor defines a set of *instance variables* which any instance of the flavor will possess. The values of the instance variables are private to each instance of the flavor. In this respect a flavor instance is very similar to the data structures found in other high level programming languages. In addition to instance variables, a flavor also defines a set of *methods*. A method is simply a named LISP function which is run whenever the function with this name is called when the first argument of the function is an instance of the method's flavor.

There are two advantages that methods have over normal LISP functions. The method has fast access to the instance variables of the instance on which it is invoked. The method can treat instance variables as local variables without the necessity of extracting the slots in the instance data structure (destructuring). Because the methods are specific to a flavor, defining methods of the same name for multiple flavors provides an efficient mechanism for dispatching on the instance type. As an example, consider a function to display an object on a window. In COMMON-LISP it might look like:

```
(defun display-object (object window)
   (case (typep object)
      (line (draw-line window
                          (line-x1 object) (line-y1 object)
                          (line-x2 object) (line-y2 object)))
      (circle (draw-circle window
                             (circle-x object) (circle-y object)
                             (circle-radius object)))
      (text (draw-string window
                          (text-string object)
                          (text-x object) (text-y object)
                          (text-font object)))))
```

Notice that this function must know about every type of object that is to be displayed. If a new object type is introduced, the display-object function must be augmented to handle it. As the number of objects increases, so does the time it takes to determine the appropriate action to take. Of course, one can implement a more efficient dispatching mechanism; this is precisely what the flavor system provides. Rewritten in flavors, the display-object function is defined as a collection of methods, one for each type of object.

```
(defmethod (display-object line) (window)
   (draw-line window x1 y1 x2 y2))

(defmethod (display-object circle) (window)
   (draw-circle window x y radius))

(defmethod (display-object text) (window)
   (draw-string window string x y font))
```

When the display-object function is invoked, the method that runs is determined by the type of the first argument in the function call (the object to be displayed). This dispatching is is provided by the function calling mechanism on the LISP Machine using microcoded hash tables so that it is very efficient. The function display-object is termed a *generic function*, because it works for a variety of instance types. Also notice that the instance variables of the flavor are accessible in the method, and need not be extracted from the instance data structure as in the first example.

Methods are usually designed in sets that define a uniform interface to objects that have a particular behavior. Such a set of methods is called a *protocol*. A protocol can be implemented as a flavor that can be combined (mixed in) with other flavors. In using flavors one concentrates on a different set of issues than one does in more conventional programming. One thinks about the set of protocols that the objects in the world must obey, and then concentrates on creating modular flavors to form the basic building blocks of the system. If these two issues are attended to, then coding is often replaced by the simpler effort of combining already existing behaviors to form new and useful compound objects. This technique was used to create a prototype layout editor by two developers in a one week, given that we already had the flavors to implement a schematic editor.

In NS, all objects that are parts of diagrams obey protocols for displaying, highlighting, copying and moving. Each primitive type of object handles these protocols in its own way. The display-object example above illustrated how one part of the display protocol is implemented. The modularity inherent in the message based approach is made clearer when the "window" is allowed to be either a screen or a hardcopy device. As long as the hardcopy device and the window handle the same drawing protocol (using a common abstract unit system), the same display code can be used to display on either device.

As in Simula and Smalltalk, the set of messages handled by a particular flavor consists of those messages handled directly by the flavor plus those handled by any of its component flavors. In contrast to the hierarchical classes of Simula and Smalltalk, a flavor may have more than one component flavor from which it inherits methods. The existence of multiple superclasses leads to a different view of inheritance. One does not think of a flavor as inheriting behavior from its component flavors; rather, one thinks of a flavor as mixing together the behavior of its various components into a larger, more complex behavior. In fact, the natural style of using flavors is to build *mixin* flavors which capture some basic packet of behavior. These *mixins* are then combined to produce more complex objects which exhibit behaviors derived from each of its component flavors.

The advantage of the Flavor system is that it allows an elegant means for combining nearly orthogonal packets of behavior. This is brought about by the ways in which methods from separate component flavors (mixins) are combined at the time of definition of a new flavor. Consider the problem of producing textual descriptions of graphical objects such as lines, wires, and vg-logs (layout sticks). All of these objects are types of lines. However, vg-logs and lines have a width property; vg-logs and wires have signal-name properties. This leads to the flavor inheritance graph shown in figure 1.

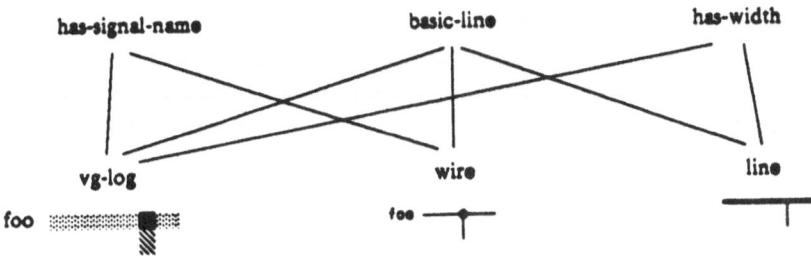

Figure 1.   Flavor Inheritance Graph

The mixin flavors has-signal-name, basic-line, and has-width (on top) are combined as shown by the connecting lines to build the graphical object flavors vg-log, wire and line. This inheritance combination is specified to the Flavor system in the following manner:

```
;; Mixin Flavors
(defflavor basic-line (point1 point2) ())
(defflavor has-width (width) ())
(defflavor has-signal-name (signal-name) ())


;; Graphical Object Flavors
(defflavor vg-log () (basic-line has-signal-name has-width))
(defflavor wire () (basic-line has-signal-name))
(defflavor line () (basic-line has-width))
```

All of these objects are lines which run from one point to another; therefore, they all want their textual description to begin as:

```
(part <type of object> :from <pt1> :to <pt2>
```

We would like to have the has-signal-name mixin insert the signal name component of the textual description, and the has-width mixin insert the width component. The Flavor system facilitates this through the use of *daemon* methods that run before or after the primary method. When two flavors are combined, the flavor system finds all the *:before* and *:after* methods contributed by any of the component flavors. It then builds a *combined method* for the composed flavor, a procedure which first calls all of the :before methods, then calls the primary method, and finally calls each of the :after methods. To obtain the desired behavior for the :text-form method, the basic-line mixin should define the primary method that prints the type of object and the two end point locations. The has-signal-name and has-width flavors define *:after* methods for the :text-form message. The *:after* method associated with has-signal-name, prints the signal-name and the *:after* method for has-width, prints the width. The LISP code for these :text-form methods is defined as follows.

```
;; primary method
(defmethod (basic-line :text-form) (stream)
  <code to print the word "part", the type of the object, and the two points>)

;; after daemon method
(defmethod (has-signal-name :after :text-form) (stream)
   <code to print the signal name on the stream>)

;; after daemon method
(defmethod (has-width :after :text-form) (stream)
  <code to print the width on the stream>)
```

2.2 A Uniform Large Scale Persistent Virtual Memory

Programming in the LISP environment bears another set of distinctions from more conventional programming. All procedures and data live within a single virtual memory which continues to exist for long periods of time.

In conventional programming environments, various procedures are compiled and linked and then are loaded into the system as a *job* (a job is a process allocated to run within its own private memory, separate and inaccessible from the memory of other jobs). Jobs run for a while and then terminate. Jobs are also typically subjected to arbitrary space limitations and compete with one another for resources of the machine. Finally, data within one job is, in principle, inaccessible to other jobs. To date, the most creative ideas on how to make the best of this situation have been found in the use of pipes in UNIX which at least provides a uniform means for jobs to communicate data. However, pipes are still remarkably limiting when they are used to convey large amounts of complex information between programs. The sequential character stream nature of pipes makes them a poor vehicle for communicating the networks of relationships between objects which are the natural representations found within CAD systems. In addition, when one wants to change an existing procedure or add a new procedure to an existing job one's only recourse is to kill the program, recompile, relink and start over.

In the LISP machine environment things are quite different. Objects and procedures (procedures are just one distinguished type of object) exist within a single virtual address space. One procedure can call another at will, passing references to whatever objects seem appropriate; since all the objects live within the same environment, there is no question of whether the access is possible. Since the procedures live within the same environment, there is no need to transform the objects into a form suitable for transmission through a pipe. The object together with all its rich interconnections to the rest of the world is simply passed onto the callee. The called procedure may in turn follow some of the references in the objects passed to it or it may modify the object. Finally, since this interaction never involves searching a large data base stored on an external medium, the programmer is not plagued by the worries of inefficiency which haunt those who try to obtain uniformity through use of relational data bases.

2.3 Persistence and Sharing

In the section on flavors, we emphasized the ability to build and share packets of behavior. The uniformity of our environment also contributes to this style of reusing software. Since procedures are persistent (that is, they stay in the environment) and the virtual address space is large and uniform, it becomes possible to provide within the system many facilities which have general utility. The construction of NS did not require a significant investment in basic user interface facilities like menus and windows simply because these are already part of the basic system. In those parts of our system which require attention to algorithmic complexity such as the layout extractor, we did not have to build our own hash-tables, priority queues (heaps) or Union-find algorithm because these are provided as existing flavors within the system. Furthermore, each application tool within the NS system does not need its own copy of these facilities; the uniformity of the environment lets them be shared. Finally, since the address space is large (28 bit word addresses) one does not have to worry about shoe-horning everything into a small place. Bit-twiddling plays a much smaller role in our style of programming.

2.4 Dynamic Linking and Garbage Collection

Procedures in our system are recompiled from within the environment. The editor and compiler exist in the same virtual address space as other system and user facilities. Whenever a procedure is recompiled, it is linked into the environment immediately and automatically; running programs call the new version of the procedure rather than the old. This dynamic linking means that program development can proceed at a much faster pace. When the system is observed to be behaving incorrectly, the developer simply jumps into the editor, fixes the offending procedures and is off and running again. Furthermore, data which stimulated the offending behavior is still around and can be used to test the change immediately. Contrast this with the more conventional approach of killing the original job, editing the source code in an environment removed from the one in which the problem was found, compiling, linking and loading and then trying to recreate the test case.

Procedures and data which are no longer accessible to anyone (the old version of a procedure which has been replaced, for example) are considered garbage. The system is responsible for periodically reclaiming all such garbage and making the reclaimed space available for new allocation. Programmers never reclaim space and never worry about deallocation. There is never the problem of incorrectly freeing space which is still in use. The garbage collector runs in its own process and does not interrupt the normal use of the machine.

3 The Organization of NS

A system design consists of a set of descriptions covering many different aspects of the design, such as the logical, electrical, functional, or physical structure. NS is an integrated design system that captures the entire description and maintains links between each of these aspects in one data base.

Some of the aspects of a design are diagrams (such as a logic schematic or mask artwork), some are textual (such as documentation or simulation programs), and some are generated by the design tools (such as mask data produced by the compactor). The aspects that NS currently supports are:

- Schematic

- Schematic Icon

- Virtual Grid Symbolic Layout

- Mask Layout

- Floorplan

- Documentation

- Electrical Network

Schematic and schematic icon aspects are for logical design. The virtual grid aspect is for design rule independent layout that is "compiled" into mask geometry placed in the mask aspect. The floorplan aspect is to plan a design physically in a top down fashion and as a specification for the automatic composition of a design into a complete chip.

A collection of aspects that represent the same subsystem is grouped together into a *module*. For example, a CMOS inverter module has schematic, schematic icon, and virtual grid aspects as shown in Figure 2. Each of the aspects of a module are different representations that should all be consistent if they exist. A collection of modules forms a *library*. Each module resides in a single library. Libraries can be used to partition a large design, or for sharing of common functionality between subsystems.

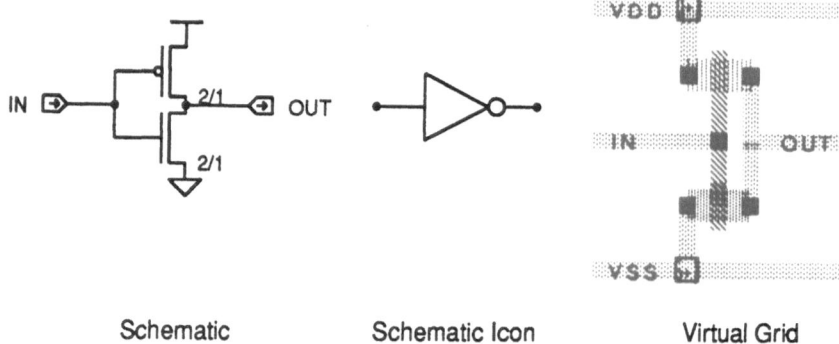

| Schematic | Schematic Icon | Virtual Grid |

Figure 2.   Aspects of a CMOS inverter

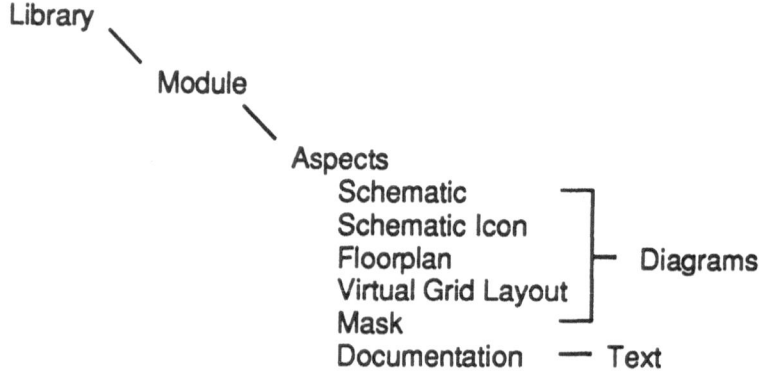

Figure 3.   Design Database Hierarchy

The user is encouraged to keep the hierarchies for the schematic and layout aspects congruent, but this is not a hard and fast rule. For example, the symbolic layout aspect for a module may have a different subtree than the logic schematic if physical constraints require the cell to be broken up differently than the logic. In this case, the logic sub-modules will not have layout aspects, and the hierarchies will split at the containing module. NS can check to make sure that the schematic and layout for a module represent the same electrical network. The advantage of making the layout and schematic hierarchies congruent is that inconsistencies between the layout and schematic are limited to the composition of the modules if each sub-module is itself consistent.

Aspects of modules may be parameterized. For example, the schematic icon aspect of a NAND gate can be parameterized to control whether it displays as an AND gate with bubble on the output or as an OR gate with bubbles on the input (its DeMorgan equivalent). A collection of variant aspects is held in a data index for fast retrieval of the appropriate aspect given a set of parameters. In cases like the NAND gate's schematic-icon, where there is little commonality between the two alternate icons, the variant aspects can be entered as separate diagrams through the graphical editor; these separate diagrams are then stored in individual permanent data structures. However, in many other cases, such as the layout of a decoder there is enough commonality to merit the use of a procedure to generate the layouts on demand. Once an aspect is generated it is stored in the aspect data index, future requests for an aspect with the same set of parameters will retrieve this stored aspect.

Most aspects are *diagrams* such as layouts and schematics. A diagram contains a set of parts which may be primitives (for example, points, lines, and text) or instances of other diagrams called *diagram-instances* (which correspond to calls in CIF). A diagram-instance contains a geometric transform, and a pointer to a diagram.

Windows that display a scaled portion of a diagram are called *views*. To facilitate fast retrieval of objects based upon geometric criteria and windowed redisplay, diagrams store their parts in quad trees [8].

Although there are many types of diagrams in NS, there is a single diagram editor. This makes it possible for the user to move freely between different aspects of the design hierarchy as well as up and down the hierarchy. It also means that the user must learn only one command interface to edit all types of diagrams. For each different type of diagram there is a corresponding mode that customizes the available commands and primitive objects appropriate to the diagram type. For example, while editing schematics the WIRE command is used to start drawing a wire. While editing a layout the same command starts drawing a layout VG-LOG (stick).

The diagram editor is organized as a basic editor that is used to draw, display, and manipulate objects. The basic graphic primitive objects are lines, points, circles, arcs, and text. These basic primitive objects have specialized behavior in the various domains. For example, the graphic primitive BASIC-LINE is an object that has pointers to its end points and knows how to stretch when its end points are moved. Schematic WIRES are BASIC-LINES that can have signal names. Layout VG-LOGS are BASIC-LINES that have a layer and a signal name and display as a stipple on a monochrome screen, or in the layer's color on the color screen. This layered approach to implementing the domain-specific primitives makes it a simple matter to implement new diagram and primitive types.

3.1 Electrical Networks

There is a single representation for electrical networks used in NS. The circuit and switch level simulators, and network comparison program supported by NS are all driven from this one representation.

For each electrically meaningful type of diagram (schematics, virtual-grid layouts, mask) there is an *extractor* which produces an electrical network corresponding to the diagram. The network is annotated with pointers to the objects in the diagram and similarly the objects in the diagram are annotated with pointers to nodes or devices in the network. These annotations allow a user to interact with the verification tools in a very natural way. Because the objects in the diagrams know which nodes in the network they correspond to, the user may tell NS to plot a node from a simulation simply by pointing at a wire in the schematic, or a location in a layout. The annotations are used to highlight nodes to a user. For example, when the network comparison program finds a node in a schematic which has no match in the virtual-grid layout, the problem is shown to the user by highlighting the unmatched node on the window containing the schematic.

Networks are composed of *nodes, devices* and *device-terminals*. A device corresponds to an electrical primitive appropriate to a target tool, such as MOSFETS in circuit level simulation. A device has a set of named terminals that connect it to nodes. A node connects together device-terminals that have the same electrical potential. Figure 4 illustrates the basic network topology. Additional data structure slots are added depending on the type of diagram being extracted to aid in mapping locations in the diagram to nodes and visa versa. Additional slots are also added for specific tools that use the network, such as the various simulators in the NS environment. A property list

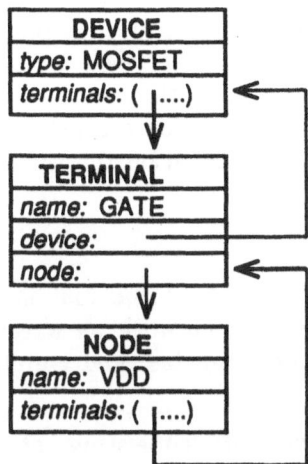

Figure 4. Network Topology

is included in all network structures so that arbitrary information can be added to them during development of network tools.

4 Design Verification in NS

In this section we will review the major subsystems that are used in the verification of a design which may be at the schematic level, layout level or both.

4.1 Network Comparison

Network comparison is used to verify that two electrical networks are identical. This is useful for checking the consistency of a module's layout with its schematic, or alternatively comparing a new implementation of a module with a known correctly constructed version.

To perform the comparison, both representations of the circuit are converted to networks, which are a flat representation of the circuit containing only nodes and transistors. The networks can be viewed as a particular type of graph. To verify that the circuits are identical, the network comparator must establish that the graphs are isomorphic.

Many variations on a graph partitioning algorithm have been described for performing connectivity comparisons. The graph isomorphism algorithm used in NS is based on the algorithms described in [9,10,11]. If there are differences in the networks, the algorithm marks a set of nodes that it regards as suspicious. Suspicious nodes are presented to the user in a graphical fashion by highlighting them on the screen.

4.2 Switch Simulation: RSIM

For chip level simulation of MOS designs, the basic tool is a LISP version of RSIM [12] which models a transistor network as a resistor-divider network, provides acceptable timing estimates, and can handle charge-sharing.

There is a procedural interface to RSIM consisting of three functions: value, to find the value of a signal or bus, set-value to cause the next simulation step to force a signal or bus to the given value, and sim-step to settle the network in response to changes in input values and then return the time it took to settle the network. Using this interface, the designer can write LISP code to generate test cases, simulate them and check that the results are as expected. The graphical interface to RSIM which is part of the NS editor provides the ability to probe layout or schematic diagrams to set and measure signals (logic and timing). As an example of such a test program, the following function simulates a 32-bit boolean function and compares the result to the expected result.

```
;; individual boolean operation test
(defun boole-test (func a b)
  (set-value 'func func)              ;set function code
  (set-value 'a a)                    ;set first input
  (set-value 'b b)                    ;set second input
  (sim-step)                          ;step simulator
  (let ((expected-value (boole func a b))  ;the right ans
    (rsim-value (value 'output)))     ;get simulated ans
    (= EXPECTED-value rsim-value))))  ;return T if OK
```

RSIM also provides a hook which lets a user-specified function be called whenever a node's value changes; the arguments are the node, the simulated time, the old value and the new value. This hook has been used to implement:

- A software "logic analyzer" which can trigger on certain events, and when triggered keep bounded or unbounded (except by virtual memory) traces of all transitions happening in the simulation. It later displays waveform traces of either all or selected signals during the whole history or a specified interval.

- A facility which keeps an event trace of nodes that change during a simulation step. The trace corresponding to the longest simulation step contains the "worst delay" path invoked by that particular set of simulation events. The nodes on the worst case path can be displayed on the layout or schematic, and frequently help in determining the critical path of a circuit.

FUNCTIONAL RSIM is a generalization of RSIM that supports mixed mode simulation at the MOSFET and functional level. FUNCTIONAL RSIM is useful for modeling the behavior of simple but large structures such as ROMS and RAMS, or a complex function prior to a gate or transistor level implementation. When a functional model is defined in the design environment for a module it is substituted during extraction of the schematic instead of expanding the hierarchy down to the MOSFET level.

The following example is the functional model for a 1K by 32 bit RAM array. The delay to data-out is specified as 40 ns, with a 7 ns rise time and 5 ns fall time.

```
(deffunctional-model 1Kx32-RAM
        (:inputs (address<9:0> write-data<31:0> write-enable)
         :outputs ((data-out<31:0> 40 7 5))
         :local-state ((ram-array (make-array '(1000)))))
    (if write-enable
        (setf (aref ram-array address) write-data)
        (setq data-out (aref ram-array address))))
```

Functional models are written in LISP. The model writer has the full power of a complete programming language at ones disposal. Incremental compilation of functional models is supported as with any other function in the LISP Machine environment. When the designer finds an error in the functional model one can correct and recompile it in a few keystrokes, and then resume or restart the simulation with the corrected model.

4.3 Circuit Simulation: SPICE

To provide electrical level simulation facilities, NS also contains an interface to SPICE [13]. SPICE can either be run on a remote machine via a network connection or it can be run locally on a LISP Machine (a FORTRAN compiler is available). SPICE has been modified to output a complete trace of the value of each node at each time step. NS collects this data for graphical plotting. NS maintains a mapping between the objects in the original schematic and their assigned SPICE node numbers. This makes it possible for the user to specify which nodes are to be plotted by pointing directly at them in the schematic. When the schematic is hierarchical the internal nodes of modules can be graphically probed by "pushing" down into the hierarchy to a lower level schematic.

5 Physical Design in NS

Physical design in NS is based on the use of the virtual grid symbolic layout methodology [2,3]. In this methodology, layout is carried out at the circuit level by placing transistors and wires on a grid that conveys relative placement information. Mask geometry is created by compacting the symbolic layout, not by the designer editing mask layers. Because the layout is specified symbolically, it is easily targeted to a variety of similar IC technologies with different spacing and width design rules. This ability is key to taking advantage of improvements in design rules as a technology matures. It also allows the design to progress without commitment to a manufacturing vendor.

5.1 Virtual Grid Compactor

Layouts are created by editing symbolic layout objects. For a CMOS process, the primitives are N and P type transistors, vg-logs (sticks), inter-layer contacts, and well contacts. These objects are placed on a virtual grid. The width of vg-logs and transistors are specified in multiples of the minimum dimension rather than in microns.

The grid upon which the objects are drawn is symbolic. The grid represents only relative placement, not physical spacing. If one object is drawn on a grid to the left of another, then this relative placement will be preserved in the final artwork. However, there is no significance to empty grid lines. It is the job of the *compactor* to turn the virtual grid into a physical one by spacing the grid lines far enough apart so that all layout design rules are satisfied. The output of the compactor is a *mask diagram*, the physical geometry corresponding to the virtual-grid layout. The user manipulates only the virtual-grid layout, remaining unconcerned with spacing geometry to satisfy design rules. The current compactor guarantees that objects drawn on the same grid line will remain aligned. We have also experimented with various forms of "grid sliding" versions of the virtual grid compactor.

The virtual grid compaction strategy is two-tiered. Leaf cells are compacted into mask diagrams. However, it is important that the assembly of leaf cells into larger modules can still proceed in the virtual-grid framework. In particular, it is important to guarantee that connectivity expressed in the virtual-grid framework be maintained in the physical mask diagram. Consider the two abutting cells shown in figure 5. The horizontal bus wires that run through both cells. If the virtual-grid layout of two cells are compacted separately they will be compressed to different sizes and the ports will not line up in the mask diagram. If the entire diagram were treated as a single compaction problem, this problem would not arise. However, in that case, aligned virtual grids from different cells which had no port in common would still be treated as a single grid, forcing unrelated objects from different cells to line up. This would result in an inferior compaction. The solution to this problem is the second tier of the compaction system, the *pitch-matcher* whose job it is to stretch the compacted leaf cells so that connection points do align in the mask diagram. This style of symbolic layout is derived from that used in the MULGA design system but the algorithms have been redesigned.

Before compaction begins the virtual grid layout is extracted to determine electrical connectivity. This allows the compactor to know that certain objects belong to the same electrical node and therefore are not subject to spacing constraints. Next, the mask rectangles for each symbolic layout object are generated with their edges expressed as offsets from a virtual grid location. Each layout primitive can generate the necessary mask rectangles, using the design rule database. The final step is to compact this set of rectangles.

The compaction algorithm implemented in NS is a left to right scan algorithm which maintains a per-layer cache of the active spacing constraints (a *fence*). As the left edge of a rectangle is scanned, its spacing from the fence is determined. The spacing between two virtual grids is the maximum of all the implied constraints. When the right edge of the rectangle is encountered, the fence is updated to include it. After left to right compaction, a bottom to top compaction is performed, taking into account diagonal interactions. A command is provided that graphically displays the horizontal and vertical constraints between virtual grids for tuning the layout for minimum size.

Supports provide a mechanism for the designer to explicitly set the minimum spacing between two virtual grids. This facility is particularly useful for designing structures where the normal process design rules do not usually apply or one wishes to "space" geometry rather than compact it. The support facility allows such structures as bonding pads and guard rings to be expressed symbolically. With supports an entire design can be described in terms of virtual grid symbolic components. The compactor deals with supports by checking to see if any supports would impose a greater spacing on the current virtual grid just after it has spaced that virtual grid against the fence.

The pitch matcher begins by flattening the hierarchical layout into non-overlapping *leaf cells* that contain primitive objects. The virtual grid layout of each of these leaf cells is then compacted. The pitch matcher does an x sweep followed by a y sweep of the global virtual grids. At each global virtual grid the pitch matcher has to determine the physical grid location of the leaf cells that intersect this virtual grid. *Pitch match points* are locations on the border leaf cells where vg-logs communicate to neighboring leaf cells. Pitch match points are kept aligned in the physical domain by grouping all cells that are locked together by a pitch match point at the current virtual grid into *gangs*. All the cells in a gang have their physical grids updated to the maximum physical grid found within the gang.

Many symbolic design systems allow compaction of primitive cells and achieve composition by river routing between connection points of adjacent cell edges. This methodology is very inefficient for regular array structures such as ROMs, RAMs, and PLAs. A key feature of the virtual grid methodology is that adjacent cells truly abut.

Abutting leaf cells can introduce design rule constraints between the interior rectangles of the abutting cells. One solution to this problem is to space all non-connection geometry from the cell edge by one half of the maximum design rule distance. Unfortunately, this approach is rather pessimistic. Designs that contain large numbers of small cells are heavily penalized. The block pitch matcher which was subsequently developed solves some of these problems.

The block pitch matcher uses the same basic pitch matching algorithm but follows this with a global compaction. This eliminates the need for half design rule spacings. To do this, the block pitch matcher needs to consider only inter-cell interactions, not just those dealing with boundary rectangles. The global compaction is done using a variant of the standard compactor.

The block compaction phase involves alternate x and y sweeps across all global virtual grids. At each virtual grid all rectangles with minimum edges (left or bottom) at that virtual grid are compared to the fence. The result of the comparison is a spacing between that edge of the rectangle and the rectangle in the fence. This spacing is recorded on the virtual grid of the leaf cell which contains the rectangle. Block compaction needs to consider the interaction of rectangles in different cells. The rectangle-rectangle spacings within the cell have already been dealt with by isolated compactions.

The block compaction algorithm is very similar to that of the normal compactor. The major difference is that in the normal compactor constraints are recorded on a per virtual grid basis for the whole cell. In the block compactor the constraints are recorded on a per virtual grid basis on the leaf cell that contains the non-fence rectangle. In the normal compactor all spacings for a virtual grid are grouped together to calculate the physical grid for the current virtual grid. In the block compactor there is no single physical grid for the current virtual grid. Each leaf cell calculates its own placement for the current virtual grid based on the spacings that have been recorded on it.

In order for the pitch match points to remain aligned, all leaf cells that intersect the current virtual grid are grouped into gangs that are locked by pitch match points on this virtual grid. All the cells in a gang have their physical grids updated to the maximum physical grid found within the gang.

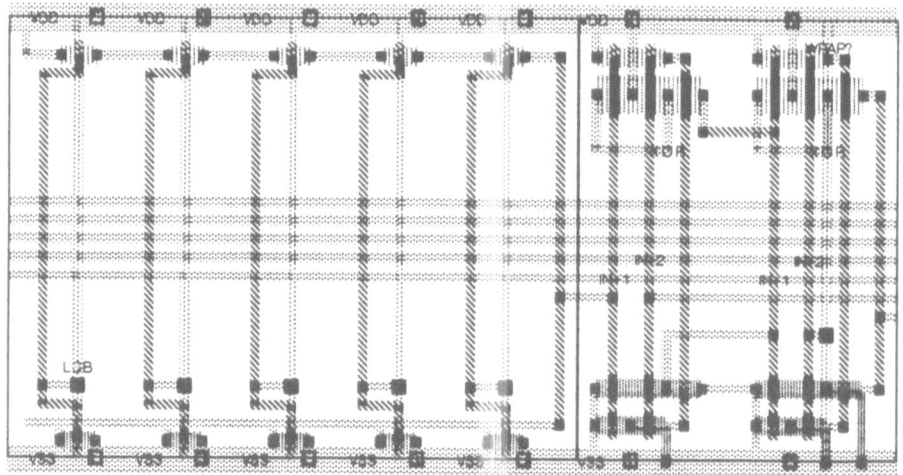

Figure 5.   Symbolic Layout for MASK-GREATERP-STAGE and MASK-XOR3

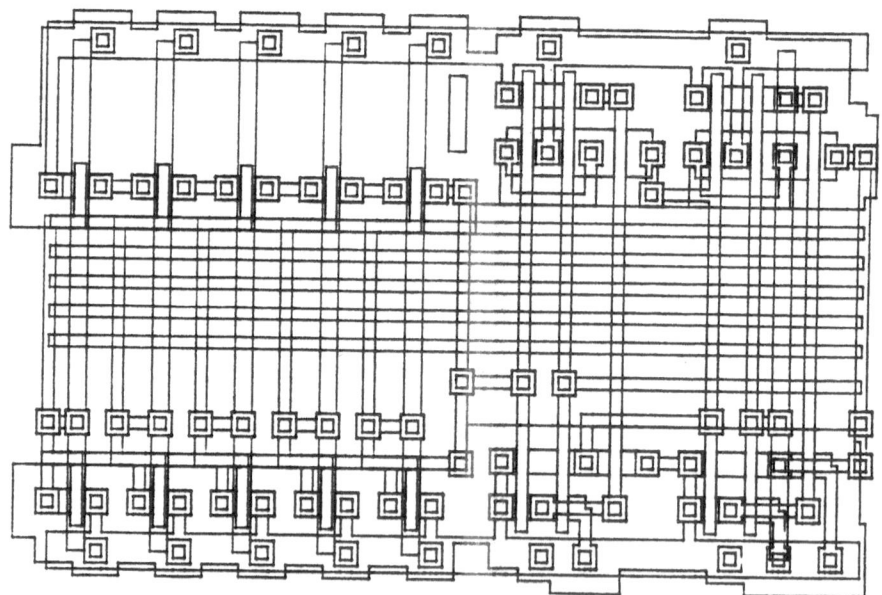

Figure 6.   Mask for MASK-GREATERP-STAGE and MASK-XOR3

Each sweep of the block compactor may alter the physical grids of a cell in the direction of the sweep. However if the virtual grid in an adjacent cell does not move by the same amount it is possible to create spacing problems in the direction opposite to the sweep. (See Fig 4) To ensure that no spacing problems are left unnoticed, the block compactor performs alternate x and y sweeps until no new spacing constraints are discovered. In most cases three sweeps are required to converge.

Figure 5 shows two abutting symbolic layouts (MASK-GREATERP-STAGE on the left and MASK-XOR3 on the right). Note that an isolated instance of the MASK-GREATERP-STAGE layout will compact to a shorter cell than MASK-XOR3 because its transistors are not stacked. The compacted and pitch-matched mask diagram for these two cells is shown in figure 6.

5.2 Floor Planning and Composition

Pitch matching is used to compose layouts that communicate by abutting ports. These cells form the major subsystems of a chip, such as RAM, ROM, or data paths. To interconnect these pitch matched blocks NS provides a set of composition tools. Placement is guided by a floorplanner which allows the relative topology of the blocks to be expressed in a floor plan diagram. Connectivity between blocks is specified by the module's schematic. The composition system uses these specifications to construct the final mask of the chip by routing the mask blocks together. Since relative rather than exact placements are specified, the composition system can guarantee one hundred percent signal routability. Power and ground signals are routed on a single layer. To aid in the layout planning phase of chip design approximate block dimensions and port locations may be specified before detailed layout has been completed. This style of chip assembly is similar to that found in the Sprint system [14].

The placement of cells in a floor plan is specified by first adding them to the floor plan by invoking a command that adds a mask outline instance (with port locations) for each icon in the module's schematic. These outlines are then arranged by moving them in the floor plan.

The composition sequence is specified by recursively grouping horizontal rows or vertical columns of blocks. Groups are specified by drawing *dividers* in the floor plan. This is a simple extension of the

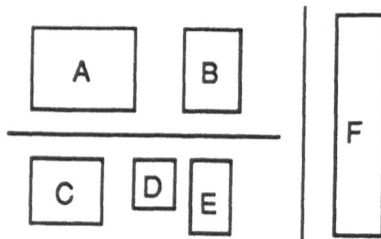

Figure 7.   Floor Plan Example

binary composition methodology (where each group composes exactly two blocks) used in Sprint.

For example, the floor plan in figure 7 specifies the following composition:

- Horizontally compose blocks A and B to form a new macro block (AB)

- Horizontally compose blocks C, D and E to form a new macro block (CDE)

- Vertically compose blocks AB and CDE to form AB-CDE

- Horizontally compose blocks AB-CDE and F

These steps are illustrated in figure 8.

Blocks are constrained to have a power bus connection on their top left corner and a ground bus connection on their bottom right corner. This constraint and the recursive composition ordering are sufficient to insure that power and ground can be routed to all blocks on a single metal layer. All of the blocks in a group share common power and ground busses. Power is routed to the cells in a horizontal (vertical) group by connecting their power bus to the group's power bus on the top (left) of the group. Ground is connected similarly, as shown in figure 8.

A channel for routing signals is inserted between each block in the group. Routing in this channel is used to connect signals between the edges abutting the channel and also to the edge of the stack.

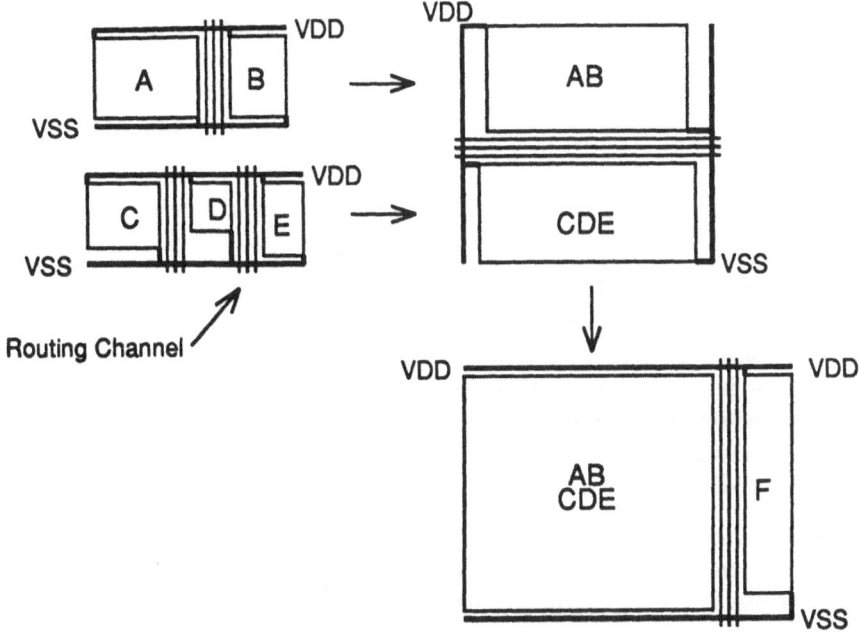

Figure 8.   Composition Example

Ports on the edges of blocks that need to connect to other blocks are extended with wire stubs across power and ground routes to the edge of the stack. Thus, the result of composing a group is a new block that has ports on its edges, and power and ground connections in the required locations.

A global routing phase determines what channels to route a signal net on before composition actually takes place. Global net assignment is is done by finding the shortest path for a net (Steiner tree) that touches all channels that have port connections to the net. Channels are routed by a simple Z channel router first. If this router fails because of cyclic dependencies, a greedy channel router is invoked.

Notice that routing channels only intersect in "T" configurations. Never having four channels intersect in a "+" configuration eliminates the need for switch box routers. This is an important advantage of this composition methodology because switch box routers cannot be guaranteed to route all nets successfully.

Composition is done by routing instances mask diagrams (generated by the pitch matcher). Virtual grid layout is not used for composition to avoid the the necessity of pitch matching an entire chip. Since the geometry used in composition is fairly simple (channel routing), there is not a lot to be gained by using the virtual grid methodology.

5.3 Standard Cell Layout Generation

ANNEAL is a facility for automatic generation of standard cell layouts from schematics. Standard cells are a convenient layout methodology for implementing control logic because it lacks structure, and hence is tedious and time consuming to layout by hand. ANNEAL is modeled after the TimberWolf3.2 package developed at Berkeley [15]. It uses simulated annealing to determine optimal placement of standard cells.

The standard cell library used by ANNEAL is captured symbolically using the virtual grid methodology. Thus, the investment in standard cell layouts is not lost when a new technology is chosen for the design.

Standard cell schematics can be drawn using the icons in the standard cell library or compiled from logic equations. We have found this capability particularly effective for generating control logic, since this logic is typically the most volatile part of a design. Estimates of control logic area are easily obtained without time consuming hand layouts.

Logic equations are defined in a simple declarative language modeled after an optimizing PAL compiler developed at Symbolics. A simple example of the syntax used is shown below.

```
(def-std-cell-schematic (control :inputs (select<1:0> A B ph1)
                                 :outputs (select=0 select=1
                                          selected-AB
                                          backup-select<1:0>))
    (setq select=0 (bus= select<1:0> 0))
    (setq select=1 (bus= select<1:0> 1))
    (setq selected-AB (mux2 A B select<1>))
    (setq backup-select<0> (d-reg select<0> ph1))
    (setq backup-select<1> (d-reg select<1> ph1)))
```

5.4 Procedural Generation of Layouts and Assembly of Modules

Procedural generation of layout diagrams and procedural assembly of layouts into larger layouts have been found useful in raising the abstraction level of design. NS incorporates a tool for supporting this style of design. Procedural generation of layout diagrams is an effective way to provide *parameterized diagrams*. Diagrams are usually parameterized for one of three reasons: Transistor sizing, flexibility of connection, and the programming of decoding logic. Below we show the generator function for a layout in one of our chips. This is an example of the decoding style of parameterization. The generator takes a single parameter (called bit). The diagram produced by the generator takes five input signals and produces an output which indicates whether the five bit input number is greater than the parameter passed to the generator. There are thirty two separate parameterizations of this layout; were it not for procedural generation, the designer would be forced to draw 32 layouts and thirty two schematics.

```
(defaspect-generator (mask-greaterp-stage :virtual-grid) (bit)
  (let ((spacing 4))
    ;; mark the LSB
    (part text :string "LSB" :height 12 :center (pt (// spacing 2) 5))
    ;; input vdd for pullups
    (part vg-log :from (pt 0 24) :to (pt 0 22) :layer 'metal)
    (part vg-log :from (pt 0 24) :to (pt 16 24) :layer 'metal)
    (part vg-log :from (pt 0 22) :to (pt 16 22) :layer 'metal)
    (loop with first-input = (how-many-low-bits-on bit)
          for i from 0 to 4
          for x = (* i spacing)
          do (if (< i first-input)
                 (part mask-greaterp-dummy :bottom-left (pt x 0))
                 (if (bit-on? bit i)
                     (part mask-greaterp-series :bottom-left (pt x 0) )
                     (part mask-greaterp-parallel :bottom-left (pt x 0))))
          finally (part mask-greaterp-final :bottom-left (pt (* i spacing) 0)))))
```

Notice that the generator code is LISP, extended by the NS function part, which constructs NS objects. The dump format for saving diagrams in the file system is also a LISP procedure, leaving no real distinction between the way generators are written and the way in which diagram themselves are described. Often a generator is written simply by modifying the code written by NS as the dump format of a non-generated diagram.

This style of parameterized layout definition was pioneered in DPL [4,5]. However, in that context the parameterization was very difficult because of the need to deal with physical dimensions and design rules. The use of the virtual grid relieves the designer of this overhead by providing an intermediate abstraction level. This simplification relieves the writer of procedural cells from incorporating the many constants and inter-layer relationships of the process in the generator. This makes it far easier to build technology independent generators.

6 History and Results

The entire NS system contains about 50,000 lines of LISP code (this does not include the FORTRAN code in SPICE, but does include the interface to SPICE). The development of NS began in the summer of 1983. The cumulative effort needed to develop the system (up until June 1986) has been about ten person-years.

NS has been used so far to design a thirty two bit (plus tag bits) data path chip, suitable for use in LISP processing applications. The chip has been designed using NS as the sole design vehicle. The chip contains a thirty two bit lookahead adder, barrel shifter, field masker, boolean function unit, interfaces to external busses, internal registers, and control logic. It contains about 24,000 transistors and is implemented in CMOS. The chip was designed primarily by three people in about nine man-months. NS is currently being used as the sole tool for the design of full custom VLSI LISP machines.

References

[1] Reference Guide to Symbolics-Lisp, available from Symbolics, Inc., 1985, Chapter 10.

[2] B. Ackland and N. Weste, "An Automatic Assembly Tool for Virtual Grid Symbolic Layout," VLSI83, F. Anceau and E.J. Aas (eds.) Elsevier Science Publishers B.V. (North-Holland), Aug. 1983, Trondheim, pp. 457-466.

[3] Neil Weste and Bryan Ackland, "A Pragmatic Approach to Toplogical Symbolic IC Design," VLSI 81, John Gray (ed), Academic Press, 1981.

[4] John Batali and Anne Hartheimer, "The Design Procedure Language Manual," Massachusetts Institute of Technology, AI memo 598, Sept., 1980.

[5] J. Batali, N. Mayle, H. Shrobe, G. Sussman, and D. Weise, "The DPL/Daedalus Design Environment," VLSI81, John Gray (ed), Academic Press, Edinburgh, 1981, pp. 183-192.

[6] R. Helliwell, "The Suds Reference Manual," unpublished but available at DEC, Stanford and MIT, among others.

[7] T. McWilliams and L.C. Widdoes, "SCALD: Structered Computer-Aided Logic Design," *Proceedings of the Fifteenth Design Automation Conference*, Las Vegas, Nev., June 1978, 278-284.

[8] Jonathan Rosenberg, "Geographical Data Structures Compared: A Study of Data Structures Supporting Region Queries," *IEEE Transactions on Computer-Aided Design*, Vol. CAD-4, No. 1, Jan. 1985.

[9] C. Ebeling and O. Zajicek, "Validating VLSI Circuit Layout by Wirelist Comparison," *Proceedings of IEEE International Conference on CAD*, Sept. 1983, pp. 172-173.

[10] R. L. Spickelmier and A. R. Newton, "Wombat: A New Netlist Comparison Program," Proc. *IEEE Int. Conf. on CAD*, Sept. 1983, pp. 170-171.

[11] Makoto Takashima, Takashi Mitsuhashi, Toshiaki Chiba, and Kenji Yoshida, "Programs For Verifying Circuit Connectivity Of MOS/LSI Mask Artwork," *Proceedings of the 19th Design Automation Conference*, 1982, pp. 544.

[12] Christopher Terman, "Simulation Tools for Digital LSI Design," MIT/LCS/TR-304, MIT Laboratory for Computer Science, Cambridge Mass., Sept. 1983.

[13] Nagel, L. W., "SPICE2, A computer Program to Simulate Semiconductor Circuits", ERL-M520, University of California at Berkeley, May, 1975.

[14] Wardle et al., "A Declarative Design Approach for Combining Macrocells by Directed Placement and Constructive Routing,"

Proceedings of the 21st Design Automation Conference, Albuquerque New Mex., June 25-27 1984, pp. 594-601.

[15] C. Sechen and A Sangiovanni-Vincentelli, "TimberWolf3.2: A New Standard Cell Placement and Global Routing Package," *Proceedings of the 1986 Design Automation Conference*, Las Vegas, Nev., June 29-July 2, 1986.

10

Trends in Commercial VLSI Microprocessor Design

Nick Tredennick

IBM Thomas J. Watson Research Center
P.O. Box 218
Yorktown Heights, NY 10598
(914) 945-1422

Abstract: Here is how commercial VLSI microprocessors have been designed, starting with the Motorola MC68000 (1977-79). I impart the design method and say what software tools were used. Here is how the chips, methods, and tools have evolved. I project the trends. In 1977, logic design was done with pencil and paper. Design verification consisted of programs simulating small sections of a chip and of TTL breadboards. Today (1986), logic design is still done with pencil and paper, but it is entered into a computer using an ordinary text editor. Design verification programs then check this text file. In the next 5 to 10 years, logic design will still be done with pencil and paper, but it will be entered into the computer with a specialized editor. Design verification will stay the same. I don't think the level of automation will increase significantly for commercial VLSI microprocessor design. The design tools will run faster because the computers they run on will be faster. There will be more computers. So there will be more instances of the use of design tools. But for commercial microprocessors, I don't think logic design tools will do substantially more than they do today. The opinions expressed are solely those of the author. These opinions do not reflect positions held by the IBM Corporation.

INTRODUCTION

This is about logic design and logic design tools for commercial VLSI microprocessors.   ("Commercial" means intended for the high-volume, commodity market.   "Microprocessor" implies single-chip.) This is a "view from the trenches" of commercial VLSI microprocessor design. I worked at Motorola (1977-1979) where I did the microcode and logic design of the MC68000 microprocessor. I went to work at the IBM Thomas J. Watson Research Center (1979-present), where I did the microcode and the logic design for the IBM Micro/370 microprocessor [Ong-86]. I will cover Motorola and IBM designs in chronological order. I tell something about: design method, verification, circuit design, layout, fabrication, initial test, manufacturing test, project staff, computers used, design tools, and problems.

I do an overview of each project in chronological order so I can point to trends in design method, design tools, and project organization. I project the trends a few years and I draw conclusions. I use the Motorola MC68000, MC68010, and MC68020 microprocessors as historical examples. I talk in detail about current practice using the Flowchart Method to design the IBM Micro/370 microprocessor. I mention current practice on the Motorola MC68030.

HISTORICAL DESIGN PRACTICE

Motorola MC68000

The MC68000 was Tom Gunter's idea. The project began (more than 1 person) in the first quarter of 1977. When I joined the project in the third quarter, there were 7 of us. The project was called "MACS," for Motorola Advanced Computer System. Tom was managing the project. Skip Stritter and David Leitch were defining the instruction set and writing the user's manual. Doyle McAlister and Richard Crisp were designing circuits. I started designing logic and microcode.   Paul Lee, temporary hire, was evaluating software performance. Our project didn't have a computer. No one had a terminal or personal computer.

Architecture: The architecture (instruction set, registers, interrupts, etc.) for the microprocessor was unknown. Defining it was part of the project. The architecture evolved considerably. The original definition had a 24-bit program counter, 8-bit condition code register, 8 32-bit address registers, and 8 16-bit data registers. The original definition evolved into an architecture with a 32-bit program counter, 16-bit condition code register, 9 32-bit address registers (counting the two stack pointers at A7), and 8 32-bit data registers. Addresses and the program counter were specified at 32-bits, though the package only had enough pins to allow 24 bits of address to make it off the chip. [Moto16-83], [Moto16-84], [Strit-79].

The architectural definition (and writing the user's manual) happened in parallel with the logic design and the circuit design. Instruction definitions changed even after the first chips were fabricated, in May 1979 (mask J5H). Second-pass parts (mask R9M), back in September 1979, contained these instruction changes.

Logic Design: Under the heading of logic design, I include: state sequencer organization, microcode, microword definition, PLA definition, PLA minimization, and general logic design. All the logic design for the MC68000 was done manually.

A processor consists of a "data path" (alu, shifter, registers, etc.) and a "controller." I use the term "execution unit" interchangeably with "data path." The execution unit elements are latent—alus and shifters only add and shift when they're told to. The controller tells the execution unit what to do when.

I use a design method I call the "Flowchart Method" [Trede-81]. It is a method I developed while working on the MC68000 at Motorola. A "Flowchart" is a bunch of boxes containing register transfers and other information (like sequencing information and external bus activity). The point of the method is to cycle through many register-transfer-level designs. By changing the register transfers (and the "other" information) and execution unit simultaneously, one can arrive at an optimum design for the controller and execution unit. The usual approach is to define (fix) an execution unit and to fix a controller structure. Fixing an execution unit defines an "instruction set." Fixing a controller structure sets the sequencing rules. The controller is then literally programmed—the execution unit operations are sequenced to achieve the functions specified by the architecture. I think this common approach is the reason people think of microcoding as programming with wide opcodes. The common approach *is* the reason microcoded implementations are slower. Using the Flowchart Method, there should be no difference in speed between a microcoded implementation and a random-logic implementation.

Back to the Flowchart Method. Each Flowchart box is a state in the controller and each box "is" one processor cycle. A sequence of boxes is an instruction execution. Box labels say which instructions use which sequences. Box contents become the control store words in a microcoded implementation. I drew Flowcharts for the state sequencer on 22"x17" vellum, in pencil. The Flowcharts for the MC68000 took 6 sheets—representing 544 states in the control store.

I reduced the Flowcharts to standard-sized copies for distribution to project members. I marked changes on a copy, then made changes to the pencil original. Flowcharts for new instructions were first done on scratch paper, then transferred to the vellum. I updated the vellum and produced new copies a couple of times a week.

The control store was compacted by splitting the contents of a control word into two parts. One part of the control word, called the microword, controlled state sequencing. The other part, called the nanoword, controlled the execution unit. Register transfers from the Flowchart boxes went into the nanoword. Sequencing information from the Flowchart boxes went into the microword. Several microwords could share the same nanoword. The MC68000 microwords were 17 bits and the nanowords were 68 bits, so the savings could be significant. The final version of the MC68000 contains 544 microwords and 328 nanowords. The microword store is 9248 bits. The nanoword store is 22,304 bits. Total control store is 31,552 bits. Without shared nanowords, the control store would have been 46,240 bits. Control store space saved is about 30%. (Potential savings using this method of compaction were greater, but changes to instruction definitions late in the design dropped the savings to 30%.) [Strit-78].

Finding microwords which might share a common nanoword was a manual procedure. This procedure is called "state minimization." Here's what I did:
- Alphabetize the register transfers in each state.
- Write the register transfers in each Flowchart box on the back of an IBM card and alphabetize the card set.
- Go through the cards, comparing each card in the deck to all the cards below it.
- Look for similar states (Flowchart boxes with similar or identical register transfers).
- On finding two similar states, look at the Flowcharts to see if the states could contain a common set of register transfers without affecting the logical correctness of their respective Flowchart sequences.
- If one set of register transfers could serve both sequences, put the register transfers into a common nanoword (the microwords had to remain distinct since they controlled the sequence of control words executing the instruction).

The bit patterns for the control words are derived from the Flowchart boxes (each Flowchart box becomes a control word). Before the MC68000 project got a computer, no translation from Flowchart boxes to control word bit patterns was attempted. Once we got the computer, I typed the contents of the Flowcharts into a text file. I proceeded to translate the Flowchart states into control word bit patterns, using an ordinary text editor. John Zolnowsky, my office mate, felt sorry for me. Or he got worried about the error-proneness of doing it that way. He wrote a program to convert the Flowchart text file into control word bit patterns.

The assignment of control words to control store locations was done manually. I did it six times. We never considered a placement program for the control store words. The problem was too complicated. (You'll see what I mean, shortly.) Some control word addresses (such as the first control word for bus error and system reset) were in fixed locations (for time-critical events). Control words with fixed addresses are placed first.

The MC68000 state sequencer allowed 4-way control-word branches. Branch target control words are related in the control store—they share the same base address. Two bits from the branch-control PLA are substituted for two bits in the microword next-address field to select the next control word. Branch-target control words had to be placed in the same column in the control store. The row they are in was determined by the two address bits from the branch-control PLA.

The MC68000 control store is divided into a microword store and a nanoword store. Look at a photograph of the MC68000. The control store is divided into 3 horizontal segments, by thin lines. The top segment is the microword store and the bottom two segments are the nanoword store. The microword store has 34 rows of 16 17-bit microwords (544 words). The nanoword store has 82 rows of 4 68-bit nanowords (328 words). (It forms a rectangle because $16 \times 17 = 4 \times 68$. Actually, 66 bits of the nanoword are used. 2 bits are unused.) The control store address enters the top of the control store, almost in the center. (There are 8 groups of 16 bits to the left of the address and there are 9 groups of 16 bits to the right.) Microwords exit the control store at the top and nanowords exit at the bottom. Each control store address produces one microword and one nanoword, simultaneously.

Several microwords can share the same nanoword. In the MC68000 control store, there is a restriction on the number of microwords which can share a nanoword. Nanowords are shared by leaving a transistor out of one row in the nanoword address decoder. If a bit is left out of a nanoword address-decode row, two microword addresses produce the same nanoword. If 2 bits are left out of the nanoword address-decode row, 4 microwords share a nanoword. If 3 bits are omitted, 8 microwords share a nanoword. Since rows in the nanoword store contain 4 nanowords, leaving bits out of a row in the nanoword address decoder creates 4 sets of microwords (one set of microwords for each nanoword in the row).

Of course, a shared nanoword could also be a branch target. Shared branch target nanowords are placed second (after control-store words with fixed addresses). After the shared branch-target control words, the remainder of the branch-target control words are placed. Next are the shared nanowords. Finally, all other control words are placed. Address assignments also attempt to balance electrical loads on the address decode lines.

I belabored the point on control word placement to show the complications in a seemingly inconsequential problem—just putting the control words in the control store. Essentially, the same control word strategy is used for all microprocessors in the Motorola family and for the IBM Micro/370 microprocessor. Details of the implementations differ. Since I have given the gory details, the rest of the examples can describe variations and discuss progress in automating what started as a manual procedure.

About a year after I joined the MC68000 project, Motorola bought a DEC PDP 11/70 computer for our use. We got one terminal for every two people initially. I never used the computer much, but others did. It quickly became saturated. PLA definition and minimization were done manually. Logic design was done manually. I used Karnaugh maps to design and to minimize the instruction decoders and the control word decoders. MC68000 instruction op codes are 16 bits (excluding addressing extensions), so the Karnaugh map was a 16-variable map occupying 20 pages. Late in the project, John Zolnowsky wrote a program to place transistors in the OR array of the instruction decode PLAs, using information in the logical description of the PLA. [Trede-79], [Zolno-79].

Design Verification: Les Crudele, who specified the MC68000 bus protocol, designed and built the TTL breadboard of the MC68000 logic design. The TTL breadboard was 13 large SSI and MSI cards. It ran at about one-fourth of chip speed. The breadboard was for logic and microcode verification. It was connected to the PDP 11/70. So test programs could be stored on the main computer and loaded into the breadboard to run. A special card in the breadboard had a socket to hold the MC68000 so test programs could be run on the breadboard or on the chip. The breadboard was the host for initial chip debugging.

Instruction set definition, logic design, circuit design, layout, and breadboard design were in progress at the same time. The breadboard was large and complex and it could not be completed until the instruction set definition and the logic design were completed. More people were working on circuit design and layout (which also could not be completed until instruction set definition and logic design were completed) than were working on the breadboard, so the completion of the breadboard was very near the end of the project. Although the breadboard ran MC68000 programs and helped debug the microcode, it was completed too late in the project to aid substantially in debugging the logic design and microcode.

Most of the microcode errors were found by Mike Spak, who did circuit simulation [Nash-79]. Mike constructed simulation models for small parts of the chip using the Flowchart states to determine input signals. The circuit simulator was not capable of simulating the whole chip or even large segments of the chip. Bill Keshlear wrote a program to simulate the Flowchart algorithms used for multiply and divide.

The MC68000 project had no Flowchart simulator and no logic simulator. We verified the Flowcharts and logic by manually going through them several times. Colleen Collins and I spent many days reconstructing designs and comparing results. The breadboard did help verify Flowcharts and logic, but the breadboard could have saved much more work if it had been available earlier.

Circuit Design & Layout: I didn't work on circuit design or layout on the MC68000 project, so I won't say much about how it was done. These are my impressions. Circuits were designed by hand with some computer simulation of small circuits. Each circuit was designed to do exactly its required task *(e.g.,* drivers were no larger than necessary to drive the expected load in a particular instance of a circuit). All layout was custom. Each circuit was laid out to fit its particular instance. Layout was done by hand on Mylar and later digitized into a Calma file. Circuit designers were assigned to the project, but layout was done by the equivalent of a large secretarial pool.

Layout checking was done by hand. One person would read the circuit diagram and a second person would crawl around on an 8-foot square plot with colored markers and trace the circuits for verification.

Fabrication: We sent the design to Motorola in Phoenix, Arizona for mask making and fabrication. The MC68000 used a 3.5-micron nMOS process. It didn't seem to take very long. I think it took about a month if we weren't in a hurry and two weeks if we were in a hurry. My notes say the design was frozen and committed to the mask shop on 24 April 1979. We got functional chips back the week of 14 May 1979. (Just 20 months after logic design began.)

Initial Test: The initial MC68000 parts were placed on a tester to see if they functioned at all. Once we found the chips were functional, a packaged part was inserted on the special breadboard card. The breadboard acted as host to the MC68000 chip. It held the program, generated the bus protocol responses, and provided controls to run, stop, and single-step the chip. Small test programs were loaded into the breadboard memory from the PDP 11/70 and run on the chip. By comparing what the chip did with what the Flowcharts said the chip would do, it was possible to verify and debug instructions. My notes record about 13 microcode, logic, and circuit errors in this initial design. Many of the errors were known and had been corrected (in the files for the next-pass parts) by the time the first parts returned.

Manufacturing Test: Motorola uses functional testing for manufactured parts. I think the goal is to take no more than one second to test each part. A packaged part is mounted on a tester and a program is run in a pass/fail functional test.

Project Summary

Project Staff: It is hard to estimate the size of a project. Projects (usually) start small, grow, fluctuate, and then shrink or diversify. I think the MC68000 project had about 10 people working for about 2 years. (The number of people does not include the people in the layout pool).

Computers: At the beginning of the MC68000 project, we didn't have a computer. The circuit designers had access to a computer for circuit simulation, but for instruction set design, microcode development, and logic design, there was no computer. After a year, our group got the PDP 11/70 computer and it rapidly got saturated.

Design Tools: Since we didn't have a computer for the first year, there weren't any design tools to aid microcode development or to aid logic design. Once we got the computer, we used it to maintain the Flowchart files and we developed programs to assemble the microcode from the Flowchart file. There were no tools for PLA design or minimization. There were no tools for logic design, for assigning control word addresses, or for verifying microcode. Circuit design was done manually. Layout was done manually on Mylar and was manually digitized into a Calma file.

Bottlenecks & Problems: A few people did a major design project very well and very quickly. Tom Gunter, the project manager, thoroughly insulated the engineers from political battles. (This is something I didn't appreciate at the time.) We worked mostly without interference and without, in my case, awareness of the political environment.

Most of the engineers on the MC68000 project were young and inexperienced. I didn't know what I was getting into. I had never designed a commercial state sequencer and I had never designed microcode. I certainly didn't know what chip design was about. [Trede-79], [Zolno-79].

The design and construction of the breadboard was a bottleneck. The breadboard was a bottleneck because it would have been useful if it had been running on the first day of the project. No matter when it was ready, it wouldn't have been soon enough.

Logic design was a bottleneck. Motorola supported the design with enough circuit designers and with enough layout designers to make me be a bottleneck. Someone was always waiting to use the work I finished.

Motorola MC68010

Architecture: The MC68010 is essentially the same as the MC68000 ([Moto16-84], [MacGr-83]). The MC68010 added a loop-mode feature and virtual memory support. Several instructions execute faster on the MC68010 than on the MC68000, notably, multiply and divide. In the MC68000, address calculation microcode was shared among all op codes. Therefore, instructions like Set According to Condition (Scc) and Clear an Operand (CLR) read the operand at the computed address even though it wasn't needed to complete instruction execution. The MC68010 did not access unused operands in most cases. The MC68010 supports virtual memory using a method known as instruction continuation (as opposed to instruction restart) [Moto16-84]. When a page fault occurs, the MC68010 can stack enough information to suspend execution till the page is available. (Execution can be suspended part way through an instruction.) When the page is available, the processor is restored to its pre-page-fault state and instruction execution continues. John Zolnowsky, who did most of the detailed instruction set definition for the MC68000, defined the instruction set changes for the MC68010.

Logic Design: The MC68010 project began in the third quarter of 1981. The project was to be a straightforward modification of the MC68000 design, to support virtual memory. The project was expected to take a year. The new loop-mode, however, was more complex than anticipated, so the project took more than a year. The circuit designers intended to do a straightforward cut and paste modification to the original MC68000 layout, to make room for added microcode and for extended instruction decoders. In the end, the major macros from the MC68000 design were reused, but the entire chip was manually placed and wired from scratch. Implementing the idea of cutting and pasting the MC68000 design became too awkward and it did not correct the existing layout imperfections. Since the project was late because of the logic designers, the circuit designers used the time to clean up the layout by doing it over. It was about 12 months from the start of the MC68010 project to first-pass parts. (Second-pass parts are considered the end of the project. The first-pass parts are expected to have enough problems to prevent sampling.)

Doug MacGregor did the microcode for the MC68010, including the Flowcharts, the state minimization, and the control word placement. The nanoword (68 bits in the MC68000) was extended to 80 bits in the MC68010. Three bits were unused (compared to two bits in the MC68000). One control bit was repeated in 2 control store columns for convenience (rather than routing it from one column to both destinations). As a consequence of extending the nanoword, the microword grew to 20 bits (from the original 17 bits). The control store is laid out with 10 groups (16 columns per group) on each side of the control store address lines. So there are 320-bits per row in the control store. There are 16 20-bit microwords per row in the microword store. And there are 4 80-bit nanowords per row in the nanoword store.

Doug did the instruction Flowcharts on paper and he entered them into the computer directly (no vellum). The computer files were the official design. A Flowchart-drawing program produced working copies in readable format. Doug did the state-minimization (sharing nanowords among microwords) and the control-store placement manually.

Some of the MC68000 logic design was reused. Bill Moyer wrote a consensus-based prime-implicants program to minimize PLAs. A PLA assembler filled in the transistors in the AND and OR arrays.

Design Verification: There was no TTL breadboard for the MC68010. Instead, an attempt was made to verify the architecture using the N.MPC simulator from Case Western. This simulator ran on the DEC PDP 11/70. A logic simulator called Simulsys was used to verify the logic for macros (small blocks) and eventually, for the entire chip. The test files for the verification programs were generated manually.

Circuit Design & Layout: Most of the cell layouts and macros from the MC68000 were adapted to the MC68010.

Initial Test: For initial test, the MC68010 was plugged into a special card on the breadboard for the MC68451 MMU. (The MC68451 is the Memory Management Unit for the MC68000; it was developed before the MC68010.) Since the MC68010 and MC68000 are pin compatible, architecture verification could be done using the MC68451 breadboard. First, an MC68000 was plugged into the special card and programs were run. Then the MC68010 was plugged into the card and the same programs were run. The memory images were compared.

Project Summary

Design Tools: Most of the design tools used on the MC68010 project were originally developed on another project. Bill Moyer's prime-implicant PLA-minimization program was an exception—it was developed on and for the MC68010 project.

Bottlenecks & Problems: The bottlenecks on the MC68010 project were in logic design and in architecture verification. Since the N.MPC and Simulsys simulators were much slower than a breadboard, verification became a major bottleneck. This experience with software simulation and verification made Motorola managers favor breadboards for design verification.

Motorola MC68020

Architecture: The MC68020 microprocessor extends the M68000 to full 32-bit external data and address buses. (The Motorola MC68000 and MC68010 microprocessors have a 24-bit external address bus and a 16-bit external data bus.) The MC68020 introduced "dynamic bus sizing" whereby the processor determines from system responses whether an attached device is operating on 8, 16, or all 32 bits of the external data bus. An instruction cache of 256 bytes is also featured. The MC68020 has new address modes, 32-bit multiply and divide instructions, new comparison and checking instructions *(e.g.,* compare and swap and compare register against bounds), new bit-field-manipulation instructions, and a new coprocessor interface. Appendix E of the *MC68020 32-bit Microprocessor User's Manual* [Moto32-84] summarizes the MC68020 extensions to the M68000 family architecture. The architecture of the MC68020 was defined by John Zolnowsky and Dave Mothersole. For additional details, see: [MacGr-84], [Moto20-84], and [Moto32-84].

Logic Design: The MC68020 project began in the second quarter of 1982 with 3 or 4 people. Unlike the MC68010 project, which began as a modification to the MC68000 design, the MC68020 was to be a complete redesign. It wasn't possible to salvage pieces of the earlier designs for the new design, because the MC68020 was converting from the nMOS of the earlier designs to CMOS. Doug MacGregor did the microcode for the MC68020, including the Flowcharts, the state minimization, and the control-store placement.

The MC68020 control store, like its predecessors, is divided into two sections. One section, the microword store, controls sequencing of the controller and the other section, the nanoword store, controls the execution unit. The microword is 36 bits. The microword store is organized as 60 pairs of rows of 8 microwords per row for 960 microwords (2x60x8). The microword store is 34,560 bits (larger than the entire control store of the original MC68000). The nanoword is 74 bits. The nanoword store is organized as 57 pairs of rows of 6 nanowords per row for 684 nanowords (2x57x6). The nanoword store is 50,616 bits. These numbers include extra control store words and spare bits. The total MC68020 control store has 85,176 bits. With no sharing of nanowords, the total control store would have been 105,600 bits—so nanoword sharing saved about 20%.

Ed Rupp wrote a Flowchart assembler and syntax checking program for the MC68020. The program also checks for assignment conflicts within states. Ed also wrote a Flowchart drawing program.

Control word placement for shared nanowords and for branch targets was done manually. Once the location of restricted microwords was fixed, a program could be enlisted to complete the control word placement. Ed Rupp wrote a program which placed unrestricted control words in the control store.

Bill Moyer's PLA generation program (which was written on the MC68010 project) was used to generate the PLAs for the MC68020.

Design Verification: Design verification used both a breadboard and logic simulation.

Circuit Design & Layout: New circuits and new layouts for all the pieces were mandated by the change from nMOS to CMOS technology.

Project Summary

Project Staff: About 8 to 10 people worked on the MC68020 project. This figure includes architecture, logic design, and verification, but not circuit design and layout.

Computers: The design team used DEC PDP 11/70 and IBM mainframe computers; and they used workstations from Apollo, Daisy, and Calma.

Design Tools: Many design tools were rewritten for the MC68020 project *(e.g.,* the Flowchart assembler and syntax checking programs). The PLA generation program was carried forward from before. A limited control word placement program was added.

CURRENT DESIGN PRACTICE DETAILS

Micro/370 Research Project

Brion Shimamoto and I started the IBM Micro/370 project in January, 1981. We work at the IBM Thomas J. Watson Research Center. Micro/370 is a research project. Our objectives were to:
1. Do a commercial quality logic design for a microprocessor.
2. Write a detailed description of the Flowchart Method.
As the project grew, we decided we wanted to see the chip fabricated. Building a chip became a project objective; the chip would be a tangible, well-documented demonstration of the use of the Flowchart Method. We got our first functional chips in the third quarter of 1985.

Architecture: Brion Shimamoto did the architecture for Micro/370. The Micro/370 instruction set is a 102 instruction subset of the IBM System/370 instruction set. Micro/370 is a 32-bit microprocessor. It has 32-bit external data and address buses. It has 32-bit registers and 32-bit internal data paths. Developing the Micro/370 architecture differed considerably from developing the M68000 architecture. The M68000 architecture started from a blank sheet of paper. The challenge to the computer architects was to create a good base for later generation microprocessors. Successive microprocessors extended the architecture. The Micro/370 architecture began with the fully defined (some say over-defined) System/370 architecture. The Micro/370 architectural challenge was to define an IBM System/370 architecture subset, in a way that permitted graceful, but optional, construction of the missing (not on-chip) parts of the System/370 architecture, at the card level. In this scenario, successive microprocessors extend the subset and come closer to the goal architecture. For M68000 microprocessors, each generation *is* the definition of the whole architecture. For System/370 microprocessors, each generation is a an implementation subset of the whole architecture. For more information about IBM System/370 and the Micro/370 design, see: [IBMop-79], [Hadse-85], [Ong-86].

Logic Design: I did the logic design and the microcode for Micro/370, except for the bus controller. Shauchi Ong and Bruce Gavril did the logic design for the bus controller. I used the Flowchart Method (the method used for all the microprocessors described in this paper). I did the Flowcharts for each instruction on scratch paper. Then I typed them into the computer. A Flowchart-drawing program produced working copies of the Flowcharts from the computer file. Richard Hadsell wrote programs to check the Flowchart syntax, to assemble Flowcharts, to index states (by location in the Flowcharts), and to draw and to print the Flowcharts from the computer files.

PLA design and minimization were done manually. The PLAs were entered into the computer in the form of text. A computer program used the text files and some cell designs to create the PLAs. Input to the instruction decode PLAs is the (pre-defined) instruction bit patterns. Since the inputs are defined by the architecture and the outputs are defined by the control word placement program, the instruction decode PLAs are good candidates for program minimization and generation. PLAs controlling the execution unit, however, have microword or nanoword fields (and other bits) as input lines and have execution unit control lines as output lines. The output lines are defined by the PLA function and design of the particular execution unit macro. The input lines are defined by the encoding of the control word field. My PLA design procedure treats the encoding of the control word field as an unknown. I *assign* control word bit patterns to best minimize the PLAs they drive. So encoding of the control word is an output of the PLA design procedure. (This is backwards from the usual procedure and a reason why PLA minimization programs weren't used. Programs which generate and minimize PLAs expect defined inputs and outputs.)

Flowchart state minimization was mostly manual. Dick Hadsell wrote a computer program to print Flowchart states in an order conducive to finding similar state-pairs. Dick wrote a program, called "Path," to find all the predecessors of a named Flowchart state. The Path program was very helpful. On the MC68000 design, I had trouble knowing the consequences of changing the contents of a Flowchart state (nanoword). Because mechanically, it was hard to find the predecessors of a given state. Flowcharts are like a river system: there are many sources (almost one per instruction), but they flow together toward the end. It is easy to follow any one instruction from start to end, but it is hard to begin at a random Flowchart state and trace all paths backward.

Control word placement was done automatically. I described the difficulties of control word placement to Dick Hadsell at lunch one day. He got interested and wrote a special program to do placement for Micro/370. (The details of control word placement for Micro/370 and for the MC68000 are different, because of differences in control store organization and control.) Unlike the MC68000, where control word placement was completely manual, I never had to do manual placement of control words on Micro/370. Ultimately, the decision was made to implement the full control store with no sharing of nanowords among microwords. (The people implementing the control store used a dense ROM from a previous design, with 32K-bits more than we asked for.) Dick modified his program to place the control store words with no sharing. Dick's program also assigned the addresses in the OR arrays of the instruction decoders.

Placement of logic macros and global chip wiring were done manually. Once we knew the major blocks in the design and their approximate sizes, we planned where they would be placed on the chip. [Tsai-82]

Design Verification: We did not build a breadboard for Micro/370. Architecture verification was done using a simulator ("Flowchart Simulator") written by Linh Lam. The simulator contains a model of the Micro/370 execution unit. It executes a System/370 instruction by driving its execution unit model with controls—Flowcharts—read from the Flowchart file. Carol Chiang, Molly Elliott, Donna Hawrot, and Linh Lam wrote test cases to verify the Flowcharts. The test cases attempted to exercise every path through the Flowcharts and to test extreme cases (like largest negative number or operands overlapping by a single byte). Linh also ran official IBM System/370 instruction test cases ("Architecture Verification Programs"—AVPs) on her simulator. Once we had Micro/370 chips, engineers at IBM Endicott ran a more complete set of AVPs on a real Micro/370, mounted on a test card they had built. Robert G. Sheldon worked on getting test cases run on a special-purpose computer meant to do fast simulation.

Logic verification was done using a simulator called "DASH," written by Dick Hadsell. DASH is a detailed, general-purpose, interactive, logic simulator. You can sit at a terminal and step through the design, one clock phase at a time. DASH uses the assembled Flowcharts and text file descriptions of the PLA logic. Dick generated most of the test cases for logic verification manually. He did use a program to automatically generate all input combinations for some of the PLAs.

Circuit Design & Layout: The circuit designers and layout designers were Hu Chao, Shauchi Ong, Jeff Tang, W. Bennett Smith, John Hou, Cindy Trempel, Kelvin Lewis, Joe Higham, and myself. Mon Yen Tsai, David Yang, Mark Birman, Vic Di Londardo, Phil Cronin, Stephen Parke, Bill Feaster, and others worked on the project too. Most of the circuit design was custom. That is, each circuit macro was specifically designed for Micro/370. Drivers were designed to drive specific loads, based on precise knowledge of what was being driven and where it was located. The layout of the individual macros was done knowing the placement of neighboring macros and knowing the global wiring. Layout was done manually, on both IBM's Interactive Graphics System and GE's Calma systems.

Fabrication: Micro/370 used a 2-micron nMOS process with two levels of metal and one level of polysilicon.

Initial Test: Warren Shih, a test engineer, was the mastermind of the Micro/370 testing [Greie-86]. Initial functional test was done on a Tektronix 3295 tester. (We used an IBM tester for a test chip.) Paul Greier and Rob Franch set up and ran the Tektronix tester, running the 40,000 or so special test patterns generated by Dick Hadsell with his logic simulator. Once we were convinced the parts were functional, we gave some packaged modules to engineers at IBM Endicott. The Endicott engineers had built an experimental card to hold the Micro/370 chip and had developed test programs. They exercised the Micro/370 chips by running diagnostics and AVPs.

Project Summary

Project Staff: The project began in January 1981 with 2 people. Over the years, the project grew to about 10 people. We always had more people actually working on the project than were formally assigned to the project. Dick Hadsell, for example, wrote several important design-aid programs before he even joined the project. (He wrote the programs to help make up his mind about whether there was enough substance in the project.) Architecture, logic design, and logic verification for Micro/370 were done mainly by people from the Computer Sciences Department. Circuit design, circuit modeling, layout, and layout verification were done mainly by people from the Semiconductor Sciences and Technology Department.

Computers: We are in the Research Division of IBM. We have lots of computer resources but we have few people to build breadboards. The division is growing fast so lab or office space is hard to get. We therefore have a strong bias towards simulators, over breadboards. Motorola builds breadboards for their microprocessors because, for them, it is cheaper than the cost to do equivalent verification by computer. A breadboard, running at a quarter the speed of the chip, is probably thousands of times faster than a simulator with equivalent diagnostic function [Carte-86]. In addition, the breadboard acts as the host processor during the initial chip testing. But, a simulator might be available earlier than a breadboard, particularly for a next generation processor. Since we were building a System/370 and our simulators were running on the same architecture, our simulators were more efficient than if they had been running on an alien architecture. (We could run a program on the System/370 mainframe, then compare the memory image to the output of the simulator.)

Design Tools: At the beginning of the Micro/370 project, we had no computer design aids. I began the way I did at Motorola, working on vellum and not using the computer. After Linh and Dick joined the project, I started using the computer. At first, I just used the computer to make nice drawings of the Flowcharts. I worked from the printed output of the Flowchart drawing program. By the end of the project, we had an impressive set of design tools. We now have: a Flowchart simulator, a syntax checker, a Flowchart drawing program, a Flowchart path program, a Flowchart indexing program, a state-sorting and printing program (aids in state minimization), a Flowchart assembler, PLA generators, control word placement programs, a logic simulator, and programs to print various design files.

Dan Beece wrote a switch-level simulator (SLS), which we used with the logic simulator to verify the layout against the logic. Gabby Silberman, Vijay Iyengar, and Leendert Huisman made major contributions to the ideas in SLS. SLS handles full-custom circuits and it is fast. Dan developed SLS at exactly the moment we needed it.

Bottlenecks & Problems: Being a research project, we got low priority for chip fabrication. Product chips come first. Micro/370 took well over 5 years to complete in this environment, so the people on the project had to be sure they really wanted to be part of the effort. Over the years we had 15 or so managers (counting two levels above us). (That's a lot of briefings!) Part of the reason Micro/370 took long is that it was only gradually, that we decided to fabricate the chip. People joined the project one at a time and just did what they happened to be good at. We had many summer students, several students who helped us part time during the school year, many temporary employees, and many people who helped us even though they weren't formally attached to the Micro/370 project. It wasn't till sometime in 1983 that we looked up, saw all these people who had signed up, and said "Hey, let's go for it."

1. On the Micro/370 project, logic design was never the bottleneck. I did the logic design, took a year's sabbatical leave to teach at U.C. Berkeley, wrote most of a book, did layout for about a year, and traveled and spoke all over the country. I never felt rushed (except when I was doing layout).

2. Once we decided to build a real chip, the bottlenecks became architecture verification and layout. The large number of verification test cases that had to be written and run, and the labor intensive nature of custom layout means you need lots of people to do this. We had few people, so it took a long time.

FUTURE DESIGNS

Motorola MC68030

Motorola is working on the MC68030 [Raju-86]. The project was begun in the second quarter of 1985. About 6 people are working full-time on the project. They are using the same tools and the same computers which were used on the MC68020 project. The MC68030 will have the same architecture as the MC68020, but it will have a data cache to complement the MC68020's instruction cache and it will have memory management compatible with the MC68851 (paged memory management unit). The bus protocol will probably be augmented with 2-cycle bus accesses (the MC68020 has a 3 cycle minimum for external accesses).

M68000 family CPU design projects beyond the MC68030 are only in the planning stage now (June 1986).

Motorola has introduced a new design in the M68000 family about every 2 years. Between 1977 and 1984, they introduced four major microprocessors in the M68000 family, not counting variations such as the MC68008. Since the MC68000 in 1979, they have designed the remicrocoded MC68000 (used in the XT/370 and AT/370 options for the IBM Personal Computers), the MC68010, and the MC68020. Expect Motorola to sample the MC68030 this year.

IBM Micro/XA Research Project

This is only a research project in logic design. We do not plan to fabricate a chip. Our objectives are to:

1. Do a commercial quality logic design for a System/370-XA microprocessor.
2. Design for a performance rating in the 5 or 6 System/370 MIP range.

Architecture: The IBM Micro/XA microprocessor will implement a subset of the IBM System/370 Extended Architecture. Micro/XA will include the basic IBM System/370 instructions, plus the decimal and the floating-point instructions. The chip will contain the System/370 control registers and the essential features of the IBM System/370 Extended Architecture.

Logic Design: We will use the Flowchart Method to design the IBM Micro/XA chip. For this design, there will be 2 logic designers instead of one. Logic designers will work directly from the English language specification of IBM System/370 Extended Architecture in the new principles of operation [IBMXA-83].

The IBM Micro/XA will be a complex design. We think the design will take about a million transistors. We plan to use extensive pipelining of the processor controller, instruction prefetch, translation and checking hardware, and the execution units. The preliminary plan for the logic design includes semi-autonomous processors for general instructions, floating-point instructions, and the storage-and-storage instructions.

Design procedures will be about the same as for the IBM Micro/370 project. Logic designers will either do initial Flowcharts on scratch paper and transfer them to computer files, or do initial Flowcharts right on the computer with an interactive Flowchart editor. PLA designs will be done manually and entered into the computer as text files.

Design Verification: Design verification takes the most people, time, and computer power. It would benefit the most from more automation. Automatic generation of test files would be helpful.

Circuit Design & Layout: We do not plan to do circuit design and layout. (The Micro/370 project began with no plan for producing a working chip—I suppose we could get carried away with this one too.)

Project Summary

Project Staff: There are 6 people working on this project, 2 full-time and 4 part-time.

Computers: We have lots of System/370 mainframe power. We plan to move some of the logic design tools to an IBM Personal Computer AT/370. Each person has a mainframe-connectable personal computer in their office and in their home.

Design Tools: W. Bennett Smith and Dick Hadsell will be working on design tools to support the Flowchart Method on an IBM Personal Computer. There will be an interactive Flowchart editor and there will probably be Flowchart-syntax checkers and a Flowchart assembler. Dick is improving his DASH logic simulator. We want to generate test patterns for PLAs automatically.

Bottlenecks & Problems: The logic design, even for a million transistor microprocessor, will not be the bottleneck. On Micro/XA, the bottleneck will be design verification. There just isn't any easy way to do that yet.

FORECAST

In the following sections I speculate about where microprocessors, design methods, and design tools are headed. These forecasts apply to commercial microprocessor designs, methods, and tools. They are my personal forecasts and do *not* reflect positions held by the IBM Corporation.

Where Microprocessors Are Going

Architecture: A few years ago Business Week published market share estimates for 16-bit microprocessors [Busin-82]. They estimated Intel and Motorola would each have 45% of the 1990 market. TI, National, and Zilog would share the other 10%. In 1986, Electronics published their estimates of the 1990 32-bit microprocessor market shares [Wolfe-86]. They gave Motorola 27%, Intel 25%, National 18%, AT&T 11%, and Zilog 9%. All others shared the remaining 10% of the market.

RISCs: RISCs are trying not to be architectures. I think they will be successful—at that. The idea of a RISC is to reduce the instruction set, make it simple enough to execute every instruction in a single cycle. If every instruction is a single cycle, there is no need for a state sequencer and microcode. In effect, the compiler compiles to the former microcode interface. Users should program everything in high level language, so there is no need for the user to know the architecture of the particular RISC implementation the program will run on. (RISCs are essentially, super bit-slices.)

System costs are significantly higher for RISC processors. Instructions are simplified and instruction throughput is boosted to compensate. RISC processors usually assume a factor of about 4 to 10 in bus bandwidth improvement over processors like the MC68000. The RISC CPU from MIPS Computer Systems, for example, claims a peak bus bandwidth of over 128 Mbytes/sec for the 16 MHz component [Mouss-86]. A 16 MHz MC68000 would have a peak bus bandwidth of about 8 Mbytes/sec. The 16 MHz MC68020 has a peak bus bandwidth of about 22 Mbytes/sec. Since the chip interface to the memory system has always been the performance bottleneck for microprocessors, significant improvements in throughput incur significant costs.

Others: I have been discussing 32-bit microprocessors. (I included 32/16-bit microprocessors like the MC68000—32-bit machines trapped in a 16-bit package.) 8-bit microprocessors are, by far, the largest share of the microprocessor market in total dollars. (If they lead in dollars, they have a runaway lead in volume, since 8-bit microprocessors are much cheaper than 16-bit or 32-bit ones.)

What about developments in the 8-bit microprocessor market? There won't be any. Economies of scale again (explained a few sections ahead). The 8-bit microprocessor market is mature. There are only a few commercially successful architectures and there won't be any more. Here's what happened:

When no one was in the market *(i.e.,* nobody was making an 8-bit microprocessor), there were some applications for which no good hardware solution existed. The existence of *any* 8-bit microprocessor conferred the competitive advantage on users almost independently of cost. The first companies to make an 8-bit microprocessor found a ready market—though part costs were high. Eventually, competition drove prices down. (I think 8-bit microprocessors in large quantities are less than $2.) Once the price is low, the development cost for creating a new part to compete in the market are again significant (even if the volume is high). Companies with established markets have amortized their development costs, so their prices depend solely on production costs. The companies making commodity microprocessors keep working on their successful parts; tuning and shrinking them to get higher performance and lower production cost. Even discounting development costs, a company introducing a new microprocessor would have a new design competing with a tuned, mature design. Production costs would be higher for the new design. Other pressures against introduction of new microprocessors include lack of a software base, lack of development tools, lack of documentation, and lack of peripheral components.

The same scenario is playing in the 32-bit microprocessor market now. The first arrivals in the market can be successful. But the market will mature quickly, permitting just a few architectures to be commercially successful.

What is on the Chip: In the past few years, there has been a controversy over what to put on a chip. I feel the controversy is about to end. The first microprocessors integrated only a basic instruction set. The next generations of microprocessors diverged as technology improved to allow more function on chip. Intel opted for operating system and memory management functions in the 80386 [Gold-85]. The Intel 80486 (or whatever it will be called) seems destined for memory management, caches, and floating-point instructions. Motorola opted for an instruction cache in the MC68020. They are building instruction and data caches for the MC68030 and and they will have an on-chip translation lookaside buffer (TLB). One might expect the follow-on to the MC68030 to have on-chip instruction and data caches, and to have floating-point instructions. The IBM Research Micro/XA (IBM System/370-XA microprocessor) design will include all the System/370-XA instructions a microprocessor needs—including the floating-point and decimal instructions.

Commercial microprocessors have always been bandwidth limited at their pins. On-chip functions can be performed more quickly than data and instructions can be transferred, so the on-chip controller is usually waiting on memory transfers. In the past, this resulted from a paucity of pins combined with simple, lengthy, transfer protocols. In the future, I think the number of pins will increase to allow multiple external buses. Designers might provide separate 32-bit instruction and data buses. Micro/370 uses the MC68000 bus protocol; this is a general, asynchronous protocol. Performance-oriented bus protocols are synchronous and use very few clock edges. Bus requests might be tagged so multiple requests could be pending and responses could come in any order.

Starting with microprocessors one or two generations beyond chips like the MC68030, we should be seeing CPU-on-a-chip architectures which are essentially complete. I will refer to these chips as MAXIs. I expect to see MAXI chips by 1990. I do not believe MAXI implementations will have significantly increased function, architecturally speaking. The successors will need improved performance or improved cost/performance. But sophisticated control and pipelining techniques will have been used in predecessor designs. So improvements (sufficient to warrant a new product) will have to come from a change in technology which the MAXI itself could not exploit. I consider this outcome unlikely. Instead, I think designs will diversify.

About 1990, I expect to see custom-order chips, in the following sense. Divide a chip into four quadrants. Put, for example, a chip like the MC68030 in one of the corners and choose the occupants of the other corners from your *Custom Mega-Cell Catalog*. Want a chip with an MC68030, two special-purpose I/O controllers, and some RAM? OK. Want a chip with an MC68030, two quadrants of RAM, and a quadrant of ROM? OK.

For microprocessors which have microcoded implementations, there is the possibility of using the microprocessor "chassis" as a universal host machine (UHM). In his master's thesis at U.C. Berkeley, W. Bennett Smith showed it was feasible to use Micro/370 as a UHM [Smith-84]. He and Robert G. Sheldon and I remicrocoded Micro/370 as a System/370 Decimal coprocessor. Similarly, Robert proposed the design of a "mail box" coprocessor. A company could publish characteristics of the chassis and let customers determine their own instruction sets. I like this idea conceptually, but the cost of supporting custom architectures will probably keep it from happening.

Once complete architectures are implemented on a single chip, it will be difficult for successor chips to show substantial improvements in performance or cost/performance. I think this will lead to greatly increased use of the microprocessor as a component on a custom IC ([Schne-86],[Buric-84]). I think the biggest future in ICs is in application-specific ICs (ASICs). When you read about new work in VLSI design (workstations, software, CAD, silicon compilers, architecture, silicon foundries, etc.), about 90% of the time the object or target is ASICs. Intel has an ASIC business. Motorola has an Application Specific IC Division. In addition, Motorola is teaming up with Silicon Compilers Inc., so designers can base ASICs on Motorola's CMOS [ElecD-86]. I think using an excellent family of microprocessors and peripherals as on-chip components, and adding one or two ASIC macros on the same chip, is a winner.

Where Design Methods Are Going

Design methods in commercial design environments show little change.

Formal Specifications: High level language specifications and formal description languages receive considerable trade press attention. If a program can specify a design, there is a potential for easing design verification. If both the specification and the design data base are set in a kind of formal language, they might be compared by a program to verify their correspondence. If the specification exists in a formal language, a program might be used to convert the specification into a logic design. Why stop at logic design? That's a formal description too. Why not just use a program to convert that into the circuit design, the layout, and ultimately, the mask patterns? That is the silicon compiler tenet.

The formal specifications I have seen look like Pascal programs. If someone gave me a processor specification written in a formal language, the first thing I would do is attempt to translate it into English. All the commercial microprocessor design projects I know anything about began with an English language specification. I don't expect this to change. Future commercial VLSI microprocessor design projects will begin with an English language specification.

A formal description seems to be the starting point for many proposed design methods. But the formal description must come from somewhere. I think it unlikely that someone would produce the formal description as the first level specification of an architecture. The English language user's manual seems a logical starting point. If the English language description is the starting point, producing the formal specification costs time and effort. I suspect the time and effort to produce an efficient formal description from an English language specification is comparable to the time and effort to produce an efficient logic design.

Silicon Compilers: Even though considerable trade press attention is given to automatic chip design programs, such as silicon compilers, these programs do not always meet expectations [Schin-86]. Converting a formal specification to mask patterns through logic and circuit design is not yet efficient. I don't think it will be any time soon. The program has to make many decisions at many levels. It is still too hard to know what a good designer considers. Figuring out the parameters of the decision process and incorporating them in a program is a formidable task. Too much time has been spent working on solutions to the problem and not enough time studying what the problem is. Specifically, I think very few people understand design method—even though they are working on automatic design systems.

I think silicon compilers are good for integrating logic from a formal description. If you have a card of TTL you want to integrate into one or a few chips, silicon compilers are appropriate. I think there is a giant market for silicon compilers for application-specific integrated circuits (ASICs). But silicon compilers are not as efficient as human designers for new designs of commercial microprocessors. Here's why.

Economies of Scale: The cost of producing an integrated circuit can be modeled as a one-time development cost and a per-part production cost. Development cost is amortized over the number of parts. Production cost is the manufacturing cost for each part. Yield, which directly affects manufacturing cost, follows a non-linear curve with respect to chip size. (Smaller chips have much better yield.) A 10% reduction in chip size might double yield—cutting manufacturing cost by two.

If the number of parts to be manufactured is small (say, in the thousands), development cost can be more significant than production cost. Companies marketing commodity microprocessors must have the lowest possible cost per part. If the number of parts is large (millions), production cost is more significant than development cost.

Suppose it costs $10 to manufacture a certain IC. (There must be some chip size you can manufacture for $10. You can buy MC68000s in large quantities for around $10, so such a chip might be just a little larger than the current MC68000 [to allow something for mark up].) If I can reduce chip size 10%, manufacturing cost might decrease to $5. If I expect to sell a million chips, I can make more money as long as the added development cost (to reduce chip size by 10%) is less than $5 million. Motorola probably spends $10 to $50 million in development cost for each M68000 family CPU design project (my own wild guess). I think this is typical for commodity market microprocessor development using current commercial design methods.

If I expect to sell a million chips, each extra million dollars I spend on development increases the purchase price by a dollar per part. If I expect to sell a thousand chips, each extra million dollars I spend on development increases the purchase price by a thousand dollars per part. Clearly, if I expect the market to be in the thousands, anything I can do to reduce development cost dramatically affects cost per part. Silicon compilers produce area-inefficient parts cheaply (and quickly). Development cost for a chip produced by a silicon compiler is probably in the tens of thousands of dollars. The resulting chip is perhaps 50 to 100% larger than what could have been produced by a development effort costing a thousand times as much. If the manufacturing cost is 10 times greater ($50 per part), it's still insignificant compared to the savings in development cost (thousands of dollars per part). I think this stuff is called economies of scale.

Design: The Flowcharts for the MC68000 through the MC68020 (four chips in the series) and for the IBM Micro/370 were done on scratch paper and (eventually) entered in computer files. I don't expect much change here. For our next project at IBM Research, I expect to be using a special interactive Flowchart editor. Designers at Motorola will either enter the Flowcharts directly using a Flowchart editor, or they will continue to use scratch paper for the initial Flowcharts [Raju-86].

The processor state minimization for all the projects I mentioned was done manually. I expect state minimization in future projects to be done manually. It's the kind of job humans are better than computers at. You need a good overview of processor operation combined with an ability to recognize significant (but not exact) similarities and subtle, but compatible (or almost compatible—with adjustments) differences. Computers can help. They will order the states, alphabetize, index, and they can identify definite incompatibilities. But they won't do the minimization (unless you don't care much about quality).

Logic Design: Logic design has been done on scratch paper and will continue to be done on scratch paper.

Motorola has gone from hand-generated PLAs to computer-generated PLAs. [Raju-86], [Zolno-86]. The IBM Micro/XA designers will still be generating PLA logic by hand. The hand-generated PLA will be represented as a text file in the computer. Computer programs, using the PLA text file as input, will generate the PLA layout.

My PLA design procedure uses output lines defined by the macro the PLA controls. (The circuit designer can tell you exactly what control lines the PLA must generate for each macro.) The Flowcharts define PLA function. Input control for the PLA might be any lines in the microprocessor *(e.g.,* control word field, instruction register, condition codes). If (at least part of) the input is a control word field, I treat the encoding of these bits as an unknown and attempt to find the encoding that produces the minimum PLA.

The chip floor plan will continue to be done manually.

Macro design, placement, and wiring will still be done manually.

Where Design Tools Are Going

There is a lot of pressure on companies making commercial microprocessors to further automate the design process. Design cycles seem impossibly long and costs too high. The trade press puts wood on the fire by headlining fantastic design tools which produce design quality equal to the best custom designer at a small fraction of the cost. Most of the reported improvements are applicable only to ASIC designs. The comparison with commercial designs is meant to demonstrate the quality of a new product against a known standard of excellence. The claims tend to be overstated—which causes immediate problems for commodity-part design managers. Corporate executives send their staffs to the trenches to hear the justification for high costs and long schedules in the face of programs reported to be fast and cheap and achieve equivalent results. The long term effect, however, is positive. The corporate executives are amenable to real improvements in design tools.

The most trade press attention is given to improvements in the function of design tools. As I said, these reported improvements apply to ASICs, not to commercial microprocessors. For commercial microprocessors, I would ask for a 10x improvement in the speed of the simulators, microcode assemblers, and other programs—with no improvement in function.

Design Verification: I think design verification has become the biggest problem in designing a single-chip microprocessor. How do you know the thing you designed is the thing you wanted? The number of test cases you have to run to determine whether the part is doing what you expect it to do is increasing rapidly as levels of integration increase. We need verification checks in two categories.

One category is checks for the pathological cases in the architecture itself. These are (architecture) verification programs for the architecture. In the case of IBM System/370, for example, the company has developed a set of programs called Architecture Verification Programs (AVPs). This set of programs is collected test cases for the architecture. Over time, the set should improve in its ability to detect variance with the architecture. The AVP test cases are independent of the implementation.

The second category of verification tests is tailored to a particular implementation. These test cases might invoke certain functions, for example, because they will cause the outputs of a certain PLA to assume all possible values. IBM developers sometimes call these Implementation Verification Programs (IVPs). The algorithm I use for checking for destructive overlap in the Move Character Long (MVCL) instruction, for example, should be checked with data that tests the boundary values for branches in the algorithm. Verification programs for the implementation have to check every path in the Flowcharts. They should check every path in the Flowcharts at the limits of that path *(i.e.,* with data that causes Flowchart branch conditions to just be met or just miss instead of data in the middle of the path). Test cases of this sort cannot be collected and accumulated over time, because they are unique for each set of Flowcharts. It would be nice to have a program to look at the Flowcharts and generate implementation verification programs (IVPs).

Table 1. Design Style Comparison

Architecture and Logic Design						
Feature\Chip	MC68000	MC68010	MC68020	μ/370	MC68030	μ/XA
Specification	English	English	English	English	English	English
TTL Breadboard	Yes	No	Yes	No	Yes	No
Flowchart Simulator	No	No	No	Yes	No	Yes
Logic Simulator	No	Yes	Yes	Yes	Yes	Yes
Self Test Logic	No	No?	Yes	No	Yes	Yes
Architecture Test File	Manual	Manual	None	Manual		Manual
Logic Test File	None	Manual	None	Manual		Manual
Flowchart Assembly	Mix	Program	Program	Program	Program	Program
Flowch. Syntax Check	Manual	Program	Program	Program	Program	Program
Flowchart Drawing	Manual	Program	Program	Program	Program	Program
Microcode Placement	Manual	Manual	Mix	Program	Mix	Program
State Minimization	Manual	Manual	Mix	Mix	Mix	Mix
Flowchart Generation	Manual	Manual	Manual	Manual	Manual	Manual
PLA Minimization	Manual	Mix	Program	Manual	Program	Manual

Circuit Design and Layout						
Feature\Chip	MC68000	MC68010	MC68020	μ/370	MC68030	μ/XA
PLA Generation	Manual	Mix	Program	Mix	Program	Mix
Macro Design	Manual	Manual	Manual	Manual	Manual	Manual
Macro Placement	Manual	Manual	Manual	Manual	Manual	Manual
Global Wiring	Manual	Manual	Manual	Manual	Manual	Manual

Table 2. Microprocessor Comparison

Feature\Chip	MC68000	MC68010	MC68020	μ/370	MC68030	μ/XA
Start of Design	3Q77	3Q81	2Q82	1Q81	2Q85	4Q86
End (1st working part)	3Q79	4Q82	3Q84	3Q85		2Q90
Technology	nMOS	nMOS	CMOS	nMOS	CMOS	CMOS
Minimum Feature(μm)	3.5		2.3	1.8		1.0
Chip Size (mm)	6.3x7.1		9.5x8.9	10x10		14x14
Transistors (k)(sites)	68		200	200		1000
Microcode (kb)	36		85	92		400
Clock MHz (Nominal)	8	8	16	20		40
Pins	64	64	114	171		264
External Address Bus	24	24	32	32	32	32
External Data Bus	16	16	8,16,32	8,16,32	8,16,32	8,16,32
Instruction Cache (B)	0	0	256	0	256?	??
Data Cache (B)	0	0	0	0	256?	??
TLB Entries	0	0	0	0		64
Floating Point	No	No	CoP	CoP	CoP	Yes

SUMMARY

Tables 1 and 2 show the development of design methods and commercial microprocessor chips for Motorola and for IBM. (Information about the MC68010, MC68020, and MC68030 comes from [Raju-86] and [Zolno-86].) There may only be one or two more generations of microprocessors until all the pieces of a complicated CPU are on one chip. Sophisticated pipelining and controller techniques are used now, so significant progress beyond the next one or two generations will be difficult. Components will diversify, moving special-purpose coprocessors, custom logic, and memory onto a chip with a microprocessor in the corner. That microprocessor will implement one of the few commercially successful architectures.

Commercial microprocessor logic design methods have been largely manual because these manual methods were the only ones available. Commercial microprocessor logic design methods will continue to be largely manual (though aided by computers for tedious, simple tasks) because manual design produces higher-speed, more compact chips. If the expected part volume is large, a 10% decrease in chip area could be worth a thousand times the development cost (the possible cost difference between a manual design method and a silicon compiler method). Macro layout, macro placement, and global wiring will be manual. Flowcharts will be manually generated and processor state minimization will be manual.

Microprocessor design projects fall behind schedule because each project is organized using previous design projects as a model. But the projects are becoming more complex—and the focus of the work is changing. Circuit design and layout has been the bottleneck. In the next microprocessors, design verification will probably require more effort than circuit design and layout.

Design tools with sufficient function to support commercial microprocessor logic design exist today. They need to be 10 times faster. (They don't need more function.) But logic design isn't the bottleneck (verification is). Tools to support design verification aren't adequate. We need programs that can automatically generate the test cases for verifying the design. But the programs must do this without putting constraints on the designers (like defining the logic or circuit vocabulary).

Acknowledgements: I thank John Zolnowsky of ViewTech and I thank Raju Vegesna and Dave Mothersole of Motorola for their help in providing historical and contemporary information on the Motorola microprocessor design projects. Forecasts are my own and do not reflect official positions or policies of either Motorola or IBM. Thanks to Sue Sanicky of IBM in Los Gatos for many hours of help preparing the document for the photocomposer. Thanks also to my usual (ghost) coauthor Ducky.

References

[Buric-84] — Misha Buric. "Microcomputers as Components of Custom ICs," *VLSI Design.* May 1984, pp 33-39.

[Busin-82] — "A Chip Waiting to Explode," *Business Week.* 26 July 1982, pp 30E-30H.

[Carte-86] — Harold W. Carter. "Computer-Aided Design of Integrated Circuits," *IEEE Computer.* April 1986, pp 19-36.

[Cole-86] — Bernard C. Cole and Charles L. Cohen. "Can Japan Catch Up in 32-bit Microprocessors?," *Electronics.* May 12 1986, pp 41-45.

[ElecD-86] — "Silicon Compilers Set to Aid NCR, Motorola ASICs," *Electronic Design.* 26 June 1986, pg 11.

[Elect-85] — "National Defends Lead in 32-bit Computer Chips," *Electronics.* 4 November 1985, pp 21-22.

[Gold-85] — Martin Gold. "After a Grand Entrance, 32-bit CPUs Start a Quest for Higher Throughput." *Electronic Design.* 17 October 1985, pp 27.

[Greie-86] — Paul Greier, R. L. Franch, W. Shih, R. W. Hadsell, J. Y. Tang, S. Ong, and H. H. Chao. "Micro/370 Functional Test," *IBM Thomas J. Watson Research Center Research Report, RC 11728.* Feb. 18, 1986.

[Hadse-85] — Richard W. Hadsell. "Micro/370," *Microarchitecture of VLSI Computers.* NATO ASI Series E, No. 96 Martinus Hijhoff, Dodrecht, The Netherlands, 1985, pp. 3-54.

[IBMop-79] — *IBM System/370 Principles of Operation.* GA22-7000, Armonk, NY: IBM Corporation.

[IBMXA-83] — *IBM System/370 Extended Architecture Principles of Operation.* SA22-7085, Armonk, NY: IBM Corporation.

[MacGr-83] — Douglas MacGregor and David S. Mothersole. "Virtual Memory and the MC68010," *IEEE Micro.* June 1983, pp 24-39.

[MacGr-84] — Doug MacGregor, Dave Mothersole, and Bill Moyer. "The Motorola MC68020," *IEEE Micro.* August 1984, pp 101-124.

[Moto16-83] — *MC68000 16-Bit Microprocessor.* Advance Information, Austin, TX: Motorola Inc., 1983.

[Moto16-84] — Barbara A. Cassel. *M68000 16/32-Bit Microprocessor Programmer's Reference Manual.* Fourth Edition. Englewood Cliffs, N.J.: Prentice-Hall, Inc., 1984.

[Moto20-84] — "32-Bit Virtual Memory Microprocessor," *MC68020 Technical Summary.* Austin, TX: Motorola Inc., 1984.

[Moto32-84] — *M68020 32-Bit Microprocessor User's Manual.* Englewood Cliffs, N.J.: Prentice-Hall, Inc., 1984.

[Mouss-86] — J. Moussouris, L. Crudele, D. Freitas, C. Hansen, E. Hudson, R. March, S. Przybylski, T. Riordan, C. Rowen, and D. Van't Hof. "A CMOS RISC Processor with Integrated System Functions," *Digest of Papers Spring Compcon 86.* March 1986, pp 126-131.

[Nash-79] — James Nash and Mike Spak. "Hardware and Software Tools for the Development of a Micro-programmed Microprocessor," *Proceedings of the 12th Annual Microprogramming Workshop.* November 1979, pp 73-83.

[Newsw-86] — William D. Marbach and Martin Kasindorf. "Computer Slugfest: Burroughs Tries to Break the IBM Stranglehold," *Newsweek.* 9 June 1986, pg. 47.

[Ong-86] — Shauchi Ong, H. H. Chao, M. Y. Tsai, F. W. Shih, J. C. L. Hou, K. W. Lewis, J. Y. F. Tang, C. A. Trempel, R. W. Hadsell, H. N. Yu, P. E. McCormick, C. V. Davis Jr., A. L. Diamond, T. J. Medve, and J. P. Higham. "A 32b Single-Chip Microprocessor," *ISSCC Digest of Technical Papers.* Feb., 1986, pp. 28-29, pg. 291.

[Raju-86] — Raju Vegesna. Motorola, Austin, Texas, *Correspondence and telephone conversations.* March to June 1986.

[Schin-86] — Max Schindler. "Silicon Compilers Travel Rough Roads to Acceptance," *Electronic Design.* 1 June 1986, pp 156-166.

[Schne-86] — Ron Schneiderman. "High-integration uPs Move Along the Road to Application Specifity," *Electronic Design.* 20 March 1986, pp 75-78.

[Smith-84] — Walstein Bennett Smith, III. *The Micro/370 Decimal Coprocessor -- A Study of Using Micro/370 as a Universal Microprocessor Chassis,* Master's Thesis, Department of Electrical Engineering and Computer Sciences, University of California, Berkeley. 23 April 1984, pp 1-64.

[Strit-78] — Skip Stritter and Nick Tredennick. "Microprogrammed Implementation of a Single Chip Microprocessor," *Proceedings of the 11th Annual Microprogramming Workshop.* November 1978, pp 8-16.

[Strit-79] — Edward Stritter and Tom Gunter. "A Microprocessor Architecture for a Changing World: The Motorola 68000," *IEEE Computer.* February 1979, pp 43-52.

[Trede-79] — Nick Tredennick. "Implementation Decisions for the MC68000 Microprocessor," *Proceedings of the 3rd Rocky Mountain Symposium on Microcomputers: Systems, Software, Architecture.* August 1979.

[Trede-81] — Nick Tredennick. "How to Flowchart for Hardware," *IEEE Computer.* December 1981, pp 87-102.

[Tsai-82] — M. Y. Tsai and Nick Tredennick. "Structured Physical Design for the Micro/370 Microprocessor," IBM Structured Logic Symposium. November 1982.

[Wolfe-86] — Alexander Wolfe. "A Three-Way Tug of War Hits The 32-bit Micro Business," *Electronics*. 5 June 1986, pp 40-41.

[Zolno-79] — John Zolnowsky and Nick Tredennick. "Design and Implementation of System Features for the MC68000," *Proceedings of Compcon, Fall 1979*. September 1979, pp 2-9.

[Zolno-86] — John Zolnowsky. ViewTech (formerly of Motorola), San Jose, CA, *Personal interviews and telephone conversations*. March to June 1986.

11

Experience with CAD Tools for a 32-Bit VLSI Microprocessor

David R. Ditzel
Alan D. Berenbaum

AT&T Bell Laboratories
Murray Hill, New Jersey, U.S.A.

Introduction

As the complexity of VLSI devices increases, so does the need to rely on computer aided design (CAD) methods. As VLSI designs grow they push the limits of the CAD tools, and in some cases require new approaches to design and verification. This paper reports on experiences with a particular approach taken to design a 170,000 transistor single chip CMOS microprocessor. The chip was an implementation of the Bell Labs C-Machine[1,2] architecture, code-named CRISP (C-Machine Reduced Instruction Set Processor) during its design. The major design tools used are described along with pleasant and unpleasant surprises in their use. Problems with more traditional approaches due to the increased size of designs are discussed.

Top Down Simulation

A first step in building a complex chip is to thoroughly understand the specification. For a microprocessor, this first specification is usually the instruction set. For our microprocessor, the instruction set was designed with an iterative cycle based upon writing a compiler, compiling many programs and measuring the output of the compiler to refine the instruction set. We then built successively more refined models of the machine until we finished with the physical implementation.

As the level of detail of simulation increases, substantially more cpu time is required and the total number of instructions one can realistically simulate grows smaller. When a bug occurs, it is often much harder to understand in the more detailed model of the machine than in a less detailed model. For this reason, we started by simulating as much as possible at the highest level. Many bugs, of course, will not be present at the most abstract level of simulation, so simulation must eventually be done at a very detailed level.

The Interpreter

Our first goal was to implement the exact specification and semantics of the instruction set. An instruction interpreter was written to execute instructions with the same semantic effect as the final hardware was to have. Using the C compiler with the interpreter allowed us to execute real programs. Two benefits from this are (1) being forced to detail the exact specification of the instruction set and (2) debugging the compiler early. Debugging the compiler early is essential so that detailed logic simulation does not appear to have hardware bugs that turn out to be compiler errors. The interpreter is also the fastest of the instruction simulation models and hence usually the easiest to debug the compiler with. Our interpreter included its own debugging system which allowed one to set breakpoints, dump memory, run for a specified number of cycles, dis-assemble instructions and other features.

Architectural Simulator

The first simulation of the hardware architecture of the processor was done with a C program called the architectural simulator. The architectural simulator modeled the hardware at the register transfer level. This simulator could run large programs, such as the C compiler, and provide an estimate of performance in terms of the number of clock cycles required. The pipeline structure and cache memory sizes could be easily changed at this stage, this allowed us to use the simulator to gather statistics on the performance sensitivity of various features. Features of the machine were only added if they could provide a sufficient performance boost to justify increased implementation complexity.

While the architectural simulator modeled all the major machine registers and datapath it was not partitioned to reflect the physical implementation. The control for the datapath elements was not explicitly modeled. Numerous details, such as the exact handling of faults and exceptions were not dealt with in the architectural simulator. The speed of the architectural simulator is about an order of magnitude slower than the instruction set interpreter.

Functional Simulator

As the final architecture of the machine solidified the architectural simulator was rewritten into what became the Functional Simulator (Fsim). The Fsim is partitioned into modules which correspond to the physical partitions of the final silicon, and all datapath and control signals are explicitly modeled. The functional simulator represents the complete design, and so is used to verify the sanity of the design, and also to provide test vectors for all later stages of debugging. CRISP uses a four phase clock, the Fsim models the hardware accurately down to the individual clock phase boundary. Each of the major pieces of layout has a corresponding C header file that details every wire entering or leaving the block and a C code file to model logic within that block. Other files contain display routines, an elaborate debugger and other support for the simulator.

We found the C programming language[3] both sufficient for describing the detailed hardware of the machine and efficient in execution speed. Figure 1a shows some edited excerpts from a header file and figure 1b shows excerpts from the corresponding C file. Each signal entering or leaving the block is described as an *input*, *output* or bidirectional *ioput*. Major internal signals are also identified. For the purposes of the functional simulator, the C preprocessor transforms these new keywords into the standard integer datatype, making this just another ordinary C program. Each signal declaration may optionally be followed by field specifying the number of bits (assumed to start at bit 0), or an explicit naming of the starting and ending bits of a bus. Signals without this field are assumed to be single bit values. The bit range specifiers are ignored by the functional simulator, but are used in backporting, which is described below.

Figure 1b shows more detail of how the functional simulator simulates the hardware. In this case the logic block is called *pdureg*, this subroutine is called once for each of the four clock phases in simulating a single clock cycle. Figure 1b shows a master/slave pipeline register clocked from the master *pdpc* into the slave *Pdpc* during clock phase one if the enable signal *ckpdpc* is active. Adders, subtracters and other bit manipulation are easily handled with standard operators in the C language. A multiplexor can be handled with a switch statement, as well as being able to check for illegal values of the mux control. In short, most hardware constructs could be easily modeled simply by writing in a stylized form. Translation of functional simulator code into schematics was usually quite straightforward.

C was chosen as the simulator language to ensure maximum simulation speed — the more code we could run through it, the easier it would be to uncover bugs. On an IBM 3081, the Fsim runs approximately 500 cycles per second, approximately 10 times slower than the architectural simulator and about 100 times slower than the interpreter. In opting for speed, we lost some advantages of specially-designed simulation languages. In particular, modelling parallel behavior in the sequential C model led to to calling order bugs that had to be ferreted out. Honesty was required by the Fsim writers to only write code which resembled reasonable circuits.

One of the practical advantages of writing the functional simulator in

```
#define input     extern int /* from fsim.h */
#define output    extern int /* from fsim.h */
#define ioput     extern int /* from fsim.h */
#define internal  extern int /* from fsim.h */
#define when      break;case /* syntactic sugar */

        /* Input output list for data path of the PDU */
input   Pdpra;   /* 0,15 */    /* instruction stream parcel from the queue */
input   Pdprb;   /* 16   */    /* instruction stream parcel from the queue */
input   eunpc;   /* 1,31 */    /* Next pc from IR stage of execute unit */
input   bradj;   /* 2    */    /* branch addition value (0,1,3) */
input   addpdu;                /* put the PDU address on the address bus */
input   ckpdpc;                /* used to enable clock of pdr.pc */

output  pipc;    /* 1,31 */    /* Tag for Instruction cache */
output  pbpbi;   /* 3,31 */    /* pbi master for prefetch buffer address */
output  irpieq;                /* high when eunpc equal to pipc */

ioput   adbus;   /* 2,31 */    /* bidirectional main address bus */

internal offset;               /* output of offset mux and pc adder */
internal tpc;                  /* target pc -- pc + offset */
internal spc;                  /* sequential pc */
```

Figure 1a. Edited portion of Fsim header file pdureg.h

```
/* PREFETCH DECODE UNIT datapath - Registers and adders and muxes */
#include "fsim.h"
#include "pdureg.h"

pdureg()
{
    /* In phase one, slaves are latched  */
    if  ( CLK1 )
    {   Pbpbi = pbpbi;          /* Latch every phase 1. */
        if  ( ckpdpc )          /* If clock enable is on */
            Pdpc = pdpc;        /* clock master into slave latch. */
    }

    if  ( CLK2 )
    {   if  ( addpdu )     /* double word align the pdu address */
            adbus = (Pbpbi & (~0x7)) | Pdusec << 3;
        irpieq = ( eunpc == pipc ); /* 31 bit compare */
    }

    if  ( CLK4 )
    {   offset += bradj << 1;   /* add in instruction length */
        tpc = offset + (pdpc & ~0x1);

        switch (pbpcmx )   /* model pbpc 3 input mux */
        {   case R_PBPCMX:  pbpbi = rpc;
            when T_PBPCMX:  pbpbi = tpc;
            when P8_PBPCMX: pbpbi = (Pbpbi + 8) & -8;
            default:    error(WARNING,"Bad PBR mux control = %d", pbpcmx );
        }
    }
}
```

Figure 1b. Section of C code of Functional Simulator.

C was that it was comfortable since all the tools and techniques for standard programming were available. The writers of the functional simulator went beyond mere simulation, and developed an extensive set of commands and debugging facilities to facilitate design and make debugging easier. One of the nicest features was an interactive display of the major machine registers. Each of the pipeline stages was shown on a CRT screen, allowing one to watch multiple instructions flowing through the machine. The practical effect was that newcomers understood the machine more easily than by looking at the simulator code, and those intimately familiar with the machine were able to understand the cause of bugs much sooner.

The functional simulator models the value of every signal at the end of a phase, but not necessarily the exact combination of logic gates necessary to produce that signal. All signals in the simulator are active high. Undefined signals, precharged nodes in dynamic logic and tri-stated outputs in I/O circuitry are not modelled. These features were left out to reduce the time it took to write the simulator and run the simulator, as well as to improve the clarity of the code. As a result some bugs (especially problems coming out of reset) were only discovered later in the backporting process.

In addition to internal consistency checks, like the validation of mux controls, the functional simulator was verified by dynamic comparison with the interpreter. No attempt was made to compare on a cycle-by-cycle basis (since the interpreter does not model cycles) or on the basis of any control signals. Stores, either to internal registers or off-chip memory, were matched to that of the interpreter. Most errors were quickly revealed this way. Since the interpreter ran about 100 times faster than the functional simulator, it was run as a background process with the Fsim, so no log files needed to be stored, and virtually every run of the Fsim could be verified this way. Errors that delayed execution but eventually produced the correct result were not automatically detected. Those that were found were uncovered by casual inspection noting that some tests seemed to take too long. A more formal regression mechanism, which compares a "good" Fsim version to a "new" Fsim, is probably required.

Schematic Logic Drawings: Draw

Schematics were used as the exact circuit and conceptual description of the machine. It is certainly possible to build complex chips without schematic diagrams. As long as some of the layout is not done automatically, schematics provide the clearest description for traditionally trained layout engineers. Engineers were comfortable with schematic representations but did not express the same appreciation for circuits described using textual representations. Some part of this is probably that schematics were able to convey 2-dimensional information, whereas text did not. We discovered and corrected many bugs at the schematic level before layout had begun.

The CRISP microprocessor was represented using 616 pages of schematics using the UNIX Circuit Design System.[4] Schematics were organized as a series of hierarchical macros. Unlike many systems, the hierarchy did not stop at the gate level, each gate or macro had to ultimately be defined from p-fet and n-fet transistors. About one third of the schematics define basic cells such as nand gates, inverters, multiplexors and latches. In this fashion, the schematics represented the exact transistor connectivity of the entire layout. Schematics could freely mix transistors in with other macros; this allowed designers to make good use of dynamic logic and to invent their own cells, rather than simply trying to design from a catalog of predefined gates.

The UNIX Circuit Design System is composed of a number of small programs that work together, rather than a monolithic schematic system. The circuit editor, called **draw**, is simply a graphics editor that facilitates the drawing of shapes, pins, wires and text. Draw produces one binary graphics file for each 8 1/2 x 11 inch schematic page. Each graphics file is first converted in to an ASCII netlist file. These individual netlist files are then run through a macro expander that produces the final netlist.

A macro is a schematic entity such as a nand gate or a multi-bit adder, that has inputs and outputs, and is composed of transistors or other macros. By convention the name of the file is the same as the type of the macro. A single macro may consist of a single or multiple schematic pages, for multiple page schematics the filename need be the same only until the first non-alphanumeric character. Since one of the main points of schematics is to make the intent of the circuit intelligible the macro expander also helps eliminate needless repetition.

Figure 2 shows an example circuit that makes use of the macro expansion facilities. This figure has 7 inputs and 1 output, all but *ck4* and *precharge* are busses. Signal *bradj<0:1>* represents the two wires *bradj0* and *bradj1*, and *tpcmx[abd]* is expanded into the three wires *tpcmxa*, *tpcmxb*, and *tpcmxd*. A 10-bit register can be similarly drawn using the macro expander. Formal and actual parameters are connected through pin names, one interesting feature is that pin names and wire names need not be the same. This is convenient when dealing with busses at high levels of the hierarchy. Consider how confusing multiplexor *abdmux* would be if all 43 inputs and output pins had to be explicitly shown in the macro call.

Each macro call has both a name (or names) and a type. Shown is a macro named *tpinc* of type *add2to10* drawn in two pieces, so that the precharge input will not impede the understanding of the function of the adder, whose internal circuitry is shown in figure 3.

One might assume that the design would progress in a completely top down fashion, where pages of schematics would be the specification handed to an engineer to layout. This was the case only for the control logic in our chip. For the remainder of the logic the functional intent was understood by the layout engineer, who then designed the circuit to execute that function according to his requirements for layout area and timing. Schematics for the bottom levels of the hierarchy were then drawn according to what the layout would be. The schematics were also driven by the layout in the sense that the physical topology was also reflected. For example, for a 3 input nand gate, we accurately reflected the order of the 3 inputs. This was necessary for later verification of the layout.

While the schematics were drawn to be as understandable as possible, there came a tradeoff in that they also had to be drawn to reflect the physical layout structure for use in verification. For example, a decoder might be too large to fit in a layout of a group of registers, but might fit conveniently into a hole in an adjacent adder; this had to be reflected in the schematics.

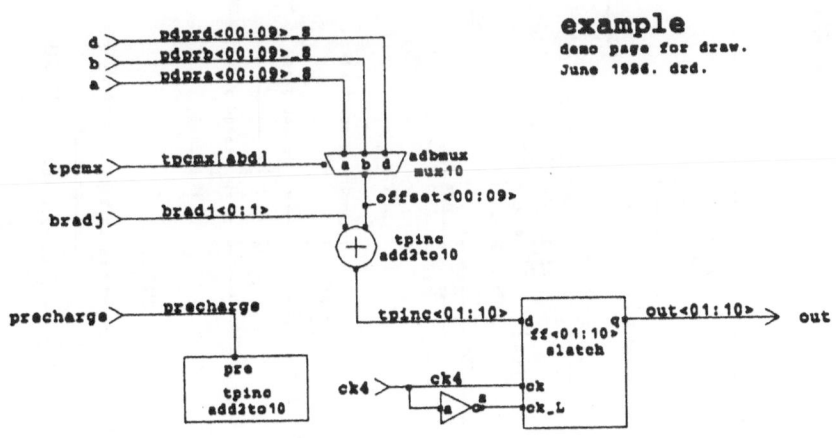

Figure 2. Circuit using macro expansion.

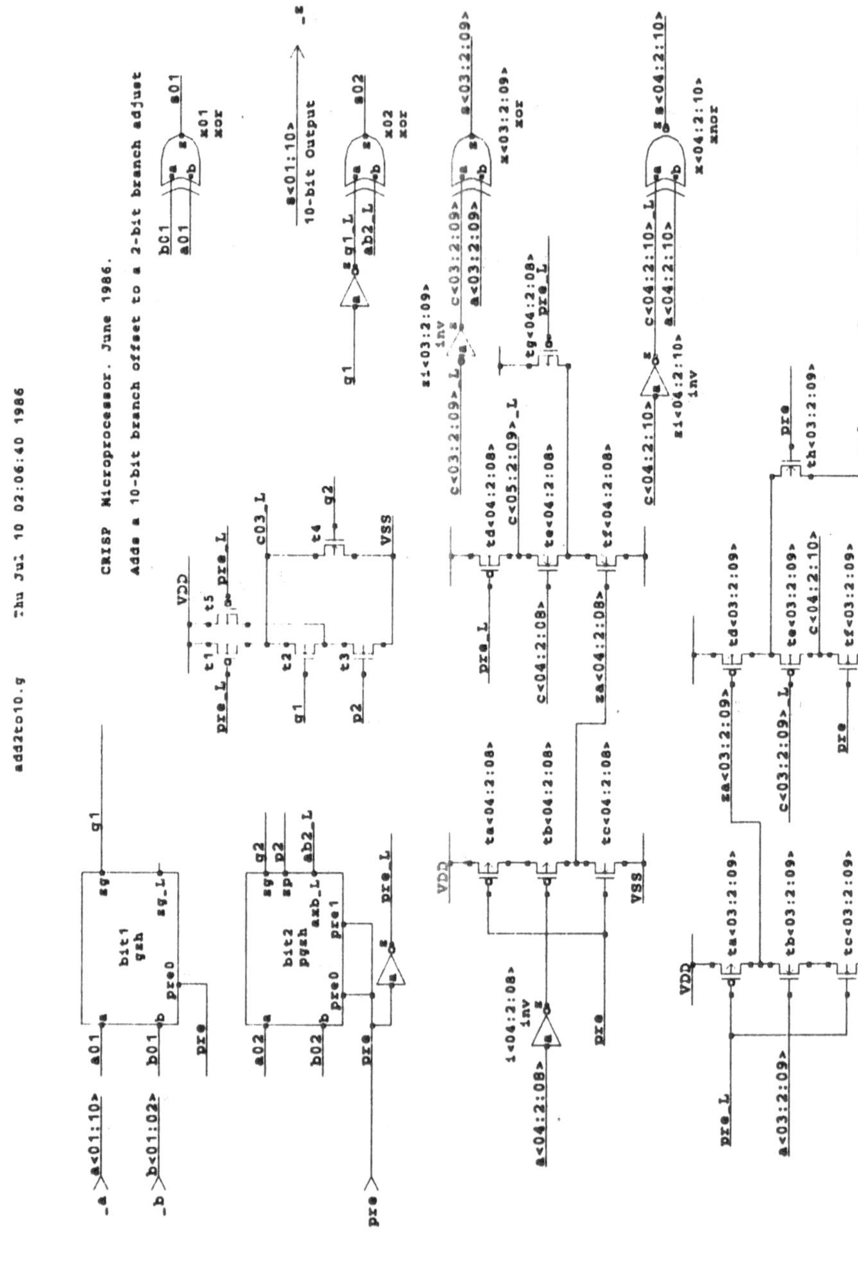
add2to10.g
Thu Jul 10 02:06:40 1986
CRISP Microprocessor. June 1986.
Adds a 10-bit branch offset to a 2-bit branch adjust
s<01:10>
10-bit Output
bit1
gzh
bit2
pgzh
VDD
VSS
pre
pre_L

Figure 3. add2to10 Macro.

Switch Level Simulation: Soisim

Transistor level simulation was handled with a switch level simulator called Soisim. The input to Soisim is a transistor level netlist containing transistor sizes and net capacitance. This ASCII input file is compiled before simulation begins. The compilation phase analyzes the circuit and produces additional information to facilitate simulation. For example, many gates are recognized and evaluated by function, rather than exhaustive transistor evaluation. Special cases usually hard for switch level simulators, such as the four transistor exclusive-or gate and memory sense amplifiers, are also recognized and handled properly. Another powerful speedup recognized by the compiler is to recognize memory cells and model rather than simulate them, as memory can be a large fraction of the total circuit. Capacitance information is used to handle dynamic logic circuits as well as detect potential problems with charge sharing. Transistor size information is used to evaluate ratioed logic.

Soisim has a number of interactive commands. Small circuits can easily be tested by setting input values, settling the circuit, and printing out nodes of interest. A benefit of the compilation process is that one may inquire as to which "gates" drive a particular node, where the notion of a gate has been automatically reverse engineered from the transistor level input netlist. The Soisim command interpreter follows the conventions of the UNIX Shell in two ways. One may traverse the hierarchy of a circuit using **cd** and **pwd** commands. Names use the character '/' as the hierarchy separator. Second, names may be specified by shell-like regular expressions using *, ?, and character classes. This is particularly convenient for dealing with groups of signals such as a 32-bit bus. Commands to the interpreter can be read from a UNIX file, and output from any command may be sent to a file rather than the terminal. The apparent command set of the Soisim interpreter can be extended with user supplied C programs compiled with Soisim, or by writing output to a file, then using combinations of the UNIX commands *awk, sort* and *grep*.

Soisim also has a non-interactive mode which can be driven from a C program. This is useful when the number of inputs and cycles of execution is too large to be conveniently typed in by hand. The values of C program variables can be bound to wires in a circuit to be simulated. Assigning a value of 0,1 or X to a variable will cause that value to be placed on the signal net before the next circuit evaluation begins. Variables must be specified as input, output, or a mixed input/output as specified by a control variable.

Backporting: Switch level simulation without vector files

The conversion from the functional simulator to schematics was a manual process. An engineer stared at the C code in the simulator and drew a series of gates to represent that code. Without another mechanism, there is no guarantee that the gates in fact produce the same behavior as the C language statements. The mechanism we use to prove that the logic is correct is called backporting. One by one, C-coded modules in the functional simulator are replaced by switch level simulation of the schematics for that module. Eventually the entire chip is modeled at the switch level by Soisim, and the Fsim is a skeleton which provides an operating environment, verification and debugging facilities. The result is an efficient mixed mode simulation capability.

Each module of the functional simulator, when called by the main driving procedure, takes all its inputs (which are C external variables) and, at the end of the phase, calculates appropriate values into its outputs (also C externals). When backporting a module, the C code is replaced by a special C code that binds a module's inputs to Soisim variables, issues a call to Soisim to settle the circuit, then maps the appropriate Soisim variables back to C variables for the module's outputs. The binding, settling and mapping function is handled automatically from the module's header file and the netlist file generated from the schematics.

The tool which generates the linking module automatically translates between the Fsim naming convention and the schematics naming convention. (The distinction exists primarily so the simulator can represent multi-bit signals as C integers, instead of a collection of 1-bit variables.) In addition, the tool provided another level of verification, since it complained if the inputs and outputs specified in the C header file do not agree precisely in number with the inputs and outputs specified in the schematics.

Backporting more than one module at a time is a simple extension of the procedure to backport individual modules. All that is required is an I/O list (which is in the form of a C header file) for the union of all the modules in question, and a netlist, which could be generated either by hand in schematic form or automatically from the constituent modules. Generating the I/O list is a manual process, but it is straightforward and eventually is double-checked against the logic. For backporting the full chip, the I/O list was the specification of the chip's I/O.

All facilities of the functional simulator are available to the backported simulator, so that the interactive debugging interface looks as it did without Soisim, all verification mechanisms function, and all Fsim variables are available for inspection by the debugger, including signals being simulated by Soisim. In addition, all nodes internal to the Soisim model could be examined or traced. The debugging paradigm is the same· as the stand-alone Fsim: run a test while comparing its results with the background interpreter, and when a conflict occurs, inspect signals until the problem is revealed. Usually, once a bug was known, its cause could be uncovered in a few minutes. As a result, we did not use any vector files until just before the chip mask date, when vector files were necessary for detailed I/O protocol verification as well as for the silicon wafer testing machines.

The speed of a backported functional simulator depends on the size and the nature of the circuit being simulated by Soisim. A small block of a couple of hundred transistors runs in nearly the same time as the bare Fsim. A full CRISP model, all 170,000 transistors, runs *very* slowly, on the order of two cycles per second of IBM 3081 CPU time. This is about two orders of magnitude slower than the bare Fsim. The process, even with the hierarchical names in the Soisim netlist flattened to save space, takes about 7 Megabytes of virtual memory space. Therefore, we ran a full-chip model infrequently, and only after we were reasonably certain that all of its sub-modules backported successfully. A full regression of the full-chip model through the 200-odd tests takes about 3 3081-CPU-days.

Layout Tools: Mulga

The layout was accomplished by three different methods. The memories and final chip assembly was done with a vlsi graphics editor called GRED. GRED allowed the editing of arbitrary polygons and mask levels in a fixed absolute geometry system. The two control PLA's were generated automatically from high level equations, and major routing channels were generated automatically using the LTX2 routing system. The remainder, and by far the majority of the drawn layout was done using the Mulga system.

Mulga[5, 6, 7, 8] provides a layout system using symbolic virtual grid layout with compaction. Layout is entered using a high resolution color display and digitizing tablet for cursor control. Transistors, wires and contacts are symbolic entities that can be placed at the crossing point of conceptual grid lines. These grid lines appear in the symbolic layout as if a piece of graph paper was superimposed on the screen. Symbolic layout still requires that the designer decide on the relative placement of transistors and wires, but actual physical design rules for spacing can be ignored.

After a symbolic design is entered, is then goes through a compaction process yielding final physical mask data. Virtual grid compaction in Mulga is conceptually accomplished with two passes, first an X-compaction step, then a Y-compaction step. In the X-compaction pass, all elements on that X coordinate are examined to see how far they might be moved to the left without violating the physical design rules to elements on the virtual grid line to its left. A single fixed mask X coordinate is then assigned to all elements on that virtual grid line which is determined by the most constraining path. This step is then repeated for the next virtual grid line to the right, and so on. The Y-compaction step is repeated similarly, with the constraints also including diagonal clearances. After compaction, all X and Y virtual grid coordinates have a corresponding physical coordinate. A program converts the symbolic data and physical coordinate information into the final physical mask data. Compaction is more a process of pushing objects closer together than one of re-arranging layout. The compaction system also adds additional mask layers not covered by the symbolic representation, such as tubs and N+.

Using symbolic layout proved beneficial in several ways. Layout of cells seemed to progress about 3 times faster for our designers with symbolic layout rather than with more traditional absolute geometry "paint" style layout editors. Symbolic layout is easier to deal with in several ways. Objects can be placed with a single atomic operation, rather than having to paint individual mask layers. For example, placing a single symbolic transistor is easier than drawing separate layers for each of diffusion, polysilicon and tubs. Time was also saved by not having to worry about particular design rules, for example the absolute length and width of transistors or the absolute spacing of wires; these would later be determined by the compactor.

Use of the compaction system also eliminated bookkeeping required for pitchmatching. Pitchmatching is necessary when many heterogeneous cells are packed in a matrix with precisely lined up interconnecting busses, as is common with datapath design. In figure 4a two cells need to be connected after layout and compaction, but the busses running though them are not aligned appropriately so that they may be butted together. If these cells are large, re-layout could be very time consuming. In a fixed geometry system one solution would be to interpose a routing channel between the blocks as in figure 4b. Such a routing channel could use large amounts of area, very unacceptable for microprocessor designs where area is always at a premium. Our execution unit datapath had 250 individual cells, each of which had to abut with cells on each of its four sides. The automatic pitchmatching used with the Mulga system aligned these cells by first compacting all cells to minimum size, then expanding the virtual grid lines of cells so that they abut exactly. Once a cell is compacted, adjacent virtual grids can be expanded without causing any design rule violations by holding the relative spacing of other grid lines fixed. Cells would now abut without the need of a routing channel between them, as shown in

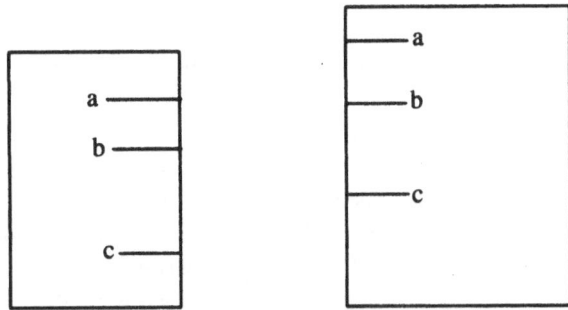

Figure 4a. Representation of 2 cells after compaction

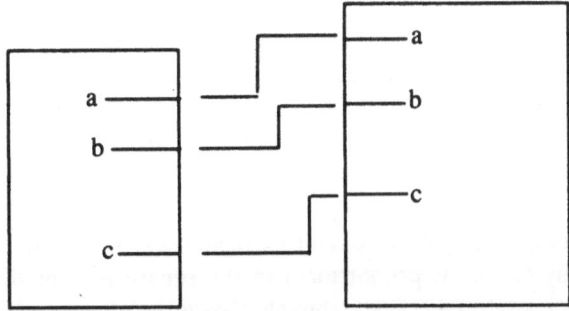

Figure 4b. Routing channel used to connect 2 cells with disjoint IO

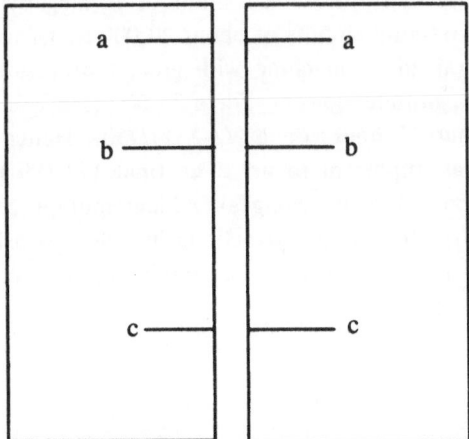

Figure 4c. 2 Cells expanded after pitchmatching now abut.

figure 4c.

Compaction allowed the layout to proceed long before the final design rules stabilized. When a design rule changed the effort was not in fixing the layout, but rather in changing some tables in the compactor. The only real concern was that all cells be compacted with the final version of the compactor. Our compactor was changed several times in the course of the layout, due both to design rule changes and to a few bugs in the compactor uncovered by use of a design rule checker. (A working compactor by definition does not produce any layout errors.) When we wished to evaluate a new circuit design for the ALU before 1.75μ processing was available, we simply took the symbolic layout and recompacted it using a compactor for 2.5μ design rules. Different compactors have also been used to create new layout based on preliminary design rules for future processes.

Circuit Extraction: Goalie

Two problems need to be solved after layout has been generated. First, is the layout correct? Second, what are the timing characteristics of the circuit? The first step in determining these answers is to automatically extract the circuit connectivity and capacitance from the final mask artwork. We used a new circuit extractor called Goalie.[9] Goalie's two most important features for a large chip design are that it is very space efficient, and also very fast compared to other circuit extraction systems. The working space required by Goalie is proportional to the square root of the size of the artwork being analyzed. Even though Goalie was space efficient, we still had problems due to the large amount of data involved. The final mask data for CRISP was 34 megabytes. The operating system for our largest IBM computer had process size limits of 10 megabytes per process, this limited us to extracting circuits of about 20,000 transistors. Larger circuits had to be extracted on machine with greater process size, but which were an order of magnitude slower. For a 35,000 transistor circuit, circuit extraction took about 12 hours on a VAX-11/750. Hence the CPU efficiency of Goalie was important to us. The final 172,189 transistor mask was extracted in about 6 hours using an Alliant multiprocessor, about 26 hours on a VAX-11/750. These extractions included parasitic capacitance and also generated intermediate files for simulation and netlist verification tools.

Netlist Comparison: Gemini

If schematics are used to specify layout, one needs a technique for certifying that the actual layout is as was intended. One might consider using simulation as a tool for verification, but this quickly becomes prohibitive as the size and complexity of circuits grow. Even if computer time were not a factor, catching a minor layout error would require perfect fault coverage, a very difficult prospect. Simulation can be considered a form a dynamic verification, we instead opted for static verification via netlist comparison. The schematics provide a transistor level netlist for the entire chip. The layout is extracted using Goalie to provide a second netlist. The two netlists are then compared for equality by a computer program called Gemini.[10] Gemini was written at Carnegie-Mellon University by Carl Ebeling and Offer Zajicek and was the only tool obtained from outside Bell Labs.

Gemini has a number of essential features for large circuits. First, its running time and space requirements are nearly linear with the size of the circuit. Second, it is convenient in that it does not depend on the actual names being equivalent; in fact, no names need to be supplied to get the program started. Third, when the circuits do not match, it does a reasonable job of producing diagnostics to help pinpoint the error.

A number of practical considerations have to be taken into account before relying on this method. One that almost caught us by surprise was the amount of virtual memory space that would be required for the finished chip simply due to the large number of transistors. Our final Gemini run took 110 Megabytes of virtual memory! This is still an average of only slightly more than 300 bytes per transistor, quite respectable. Most of the UNIX systems available to us provided for only 2 to 32 megabytes of virtual memory space per process. We were fortunately able to borrow a recently purchased Alliant mini-supercomputer which allowed for up to 256 megabytes per process. Our project could have been in serious trouble without the use of this machine. How big a process does your machine really allow? Try the C program in figure 5.

The sizes of files also became almost unmanageable. The Gemini input format was a flat ASCII database, with one line per transistor. Each line contained the device type (p or n), the names of the nets on the gate, source and drain, the length and width of the transistor, and a unique name for the transistor. For the extracted circuit the net names and transistor name contained no hierarchy, and were typically 6 characters in length (the default name for nets with no identifying text was "N" followed by a net number). The size of the Gemini file for the extracted chip was about 9 megabytes. The Gemini file generated from the schematics was over 25 megabytes, as the hierarchical names were each often long, for example **crisp/pfdata/pbufdata/array_d/nyble00/z/bit_L**. Other than disk space, one problem with large files was that they occasionally had to be moved between machine and few networks could transfer files that large in reasonable amounts of time. Magnetic tape ("tapenet") was used all too often.

It is best to start verifying small pieces of layout first, then gradually work to larger pieces. The intent is to verify a circuit where there are never more than a few errors, otherwise the diagnostics are not likely to be very useful. This incremental debugging approach required us to carefully draw the schematics so that there would be exact correspondence with layout blocks at various stages of the hierarchy. This required many tradeoffs between making the schematics easy to understand versus making them model some physical layout.

Although Gemini could often describe an error in terms of connectivity, finding the error in the physical layout could sometimes be difficult. Consider the case where a long wire is broken in two pieces by accident. Looking for the error with the naked eye may be impossible at any resolution large enough to see the entire circuit. Several of our plotting programs had the ability to highlight a particular net, thereby making the physical location of many problems plainly visible. The most difficult problem to identify turned out to be when power was accidentally shorted to ground creating a single power net. We found no easy way to detect the physical location of this particular error.

Gemini was extremely successful at finding bugs in what we would otherwise have thought was "perfect" layout. One of our strongest recommendations is that netlist comparison be a fundamental part of chip design methodology.

```
#include <stdio.h>
main()
{ int i = 0,meg = 0;
    while( ( i= malloc( 1000000) ) != NULL ) {
        printf("space for a meg at %d\n", i ); meg++; }
    printf("allocated a total of %d megabytes\n", meg );
}
```
Figure 5. C program to determine maximum process size.

Timing Analysis: ADVICE and Leadout

Detailed timing analysis was first done with a detailed circuit analysis program called ADVICE,[11] a Bell Labs proprietary version similar to the more commonly available SPICE.[12] ADVICE models the device physics of fets to produce highly accurate analog simulation based upon the actual process parameters. Input decks for ADVICE were initially hand extracted from plots, measuring transistor sizes and wire lengths via ruler and entering best approximations of values. Goalie was later modified to automatically extract an input deck for ADVICE, but it was still necessary to type in input stimulation waveform information. While this was merely tedious for small circuits, the job quickly became difficult for large circuits. Simulations of this kind had absolute limits of a few thousand devices due to both program memory requirements and excessive amounts of CPU time. ADVICE use concentrated around the design of the memories, drive and noise analysis in the IO frame, and on a few expected critical areas such as adder and ALU design.

One option we had available to us for large scale chip timing was to use a less accurate multiple delay simulator such as MOTIS.[13] This could have been used to simulate the execution of the machine for many clock cycles at a given clock frequency, i.e. a dynamic timing analysis. In addition to the still massive CPU requirements this would have cost, we saw several more basic problems with this approach. First, errors caused by signal paths running too slow would likely be very difficult to find. One might have to examine hundreds of nodes to determine which signal was at fault; this seemed unreasonable for large complex circuits. Second, this type of vector driven simulation would have no guarantees of exercising and finding all worst case signal paths. Third, the information conveyed by a timing failure would show only the last gate in the chain. The designer needs to be told the entire path of a signal in order to fix it. For these reasons, we abandoned this technique of dynamic timing analysis in favor of static critical path analysis.

For static critical path analysis we used a new program called Leadout.[14] Leadout's main job is to tell the user what paths do not arrive at their destination soon enough for proper operation at a given clock frequency. Leadout analyzes a circuit over a single clock cycle in a way that is independent of the logical values of most nodes. It does take into account and exploit information about clocks, invariant signals and other controlling signals whose timing behavior is specified. Leadout assumes that a circuit will run correctly if clocked slowly enough. This correct behavior is used to produce a set of internal equations that describe the time of occurrence of signal changes for any frequency, and a set of constraints on these times that must be satisfied in order for the circuit to operate correctly. The result is a fast timing static timing analyzer that could handle our entire chip in reasonable time (minutes) on a VAX-11/750. Accuracy of Leadout is within 10% of the time predicted by ADVICE.

Because Leadout was fast and easy to use it was convenient in sizing transistors. Gates driving signals internal to small blocks of logic were first laid out using minimum size devices. This provided a benefit by reducing the gate loading for most signals. Leadout was then used to selectively identify slow paths, these were then resized with larger transistors. Leadout would provide immediate feedback on timing improvements from transistor sizing. Designer time was probably saved as no time was wasted optimizing non-critical paths. Before Leadout was available we had done transistor sizing based on fanout, assuming that a gate with a large load needed more drive. If a gate had more than an ample timing budget (slack), then this unnecessarily increased both the area and capacitance on its inputs.

Leadout was not a substitute for ADVICE, but a tool that worked well with it. In fact, one of Leadout's commands is to produce a complete input deck for ADVICE for a critical path or other net, including input stimulus and producing plots. Leadout could not be used for circuits of a more analog nature, such as memory sense amplifiers.

Naming Conventions

What's in a name? Headaches. One of the biggest factors in making all the tools work together smoothly was in defining names. Our set of tools came from many sources, each with its own history for why what it had done was the "right" choice. The tools themselves were not even as much of a problem as the languages they had to deal with. The oldest of these was the XYMASK physical mask layout language, which had descended from the days of making printed circuit boards. XYMASK names could be a maximum of 8 uppercase alphanumeric characters, officially no special characters such as underscore were allowed. The C compilers on the large IBM machines initially also had restrictions of 7 characters on names, but would allow both upper and lower case, and underscore. Seven characters in a flat name space (i.e. no hierarchy) is just not enough. Fortunately the schematic system allowed hierarchical names of essentially unlimited length. In the end we decide not to try to force all cad tools to use exactly the same names. We cared mainly about the schematics and layout, Gemini provided the key by producing a dictionary of equivalences as it matched the two nets. In practice, using a dictionary turned out to be only a minor nuisance, perhaps no worse than would have been the bookkeeping to try to keep 68,000 unique 8 character net names.

The first fight among any group of engineers is how to represent active low signals in their schematics. Here is how we solved the problem. Names could optionally be followed with an underscore and *tag* field. If the tag was the letter *L*, the signal would be considered active low; *xyz_L* is an example of an active low signal. There were often many variants of the same signal, a common variant was to distinguish the master and slave nodes of a latch. Slave nodes would contain the character *S* in the tag. Buffered versions of the same signal could contain a number in the tag, hence *xyz31_2SL* would be the second slave latched version of bit 31 of bus *xyz* active low. Tags could in practice contain any legal characters, and their use was merely a convention extendible whenever the need would arise.

Most of the tools had their own ASCII intermediate language. Interfacing different cad systems could quite often be done with a relatively trivial translation program written in awk or C.

Our use of Gemini to match the schematics against the layout required that the nets match exactly. In particular, the order of inputs to gates had to match, so we needed a naming convention for gates. Our convention was that lower valued pins would be nearest the Vss and Vdd supply rails, higher valued pins would be nearest the output. For this definition, the value of 1 was less than 2, and the value of A was less than B. This convention seemed to work well even for gates such as the 3-2-1 input and-or-invert gate of figure 6.

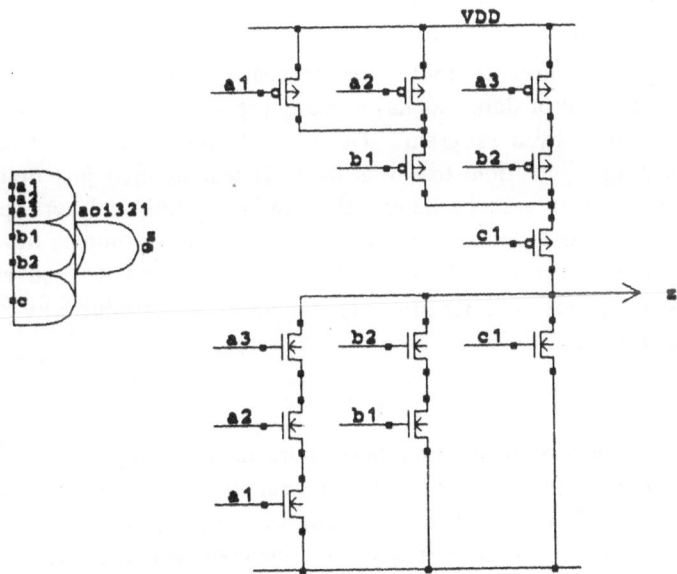

Figure 6. Example of naming convention for an And-Or-Invert gate.

Control the Source Code

In almost every case we obtained copies of the source programs for our tools. In retrospect, this was not only a wise choice, but essential in making the tools effective. Our tools came from several different sites, each with its own view of a particular problem. As a result, few of the tools would interface with each other in their original form. For example, our schematic entry system was originally designed to handle wire-wrapped circuit boards of a small number of TTL chips, not a large scale VLSI design. The wire-list format for TTL chips was not used directly by any of our tools, but a simple program converted the file into the format required by the Soisim simulator.

Few of the tools were equipped to handle the massive size of the circuits used in the chip. In many cases a program would halt with a diagnostic saying that the size of our circuit exceeded the allowed size of some internal data structure. By having the source locally available we were able to quickly identify the problem, change our copy of the source, recompile, and be in operation again in only a few minutes. These errors were too frequent to have the problem fixed by some separate support organization, and of course occurred on nights and weekends when no official support would have been available. The operating system also required changes to support our tools, primarily in adjusting process size limits and adding device drivers.

Having the source to tools helped create dozens of spinoff tools. Because the amount of data was large, many repetitive tasks could be more efficiently handled by a program. The tradeoff was always to determine whether writing a one time tool was more expedient than just doing the brute force repetitive task by hand. By stealing lexical analyzers, parsers and code to build data structures from an existing tool, complex new tools could be put together in only a few hours. In the end we had created many new programs that significantly enhanced the productivity of our CAD environment.

UNIX

The CAD environment consists of more than just layout and simulation tools. For us, the best foundation for our tools was the UNIX operating system. UNIX was useful in several ways. First, it provided a number of support programs, we most frequently used *grep*, *awk* and *sort*. Second, its ubiquity was important in that we had to deal with a large number of different computers. The last thing an engineer wanted to have to deal with was a new operating system. In our particular case, VMS was favored by some of the CAD tool writers as its FORTRAN compiler ran somewhat faster than that offered by UNIX for the same computer hardware. In no case was this change to a non-UNIX operating system appreciated by our designers. Early in the project we had available an order of magnitude increase in performance available on a non-UNIX machine and turned it down.

Some Statistics

Numerology is often interesting, table 1 contains some of the more relevant statistics gathered during the design of CRISP. The CRISP microprocessor was designed in 36 months by a team that averaged about 6 full time engineers. Additional support staff also made contributions. The sizes of programs are listed to give some feel as to the amount of effort required to install and maintain them. Some of these, such as Mulga, are actually a suite of several programs. In this case the total lines of code for all programs in the suite is given whether we used them all or not.

1	1.75μ CMOS Microprocessor
172,189	transistors
68,341	nets
616	schematic pages
2,892	files in the mulga subdirectories
3,507	files in the draw subdirectories
26	VAX-11/750 hours to extract with goalie
370	megabyte directory needed to extract chip
5	VAX-11/750 hours to rasterize a 200μ/inch color versatec plot
240	disk megabytes reserved for rasterizing plots
110	megabyte process size for final gemini run
25,500,634	bytes schematic generated gemini file
9,032,022	bytes goalie extracted gemini file
34,205,222	bytes final mask data
3,295	interpreter lines of C
4,308	architectural simulator lines of C
4,654	gemini lines of C
16,670	fsim lines of C
20,023	draw suite lines of C
21,699	Leadout lines of C
36,738	goalie/soisim suite lines of C
119,401	mulga suite lines of C
3	IBM-3081 CPU days for full backport regression
0	card images

Table 1. Some statistics gathered during the design.

Weak Spots

On the whole we were very pleased with our methodology and the tools. One area for improvement would be in the layout system. Layout was done in two vastly different styles, symbolic layout with compaction and a physical layout editor; a single integrated layout system would have been vastly superior. Mulga was excellent for designing small cells, but not very good for large blocks and final chip assembly. Symbolic layout was very time consuming as the size of blocks got large, even for a tiny change as layout verification was necessary every time. Recompacting, then circuit extracting in order to verify with Gemini could take several hours for a large circuit. Many hours could have been saved had the layout editor kept electrical net information and been able to generate a Gemini netlist directly.

A small change in a symbolic cell might change the overall size of a layout block after compaction, as well as the position of its inputs and outputs. This greatly interfered with inserting the compacted layout into the physical layout system during chip assembly. A single integrated system allowing both compaction and physical placement might alleviate this problem.

Results

There now exist working CRISP chips that run at speeds at or above the design frequency. As of this writing we have found only two bugs in the initial silicon. One bug was a charge sharing problem, another was due to Miller effect problems involving a dynamic node. The charge sharing problem was at the interface of two blocks, and would not have been a problem if the timing interface had been constrained in a slightly different way. This bug could have been caught by our Soisim backporting, but due to either haste or lack of communication this simulation was done without ever having included extracted capacitances. The Miller effect problem was a clear design bug, obvious when one looked at the schematics. We had no tools to effectively look for this type of analog design problem. The Miller effect bug could only have been seen through detailed ADVICE simulation, which was of little use unless one knew exactly where to look. If we had known where to look, we wouldn't have needed simulation to see the bug. A few of our first wafers had been held back from final metalization, we were able to make a minor change to the metal mask to correct these two bugs and obtain working chips from our first silicon wafers.

Conclusion

While many of the CAD tools have not been described, the overall methodology and experiences give a good view of how the chip was constructed, and the obstructions along the way. We urge designers of large chips to follow a few key recommendations. First, use tools that are capable of handling circuits as large as the entire chip. Piecepart verification would have left us with many bugs. Second, verification efforts should start as early as possible, and be capable of incremental verification. Multiple levels of simulators worked well for us compared to the more traditional approach of not simulating until extracted layout was available. Third, static timing and layout verification is vastly superior to dynamic verification. Apart from the tools themselves, the hardware resources were never sufficient or perfectly reliable. Much time was spent dealing with computer issues of networking, broken hardware, porting software and a constant stream of bugs that stole time that might have been used for useful design.

Acknowledgements

Many people contributed to the tools that made CRISP a working chip, and to the incremental improvement of our CAD environment. Rae McLellan was a fundamental contributor to CRISP and pushed the tools to their limits. Kevin O'Connor was quick to pick up new tools, find their bugs, and request fixes and improvements. T. Szymanski wrote a vast quantity of our tools and was always willing to solve special requests. N. Weste and B. Ackland and the rest of the Mulga crew provided much assistance in getting Mulga running. A. Fraser is thanked for initiating Draw and J. Condon for supporting and improving it over the years. R. Cmelik rewrote the Draw macro expander and worked on many of the early C Machine tools. In addition to his layout responsibilities R. Freeman provided valuable software support for many of the layout tools during the project. S. Eggers is thanked for the meticulous reading and insightful comments on the paper as well as moral support. M. Kahrs and R. Freeman are thanked for the comments they made on this paper. Without the numerous meetings and position papers we could have finished the design much sooner. A.G. Fraser, V.A. Vyssotsky and A.A. Penzias are thanked for their constant moral support, encouragement, and making available the amazing resources of AT&T Bell Laboratories which made our job easier.

References

1. D. R. Ditzel and H. R. McLellan, "Register Allocation for Free: The C Machine Stack Cache," *Proc. of Symposium on Architectural Support for Programming Languages and Operating Systems*, Palo Alto, California, pp. 48-56 (March 1982).

2. D. D. Hill, "An Analysis of C Machine Support for Other Block Structured Languages," *Computer Architecture News* **11**(4), pp. 6-16 (September 1983).

3. B. W. Kernighan and D. M. Ritchie, *The C Programming Language*, Prentice-Hall (1978).

4. A. G. Fraser, "Circuit Design Aids," *The Bell System Technical Journal* **57**(6, Part 2), pp. 2233-2249 (July-August 1978).

5. N. H. E. Weste, "MULGA - An Interactive Symbolic Layout System for the Design of Integrated Circuits," *The Bell System Technical Journal* **60**(6, part 1), pp. 823-857 (July-August 1981).

6. N. H. E. Weste , "Virtual Grid Symbolic Layout," *Proceedings of the 18th Design Automation Conference*, pp. 225-233. (June 1981).

7. N. Weste and B. Ackland, "A Pragmatic Approach to Topological Symbolic IC Design," *Proceedings of the 1st International Conference on VLSI* , Edinburgh UK , pp. 117-129 (August 1981).

8. B. Ackland and N. Weste, "An Automatic Assembly Tool for Virtual Grid Symbolic Layout," *Proceedings of the 2nd International Conference on VLSI*, Trondhiem Norway, pp. 457-466. (August 1983).

9. T. G. Szymanski and C. J Van Wyk, "Goalie: A Space Efficient System for VLSI Artwork Analysis," *IEEE Design & TEST* **2**(3), pp. 64-72 (June, 1985).

10. C. Ebeling and O. Zajicek, *Validating VLSI Circuit Layout by Wirelist Comparison*, Proceedings of the Design Automation Conference (1983), pp. 172-173.

11. L. W. Nagel, "ADVICE for Circuit Simulation," *Proceedings of the 1980 Symposium on Circuits and Systems* (April 28, 1980).

12. L. W. Nagel, *SPICE2: A Computer Program to Simulate Semiconductor Circuits*, University of California, Berkeley (May 1975).

13. B. R. Chawla, H. K. Gummel, and P. Kozak, "MOTIS - An MOS Timing Simulator," *IEEE Transactions on Circuits and Systems* **22**(12) (December, 1975).

14. T. G. Szymanski, "LEADOUT: A Static Timing Analyzer for MOS Circuits," *Proceedings of the ICCAD Conference* (October 1986).

12

OVERVIEW OF A 32-BIT MICROPROCESSOR DESIGN PROJECT

Pat Bosshart

Texas Instruments Incorporated
Dallas, Texas 75265

INTRODUCTION

This paper will attempt to give an overview of a large microprocessor design project. Its purpose is not so much to describe a particular machine, but rather to describe the design process. It will present the design approaches taken and the design tools used or created during the project, and will cover topics such as the division of work, problem areas and important results. However, in order to properly set that discussion in perspective, a brief description of the microprocessor will also be given.

The first section will provide an introduction to the architecture of the LISP processor chip Next some of the external constraints and high level decisions which affected the overall shape of the design task will be discussed. The following section will describe the design flow, with emphasis on the design tools which were used or created in the design. Tracing the flow of design information will provide the greatest detail in understanding the design process. The next section will discuss the division of labor. Finally, some of the problem areas in the design task will be described, and the paper will conclude with discussions of some of the important results from the project.

PROCESSOR DESCRIPTION

The chip is a 32-bit LISP microprocessor which is upwardly compatible at the microcode level with an existing LISP machine, the Texas Instruments Explorer-I. Microcode is contained off-chip in a writable control store; the chip implements a fairly simple microengine.

A simplified block diagram of the processor datapath is shown in Figure 1. The machine has a 3-address architecture: microinstructions may specify two sources and a destination. One of the two sources is the A-memory, a 1K-word on-chip memory intended for use as a microcoder's scratchpad. A variety of inputs may drive the other source, the M-bus. M-memory is a 64-word microcoder's scratchpad, and the 1K-word PDL-memory is used as an on-chip stack cache. The virtual machine implemented by the system microcode is stack-oriented, so the presence of the top-of-stack on-chip reduces off-chip memory references. There are numerous other inputs to the M-bus. Among them are pointers to address the PDL-memory, one of which is maintained to point at the top of stack, and registers which implement the interface to off-chip virtual memory. There are also other registers, such as the macroinstruction program counter, whose purpose is to aid the emulation of macroinstructions, and also connections to load and read other local memories, such as the microprogram stack memory. All microinstruction destinations may be reached through the O-bus.

All sources feed the execution unit, which consists principally of an ALU and barrel shifter/masker. The ALU performs all arithmetic and logical operations, in addition to atomic operations necessary for multiplication and division. The shifter performs rotation and arithmetic or logical shift operations. The masker allows a rotated field from the M-source to be merged into the A-source operand with single-bit resolution.

A simplified block diagram of the processor's control paths is shown in Figure 2. At its heart is the microprogram counter or PC. Jump and call instructions obtain the next-PC directly from a field in the microinstruction. A call instruction will push a return-PC onto the micro-stack (UPCS) memory. Microcode can implement either normal or delayed branches. A set of hardwired trap locations form another PC input, while the last input comes from Dispatch memory, and is used for multi-way branch instructions. A dispatch instruction allows any field from any source to form the index into a dispatch table whose base is specified in the microinstruction. The dispatch entry retrieved from the 45K-bit dispatch memory contains both the target address and the transfer type (jump, call, return, or no-branch).

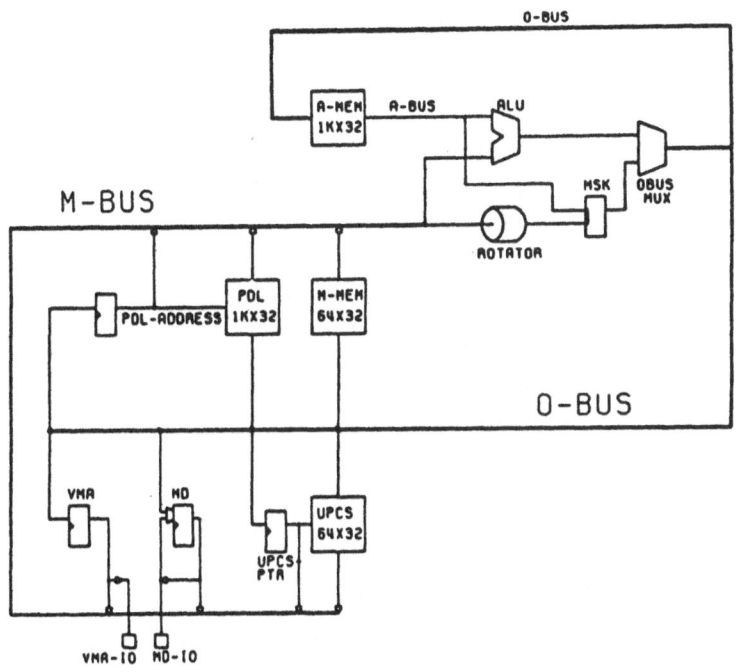

Figure 1. Processor datapaths.

The machine has four types of microinstructions; ALU, Byte, Jump and Dispatch, which differ in format only slightly from one another.

The character of this processor can therefore be briefly summarized as follows: It is a simple micro-engine with a fairly regular structure and simple datapath. It processes microinstructions, all of which are single-state instructions and are fairly horizontally encoded to minimize the amount of control logic required on-chip. In particular, there is no extensive control logic to interpret complex macro-instructions in hardware. The processor is memory-dominated; it contains 114K bits of on-chip RAM, and 16K bits of on-chip microinstruction ROM, used for self-test and boot load. Its large pin count of 224 is necessitated by the use of off-chip microcode. Chip area is dominated by three large RAM's, with a regular datapath structure occupying most of the rest of the chip area. Only a small amount of control logic exists to provide detailed control of the datapath. The processor contains about 550,000 transistors.

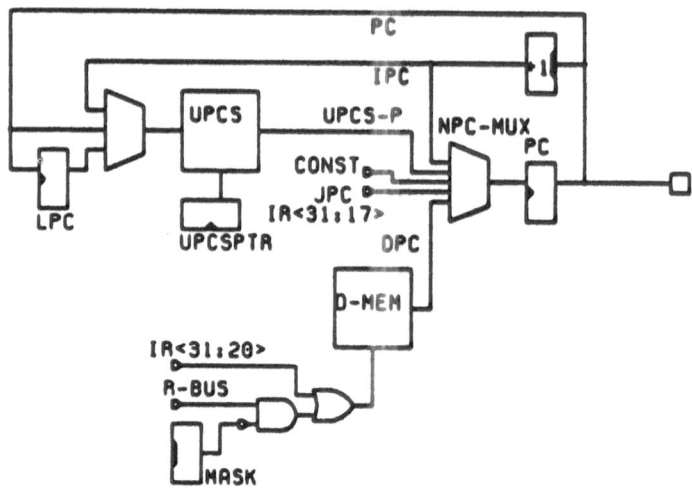

Figure 2. Processor control paths.

HIGH LEVEL DECISIONS AND INFLUENCES

This section will describe some of the outside influences and technical constraints which determined some of the global aspects of the chip design. It includes marketing, financial and schedule considerations. The most important external factor in the entire project was the source of funding; this project was done under a grant from the Defense Advance Research Projects Agency (DARPA) for the purpose of developing a compact LISP machine to be used in future embedded military applications. While the contract would provide for the development of the processor hardware, software development would be in common with TI's commercial LISP machine family. In any case, without that external funding, the project would not have been possible.

Since this processor chip was meant to be a new member of an existing computer family, opportunity for architectural experimentation was limited. Microcode compatibility was required. Another reason to restrict the range of architectural innovation was the fact that the group doing the design had never before designed a microprocessor. Though the laboratory had extensive experience in RAM design, its experience in designing logic devices was limited to a few projects of a much smaller scope. In fact, almost no one on the design team including the team leader, had ever participated in a microprocessor design A third reason for close adherence to the Explorer-I implementation was schedule It was felt that it was more important to be able to introduce the product on time than it was to achieve revolutionary improvements in functionality or performance Therefore the aim of the project was to produce a relatively faithful implementation of the earlier machine, relying heavily on IC technology for cost/performance improvements.

The large number of on-chip memories meant that the entire processor had to go on a single-chip; connecting to multiple external RAM's would have cost far too many chip pins. Likewise, there was no easy partition which would allow a multi-chip implementation. Once all the memories could fit onto the chip, their effective access times could be dramatically reduced because signals did not have to traverse chip boundaries, allowing much faster microinstruction execution rates than before. The Explorer-I executed microinstructions at a 7MHz rate; a 20MHz rate was (rather arbitrarily) chosen for the LISP chip. This proved to be an aggressive goal, for an instruction included not one but two memory access times.

Writable control store relieved the design team from the requirement for correct system microcode before the end of the chip design cycle. It also meant that simulation of all the high-level system microcode routines was not considered part of the chip design. The job of the design team was reduced to designing a machine which would correctly implement the four above-mentioned microinstruction types. This is perhaps a factor of two smaller job than designing a machine with a large amount of built-in microcode.

The design team leader had 6 months before the start of the design to learn enough about the processor so we would know what it was we were supposed to build. During that time there was a large amount of interaction with the designers of the Explorer-I. Materials provided for study included the Explorer-I processor and virtual machine specification documents, system microcode, processor schematics and PAL equations, microcode diagnostics and an architectural simulator. This time coincided with the later stages of the Explorer-I design, so design reviews, bug reports and design change notices all helped fill in details of the workings of that machine.

DESIGN TOOLS

This section will describe the design tools used during the course of the project, and how they related to each other. This will provide the most detailed account of the design process. The following design steps will be included in the discussion: computing environment, schematic entry, electrical simulation, logic simulation, functional simulation, functional test generation, RTL design, control section layout, timing verification, floor planning, manual layout entry, layout verification, datapath assembly, chip assembly object-oriented database, structural test pattern generation, fault grading, test pattern translation, and chip debug.

Computing Environment

About 9 months before the official start of the design project, an evaluation was made of commercially available design tools. Our group eventually concurred with the decision made by our Design Automation Department (DAD) to use the MENTOR system for schematic entry. We also chose at that time to buy a ZYCAD logic simulation accelerator. At the start of the design project, our local computing environment consisted of the following machines: 1 VAX 11-750 and 2 VAX 11-780's connected via DECNET (also used by other groups in the laboratory); 4 APOLLO's (2 DN600's and 2 DN660's) connected together via Apollo's DOMAIN network, but not connected to the VAX'es by any network; One Symbolics-3600 LISP machine connected to a VAX via CHAOSNET; 1 ZYCAD LE-1002 hardware simulation engine interfaced to one of the VAX'es; one Fairchild SENTRY-20 64-pin chip tester, not interfaced to any other machine; one AMRAY Scanning Electron Microscope with a LINTECH waveform sampling unit to do SEM probing; and an RJE link to more IBM mainframes than we could ever possibly afford to use.

During the course of the design project, we acquired a VAX 11-785 and transferred the ZYCAD interface to it. This machine was mostly used for running large batch jobs, such as SPICE and the ZYCAD flattening software. All the VAX'es were clustered during the course of the design. We acquired 4 more Symbolics 3600 LISP machines, and when the TI EXPLORER LISP machine became available, 6 of those. These LISP machines were connected to the now ubiquitous Ethernet which linked all our systems, including the APOLLO's. We acquired a CALMA system, mostly out of fear that the APOLLO-based in-house layout editor would not be able to handle large databases. That fear turned out to be unfounded, and the extra layer of data transmission and translation made the CALMA so inconvenient to use that we hardly used it at all for layout. It was used as the driver for the color VERSATEC plotter we acquired. For testing we acquired a MEGATEST MEGAONE tester with 220 pin capability, also interfaced via Ethernet; and a MICROVAX to interface to the SEM prober. It took a long time into the design cycle before TCP/IP Ethernet was available to interface APOLLO's to VAX'es; during the interim we had to make do with a file transfer interface routine built on top of an RS232 link. During the course of the project, an SNA link was also added between the VAX'es and the central IBM mainframes.

Schematic Entry

Schematics were entered using the MENTOR schematic editor resident on APOLLO workstations. We wrote our own software to interface MENTOR to our design system; MENTOR is a closed-architecture system which makes this task more cumbersome by not allowing the user to directly get at the internal database. Instead, a service routine outputs MIF (MENTOR Intermediate Format), a text file expressing the connectivity of the design block. Unfortunately, this routine wanted to flatten hierarchical schematics before producing MIF, rendering it useless for all but very small circuits. It was a small matter to defeat the flattening mechanism to output a hierarchical description in MIF. MIF was then translated to HDL, TI's Hardware Description Language, and hierarchical SPICE descriptions. HDL is the hierarchical description language used for schematic verification and for the centrally-supported logic simulation programs. There was no automatic provision for updating old MIF and HDL files; updating was initiated manually.

The largest problem we had with the MENTOR system was that it was very inconvenient to enter schematics of very large blocks; a symbol for the datapath would have had over 400 interface signals. There was no facility which would allow a user to write software to draw schematics or manipulate the internal database in any way. Another problem we had with MENTOR was that while HDL and SPICE use call-by-position syntax, MIF uses call-by-name syntax. Therefore, in a MIF subcircuit call, the order of the interface signals carries no meaning, while that is not true in HDL and SPICE. After one of MENTOR's new software releases, we found that the MIF signal ordering had changed from reverse alphabetical order to alphabetical order, with the result that all the HDL and SPICE data had to be regenerated to restore consistency.

Electrical Simulation

For many performance-critical circuits, we performed electrical simulation using TI-SPICE on our VAX 11-780 or 11-785. These blocks included the RAM's and ROM, the ALU and shifter, the clock generator, and most of the library cells used in the datapath, control section, and pad frame. Schematics were drawn on MENTOR to provide the circuit description, and MENTOR logic simulation log files could be automatically translated to provide input stimuli, or the stimulus control text files could be written by hand. A variety of output formatting tools were used, most commonly waveform plot routines which could add additional information such as signal transition times. These were viewable on screen or as hardcopy. Other formats included outputs where the parameters of interest such as delay times, rise and fall times, and voltage levels were automatically extracted.

Before running SPICE, circuit files were sent through a program which added estimated parasitic diode areas. A simple formula was used which related average diode area to transistor width. The formula was arrived at by processing layout extractions of previous designs, producing statistics and adding a dose of conservatism. In post layout simulation, we never had to change a circuit design due to layout parasitics except in a few cases where the designer had forgotten to include the effects of long metal wire capacitances.

A program called SPYE [1] (Statistical Parametric Yield Estimation) was used to determine the parametric yield of a circuit vs. process statistics. For example, this was used to compute the yield of the RAM's in meeting speed requirements given typical process variations in the transistor performance. Another program, OPDIC, was used to automatically choose ciruit transistor sizes based on desired performance parameters. Both of these programs are based on SPICE.

Statistically accurate transistor parameters were automatically extracted from test chip devices using the AMSAM (Automated MOS Statistical Analysis and device Modeling) [2] system, which provided SPICE models for mean, +sigma and -sigma devices over temperature. Large amounts of statistical process data were collected and analyzed using the SPADS (Statistical Process Analysis and Design rule Synthesis) [3] system to validate process integrity and accurately model other process parameters.

Logic Simulation

Very small logic simulation jobs were initially run on MENTOR's resident logic simulator. That approach was extremely limited both due to the inconvenience of entering stimuli manually, and due to the need to visually check outputs for correctness. However, the most important limitation was due to the long conversion (EXPAND) times from schematics to the logic simulator circuit structure database. At one point, the entire ALU was simulated there, but the EXPAND time was measured in hours.

For most of our logic simulation jobs, we used a ZYCAD LE-1002, a hardware accelerator for logic simulation. The ZYCAD was resident in our local computer facility, and was interfaced first to a VAX 11-780, and later to an 11-785. Our design group wrote all our own interface software for the ZYCAD. We output a simple hierarchical description language called HIF (Hillcrest Intermediate Format) from MENTOR's MIF to use as the input language for the flattener, the program which would produce the flat (non-hierarchical) circuit description in the required binary ZIF (ZYCAD Intermediate Format). Our first generation flattener was so I/O bound that it would take 3 hours to flatten the datapath, even though the CPU time was considerably less. The second generation flattener eliminated the I/O bottleneck and could flatten the entire chip into about 60K ZYCAD primitives in about 35 minutes. Moving the interface to the VAX 11-785 reduced flattening time to under 20 minutes.

Our third generation flattener stores Pascal binary record versions of each block, so time spent in parsing input files and creating the internal data structures for the flattener is minimized. This system, which was not available in time for the LISP chip PG release, can typically flatten the entire chip in about 4 minutes of 11-785 time, if not many newly changed blocks must be parsed from MIF. It also eliminates the extra step of translating from MIF to HIF.

We also defined formats for stimulus, expected response, and signal and memory initialization files. These files were normally produced by the functional simulator. When running, the ZYCAD would produce binary files storing output signal values, and these would be compared with the expected results on the VAX. The user output would be an error list, displaying for each time which signals had incorrect output values. Other output interface routines allowed probing any internal ZYCAD signals.

The ZYCAD was also used to generate long tests which were used at wafer probe. In these cases, the functional simulator was not run. The input stimulus files were trivially simple, and the signals of interest were probed inside the logic simulator. The ZYCAD would simulate the entire chip at about 15 clock cyles/second. When running tests generated by the functional simulator, typically a few thousand signals would be probed, so the ZYCAD execution rate would decrease by about 50% due to the increased I/O activity. No attempt was made to decrease the level of detail in the ZYCAD simulation in order to speed execution; many domino circuits were modeled at the transistor level.

Functional Simulation

Our functional simulator had two conflicting goals. The first was to provide a high-level architectural simulation of the processor, while the second was to provide a system which would produce test patterns for logic simulation, so that the logic simulation results would never have to be checked manually. The problem with these goals is that the very same decisions which lead to a simple, efficient architectural simulator also make it unlikely that it will track the operation of the logic model closely enough to allow automatic test pattern comparison at every clock cycle.

In our functional simulator, the control section was modeled at the Register Transfer Level (RTL), but the datapath and memories were modeled functionally. This turned out to be a mistake; the datapath should also have been modeled in RTL. We had inherited an architectural simulator from the Explorer-I effort, and instead extended its modeling of the datapath until it had sufficient accuracy for test pattern generation. This turned out to be a time-consuming, error-prone process; the architectural simulator grew unwieldy as the level of detail it carried increased. When using test patterns to debug logic simulations, for every logic bug which was found, several functional simulator bugs of various sorts had to be repaired.

The original Explorer-I architectural simulator and our functional simulator were both implemented on a LISP machine, and extensively used capabilities inherent to the LISP language which are not available elsewhere. Typically this meant using the capability of LISP to efficiently represent LISP programs as data. Programs would be written to write custom programs which would run very efficiently after compilation.

In a normal hi-level simulator, there is some type of decoding for each microinstruction field in order to determine what operation to perform. This decoding activity is often far more expensive than modeling the microinstruction itself, which may simply add two registers together and put the results in a third. Instead of decoding instructions at run time, the simulator decoded each of them at load time, producing a LISP function which would express exactly the intended effect of that microinstruction. These were then compiled, and executing only these at run time eliminated the extra overhead. The architectural simulator ran at an execution rate of 1000 microinstructions per second.

The control section of the processor chip was written in RTL equations, which drove both the functional simulator and the control section layout synthesis program. The control section RTL simulator was quite simple, in that it needed neither hierarchy nor the ability to represent signals more than one bit wide. It also only needed to model static gates and registers; all other more detailed timing information was suppressed, so it did not need to model events at a finer grain than a complete clock cycle. The RTL module given to the simulator contained both the RTL equations and a list of signals meant as either test pattern inputs or outputs. Starting from these signals, equations were included for simulation until there was a consistent set, with input signals coming from the functional simulator. In this way, unneeded equations were not simulated. These included equations were ordered for execution, so that each equation was evaluated exactly once per clock cycle. LISP machines have an internal stack which can hold up to 128 function arguments and local variables. During execution, the 1-bit wide control equation signals were packed many to a word and pulled onto the local stack, so that all the logical operations worked off the stack, eliminating data memory references. The equation translator wrote a LISP function to copy values onto the stack, do all the single bit logical operations and then return values to storage. This function was compiled, and would typically run at 20,000 gate evaluations per second. Custom functions were also written (by LISP programs) to write test patterns, so that pattern generation runtime overhead was minimized.

The main problem with this approach of building custom functions was the time required to compile them. Depending on the size of the control equation set being run, the turnaround time to make an equation change varied from 3 to 12 minutes. Fortunately, most equation development was done with small simulation modules, where changes could be implemented in under 5 minutes. During actual execution, the entire simulator with its RTL and functional parts could simulate the processor at about 3 clock cycles per second, with most of the computation time spent in the functional portion of the simulation. When generating large test pattern sets, execution speed would drop to 2 cycles/second due to the file system activity.

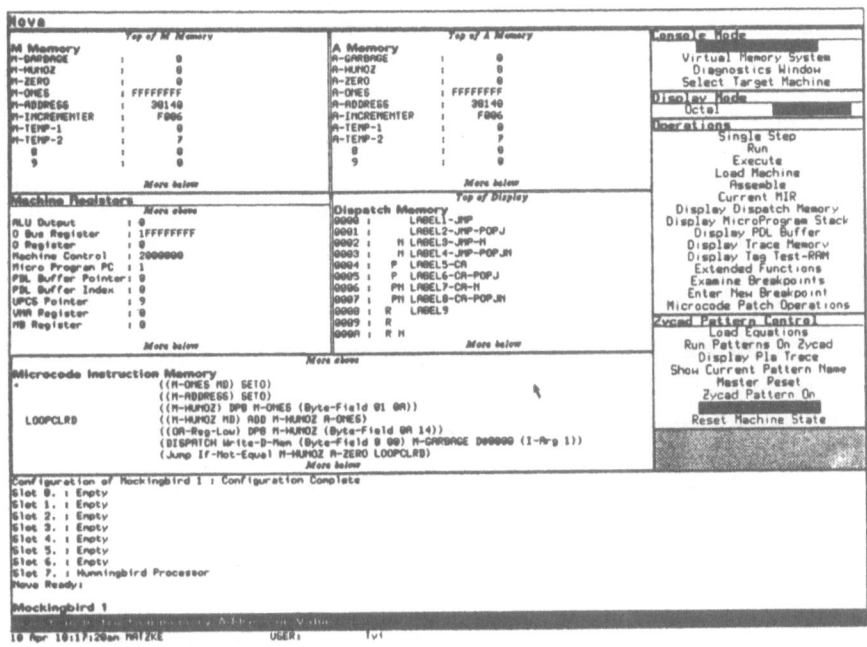

Figure 3. Functional Simulator Window Interface.

The window interface to the functional simulator is shown in Figure 3 for a typical configuration. Included are panes to display various memories, including contents of microinstruction RAM in a disassembled format. A group of registers to be displayed can be selected from a pop-up menu, and a group of control equation signals can likewise be selected. While only the current value of registers and memories are displayed, a trace history of RTL variables is kept for display, and the display window may be scrolled to show older values. The interface also includes menu items to select memories for display, load and compile control equations, enable test pattern generation, and output test patterns to the ZYCAD. Other windows show virtual and physical memory, the cache and memory map for system simulation. A window interface generator package was written which let the user generate custom interfaces with only a few pages of code.

Functional Test Generation

Functional tests were written in microcode. These could be used to exercise the whole chip or any smaller module; the correct inputs and outputs for a chip partition simply had to be listed together to create test pattern files. There were automated ways of generating input and output signal lists for almost any chip module, including various pieces of datapath, control sections, clock generators or bondpads, together with combinations of these. Some of the functional tests were self-checking; these "good" diagnostic routines had code built in to independently ensure that the various operations being tested worked correctly. Other test results had to be visually checked on the functional simulator, a clearly less desirable situation. Ultimately, good self-checking microcode diagnostics should be the final test of system functionality. Altogether there were some two dozen functional tests totaling about 25,000 clock cycles. The simplicity of the machine and the lack of a requirement to check out any macrocode greatly contributed to the small size of the microcode test suite.

RTL Level Design

Very early in the design it was recognized that even with this simple processor, the amount of control logic present prohibited gate level design of the processor control section. Instead, arbitrary boolean equations were written to drive the simulator and control section autolayout system. A typical equation is written below:

$$(\text{DEFEQN A}\quad \text{F}\ //\text{G} + (\text{B} + //\text{C})\ \text{D E})$$

This would define a complex static gate producing the signal A as an OR of two terms: the first is an AND of two signals, F and NOT G (// means complement), while the second is more complex. The boolean equations can also express dynamic domino gates of various clock phases, registers, and latches of different phases. In general, the keyword replacing DEFEQN in the example is used as a directive to the auto-layout system, and as the design progressed, other types of gates implying specific layout styles also were included.

As the equations were processed to assemble a layout module, the equations were manipulated in several ways. Some complex equations were factored into simpler ones, sometimes default terms (such as scan connections) and additional equations were added mechanically. For conversion to simulation, similar processing was done. For example, equations were added to check multiplexer control signals for conflicting inputs.

For simulation, all equations types were also compressed into either static gates or registers. As a result of this loss of detailed timing information, this timing had to be verified in a different way. A small rules-based Timing Syntax Verifier was used, which ensured that gates were not connected together in illegal ways. For example, a domino gate of a particular clock phase can drive another identically clocked domino gate, but not through an inverter. This verification approach was better than simulation because it gave instant diagnostics without having to simulate a particular errorful circuit. All of the control equation software was written in LISP.

Control Section Layout

Once control equations were produced and collected together into a layout module, the RESCUE (Rtl Equations Synthesize Control Unit gEometries) [4] system automatically synthesized the control section layout. There were actually three generations of the control section synthesis program produced during the course of the design. Each generation allowed more layout flexibility, produced a more complete layout solution with less manual intervention,and increased the layout density, with perhaps 1/3 of the total 1.5 man-year software effort specifically directed at achieving enough layout density to allow the control logic to fit into the allotted area.

The control section was laid out as a regular array of static and dynamic domino gates, river-routed to an overhead section where output buffers and registers were placed. Outputs from the overhead section could be routed back into the array, so that several logic levels could be implemented per clock phase. A layout block diagram of this arrangement is shown in Figure 4. An extensive analysis of charge sharing effects in the domino array was done; this sometimes resulted in manual directives to the array folding program to minimize charge sharing effects, in addition to other corrective measures. It was also found that by manually altering the folding of the array, about 10-15% size reduction could be achieved. This manual folding manipulated intermediate folding files and required a few man-weeks of effort at the end of the design cycle.

Part of the layout synthesis effort included automatically writing HDL, our logic description language. This output was used for logic simulation and schematic verification (checking of layout topology vs. HDL).

Timing Verification

While SPICE was used to check the speed of small blocks and some functional units like ALU's, RAM's, ROM and clock generator, it was incapable of checking the performance of the entire chip. In the datapath, there were no formal tools to support timing verification. The execution unit of the datapath (the module including the ALU and shifter) was extensively SPICE'd, but most other datapath sections were not. Since the datapath was composed mostly of a relatively small number of library cells, those cells were characterized and used to manually calculate signal timings. Datapath cells were characterized using loads large enough to include standard datapath wiring. While the architecture of the datapath tended to concentrate critical paths in a few well-known places, mostly involving RAM accesses, the lack of true timing verification allowed a critical timing path to escape until almost the very end of the design cycle

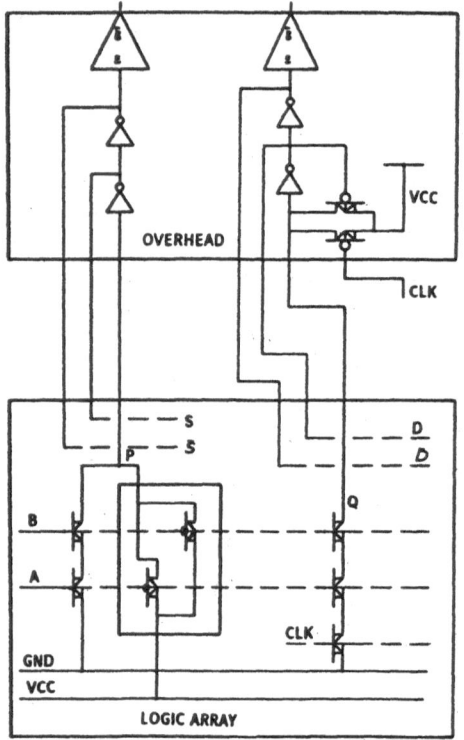

Figure 4. Control section layout block diagram.

The control section had many critical paths, so automated verification of timing [5] was essential. A small custom timing verifier was written to check the speed of operation of the control section; since it only had to deal with a few types of logic gates, simple equations could be written for gate delays and calibrated using SPICE runs. These equations worked for any synthesized gate, and included effects due to fanout and actual wiring lengths. A static timing analysis was then done to compute delays of all signals. The timing verifier was integrated into the LISP-based equation processing package, and gave turnaround times of under a minute, allowing rapid interactive timing optimization.

Floor Planning

Chip floor planning was made easy by the simple chip architecture and large amount of memory. Arranging the large datapath modules to communicate effectively with the RAM's above and the bondpad connections at the left and right ends was the only requirement. The larger floor planning task was estimating datapath length. Very near the start of the design, about one man-month was spent doing a coarse datapath design, and performing quick trial layouts of various blocks to estimate area. That estimate proved to be about 17% optimistic, partly because extra functions were added through the course of the design, and partly because the overhead circuit areas of the small RAM's and ROM which lay in the datapath were underestimated.

Manual Layout Entry

Manual layout entry was done on an in-house layout editor ICE (Integrated Circuit Editor) running on APOLLO workstations. Our layout assembly tools also output to this database, and the final chip layout database resided there. The fact that our layout editor used a LISP-readable text file format (LAFF – LISP Archival File Format) as an ASCII transmission standard made it much easier to interface to our LISP-based layout tools.

With the exception of the RAM's and ROM, all layout and chip assembly was done in a coarse grid sticks layout style. A locally-generated software system called INCOGNITO (INtelligent COarse Grid Non-compacting InstantiaTOr) was used to instantiate these symbolic layouts into full geometric realizations. The resultant geometric layouts were read back into a different layout database. Schematic verification programs could be run off the symbolic layout database before instantiation, or off the geometric database afterwards.

INCOGNITO was resident on a LISP machine connected through Ethernet to the APOLLO'S. The program did not do compaction, but it did perform some local movement of geometries to satisfy design rules. In other words, symbolic grids were directly tied to geometric coordinates through a scale factor, rather than being much more loosely coupled as in compacting [6] systems. The program allowed some of the layout productivity improvements due to working with symbolic figures and working on a coarse grid while avoiding more complex issues such as pitch-matching that true compactors face. Sample symbolic and geometric layouts are shown in Figure 5.

Layout Verification

Schematic verification programs were run to ensure that the layout topology matched that of the schematics, represented by HDL. Schematic verification was performed by running batch jobs on remote mainframes. For small jobs the layout and HDL databases were shipped with the job, but for larger tasks each transmission and processing step could be run to completion separately, with the intermediate databases kept on the mainframe. For large tasks, a separate program was used first which checked the layout for opens or shorts in the signal names placed in the layout. The diagnostics from this program were so much better for high-level

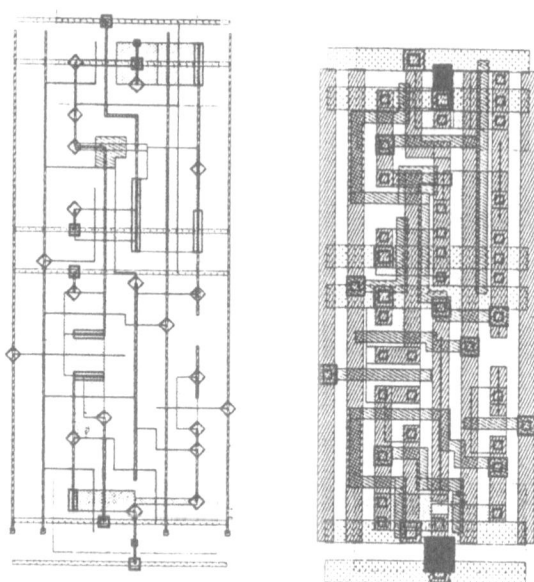

Figure 5. Sample symbolic and geometric layouts.

interconnect errors that it was generally useless to run the SV program until the open/short checker came back error-free. A liberal sprinkling of (automatically generated) global signal names made this program very useful. When the whole chip was verified, the inner core cells were removed from the large RAM's, but the core cells of the small RAM's and ROM were left intact.

The SV system suffered from productivity problems. The fact that it was a batch program on a distant mainframe gave tasks a minimum overhead time which lowered productivity on the numerous small SV jobs. The fact that it didn't use hierarchy made turnaround times unavoidably long for big jobs, such as verifying the whole chip.

Small design rule verification tasks could be run locally on the APOLLO, while large jobs were run on remote mainframes. The presence of local verification improved productivity on small jobs, but the non-hierarchical nature of the tool made turnaround times long when verifying the whole chip, or large fractions of it. We attempted to run medium-sized DRC jobs on our resident CALMA, but the overhead of translating the layout data from the APOLLO-based system negated any productivity gains, and the checker was incapable of verifying some of our more complex rules. The diagnostics were also poorer, so this effort disappeared shortly. We had also attempted to use the CALMA for layout entry, but again the database translation and transmission times made it so inconvenient to interface to our design tools that we soon stopped.

When verification jobs were run on the centralized IBM mainframes, until we were verifying the entire chip the layout data was sent over the SNA link. When the layout database for the entire chip had to be sent, it became faster to write a tape and carry it the few miles to the main site.

CIDER, (Circuit Interconnect and Device Element Recognizer), the circuit extractor used for schematic verification, was also used to extract layouts with parasitics for post-layout SPICE simulation. It was capable of extracting area and fringing capacitances for either node to substrate or node to node capacitances, but was incapable of extracting fringing capacitance between two adjacent wires on the same layer. It could also extract resistances, but none of the circuits which were of a scale small enough to simulate on SPICE had resistances of interest. The resistance extraction program was part of a larger timing verification system which at that time was not sufficiently mature to use in a production mode.

Datapath Assembly

While small cells were laid out manually, all datapath and chip assembly was done via software, so that the final chip layout was intended to be "untouched by human hands." The only point at which this was violated was by some final manual edits which widened some power supply wires to use up available space around the pad frame. Manually laid out cells included library cells for the datapath, control section and bondpad, and various larger special purpose cells in the datapath, such as the ALU. The execution unit, RAM's and ROM were also assembled by hand. The datapath placement and routing, pad frame and final chip assembly were all accomplished using special software tools, all written in LISP.

An initial file to list datapath components in order to estimate datapath length grew into a full datapath floor plan, which usually simply placed cells by abuttment. Datapath library cells had a standard geometry placed in them which indicated the length of datapath they would consume. The datapath cells were placed by processing these two sources. All datapath cells also had their interconnect points indicated by geometries in the cell, so these could be extracted by the datapath router. Datapath routing between cells was done horizontally using the second metal level, and normally the I/O points in library cells were run on vertical first metal wires which attempted to span as much vertical distance as possible in order not to constrain the 2nd metal routing interconnect points.

The datapath HDL was read to determine the high-level connectivity, and the datapath placement and cell I/O position information were collected. The datapath router then performed a routing, where the simple wires were routed correctly, but in difficult situations would allow wires to overlap and cross in order to complete the routing. We then used a routing editor on the LISP machine which would allow the user to move wires in order to eliminate overlapping and crossing wires, but would not allow any required connections to be broken or additional connections to be made. This eliminated the potential for the manual datapath routing effort to introduce connection errors in the datapath layout.

The routing editor would choose the most complex bit of the particular datapath section for the user to manually route, and then use that as a template to route the other bits, presenting the user bits which still were not correctly completed for further work. Signal names were placed on all connection points and connecting wires to aid in the open/short check and in schematic verification. Intermediate results were stored a format which allowed the wiring to be reconstructed automatically when small changes to the placement occurred, for example, when the size of a particular datapath layout cell changed.

Chip Assembly

The pad frame was laid out using a manual software approach. This program simply started with an ordered listing of bondpads and the layout cells which implemented them, together with the cell sizes. It constructed the frame and used the resulting terminal positions in the routing software. A simple floor plan of the entire chip allowed assembly of the top level layout cells to be relatively independent of final sizes.

The channel routing between the control section and the datapath used three level routing; poly and 2nd metal ran horizontally, while 1st metal ran vertically. YACR, (Yet Another Channel Router), a two-layer channel router from the University of California at Berkeley, was used to do the routing, given the absence of a three layer router. A program divided the routing into two separate jobs, with some time-critical signals constrained to route on 2nd metal, and others free to choose either layer. The two separate routings were afterwards combined, with post-processing to eliminate any 1st metal conflicts between the two routings. This worked fairly well, since the 1st metal layer was quite unconstrained in the layout, whereas the other two were naturally much more critical. In the end, however, our software couldn't always eliminated the conflicts, and about half a dozen places had to be manually edited to finish the job.

Once the entire chip layout had been assembled and was undergoing final touch-ups, there were still a few late logic simulation bugs which had to be corrected. Rather than run the entire control section, datapath or channel routing software over again, the last few small changes were entered into the layout by hand.

Object-Oriented Database

A design tool which was used in several different places was DROID, (Design Representation using an Object-oriented Integrated Database), our object-oriented database system. This LISP tool implemented a data structure which could represent schematics and some aspects of layout in an internal Flavor-object based representation. The representation was documented for users and easily extensible. Interface routines were provided to read in and write out common design languages, and many utilities were included which allowed users to write software which moved around in or manipulated this database.

The effort to write HDL for the control section of the chip actually created this database, and then used a standard utility to output the HDL text. The datapath layout system used DROID to represent the layout and HDL data in a common form. There were also other manipulations of the HDL representation used for simulation, schematic verification and test pattern generation, which were implemented with DROID.

Structural Test Pattern Generation

We committed very early to using a Scan design, as well as several other design for testability strategies [7]. While the control section uses serial scan, the datapath uses a different arrangement designed to make use of the data buses already present. The scan system only added 2.5% to the total chip area. One reason that this cost is so small is that a large fraction of the chip area is RAM, which incurs no scan overhead. The most important benefit a Scan design gave was the ability to use automatic test pattern generation software. We used a commercially available system, AIDSTG, which can generate tests for static combinational circuits, including transistor level faults. The HDL schematic description for AIDSTG had to be topologically manipulated in order to be acceptable as input; this manipulation was done using DROID.

AIDSTG performs its own fault grading; however, we evaluated its test patterns on FMOSSIM, a switch level fault simulator, and on the ZYCAD run in serial fault simulation mode. Only after resolving the differences between the three systems were we willing to believe any one of them.

Other design-for-test features implemented include special microcode extensions for testing the on-chip memories at full speed, and enough space in the on-chip microcode boot ROM to execute a self-test routine. Also included were signature analyzer registers on the main data buses for compressing test results, use of a scanring or boundary scan to enable wafer-probe testing of this 224-pin chip using only 13 signal pins, two-pattern test capability, and zero static power for current-mode testing.

Chip Debug

Wafer probing was done on a MEGATEST MEGAONE tester connected via ETHERNET to the rest of our machines. Functional test patterns were generated either by the functional simulator or by the ZYCAD logic simulator. A LISP-based facility translated these HIF format vectors to the VPL format required by the MEGAONE. This translator could also take functional patterns and expand them out to scan mode, or take structural test patterns generated by AIDSTG and produce the correct scan sequence to set up the required state and observe the output state. This translator is part of a larger system, currently in development, wherein the functional simulator can be run to produce tests which are translated to VPL and interactively run on the MEGAONE across the ETHERNET. Test results are brought back to the functional simulator for display. This will allow the design engineers to run tests directly on the finished chips in the same way the functional simulator was used during the design stage.

Early in the design cycle, a test chip was produced which contained the large RAM's and process test structures. This was tested on a FAIRCHILD SENTRY-20 tester with 64-pin capability.

During initial chip debug, the MEGAONE was used to run functional tests, and an Electron-Beam Prober was used to measure analog waveforms on internal nodes. Many nodes had small pads with protective oxide openings to facilitate SEM probing. This was used to verify that all the clock generator delays were in spec, for example, and to chase down some problems with voltage margin. It was also extensively used to characterize the RAM's on the test chip.

DIVISION OF LABOR

This section will describe how the design was apportioned among the various members of the design team. Perhaps the most striking feature of how this design was accomplished was the large amount of work which went into custom tool development, and the large fraction of designers who were also required to be expert or at least adequate programmers. Often custom tools can be implemented quickly enough and have a large enough impact on design productivity that the payback period is considerably shorter than the design cycle. An additional advantage to this approach is the improvement in the design infrastructure remaining at the end of the design cycle. Having a large fraction of the design team actually doing tool development means that fewer people are actually involved in the details of the design, with the result that design assumptions and changes have to be communicated among fewer people; the number of people who actually have to know how the processor works, in whole or in part, is significantly reduced. It is an efficient way to structure a large design project, given that productivity is often inversely proportional to the number of people involved in any given task.

Of the 14 engineers working on the design, only 4 did not do significant software development. Of the remaining 10, 6 spent literally all their time developing design software, with 3 of the other 4 spending large amounts of time in software development. As a result, the productivity of 7 of the team members was completely determined by their software development skills, rather than by their IC design ability, and the productivity of 2 more team members was significantly influenced by their software skills.

Another way of viewing this division of labor is to examine what various team members needed to know about the design in order to do their job effectively. Three team members who worked on basic tools needed to know almost nothing about the design; it was almost impossible for any design change to cause them to do any extra work. Three more team members needed to know only something about the design style of a particular part of the chip; as long as the design style remained within bounds, their programs would work and individual design changes were transparent. Two more team members needed to know more about the chip design, but still only in a limited way. These people were designing well-defined parts of the chip which were cleanly interfaced to the rest of the design, so they could proceed quite independently of the others. This meant that 8 of the 14 engineers were separated from the detailed design of the main body of the chip.

This left only 6 people who had to be directly involved in the main design flow, who might potentially have to respond to ECN's of one sort or another. The tasks of these 6 included lead designer, execution unit design, cell library design, chip schematics, logic simulation, functional simulator design, datapath layout assembly, and whole chip assembly and verification. The datapath assembly task was actually a software development job which was difficult enough that the program could not be made powerful enough to be completely insulated from design changes, but it was decoupled from most of them. In practice, not all of these 6 were at any one time involved in the mainstream of design. It was highly unusual for a design change to involve more than four people, with most changes involving two or three people.

In addition, three technicians worked on the project, mostly performing chip layouts and doing layout verification. One of these was promoted to engineer status during the course of the project. The small number of dedicated layout people was mostly due to the automated layout methodologies used in the project.

Table 1 lists several categories of tasks in the design, together with the people who worked in those areas. Table 2 lists the various team members, together with the areas they worked in. Work items marked with an asterisk indicate software development activities, where productivity was mainly limited by the rate at which software could be written and debugged; it does not include simply using existing tools.

TABLE 1 TASK COMPOSITION

DESIGN MANAGER	CRH
DESIGN LEADER	PB
ARCHITECTURE AND HI LEVEL DESIGN	PB
MEMORY DESIGN, LAYOUT	TH, ey
LOGIC & ELECTRICAL DESIGN, LAYOUT	MCC, KC, DS, TH mo, sy
LOGIC SIMULATION/CHIP SCHEMATICS	KR
ZYCAD INTERFACE SOFTWARE	RLS, MCC
FUNCTIONAL SIMULATOR	CH, PB
RTL SIMULATOR, ZYCAD TEST PATTERNS	DM, CH, PB
OBJECT ORIENTED DATABASE	SL, VK
CONTROL SECTION AUTOLAYOUT/HDL	CHS, VK
DATAPATH AND CHIP ASSEMBLY	SSM, CH
CHIP LEVEL LAYOUT VERIFICATION	SSM, KC
TESTABILITY	TS

TABLE 2 DESIGN TASKS

CRH: DESIGN MANAGER

PB: DESIGN TEAM LEADER
        FLOOR PLANNING
        HIGH LEVEL DESIGN
        RTL CONTROL SECTION DESIGN
        RTL DESIGN SYSTEM *
        MICROCODE TEST DEVELOPMENT *
        FUNCTIONAL SIMULATOR *
        FUNCTIONAL/LOGIC SIMULATOR INTERFACE *
        DATAPATH WIRING *
        LOGIC SIMULATION DEBUG *

MCC: EXECUTION UNIT DESIGN
        APOLLO GURU *
        ZYCAD INTERFACE SOFTWARE *
        MICROCODE TEST DEVELOPMENT *

KC: DATAPATH CELL LIBRARY DESIGN
        BONDPAD CELL LIBRARY DESIGN
        DATAPATH DESIGN
        CLOCK DESIGN
        LARGE LAYOUT VERIFICATION

DS: LOGIC AND ELECTRICAL DESIGN

TH: MEMORY DESIGN
        TEST CHIP DESIGN
        VERIFICATION SOFTWARE TECHNOLOGY FILES
        PG
        CHARGE SHARING, TIMING ANALYSIS

KR: DATAPATH DESIGN
        DATAPATH SCHEMATIC ENTRY
        LOGIC SIMULATION
        ZYCAD SOFTWARE*
        ELECTRICAL DESIGN

* INDICATES SOFTWARE DEVELOPMENT TASK

TABLE 2 DESIGN TASKS, CONTINUED

TS: DESIGN FOR TEST, CHIP TESTING

SSM: DATAPATH ASSEMBLY *
           CHIP ASSEMBLY *
           LARGE LAYOUT VERIFICATIONS

RLS: ZYCAD INTERFACE SOFTWARE *
           CALMA SOFTWARE *

SL: OBJECT-ORIENTED DATABASE *
           FULL-CHIP HDL *
           TESTABILITY RESEARCH *

CH: FUNCTIONAL SIMULATOR *
           MICROCODE TEST DEVELOPMENT *
           BOOT ROM PROGRAM *
           CHANNEL ROUTER *
           PAD FRAME PLACE/ROUTE *

DM: FUNCTIONAL SIMULATOR ENVIRONMENT *
           FUNCTIONAL SIMULATOR TEST PATTERN SOFTWARE *
           RTL SIMULATOR *
           CHARGE SHARING ANALYSIS *
           TEST PATTERN TRANSLATION *
           TESTER INTERFACE *

CHS: CONTROL SECTION LAYOUT SYNTHESIS *

VK: CONTROL SECTION HDL SYNTHESIS *
           OBJECT-ORIENTED DATABASE *

* INDICATES SOFTWARE DEVELOPMENT TASK

PROBLEMS

This section will describe some of the problems we had in the design. In many cases these problems were just the inevitable result of inadequately planned tool capabilities; other times they were the problems we faced in dealing with early versions of our own tools, which weren't quite up to the job. Other problems just seemed to arise spontaneously.

The biggest problem area in the entire design was the functional simulator. Recall that it was composed of two segments: an RTL simulator for the control section and a functional part for the datapath. The RTL simulator worked extremely well, but the functional part was a constant source of problems. It was a special piece of code to model the datapath, which didn't always model it with sufficient accuracy. When errors occurred in the ZYCAD logic simulation, often it was the fault of the functional simulator. Sometimes the error was in the functional simulator itself, and sometimes it was in the mapping between functional simulator variables and logic simulation signals. We attempted to cover up for the simulator's deficiency by allowing extra control signals in the RTL simulator to compute on a cycle by cycle basis whether any particular datapath signal should be compared with the ZYCAD output or not. But this was not adequate either. Of the 91 errors found during logic simulation, 50 were in the functional simulator and its associated test pattern generation modules, 20 were control equation bugs, with only 21 being true logic errors. The person debugging the logic simulation had to know the chip specification, the logic design of the entire chip, the control equations, and the functional simulator and test pattern module code. Unfortunately, there was only one such person on the team, and he was already quite busy enough.

The obvious better choice would have been an RTL simulation of the datapath, but by the time we realized this, we had no additional resources to develop it.

Another problem area was the first generation ZYCAD flattening software. It was slow; requiring 3 hours to flatten the datapath. As a result, logic simulation productivity was very poor until the second generation software emerged about 4 months before the end of the design cycle. This problem, coupled with the functional simulator problems, made the logic simulation task constantly on the critical path, with the result that in the last month before PG there were 8 (small) logic fixes to the chip, just as the chip layout was going through all its final verification stages.

During early tests of our schematic verification capability we wanted to ensure that there would be no problems of scale in trying to verify the topology of a circuit containing about 100K devices (RAM cells in the large memories were omitted).

We found that the HDL we were producing would run much too inefficiently on that portion of TI's SV program which flattened hierarchical HDL into an internal database. The problem was that while HDL and the TI design system in general makes use of arrayed signals and sub-component calls for the sake of efficiency, the MENTOR system didn't preserve the array information. Each member of an arrayed component call would be called separately in MIF. In order to get the HDL input routine to work, we would have to reconstruct the arraying information in our HDL, by combining separate calls to arrayed sub-components into a single arrayed call, and combining arrayed signals likewise. Even if we would have wanted to do this manually, there were still many design changes filtering in, so there was no choice but to do this with software, so updates could be readily included. The DROID database was used for this task. In about one man-month, a utility was written which took the internal representation and combined together all signals and components which could be arrayed. The signal or component name and subscripts were used to guide the arraying process, so that $X<1>$ and $X<0>$ could be combined into a single signal $X<1:0>$, for example. This would not have been a large problem except that we couldn't start large scale schematic verification jobs until the HDL had been arrayed, so it was on a critical path.

Another problem with large scale HDL was one we brought upon ourselves. It was sufficiently inconvenient to draw schematics of large scale blocks on the MENTOR system that we didn't. Instead we made all I/O signal names global on the top level drawn schematics, and connected them together by name. The top-level HDL was created by loading all these blocks into DROID and creating the top-level blocks which would pass all the right signal connections down to them. This was another man-month job on DROID, but here the problem was getting consistent I/O signal lists from all the required blocks in the design. For example, the datapath control signals specified by the RTL control section design had to match the control signals output by the control section layout synthesis program, and had to match the control input signals in the datapath schematics. Other high level blocks were also involved. The main problems here were that there was no central design representation where these different types of information could be kept and resolved earlier, and that we weren't very systematic about deriving that information and making sure it was correct from the start. This was an unpleasant exercise to go through because this top-level HDL was required to do whole-chip simulations, with the result that for a while, the simulation effort had to grind to a halt until these problems were resolved. Actually, in our very first design review, this issue had been brought up by a visitor from another part of TI, but somehow we forgot to respond to it until it was promoted to emergency status.

A smaller problem which was felt earlier in the design cycle came from conflicting signal name formats in the different design languages. Different signal formats which might be acceptable in different places were foo-bar<1>, foo_bar<1>, foo-bar<01> and foo_bar<01>. It took us awhile to get all the permutations correct and have all the translators in place. This is just a small example of what happens when a seemingly unimportant detail is not organized effectively.

The RESCUE control section layout synthesis program was constantly on the critical path, due to a large degree to the magnitude of the software development effort required to reduce the layout area so the control section would fit in the allotted space. Most of the last 6 months of RESCUE development were devoted to this problem, and until close to the end of that time, it was difficult to predict how long it would take to make the control section fit. If it weren't for the fact that logic simulation was going so slowly, we would have been much more concerned about this delay. Not until a few months before PG was this issue completely laid to rest.

Not until late in the design of the RESCUE system did we get serious about modeling charge sharing in the domino gates implementing most of the control section. Fortunately, some team members were finishing their portions of the design so people were available to work on this problem, but several steps had to be taken to mitigate charge sharing effects, and the worst problem was that not until we had shown the problem to be eliminated for all gates were we sure that we could actually accomplish that goal. This was another problem area which was not completely resolved until about 3 months before PG.

One somewhat more legitimate problem area concerned the mapping of signal names from RTL to schematics. Often one RTL signal name mapped to more than one signal name in the logic model; this usually was due to differently buffered versions of a signal. In other places, the node which was required to have the name that matched the RTL signal name was not the node which would have to be initialized for logic simulation; it may have been one buffer downstream from the initialized node. These are a few examples of the problems which occur due to the sometimes loose mapping between RTL and schematics. We handled most of these classes of problems on an individual basis, but we recognize that a more formal mechanism was required.

After careful datapath floor planning and routing analysis, the height of each datapath bit was set at 12 2nd metal pitches. Fifteen months later we found the part of the datapath where the 13'th 2nd metal wiring track per bit was required. Since the region involved was rather small and was populated with regular circuits, we did a relayout of that area which allowed some of the more local connections to be routed internally to the cells on the poly layer. The problem was relatively quickly taken care of, but it did create quite a bit of worry while it lasted.

About 6 months into the design, we released PG for a test chip which contained the three large RAM's and several process test structures. This would allow us to verify the performance of a critical part of the design, while giving the production fabrication facility a vehicle to test yield on. After PG we suffered a long series of delays. All along we had been working on getting a commitment from a production fabrication facility to process the both the test chip and the processor. Just as the test chips were finally about to start in the process facility after some delays due to reticle sizing problems, that commitment evaporated. So we ran PG again and produced reticles for our process development laboratory, where a few weeks later the test chip was again started, only to have the lot scrapped a week later due to resist problems. When the lot started again (this time for real!), 12 weeks had elapsed from the original test chip PG.

Several of our tools were in a state where they could accomplish 95% of the job, but the remainder had to be done manually. This was true of the control section array folder, the 3-layer channel router, and the standard cell layout tools we used for the clock generator. With this approach, when a small change occurs it is easier to update the layout manually than it is to rerun the program and manually re-do that last 5% of the job. Unfortunately, if there are ten changes, as each one occurs it is easier to enter it manually, but in the end a lot of manual work has been done. With fully automatic tools changes are inexpensive, but with almost automatic tools changes are only inexpensive if they don't come at the end of the design.

IMPORTANT RESULTS AND CONCLUSIONS

In spite of all the problems and delays in accomplishing the design, there were many advances in design technology which emerged from this work. Some of them emerged early in the design cycle and resulted in areas of the design which went smoothly, while others emerged only as hindsight.

One task which was very pleasant due to the available tools was the design of the processor control section. Writing control equations was a powerful way of expressing a very flexible form of logic. The RTL simulator, other than its long compile time, was an almost ideal tool on which to debug these equations. The internal representation of the equations was simple to manipulate, so various analytical tools, such as a timing syntax checker and a timing verifier, could be added to provide early and thorough checks of important aspects of the design. The existence of the RESCUE layout synthesis tool meant that changes could be entered with impunity, since no one else was involved in individual changes. Over the course of the design, ten control equation releases for layout were made, some of which were for RESCUE development purposes.

The approaches used to ensure a testable design have proven so worthwhile that it is difficult to see how anything can be accomplished without them. In particular, our recommendation is to never, ever, ever do a processor design which does not use scan. To avoid a scan system is to avoid being serious about quality. Scan design provides a basis for a large range of capabilities; it is the foundation upon which more difficult test problems can be solved.

The goal of producing a final chip layout "untouched by human hands" lead to several software tools for layout assembly, which provided far more flexibility than manual layout methodologies would have. This was true of the datapath placement and wiring programs as well as the full chip assembly tools. Changes would be implemented easily by changing and re-running programs, whereas in manual approaches, once work had been invested in doing the job once, it would have to be thrown away and re-done to incorporate changes. Small details of the datapath floor plan evolved continually as better ways were found to place and route components, resulting in more efficient layouts. Without our software-intensive layout methodologies, these improvements would have been too expensive to implement.

The initiation of the DROID object-oriented database effort was a very important result. It provided a quantum leap in the ability of the designer to manipulate design data. Integrated databases are clearly the wave of the future, and placing this development in the middle of a design project gave the opportunity to use this new tool to solve some very real-world problems. With this tool as well as with others, initiating new tool efforts within design projects provides quick feedback about real issues and problems, and serves as an effective guide to keep tool development moving in the right direction. Of course, the problems inherent in trying to use tools as they are developed are not at all trivial, but they are minimized by maintaining close communication between the chip builders and the tool builders.

The demonstration of LISP-based design tools was an important result. Some of the capabilities of the language and the programming productivity provided by the LISP machine environment are markedly superior to any competing systems; but that can only be proven by results. We feel that some of the tools we have produced are a step towards demonstrating those results. In any case, tool development in our laboratory will continue to move solidly into the LISP environment.

The goal of requiring almost every designer to be a programmer (usually a LISP programmer) produced the most important results of the entire project. After awhile, software infrastructures were built up so that it was easy for designers to accomplish small tasks by writing programs. Once the designers got used to the idea of writing software to accomplish design tasks, they felt that without that capability they couldn't get anything done at all. This software-based design methodology caused some painful learning curves, but it was a good investment. While first generations of home-grown tools often often didn't quite do the job, the next versions were much better, and now the tool effort has taken on a momentum of its own. After all, there are no more dedicated CAD people in the world than those who have just recently suffered through a design.

REFERENCES

[1] P. Yang, D. Hocevar, P. Cox, C. Machala, and P. Chatterjee, "An Integrated and Efficient Approach for MOS VLSI Statistical Circuit Design," IEEE Transactions on Computer-Aided Design, VOL CAD-5, No. 1, Jan. 1986, pp 5-14.

[2] P. Yang and P. Chatterjee, "An Optimal Parameter Extraction Program for MOSFET Models," IEEE Transactions on Electron Devices, VOL ED-30, No. 9, Sept. 1983, pp 1214-1219.

[3] A. Nishamura, S.S. Mahant-Shetti, J. Givens, E. Born, R. Haken, R. Chapman, and P. Chatterjee, "Multiplexed Test Structure: A Novel VLSI Technology Development Tool," Proc. IEEE Workshop on Test Structures, Long Beach, CA., Feb. 17-18, 1986, pp 336-355.

[4] C. Shaw, P. Bosshart, V. Kalyan, T. Houston, and D. Matzke, "RESCUE: A Comprehensive Control Logic Layout Synthesis System," 1986 IEEE International Conference on Computer-Aided Design, Santa Clara, Ca., to be published.

[5] J. Ousterhout, "Crystal: A Timing Analyser for NMOS VLSI Circuits," Proc. 3rd Caltech Conf, on VLSI, March 1983, pp 57-69.

[6] N. Weste, "MULGA, An Interactive Symbolic Layout System for the Design of Integrated Circuits," Bell System Technical Journal, VOL 60, No. 6, July-August 1981, pp 823-858.

[7] P. Bosshart, and T. Sridhar, "Test Methodology for a 32-Bit Processor Chip," 1986 IEEE International Conference on Computer-Aided Design, Santa Clara, Ca., to be published.

13

Architecture of Modern VLSI Processors

Priscilla M. Lu
Don E. Blahut
Kevin S. Grant

AT&T Information Systems
Holmdel, New Jersey

1. Introduction

In recent years, the focus of VLSI architecture effort has been primarily on the tradeoffs possible in new microprocessor instruction sets. The result has been a collection of machines with new streamlined instruction sets, and new hardware subsystems tuned to maximize performance. This leaves many designers with a difficult problem: how to apply these new ideas within the constraints of an existing instruction set. Moreover, as the industry converges on faster internal architectures for microprocessors, the design problem changes to address more system-level issues, such as caching structures, I/O, memory interfaces, and peripherals. Traditionally, it has been difficult to analyze these system-level issues in detail, and as a result, many machines have been built based on intuition or incomplete data. However, the availability of existing microprocessors, and rapid advances in CAD techniques, have made possible experiments that help guide design decisions with more solid data.

This chapter discusses how these issues affected some of the design decisions and tradeoffs made in the development of several generations of AT&T's 32-bit microprocessor chip-set. The first section reviews the considerations involved in instruction set design, including the issues involved in speeding up an existing instruction set. Crucial issues here include pipelining, caching, and detailed formatting of the instructions. Section 3 deals with memory management architecture, and the tradeoffs involved in caching, segmentation and page fault systems. Section 4 describes another peripheral subsystem that accelerates I/O. In each case, the design decisions discussed were based on measurements made on the existing chips (where possible) combined with detailed simulations of the proposed architectures. These simulations make use of custom-coded C language programs that model the architecture of the chip in great detail, so that cycle-accurate, or even phase-accurate performance data can be obtained.

2. Microprocessor Architecture Design Considerations

Microprocessor performance is dependent on the following aspects of the architecture:

1. Efficient encoding of the instruction set - The instruction set encoding should be regular, simple to decode, with predictable instruction lengths and addressing modes.

2. Highly-tuned pipeline decode and execution units - A well-engineered pipeline reduces cycle overhead due to *discontinuities*, such as branches, as well as overhead from *hazards* such as the back-to-back access of the same register or memory location.

3. Efficient I/O architecture - The I/O subsystem should provide sufficient throughput for the processor, so that "idle" or "blocked" time within the CPU is reduced to a minimum. In addition, the I/O to memory protocol should maximize the time allotted for the memory system to respond to a read or write operation without incurring additional wait states.

The original AT&T WE32000 was designed in 1977 when the most widely accepted metrics for an efficient instruction set were based on size of the object code, and the ease with which one could program the machine (for the compiler writers and the assembly programmer). At the time the size of the object code was a concern due to the cost of memories.

Today, we find that the focus of concern is in achieving the highest performance at the lowest system cost. This has lead to the popularity of the

"RISC" architectures[1],[2]. By using the simplest instructions, they encourage the use of efficient decoders and fairly simple execution units. The tradeoff here is to use more advanced software/compiler technology in place of complexity in the hardware. Some measurements show that with the simpler instructions, code size could increase 25% to 50%, compared to more traditional instruction sets like the VAX 11/780. Likewise, the number of instructions executed would also be greater. However, the resultant overall performance of the simpler machine could be much higher. Furthermore, some people hold that simplicity in the VLSI hardware has the additional potential advantage that it can be easily updated to a more advanced technology, and thus can run at higher clock speeds.

In evaluating and redesigning the WE32000 microprocessor, the sensitivity of performance to the instruction set architecture was carefully studied. The following describes some of the analysis that were done, and compares it to the designs of RISC-like machines.

2.1 Efficient Instruction Fetch Units

The recent "RISC" machines[3],[4],[5] have all adopted a 32-bit wide instruction format, in which the instructions are word-aligned and therefore decodable with one instruction fetch. Many of the more traditional machines allow variable-length instructions, and thus make it more difficult to decode a whole instruction in one cycle. Although variable-length instructions offer more compactness in the object code, they can fall across word boundaries and therefore require more than one fetch before the instruction can be fully decoded, as shown in Figure 1.

2.1.1 Instruction Caching
In the case of AT&T's WE32000 microprocessor, instructions are variable length and can fall on arbitrary byte boundaries. An instruction queue was added to provide for alignment and queuing of prefetched instructions, so that the additional cycles that would have been required for fetching across word boundaries and alignment can be overlapped with the previous instruction decode. To further improve the throughput of instruction fetch so that the decode unit can maximize its efficiency, an on-chip prefetch buffer/instruction cache (I-cache) was added. In the WE32100 microprocessor, a 64-word instruction cache provided a performance boost of about 15% to 20% for typical programs, when measured with zero-wait-state external memory. The size of this case was determined by software simulations, which showed that the performance improvement that could be obtained by doubling the I-cache to 128 words, was only an additional 5% at zero wait-states, or about 7% at three wait-states. These were measured for typical UNIX C programs that had hit rates at around 50%-60% for the 64 word I-cache.

In a study of a more aggressive pipeline design proposal of the WE32200, we found the on-chip I-cache to be less significant in its contribution to the overall performance of the processor. The performance improvement was only about 3% for a 64 word I-cache, and the overall sensitivity of caching instructions on-chip was no more than 7% to 8%. The reason for this is that the decode unit was already operating at, or near, it maximum efficiency, with the just the help of a simple instruction fetch queue. This shows that the effectiveness of an I-cache is highly dependent on the overall cycle structure of the machine, and that the need for an I-cache may be questionable for machines that have an efficient I/O subsystem and instruction queue that can maintain a high occupancy of the decode unit. This is particularly true if the machine is memory-bound on data fetches. In some applications this is unavoidable, but often the effect can be reduced by improving the compiler's use of the machine's registers or its stack/data cache.

In summary, the need for an I-cache is a function of all the stages in the pipeline. If the execution unit, or the instruction fetch/decode unit is operating at maximum utilization, then the processor will probably be less sensitive to caching. For example, if the decode unit is approaching maximum utilization, further improvement of I/O throughput would have little performance impact on the overall processor. We have found that discontinuities due to branches and multiple-cycle instructions usually cause the decoder of an efficient machine to operate at slightly better than half its occupancy rate. This is because one third to one fourth of all instructions are usually branches; multiple-cycle instructions and pipeline hazards account for the rest of the delays.

2.2 Data Caching

There are several options for reducing the delays for data fetches. For example, the BELLMAC-8[6] used a register-file scheme for this purpose, as did subsequent machines such as the RISC microprocessor. In the case of BELLMAC-8, register sets in memory are pointed to by a register pointer. The register set is just a cached representation of the memory-based locations. A write-through convention is used for updating the registers. The register can be treated as a stack by using it as a circular register file. This provides an efficient means of storing local variables on the stack, without incurring the overhead of memory accesses.

More conventional machines have explicit register-based operands. For example, the WE32100 provides 16 general registers. The WE32200[7] CPU has increased this number to 32. Since these registers are not designed as a contiguous register set, efficient usage of these registers is dependent on an efficient optimizing compiler. With proper allocation of local and global registers, and user versus privileged registers, it is believed that can reduce the number of saves and restores across subroutines and system calls.

Still another approach is illustrated in the C Machine[8] which makes use of stack cache keyed to subroutine entries and exits. In the analysis of one model of the C Machine, we have found the optimal stack frame size to be between 32 and 64 entries. Doubling the stack would contribute to only about 3% to 4% more performance compared to a 32 entry stack cache, depending on the load and external memory speed. A stack cache has the advantage of providing a simplified model for compiler optimization, since the cache is treated like memory. However, in this case the compiler has to provide efficient algorithms for stack compaction to assure efficient usage of stack space.

The use of an on-chip data cache may enhance operand accesses, but it also incurs additional complication of cache coherency for shared memories. If the on-chip data cache operates in virtual address space, it would be difficult to provide physical address bus monitoring to guarantee on-chip cache flushing if updates of the same physical memory was made over a back plane bus. The speed advantage of on-chip cache accesses over even the best external memory (zero wait-state) is almost one cycle, and has an additional advantage in that it reduces bus utilization. At the same time, caching is no substitute for fast memory: zero wait-state memory has the advantage of minimizing contention in the I/O control, thus reducing conflicts between instruction accesses and data accesses. We found that speeding up instruction fetches by using quad-word fetches, in some designs, actually contributed to degrading or having negligible performance impact on the overall processor at zero wait state. This is because quad word fetches could block operand accesses needed for the execution unit. In machines that have a lower number of data accesses, such as the C Machine, additional on-chip data caches would not impact performance significantly (5%).

In the new WE32200 Memory Management Unit, which has a physical data cache on-chip, we found that a minimum of 4 Kbytes of data cache was needed to obtain a performance gain of about 10-15% in the WE32200 chip set environment.

2.3 Accelerating the Decode Unit

In order to execute one instruction per cycle, it is usually (but not always) necessary to decode instructions at that rate. In general, an instruction contains several fields. In the WE32100 instruction format, the opcode field specifies which operation is to be performed and what resources will be used. If the operation requires data, it can be implicit in the opcode (i.e. POP the stack) or it can be explicitly described in the operand fields. Each operand in a WE32100 instruction contains a descriptor which indicates which addressing mode is being used, and zero to four data bytes.

Decoding an instruction requires the decoding of each of these fields. It is evident from the instruction formats used in RISC machines that fixed-length instructions with independent fields and few formats are easiest to decode. Since the WE32100 instruction set does not possess these characteristics it decodes each field of an instruction serially. The serial decoding of a simple dyadic register-to-register instruction (i.e. ADDW2 %r0, %r1) requires three cycles to decode (one cycle for each field). The actual execution of this instruction requires only a single cycle! This speed mismatch between processor units has driven us to explore decoding schemes with more parallelism.

To perform a single-cycle decode, the entire instruction must reside in the instruction queue. Unfortunately, the maximum length WE32100 instruction is 25 bytes long. Single-cycle decoding for all instructions is therefore impractical because it is difficult for the instruction fetch unit to load such large numbers of bytes into the queue in a single cycle. The processor can fetch at most four bytes at a time from off-chip memory. Even if an on-chip cache is organized as an array of double or quad words, it would still not be sufficient to fetch every instruction in a single cycle. Rather than allowing the decode unit to idle until the entire instruction arrives, it is advantageous to decode instruction fields as they arrive in the queue. Furthermore, it is difficult and chip area intensive to implement a maximum size (25 byte) queue and to provide all the logic and control necessary to decode every possible instruction format which can occur.

2.3.1 Encoding Format Regularity of instruction encoding is necessary to achieve a one-cycle decode. For this purpose, it would be desirable for the instructions to be word-aligned, with predictable instruction lengths and simple addressing modes. However, in the WE32100, variable-length encoding was adopted to reduce bandwidth to memory and also to reduce program size. The resulting variable-width instructions complicated the instruction fetch unit in several ways. To retain a one cycle per instruction execution, the instruction fetch unit's output register would have to be sufficiently wide to contain the longest instruction. Instructions are not word-aligned, introducing an additional requirement for aligning instructions. Also, the instruction fetch unit must provide status indicating the amount of valid data in its output register. To obtain the absolute maximum performance, alignment and instruction fetching would all be implemented within the one-cycle per instruction constraint.

2.3.2 A Revised Instruction Format In the case of the WE32100, it would have been difficult to decode all the instruction fields in parallel because the variable length of the operand fields makes their location within the instruction difficult to determine. For example, although operand one always begins at the second instruction byte, the beginning of operand two is unknown unless the length of operand one is known (this is the essence of the serial decode problem). With the current instruction format, it is not difficult to decode the opcode field and the entire first operand in parallel. Additional operands would require an additional cycle for each one. However, decoding additional operands at the same time would be difficult.

In order to simplify decoding, it has been proposed to re-order the instruction format such that all operand descriptors follow the opcode. Hence descriptor one is contained in byte 2, descriptor 2 is contained in byte 3, etc. The part of the instruction containing the opcode and operand descriptors will be called the base instruction. The format of the base instruction is simple enough to allow the decoding of each field to proceed in parallel. Instructions limited to the base instruction format would include: register, register deferred, positive and negative literal[1], and argument pointer and frame pointer short offset[2] addressing modes.

2.3.3 Look-Ahead Decoding In decoding the fields of the base instruction in parallel, one also has to deal with problems related to the lack of orthogonality of the fields. The number of operands, and hence the length of the instruction, is unknown until the opcode is decoded. Also, the descriptor for operand one may indicate a register- displacement addressing mode. An effective address will be formed by adding the displacement to the contents of the specified register. If the opcode is MOV, the address must be issued to the memory system to fetch the instruction's data. However if the opcode is MOVA (move address) the effective address is itself the desired data. Worse still, if the opcode is BCCB (branch on carry clear with byte displacement) there is no operand descriptor, just the displacement.

To simplify the decoding of operands it may be advisable to use a look-ahead decoding scheme also known as "pre-decoding". A look-ahead decoder can decode the next instruction's opcode while working on the current instruction. It actually takes two cycles to decode the instruction but the

---

1   A literal for the WE321000 is an immediate between -16 and +63.

2   The address of the operand data is formed by adding a literal to the contents of the argument or frame pointer register.

process is pipelined to achieve a one cycle throughput. During the pre-decode, the number and type of operands can be determined. This may simplify the decoder but not without some penalty. Although the two stage decode is pipelined to achieve a one-cycle rate, a one-cycle penalty is encountered any time the pipeline breaks because of program discontinuities or stalls (i.e. the instruction fetch does not deliver enough instruction bytes). Under these circumstances the decoder cannot overlap the look-ahead with any other useful work. The penalty on discontinuities is mitigated somewhat by the nature of instruction fetching in variable length instruction set. In an instruction set like the WE32100 where instructions may start on arbitrary byte boundaries but instruction fetches always occur on word boundaries, the first fetch often will not acquire enough bytes to decode a whole instruction. However only one byte is needed for the pre-decode. Hence the pre-decode can sometimes be hidden in the instruction fetch delay.

2.3.4 Hardware Tradeoffs in the Decode Unit In order to achieve the parallel decode of all operands, the necessary hardware resources must be accessible to each operand. In load/store architectures the sharing of hardware for ALU operations and address computations is reasonable. Only one address is generated in a single instruction. Also, memory accesses can be scheduled to occur before the data is actually used in the pipeline, avoiding data dependency delays (hazards). However in memory-based architectures, there is a significant benefit to dedicating hardware to perform address computations as early in the pipeline as possible.

The WE32100 instruction set has 17 different addressing modes. Six of these modes can be encoded in the base instruction's operand descriptors without needing additional data bytes (register, register deferred, positive and negative literals, and two short-offset addressing modes). In order to support the single-cycle decode of the base instruction, three read ports to the register file are necessary for the register modes. Adders are necessary for the short offset addressing modes to add the offset to the argument pointer or frame pointer. The remaining 11 addressing modes involve generating addresses or data using additional bytes in the instruction queue and adding a displacement to the contents of a register to form an address.

There are a number of design tradeoffs to be examined here. A cycle-accurate behavioral simulator is an invaluable tool in analyzing the performance impact of these design decisions. It would quantify the performance benefit of particular implementation choices. These choices must be carefully considered. Conceptually, the decoder could be implemented as four interacting state-machines, one for each field.[3] In attempting to decode the three operands in parallel, it may be simpler to design three independent address arithmetic units instead of one common unit. Adding hardware increases the size of the decoder. Certainly chip area is another important consideration, since the size of the decode unit is bounded. There are other considerations related to address generation. Although two short-offset addresses could be decoded in the same cycle, one of them will go unused in that cycle unless a multi-ported memory interface exists for operands. In all other cases of address generation, additional bytes are necessary beyond the base instruction. It would be costly in hardware to attempt to generate more than one address per cycle in these cases.

Machines have been designed to decode both opcode and first operand in parallel. This frequently would require look-ahead decode to determine point of termination of the instruction, unless the instruction length is fixed. This improved decode rate does not always translate directly into an overall instruction execution improvement because of the interaction of the various pipeline stages. The execution unit has to be able to keep up with the decode unit.

2.3.5 Discontinuities A discontinuity is the result of writing a new target address in the program counter (PC) that breaks the sequence of instructions fetched by earlier stages of the pipeline. In the case of conditional branches, condition flags are set at the last stages of the execution unit. If a new target branch address is loaded, a latency of several cycles could result because of the discontinuity. At least one additional cycle is imposed for fetching the next instruction, with additional cycles for decoding and fetching of the operands for the new instruction (if the instruction fetch unit did not anticipate the branch). Return and indirect addressing modes would require even more additional memory accesses.

As the pipeline becomes more efficient the delay associated with discontinuities becomes more pronounced. Since 20 - 30% of the instructions executed are discontinuities, the delays associated with them are significant. Reducing the number of pipeline stages will minimize the penalty associated with discontinuities, but more than this can be done.

---

3 At first glance, it seems the opcode decoding is a simple one-to-one mapping. However, instructions such as CALL, invoke a micro-sequence that requires the decoder to generate several internal instructions.

The latency of unconditional transfers can be reduced by informing the instruction fetch unit to prefetch from the anticipated branch-target address as soon as it is computed. For conditional transfers, condition evaluation is typically performed in the execute unit. Evaluation of the condition frequently results in delay of the pipeline. During that delay, instructions can be inserted in the pipe consistent with some predicted outcome.

There are two techniques for optimizing on branches: static branch prediction or dynamic branch prediction. Several branch prediction strategies were discussed in a study by Johnny K. F. Lee and Alan J. Smith [9]. As was discussed in the study, dynamic branch prediction can be done by a "look-ahead" technique, where the earlier stages of the pipeline may be able to resolve condition codes affecting the branches, and essentially prefetch the target branch address, thus reducing the overhead due to delays in discontinuities in the pipeline.

There are four ways to implement static prediction:

1. always predict no discontinuity,

2. always predict a discontinuity,

3. always predict the same for a given opcode, where the prediction is statistically determined and built into the hardware.

4. Provide two versions of each conditional instruction; one predicts branch, the other no branch. Prediction can be done intelligently by the programmer/compiler.

In trying to optimize branches in the WE32100, branch prediction was found to be worthwhile. A static branch prediction algorithm based on the branch opcode (i.e. branch on overflow is always predicted not taken) was found to be over 75% accurate on series of C benchmarks representing UNIX programs. Since this branch-prediction algorithm is built into the machine, it cannot be expected to perform equally well over all applications.

Branch prediction can be used to avoid effectively the latency between instruction decode and execution during discontinuities. However it does not always avoid the instruction fetch delay. Even if a branch is predicted correctly, the instruction queue may have to be flushed and refilled from the branch target address.

As mentioned earlier, the WE32100 instruction set was optimized for compactness of object code. The I/O limitations of machines like the WE32100 are less than those of a RISC machine. The degree of bottleneck can be measured by determining the frequency of I/O contention between different stages in the pipeline. In a RISC machine, I/O bandwidth can be crucial to achieving higher performance, especially if there are no on-chip caches. For these machines, a high performance I/O protocol is essential to achieving one cycle per instruction execution.

Experiments with various I/O protocols showed that an instruction cache may not always be necessary to sustain best throughput in instruction decoding. Nearly all the delays in an efficient instruction fetch unit in a highly tuned pipeline processor were attributable to discontinuities. In these cases, efficiency of the machine could be improved by adding branch target address caches. A branch address cache can help reduce the penalty associated with discontinuities. Whenever a new program counter value is sent to the instruction fetch unit, the instruction bytes fetched at that address are cached. When that discontinuity occurs again, the target address will hit in the branch address cache. The instruction bytes can be loaded into the queue immediately while an incremented version of the address is sent to the I/O to fetch along that path. The number of bytes cached with the branch address would depend on the results of performance analysis and chip area considerations.

2.3.6 Decoded Instructions Queues and Caches Some processors, such as the Intel 286 and 386, contain a decoded instruction queue between the decode unit and execute unit. The queue is loaded with all the control information, and possibly the data, necessary to execute the instruction. The queue is capable of storing several instructions. The latency of the queue when empty should be minimal since the longer the pipeline becomes, the greater the penalty for discontinuities.

One advantage of a decoded instruction queue is that the buffering it provides can reduce the performance penalty caused by speed mismatches between the decode and execute units. If the execute unit slows down, the decode unit can insert several instructions into the queue. If at a later time the decode unit slows down, the execute unit can continue to operate at its peak rate until the queue is emptied.

Another advantage of a decoded instruction queue involves memory-based data. In accessing data from memory, the latency between address generation and receiving the data is generally greater than one cycle. If addresses are generated in pipeline stage i and data is collected in stage $i + 1$, there will be a i cycle delay each time an off-chip operand is fetched. If the decode unit issues an address, it must wait before decoding the next instruction because the next pipeline stage is busy waiting for the data to return. If the I/O protocol is pipelined to maximize throughput, then the internal pipeline can be designed to take advantage of it. The decoded instruction queue can be used to collect data and pipeline the memory access latency while the decode unit continues to work. For an n-cycle memory access latency, the queue must be able to store n instructions.

A logical extension of the decoded instruction queue is a decoded instruction cache. Just as a regular instruction cache reduces the delay of instruction fetch, a decoded instruction cache will reduce the delay of instruction fetch and decode. In the ideal case of a 100% hit rate, the processor would be limited only by the speed of the execute unit and operand fetch delays. Some of the considerations in implementing such a cache are discussed below.

Decoded instructions would be cached based on their address. The cache would be organized as an array of n bit elements where n is the length of a decoded instruction. Since decoded instructions are generally much longer than their encoded form, the size of the decoded cache would be much larger than a normal cache with a comparable hit rate. Studies are necessary to quantify the performance gain as a function of the size of the cache and evaluate the effectiveness of such a scheme.

A policy must be established to handle discontinuities. Some unconditional transfers can be followed by the decoder. However transfer addresses which are runtime dependent cannot be followed. The decode unit must stop at this point until the execute unit can compute the next instruction address. Conditional transfers can also be followed by the decoder if a prediction scheme is used (provided of course that the target address is not runtime dependent). The execute unit must be able to indicate when the prediction is incorrect and reset the decode unit to the correct path. Once the decoded cache has been filled, many program control transfers will incur no more delay than sequential code. This could prove to be a significant benefit.

The insertion of a cache between the decode unit and execute unit, allows the two units to operate even more independently. If the execute unit is faster it will have to wait for the decode unit to decode the next instruction. However, if the decode unit is faster it can start to thrash the decoded instruction cache. The decode unit may be adding an instruction to the cache which replaces a previous instruction before the execute unit gets to it. The execute unit would then have to reset the decode unit to fetch and decode the missing instruction again. The probability of thrashing is related to the size of the cache. This could be avoided by devising a tighter synchronization scheme.

Another consideration particularly relevant to an instruction set like the WE32100 is how to handle micro-sequences. Certain instructions, such as "process-switch," generate internal sequences of instructions within the processor. Each micro-instruction is associated with the same instruction address and cannot be cached in the normal manner. Furthermore, a significant portion of the cache would be used up for each micro-sequence.

This would crowd-out other instructions. The alternative is not to cache each of the micro-instructions but to re-locate a portion of the decoder on the execute unit side of the cache in order to generate the micro-sequences.

One final consideration is the amount of decode unit functionality that must exist on the execute unit side of the cache. As previously mentioned, run-time dependent information cannot be cached. This includes register contents and operand addresses formed using register values. A portion of the operand decoders must be present in the stage following the cache in order to handle the run-time dependent decoding. This lengthens the pipeline and increases the chip size.

2.4 Pipelining

The efficiency of a pipelined implementation depends on many factors: the interface and control between stages in the pipeline, the handling of anomalous conditions (hazards) in the execution sequence, and the efficiency of code generated by the compiler, which could reduce contention or conflicts in resource (I/O, registers) accesses during execution.

This section briefly reviews some of the considerations in designing a pipeline, how performance can be impacted, and how a design might compensate for these performance degradations.

2.4.1 A Basic Pipeline Figure 2 shows a 4-stage pipeline in a simple register-based machine. Suppose an instruction is fetched at each cycle, and an instruction completes at each "store" cycle. The execution time per instruction over the length of the pipe would be four cycles. Even this over-simplified model exhibits some of the pipeline problems. For example, in an efficient pipeline design, the propagation delay for each pipeline stage should be approximately equal. Therefore, in order to retain a full pipe, the instruction fetch stage must fetch and assemble its output, ideally, at one instruction per cycle. Each subsequent stage should require only one cycle. If more cycles are required, due to operand resolution or conditional flag computation, the pipe would have to be halted. This would incur additional complexity, as well as performance penalty. Increasing the number of stages could reduce this impact for code with no branches. However, for branch instructions, additional stages in the pipe could result in delays due to pipeline flushing and overhead due to refilling of the pipeline stages.

2.4.2 Managing Data Dependencies Data dependencies in the pipeline result from anomalous conditions. They are a result of operand access conflicts, which introduces delays in the pipeline. A data dependency in the pipeline results when an operation to be performed requires a variable, where that variable has a pending change from a previous instruction. Three types of data dependencies are examined: register-operand dependencies, base-register dependencies and memory-based operand dependencies.

1. Register-Operand Dependencies - In **Figure** 2, if the result of instruction 1 is the same register as the one identified as a source of instruction 2, the operand fetch in clock cycle 3 will fetch the incorrect data. A pipeline implementation must detect the occurrence of such a dependency and either halt the pipe's advance, or appropriately manage the dependency with bypass control between stages in the pipeline.

    One solution to avoiding data dependencies is to require that the compiler guarantees that no data dependencies exist, by reordering instructions or by inserting sufficient no-op instructions between instructions that would otherwise result in a dependency. This is the approach taken by MIPS. It requires the compiler to include knowledge of the cycle architecture of the processor. Runtime conditions in the pipeline may prevent a potential data dependency from taking place. In these cases, compiler inserted no-op's would be unnecessary and would reduce the performance of the processor. Since all memory addresses cannot be known at compile time, the occurrence of memory-based operand dependencies can not be totally eliminated, and so some hazard detection hardware is still required.

    As mentioned in the previous section, the pipeline advance can always be stalled until the operand conflict is resolved (i.e., valid or available). The occurrence of the dependency, however, must always be detected, so that advances in the pipeline can be controlled.

    Figure 3 shows a bypass which manages the register-operand dependencies without imposing any cycle delays. The multiplexers at the inputs to the execute unit can select the execute unit's result from the previous cycle. To use this technique, a bypass must be provided to each stage between the registers and the execute unit. In fact, the equivalent of the bypass is also required within the register file (i.e. write before read each cycle).

2. Base Register Dependencies - The second type of dependency involves data accesses to memory. Figure 4 shows the memory interface as well as a separate address arithmetic unit, and an added pipe stage. All addresses are assumed to be dyadic, formed by adding an immediate field, contained in the instruction, with a base register. By zeroing the base register or the offset, the AAU input provides an absolute address, or register indirect, respectively. Note that the use of a register value yielded a potential hazard. A bypass expedites updating of AAU inputs. However, an inhibit of the pipe's advance is still required. Otherwise, the AAU add, as well as the memory access, would be required within

the single cycle. Assuming a single cycle access and a single cycle AAU, the pipeline would be delayed by one cycle. Memory wait-states would increase this delay cycle for cycle. An alternate approach, as shown in Figure 5, includes the AAU operation as part of the path to memory instead of an additional pipe stage. Although register-based instructions now execute faster, memory-based fetches require the pipe to be delayed by a minimum of two cycles.

3. Memory-Based Operand Dependencies - The third type of dependency involves memory-based accesses. Any memory based operand read can result in a hazard if a memory write is pending. If the addresses match, the hardware must inhibit the data read access until the store completes.

2.4.3 Special Cases Multi-cycle instructions - Most ALU operations complete in a single cycle; exceptions are integer multiply, divide and modulo as well as floating point, which are typically implemented as micro sequences. In the WE32100, a Macro ROM was implemented to simplify implementation of macro instructions that can be decomposed into micro instructions. Here, the macro instructions consisted of process switch sequences, interrupt and exception sequences of the processor.

Faults - Faults are handled in sequence, even though different stages in the pipeline could potentially cause a fault to be recorded out of order. Complexity in handling faults result from the need to determine validity of the fault depending on the status of execution of latter stages in the pipeline. For example, a memory fault may result for an instruction following an inaccurate branch prediction. That instruction will not be executed, and, therefore, the fault should be ignored. Another example is a memory fault that was preceded by a divide-by-zero fault.

Faults can be managed by accumulating all reported fault indications in the pipelined representation (microinstruction format) of the associated instruction. Each instruction, on reaching the execution stage, will either be executed or cause a fault.

Restartability or Resumability of Instructions - In order to recover from a fault, the address of each instruction, or at least the ability to compute that address, must accompany the pipelined representation of the instruction.

2.5 I/O Architecture

One calibration of the efficiency of an I/O architecture is to what extent the internals of the CPU are blocked waiting for I/O. This is dependent on the number of instruction and data fetches needed by the microprocessor when executing a program, and can vary depending on the efficiency of the compiler in utilizing on-chip registers or caching. For the same machine, with different compilers, we have observed a difference of 10% or more due to differences in compiler register usage.

We have also observed a difference of a factor of two in the number of data fetches in the same program, when comparing a register-based versus a stack-cached-based architecture. One reason for such a dramatic difference is that many local variables, such as arrays or structures that could be pushed onto the stack, cannot easily be stored into registers (unless the registers are organized sequentially as register files).

The amount of I/O required for a given program determines to what extent the processing units in the CPU are dependent on the efficiency of the I/O subsystem. In analyzing the different I/O architectures that can be applied to a given microprocessor, we have observed that the sensitivity of performance to the different I/O architectures was not more than 15%, when comparing the most efficient I/O to the more traditional standard I/O, if the machine was highly pipelined. In this study, "standard I/O" was taken as one with demultiplexed instruction and data buses used for both data and instruction fetches, 2-cycle overlapped I/O (which effectively provided 1-cycle access), and with all memory hazards resolved in the I/O frame, as opposed to the memory system.

The following options in I/O architectures were considered:

1. No I/O bottlenecks - (5 buses) with separate instruction-address bus, instruction-data-fetch bus, data-address bus, data-fetch-write bus and data-fetch-read bus.

2. Split address/data buses for instruction fetch and separate address/data bus for data fetch (4 buses).

3. Separate address bus, with separate read/write data bus (3 buses).

4. "Standard bus" - demultiplexed address and data bus used for both instruction and data fetches (2 buses).

The above options resulted in the following observations. Using the "standard bus" architecture as baseline:

- There was a 10% - 12% improvement with the 5-bus system, due to splitting address and data (with separate buses for reads and writes).

- There was an 8% - 10% improvement with the 4-bus approach - i.e., separate address and data (no split data buses for reads and writes) for instruction and operand fetches.

- There was a 5% improvement if memory hazards were checked and handled by the memory controllers, as opposed to the I/O frame.

We found that in this particular design, the maximum performance gain, assuming best-case I/O, would only be about 17%. Since increasing buses on the microprocessor would increase the pin count, and thus the cost of the system, the standard bus architecture was chosen.

3. Memory Management Architectures

The design of a memory management unit must be tightly coupled with the design of the operating system. AT&T's WE32101 Memory Management Unit[10] was derived from evaluation of the needs of the UNIX System V operating system. Special OS considerations include:

1. Memory partitioning conventions -

   - Paging - The page sizes may be variable or fixed, and they usually range from 512 bytes up to 4 or 8 K bytes in some systems. The page-size variation is dependent on the available physical memory size of the target system, and also on the frequency of process switching and the number of simultaneous processes. The objective is to find an optimal size that minimizes thrashing, and maximizes efficiency of the physical memory. Today, hardware support for page replacement procedures is common. This requires the hardware to provide automatic update of indicators for Reference or Modified bits.

   - Segmentation - Segments can range from a few bytes long to several thousand bytes long. The smaller segments can be used for message passing, or for special data shared amongst multiple processes. The larger segments are a means of managing categories of text or data under one common access protection scheme and within a contiguous memory (e.g., stack, kernel text, shared libraries, etc.)

     Shared segments are supported in WE32101 so that multiple processes can access the same physical memory space with different virtual address mapping and under different access protection schemes. (That is, one process may be allowed read-only access, while another may be allowed to write into the segment space).

     The convention used in the System V UNIX operating system, supported on AT&T's 3B machines, is to use the segment as the logical partition of memory space that is visible in the OS architecture. For example, this is used separating shared libraries, kernel versus user text and data. Pages are used as an internal memory management scheme as the structure for managing virtual memory.

2. Memory protection schemes - hierarchical protection structure for user/kernel access on any combinations for read/write/execute/no access.

3. Multiple process support - The WE32201 memory management for AT&T's third generation 32-bit chipset provides an automatic process tracking mechanism, whereby processes are uniquely identified by the base-table pointers of the page or segment tables. This tag is used to identify uniquely the process id associated with each entry in the translation buffer cache.

    In systems that rotate among a few processes (say no more than 3 or 4), such as real time systems, the memory management hardware can provide automatic replacement of translation descriptors, without the need for the operating system to flush the cache entries at process-switch time. Each translation descriptor will be tagged so that on returning to a recently executed process, the overhead to reload descriptor will be minimized.

    Interprocess communication can be achieved in the WE32201 by loading the active process tag with the target process id (with privileged instructions), and writing into the target process's segment or page, if the target process space is defined to be accessible by the active process.

4. Virtual machine support   The IBM/RT machine[11] supports the concept of virtual machine. This gives users the ability to switch from one OS environment to another. To facilitate this, the IBM/RT has increased the addressing space from the more popular 32 bits to 40 bits of addressing. With the additional bits, one could switch from one kernel space to another, while executing under the same user process.

The WE32101 supports systems that require contiguous segments, or paging, or both in a mixed mode. The tradeoff between segmentation versus paging is external fragmentation (gaps between segments when doing segment placements) versus internal fragmentation (wasted memory space between end of text or data to end of page) of memory. In systems (like System V Release 2.0 UNIX), where demand paging is not needed or supported, segmentation is more desirable. This is typically used in very small systems (turn-key applications) with small programs that do not need virtual space.

3.1 Tradeoffs in On-Chip Memory Management Unit

There are several memory interface functions that fit naturally into the memory management unit. They include virtual-to-physical address translation, structuring of memory partitioning, memory-access protection checking, translation buffer cache control (miss processing), memory access fault detection, memory control interface, direct memory access control for accelerated block moves. In evaluating the functions of an on-chip MMU, analysis was done in the following areas to determine the tradeoffs between feature versus area. This analysis uses the WE32101 MMU as baseline.

1. Size and types of translation buffers - We found that fully associative buffers for page descriptor caches at 64 entries would provide a 99% hit rate for typical UNIX programs. This compares with 96-97% for two-way set-associative caches of the same size. We also found that a degradation of 3-4% in hit rate reduced overall system performance by 10-12%. This is based on a miss processing overhead of 15 to 30 cycles (for systems with a few wait states), depending on whether there was a miss in both segment as well as page descriptor caches.

2. Hardware versus software support of miss processing - Software miss processing could take 150 to 300 cycles. By supporting miss processing in hardware, we can improve performance of the overall system by 15-25% depending on the miss rate.

By integrating the MMU with the CPU, one can save the translation overhead by overlapping translation with prefetching. This would provide a 1 cycle advantage over a separate MMU. In terms of performance, this would be an 8-15% gain for typical programs. However, the impact of this savings is reduced when examining the WE32201. We find that with an integrated data cache on the MMU, like the WE32201, translation overhead can be reduced, even though the MMU is separate from the CPU. This is because a data cache hit on the MMU would be equivalent to a zero wait-state access. In this case the 4K bytes of data cache on the MMU, and the 64 word I-cache on the CPU provided a better than 60% hit rate for most memory accesses.

In this case, if one were to put the MMU, without the 4K bytes data cache on the CPU, and reduce the translation buffer to a 2 way set associative convention, as oppose to the fully associative cache, the loss in system performance would outweigh the benefit of the 1 cycle savings. This is further amplified as one increases wait states in the interface to memory.

The integrated CPU/MMU approach does provide cost savings, and would be attractive for low end systems that do not require elaborate memory management features and well-behaved locality of program execution.

4. Memory Interfacing Peripherals

In evaluating UNIX program behaviors we found that about 10% of its time was spent in memory moves. Overall system throughput could be improved by hardware support of memory to memory copies and memory fills. UNIX system routines, such as Fork, Exec and buffer copies between user and system space could be accelerated by a fast intelligent DMA capability.

The WE32104 DMAC was designed to address some of these needs. This provided an efficient interconnection between a 32-bit system bus and the byte oriented peripheral devices, such as UARTS, disk controllers and network interfaces. We found that typically 4 channels could be used simultaneously in a fully configured system. This provided a maximum throughput of 11.2MB/sec for memory copies, and 20.3MB/sec for memory fills when running in burst mode. This is about 5 times faster than the CPU for large block moves. For transfers over the peripheral bus, a maximum throughput of 6MB/sec could be achieved. To provide maximum flexibility in transfer modes, byte, halfword, word, as well as double and quad word transfers were supported. In quad word transfer mode, bus utilization is reduced by a factor of ten compared to the byte mode transfer.

The WE32103 DRC was designed to support efficient memory accesses. Special features such as options to select page and nibble mode for fast double and quad word memory accesses, pre-translation mode for improving access time to paged memories. The pre-translation mode reduces access time by overlapping the row portion of a memory access with the address translation time performed by the memory management unit. Overall, an efficient memory interface can improve overall performance by over 5% if these features are properly utilized.

5. Summary

The overall architecture of a chipset contributes to the total performance of a system. At the core of this is the CPU. Central to the consideration, beyond the instruction set architecture and implementation, is the interface to the peripheral chips. We have found that CPU performance is reduced when it measured in a system configuration. Performance loss of 20-40% could be recovered if the peripherals are configured and used properly to maximize throughput.

6. Acknowledgements

The above analysis reflects the results of work done by the many architects of the WE32000 family. Specifically, we would like to acknowledge the work of the architecture team of the AT&T Microsystems Engineering

Department: Pamela Kumar, Mike Kolodchak, Z. J. Mao, Matt Nelson, Ben Ng, Steve Pekarich, Rich Piepho, William Wu, from which the material of this paper was based.

REFERENCES

1. Patterson D., "Reduced Instruction Set Computers", *CACM* Vol.28 No.1, January 1985, p.8-21.

2. Patterson D. A., and C. Sequin, "A VLSI RISC," *Computer*, Vol. 15, 9 (Sept. 1982) p.8-21.

3. Moussouris J., et al., "A CMOS RISC Processor with Integrated System Functions - The MIPS Microprocessor," *Spring COMPCON '86*, p. 126-131.

4. Birnbaum J. S., W. S. Worley, "Beyond RISC: High-Precision Architecture - The Next Generation of HP Computers: The Spectrum Program," *Spring COMPCON '86*, p. 40-47.

5. Radin G., "The 801 Minicomputer", *IBM Journal of Research and Development*, Vol 27, 3, May 1983, p. 237-246.

6. Campbell S. T., "MAC-8 Microprocessor Summary", *MAC-8 Systems Designer's Handbook*, Bell Laboratories, March 8, 1976.

7. Nelson M. S., B. Ng, W. Wu, "Architecture of the WE32200 Chip Set" *COMPCON '87 Spring*

8. Ditzel D., R. McLellan, "Register Allocation For Free: The C Machine Stack Cache," *Proceedings Symposium on Architecture Support for Programming Languages and Operating Systems* March, 1982, p.48-56.

9. Lee J. K. F., A. Smith, "Branch Prediction Strategies and Branch Target Buffer Design," *Computer*, Vol 17, No. 1 (January 1984), p.6-22.

10. P. M. Lu, et al., "Architecture of a VLSI MAP for BELLMAC-32 Microprocessor", *COMPCON Spring 83*, p213-217.

11. *IBM Personal Computer Technology Handbook*, 1986.

ENCODING FORMAT

WE32100

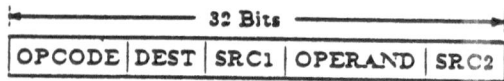

6 Bits	4 Bits	4 Bits	0/8/16/32 Bits	4 Bits	4 Bits	0/8/16/32 Bits
OPCODE	MODE	REG	EXTENSION	MODE	REG	EXTENSION

FIRST OPERAND   SECOND OPERAND

RISC

|←──────── 32 Bits ────────→|
| OPCODE | DEST | SRC1 | OPERAND | SRC2 |

FIG . 1

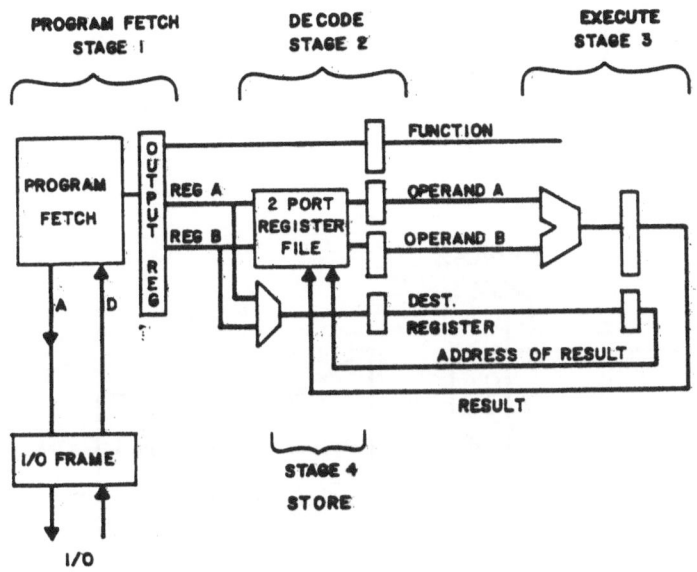

CYCLE		1	2	3	4
INSTRUCTION	1	INSTRUCTION FETCH	DECODE & FETCH OPERANDS	EXECUTE	STORE
INSTRUCTION	2		INSTRUCTION FETCH	DECODE & FETCH OPERANDS	EXECUTE
INSTRUCTION	3			INSTRUCTION FETCH	DECODE & FETCH OPERANDS

FIG. 2

EXECUTE STAGE

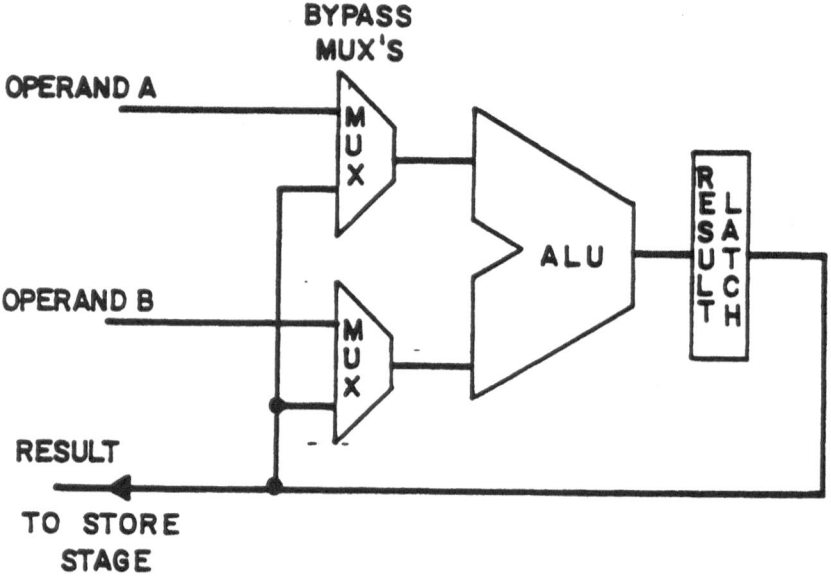

FIG. 3

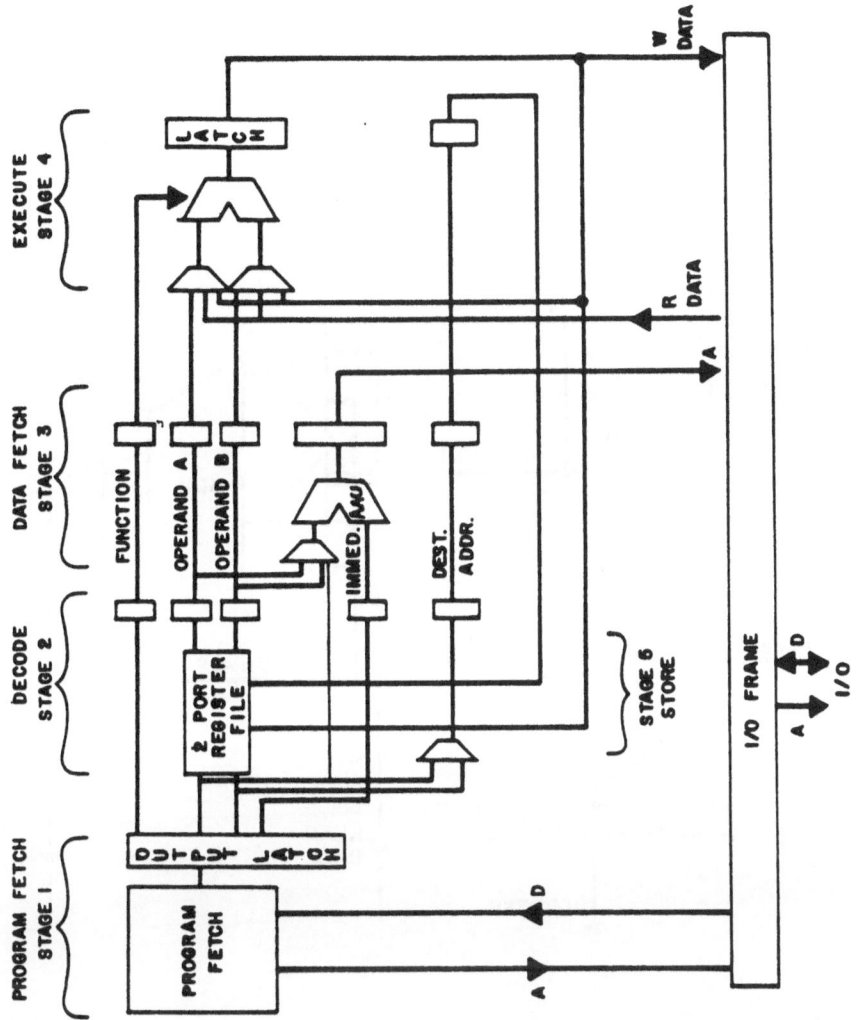

FIG. 4

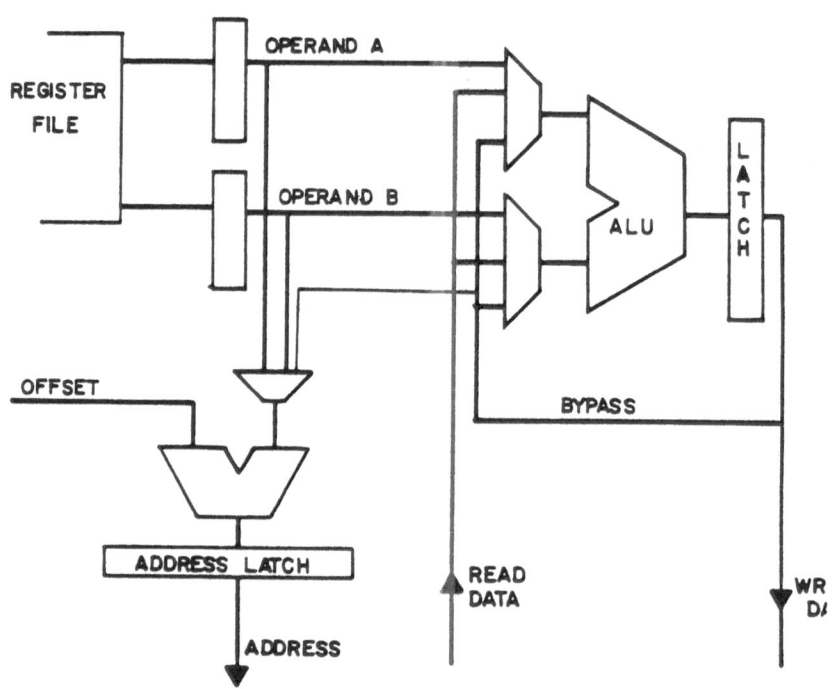

FIG. 5

14

A Comparison of Microprocessor Architectures
in View of Code Generation by a Compiler

N. Wirth

Abstract

A high-level programming language mirrors an abstract computing engine, implemented by a combination of a concrete computer and a compiler. The pair should therefore be carefully tuned for optimal effectiveness. Otherwise, compromises between more complex compiling algorithms and less efficient compiled code are inevitable. We investigate three processor architectures and analyze their effectiveness for use with a high-level language. The conclusion: neither particularly sophisticated nor drastically "reduced" architectures are recommended. Instead, the proven and pivotal mathematical concepts of *regularity and completeness* hold the key to performance and reliability.

Introduction

To a programmer using a high-level language, computer and compiler appear as a unit. They must not only be regarded, but also designed as a unit. Most computers, however, display a structure and an instruction set - an architecture - that mirrors the metaphor of programming by assembling individual instructions. More recent designs feature characteristics that are oriented towards the use of high-level languages and automatic code generation by compilers.

By orienting an architecture towards high-level languages two principal goals are pursued:

- *Code density.* Densly encoded information requires less memory space and fewer accesses for its interpretation. Density is increased by providing appropriate resources (e.g. fast address registers), suitable instructions and addressing modes, and an encoding that takes into account the instructions' relative frequency of occurrence.

- *Simplicity of compilation.* A simple, compact compiler is not only faster, but more reliable. It is made feasible by regularity of the instruction set, simplicity of instruction formats, and sparcity of special features.

In this paper, we make an attempt to measure and analyze the suitability of three processors in terms of the above criteria. In general, this is a difficult undertaking, because three variables are involved, namely the computer architecture, the compiler, and the programming language. If we fix the latter two, we have isolated the influence of the architecture, the quantity to be investigated. Accordingly, we shall involve a single language only, namely Modula-2 [1]. Unfortunately, fixing the compiler variable is not as easy:

compilers for different processor architectures differ inherently. Nevertheless, a fair approximation to the ideal is obtained, if we use as compilers descendants of the same ancestor, i.e. variants differing in their code generating modules only. In particular, we have designed compilers that use the same scanner, parser, symbol table and symbol file generator, and - most importantly - that feature the same degree of sophistication in code "optimization".

It is reasonable to expect that a simple and regular architecture with a complete set of elementary operations corresponding to those of the language will yield a straightforward compiling algorithm. However, the resulting code sequences may be less than optimally dense. The observation that certain quantities (such as frame addresses) occur frequently, may motivate a designer to introduce special registers and addressing modes (implying references to these registers). Or the observation that certain short sequences of instructions (such as fetching, adding, and storing) occur frequently, may spur the introduction of special instructions combining elementary operators. The evolution of more complex architectures is primarily driven by the desire to obtain higher code density and thereby increased performance. The price is usually not only a more complex processor, but also a more complicated compiling algorithm that includes sophisticated searches for the applicability of any of the abbreviating instructions. Hence, the compiler becomes both larger and slower.

Whereas for decades the future was seen in more baroque architectures (the huge number of instructions being a favourite item for advertisements), the pendulum now appears to swing back towards the opposite extreme. The ideal machine is now said to have few, simple instructions only [2]. Quite likely the optimal solution is to be found in neither extreme.

The processor architectures chosen for this investigation are: Lilith [3,4], National Semiconductor 32000 [5], and Motorola 68000 [6]. (To denote the latter two, we shall use the abbreviations NS and MC). Lilith is a computer with a stack architecture specifically designed to suit a high-level language compiler, i.e. to obtain both a straightforward compiling algorithm *and* a high code density. Both the MC and in particular the NS are claimed to be designed with the same goals, but feature considerably more complex instruction sets. The same observations hold for the DEC VAX computer family.

Although the above considerations are correct in general, our empirical results do not support them. It is disappointing that the more sophisticated architectures of the NS and MC do not only demand a more complicated compiler, but also result in considerably less dense code. We shall try to pinpoint some of the causes for this negative achievement.

The Target Architectures and their Instruction Formats

We first present the essential and relevant features of the three considered architectures in a comparative way. For further details the reader is referred to descriptions of the specific processors. All three mirror a run-time organization tailored for high-level languages involving a stack of procedure activation records. Lilith and NS feature three dedicated address registers for pointing to the frame of global variables, to the frame of the most recently activated procedure, and to the top of the stack. In the MC three of the seven general purpose address registers are dedicated to this purpose.

For expression evaluation and storing intermediate results, Lilith features a so-called *expression stack,* i.e. a set of fast registers that are implicitly addressed by an up/down counter whose value is automatically adjusted when data are fetched or stored. The expression stack logically constitutes an extension of the stack of procedure activation records. It is empty at the end of the interpretation of each statement. Therefore, the difficulties inherent in any scheme involving two levels of storage are minimized: the expression stack need be unloaded (from the registers) into the main stack (in memory) only when context is changed within a statement, i.e. only upon calling a function procedure. In contrast, the other processors offer a set of explicitly numbered data registers. The run-time organizations of the three processors used by the Modula-2 system are shown in Fig. 1.

The processors' instruction formats are shown in Figs. 2, 3, and 4. Lilith and NS instructions form byte streams, whereas the MC instructions form a stream of 16-bit units. Lilith is a pure stack machine in the sense that load and store instructions have a single operand address and actual operators have none, referring implicitly to the stack. Instructions of the NS and MC mostly have two explicit operands. Their primary instruction word contains fields a1 and a2 indicating the addressing mode (and a register number), and frequently require one or two extension fields containing the actual offset value (called *displacement*). In the case of indexed addressing modes the extensions include an additional index byte specifying the register to be used as index.

The so-called *external* addressing mode of Lilith and NS deserves being mentioned specially. It is used to refer to objects declared in other, separately compiled modules. These objects are accessed indirectly via a table of linking addresses. The external addressing mode, when used properly, makes program linking as a separate operation superfluous. This is a definite advantage whose value cannot to be overestimated. In the case of Lilith, module linking is performed by the loader. The use of a single, global table of module references makes it necessary that the loaded instructions are modified. Module numbers generated by the compiler must be mapped into those defined by the module table. The NS system eliminates the need for code modification by retaining a local link table for each module. The loader then merely generates this link table.

Another difference worth mentioning concerns the facilities for evaluating conditions. Lilith allows to treat *Boolean expressions* in the same way as other expressions. Each relational operator is uniquely represented in the instruction set and leaves a Boolean result

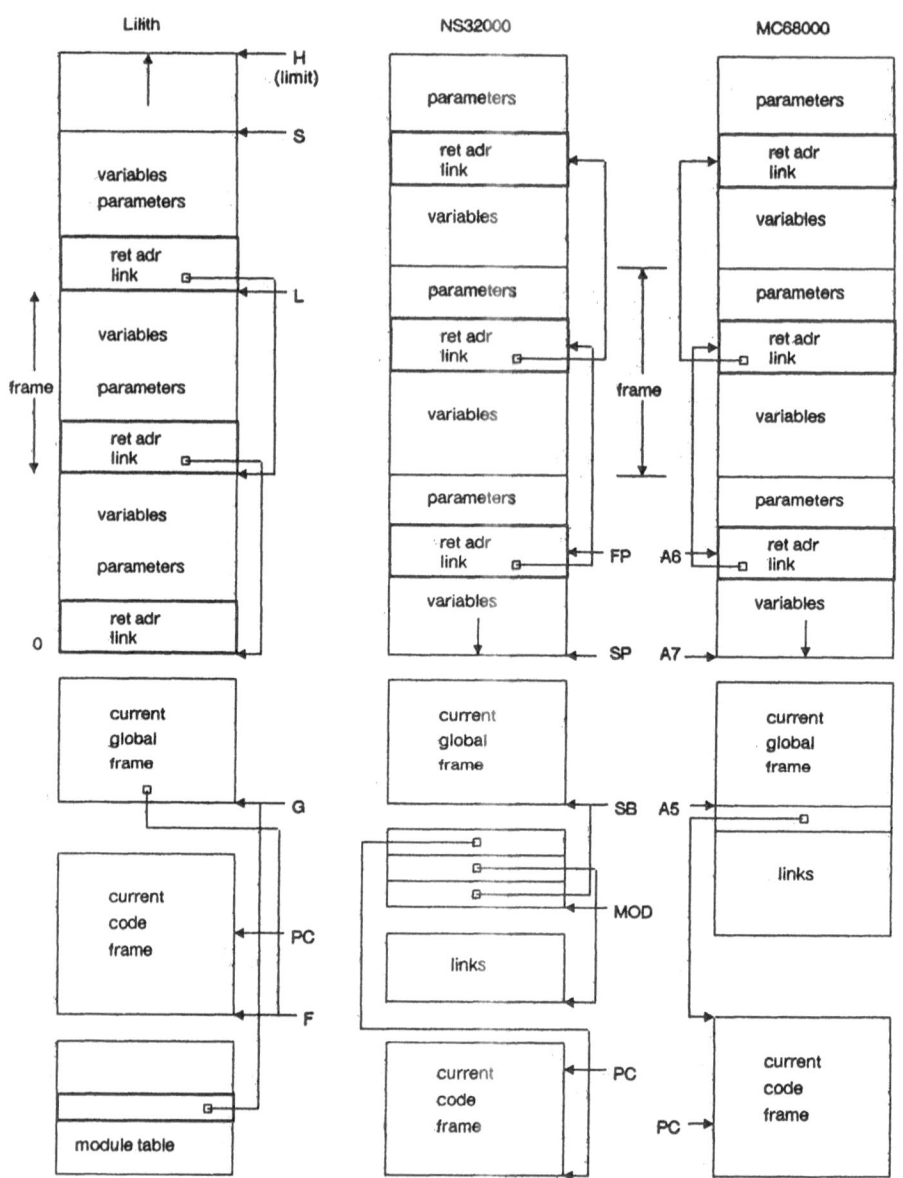

Fig.1   Run-time Organizations

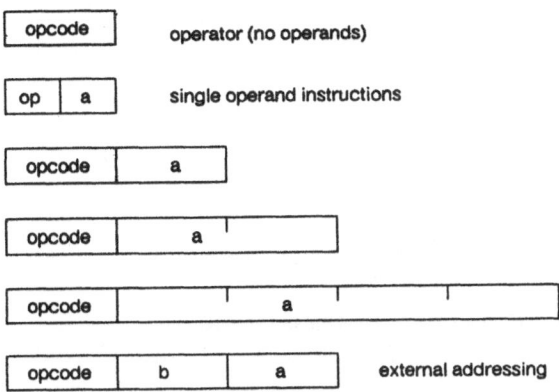

Fig.2   Instruction formats of Lilith

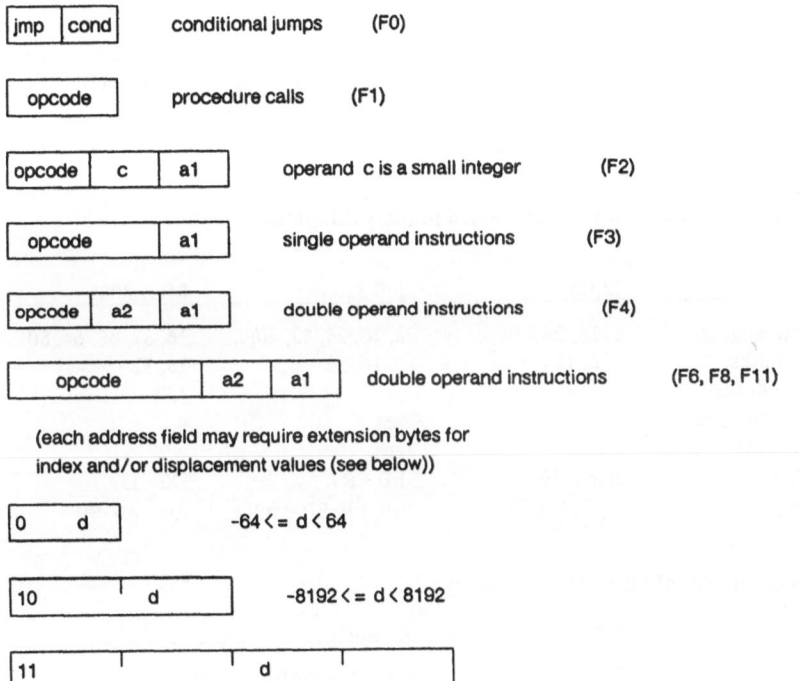

Fig. 3.   Instruction and displacement formats of NS 32000

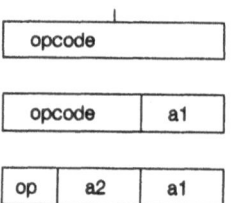

(each address field may require additional words
for index and displacement values)

Fig. 4.   Instruction formats of MC 68000

on top of the stack. In addition, there are conditional jumps corresponding to the AND and OR operators; they are suitable for the abbreviated evaluation of expressions: if the first operand has the value FALSE (TRUE), this value is left on the stack and the processor skips evaluation of the second operand.

By contrast, the NS and MC architectures offer a single comparison instruction leaving its result in a special *condition code* register. The distinction between the various relational operators is established by the use of different condition masks in a subsequent instruction that converts the condition code into a Boolean value. As a result, the compilation of Boolean expressions differs from that of arithmetic expressions and is more complicated. The condition code register is an exceptional feature to be treated differently from all other registers (see Examples 1 and 5 in the following Section).

The following tables summarize the mentioned primary differences.

	Lilith	NS 32000	MC 68000
Instruction lengths	8, 16, 24	8, 16, 24, 32, 40, ...	16, 32, 48, 64, 80
Address lengths	4, 8, 16	8, 16, 32	16, 32
Addresses per instr.	0, 1	1, 2	1, 2
External addressing	yes	yes	no
Condition code	no	yes	yes
Data registers	stack (16)	R0 - R7	D0 - D7
Address registers	G, L, S, (H)	SB, FP, SP, MOD	A0 - A6, SP

Data addressing modes of Lilith (M = memory)

mode	operand	comment
stack	T	top of expression stack
local	M[L + a]	a = offset
global	M[G + a]	

external	M[M[t+b]+a]	t = module table origin
indirect	M[T+a]	
indexed	M[T+T']	base adr + index value
immediate	M[PC]	for literal constants

Notes:

0. M denotes memory.
1. capital letters denote resources of the processor, small letters parameters of the instruction
2. T denotes the top of the expression stack, T' the next to top operand.

Data addressing modes of NS and MC

mode	operand (NS)	operand (MC)
register	R[n]	D[n]
address register		A[n]
register indirect	M[R[n]]	M[A[n]]
autoincrement	M[SP]; INC(SP)	M[A[n]]; INC(A[n])
autodecrement	DEC(SP); M[SP]	DEC(A[n]); M[A[n]]
direct	M[SB+d]	M[A[n]+d]
	M[FP+d]	
	M[SP+d]	
indirect	M[M[SB+d1]+d2]	
	M[M[FP+d1]+d2]	
	M[M[SP+d1]+d2]	
indexed	M[SB+d+R[x]*s]	M[A[n]+d+D[x]]
	M[FP+d+R[x]*s]	M[A[n]+d+A[x]]
indirect indexed	M[M[SB+d1]+d2+R[x]*s]	
	M[M[FP+d1]+d2+R[x]*s]	
external	M[M[M[MOD+8]+d1]+d2]	
immediate	M[PC]	M[PC]

Notes:

0. capital letters denote resources of the processor, small letters parameters of the instruction
1. n, x are register numbers (0 ... 7), d, d1, d2 are displacements (offsets)
2. autoinc and -dec modes are called stack mode on the NS and apply to the SP register only.
3. s denotes a scale factor of 1, 2, 4, or 8
4. MC's term for "direct" is "register indirect with offset".

Code Generation

The three compilers not only use the same scanner, parser, table handler, and symbol file generator modules, they also use the same method for code generation [7]. This is a straightforward technique based on the premise that each syntactic construct be represented by a (recursive descent) procedure and by its result parameter. This parameter (defined to be of type *Item*) contains the various attribute values describing the parsed construct. The method is further based on the premise that the values of the construct's attributes, and the corresponding code generated, are to be determined exclusively by the attribute values of the construct's constituents, i.e. that they are context-free.

Before we proceed to demonstrate the method by a few characteristic examples, we need to know what attributes might be involved in describing constructs and what determines their choice. When compiling an expression, for instance, we wish to distinguish whether the expression represents a constant or a variable, because if an addition is compiled, the compiler may add directly, if both operands are constants; otherwise it issues an add operator. In order to allow (constant) expressions to occur in declarations, the compiler's ability to evaluate expressions is indispensible. In essence, we wish to distinguish between all modes of operands for which the eventual code might differ. Code is emitted whenever a further deferment of code release could bring no advantage. The following table displays the modes of item descriptors and their attributes chosen for the three processors.

Lilith		NS 32000		MC 68000	
conMd	value	conMd	value	conMd	value
dirMd	adr	dirMd	adr		
indMd	offset	indMd	adr, offset		
		indRMd	R	indAMd	adr, A
inxMd		inxMd	adr, RX	inxAMd	adr, D, A
		inxiMd	adr, offset, RX		
		inxRMd	R, offset, RX		
stkMd		stkMd		stkMd	
		regMd	R	AregMd	A
				DregMd	D
		cocMd	cc, Tjmp, Fjmp	cocMd	cc, Tjmp, Fjmp
typMd	type	typMd	type	typMd	type
procMd	proc	procMd	proc	procMd	proc

Note: The value adr is actually a triple consisting of module number, level, and offset.

The "original" modes are *conMd, dirMd, indMd, typMd,* and *procMd.* They are the modes given to a newly created constant factor, variable, var-parameter, type transfer function, or procedure call respectively. The other modes emerge when appropriate constructs are recognized: for instance, an item is given *inxMd* (or *inxiMd*), when an item is combined with an index expression to form an indexed designator (see Example 4 below). Or an item obtains *indMd,* if a pointer variable (dirMd) followed by a dereferencing operator and a field identifier have been parsed (see Example 3 below). In general, the more complicated modes originate from the reduction of composite object designators.

Evidently the modes of an item are determined largely by the available addressing modes of the target processor. The more addressing modes, the more item modes, the larger the state space of the items to be compiled, and the more complicated the transformation and code selection routines. This comes as no big surprise; the benefit of a complex instruction set is, alas, not expected to lie primarily in a simpler compiling algorithm, but rather in shorter and more efficient target code.

The following examples show the parsing steps for several simple assignment statements. At the left the syntactic reduction step determined by the parser is indicated. Then follows the mode of the attribute (item) associated with the syntactic unit resulting from the step. The next column displays the code generated by this step, if any; and lastly, for clarification, the column labelled *stack* indicates the sequence of values stacked at this point during the execution phase.

Example 1: x := y+z

	Lilith			NS 32000 and MC 68000	
reduction	mode	code	stack	mode	code
var0 ← x	dir x			dir x	
var1 ← y	dir y			dir y	
fact1 ← var1	stack	LLW y	y	dir y	
exp1 ← fact1	stack		y	dir y	
var2 ← z	dir z		y	dir z	
fact2 ← var2	stack	LLW z	y, z	dir z	
exp ← exp1 + fact2	stack	ADD	y+z	reg 0	MOV y,R0; ADD z,R0
stat ← var0 := exp	–	STO x	–	–	MOV R0,x

Example 2: x := 3+5

reduction	mode	code	stack	mode	code
var0 ← x	dir x			dir x	
fact1 ← 3	con 3			con 3	
exp1 ← fact1	con 3			con 3	
fact2 ← 5	con 5			con 5	
exp ← exp1 + fact2	con 8			con 8	
stat ← var0 := exp	–	LIT 8; STO x		–	MOV 8,x

Example 3: x := r↑.f

reduction	mode	code	stack	mode	code
var0 ← x	dir x			dir x	

reduction	mode	code	stack	mode	code
varl ← r	dir r			dir r	
varl ← varl ↑	indir 0	LLW r	r = @r↑	ind reg 0	MOV r,A0 (MC)
				indir r, 0	(NS)
varl ← varl . field	indir f		r	ind reg 0, f	(MC)
				indir r, f	(NS)
exp ← varl	stack	LSW f	r↑.f	ind reg 0, f	(MC)
				indir r, f	(NS)
stat ← var0 := exp	-	STO x	-	-	MOV f(A0),x (MC)
				-	MOV f(r),x (NS)

Example 4: a[i] := b[j] (without index checks)

reduction	mode	code	stack	mode	code
var0 ← a	dir a			dir a	
var0 ← var0 [	indir 0	LLA a	@a	dir a	
varl ← i	dir i		@a	dir i	
expl ← varl	stack	LLW i	@a, i	dir i	
var0 ← var0 expl]	index		@a, i	inx a R0	MOV i, R0
var2 ← b	dir b		@a, i	dir b	LEA a,A4 (MC)
var2 ← var2 [	indir 0	LLA b	@a, i, @b	dir b	
var3 ← j	dir j		@a, i, @b	dir j	
exp3 ← var3	stack	LLW j	@a,i,@b,j	dir j	
var2 ← var2 exp3]	index		@a,i,@b,j	inx b R1	MOV j, R1
exp2 ← var2	stack	LXW	@a, i, b[j]	inx b R1	LEA b, A3 (MC)
stat ← var0 := exp2	-	SXW	-	- MOV b(R1), a(R0) (NS)	
				MOV (A3,R1),(A4,R0) (MC)	

Example 5: x := y < z

reduction	mode	code	stack	mode	code
var0 ← x	dir x			dir x	
varl ← y	dir y			dir y	
fact1 ← varl	stack	LLW y	y	dir y	
expl ← fact1	stack		y	dir y	
var2 ← z	dir z		y	dir z	
fact2 ← var2	stack	LLW z	y, z	dir z	
exp ← expl + fact2	stack	LESS	y<z	coc <	CMP y,z (NS)
stat ← var0 := exp	-	STO x	-	-	SCC < x (NS)

(for MC: MOV z,D0; CMP y,D0; SCS D0; NEG D0; MOV D0,x)

The following further examples are included to provide additional insight into specific areas of code generation and the processors' influence on it.

Procedure declaration:

```
PROCEDURE P(x, y: INTEGER; VAR z: INTEGER);
    VAR i, j: INTEGER;
    BEGIN ...
    END P
```

ENT	5	ENTER 0 4		LINK	A6, 4	
SLW	z					
SLW	y					
SLW	x	EXIT	0	UNLK	A6	
RTN		RTN	8	RTD	8 (NS)	
				MOVE	(A7)+,A4	(MC)
				ADDQ	#8, A7	(MC)
				JMP	(A4)	(MC)

Procedure call:

```
P(17, k + 5, k)
```

LIB	17	MOVW	17 TOS	MOVW	17 -(A7)
LGW	k	MOVW	k(SB) R0	MOVW	k(A5) D0
LI5		ADDW	5 R0	ADDW	5 D0
UADD		MOVW	R0 TOS	MOVW	D0 -(A7)
LGA	k	ADDRD	k(SB) TOS	PEA	k(A5)
CLL	P	BSR	P	BSR	P

Procedure parameters are passed via the stack of activation records. The NS/MC processors deposit the parameters' values or addresses on top of the stack (allocated in memory) before control is transferred to the procedure. Since parameters are addresssed relative to the local frame base, they are already in their proper place when the procedure is entered. In the Lilith computer, parameters are also put on the stack. However, because the top of the stack is represented by fast registers (the expression stack), and because this stack is reused in the procedure for expression evaluation, the parameters have to be unstacked into the memory frame immediately after procedure entry. This complicates code generation somewhat, but in general shortens the generated code, because the unstack operations occur in the procedure's code once, and not in each call. The fact that the NS/MC architectures include a move instruction alleviates this advantage of Lilith, because the move instruction bypasses registers (which play a role corresponding to the Lilith expression stack).

Indexed variables:

```
VAR a: ARRAY [0 .. 99] OF INTEGER;
    b: ARRAY [-10 .. +10] OF INTEGER;
    c: ARRAY [0 .. 99], [0 .. 15] OF CHAR;

u := a[9]
```

LGW	a	MOVW a-18(SB) u(SB)	MOVW a-18(A5) u(A5)
LSW	9		

```
        SGW   u

u := a[i]

    LGW   a
    LGW   i                                                    MOVW   i(A5) D0
    LIB   HIGH(a)                                              CHK    99 D0
    CHKZ              CHECKW R0 [0,99] i(SB)                   ASLW   1 D0
    LXW               FLAG                                      LEA    a(A5) A4
    SGW   u           MOVW [R0:W] a(SB) u(SB)                  MOVW   0(A4,D0.W) u(A5)

u := b[i]

    LGW   b
    LGW   i                                                    MOVW   i(A5) D0
    LIW   -10                                                  ADDW   10 D0
    ISUB
    LIB   20                                                    CHK    20 D0
    CHKZ              CHECKW R0 [-10,+10] i(SB)                ASLW   1 D0
    LXW               FLAG                                      LEA    b(A5) A4
    SGW   u           MOVW [R0:W] b(SB) u(SB)                  MOVW   0(A4,D0.W) u(A5)

u := c[9, 9]

    LGW   c
    LSA   216
    LSW   9
    SGW   u           MOVB [R0:W] c-450(SB) u(SB)              MOVB   c-450(A5) ch(A5)

u := c[i, j]

    LGW   c
    LGW   i                                                    MOVW   i(A5) D0
    LIB   103
    CHKZ              CHECKW R0 [0,99] i(SB)                   CHK    99 D0
    LIB   24          FLAG
    UMUL                                                       ASLW   4  D0
    UADD                                                       LEA    c(A5) A4
    LGW   j                                                    MOVW   j(A5) D2
    LIB   23          CHECKW R1 [0,23] j(SB)                   CHK    15 D2
    CHKZ              FLAG
    LXW               INDEXW R0 23 R1                          LEA    0(A4, D0.W) A4
    SGW   u           MOVB [R0:W] c(SB) u(SB)                  MOVB   0(A4,D2.W) u(A5)
```

Because indexed variables occur very frequently, the resulting code should be short. All three processors therefore include special instructions for indexed address computation. They include the validation of array bounds, i.e. they check whether the index value lies within the bounds specified by the array variable's declaration. In the case of Lilith, the code differs in the case of the low bound being zero. Although this may appear as an insignificant peculiarity, it contributes to the effectiveness of the architecture due to its high frequency of occurrence.

Arithmetic expression: (assume global variables a, b, and local variables i, j)

$(a + 10) - ((i + b * 5 + j * 2) \, DIV \, 4)$

```
LGW    a          MOVW a(SB) R0          MOVW  a(A5) D0
LI    10
UADD             ADDW 10 R0             ADDIW 10 D0
LLW    i
LGW    b          MOVW b(SB) R2          MOVW  b(A5) D2
LI     5          MEIW 5 R2
UMUL             MOVW i(FP) R1          MULS  5 D2
UADD             ADDW R2 R1            ADDW  i(A6) D2
LLW    j          MOVW j(FP) R2          MOVW  j(A6) D4
LI     1
SHL              LSHW 1 R2            ASLW  1 D4
UADD             ADDW R2 R1            ADDW  D4 D2
LI     2                               EXTL  D2
SHR              LSHW -2 R1            DIVS  4 D2
USUB             SUBW R1 R0            SUBW  D2 D0
```

The NS/MC compilers utilize the data registers in a manner similar to a stack. The compiler does not keep track of what was loaded into these registers. Hence it is clear that the registers are not used in an optimal fashion; but any further improvement increases the compiler's complexity considerably. However, multiplications and divisions by integral powers of 2 are (easily) recognized and represented by shift instructions.

Boolean expressions:

Boolean expressions require special attention. Although they are specified by the same syntax as other expressions, their evaluation rules differ. In fact, the definition of the semantics of Boolean expressions is inconsistent with their syntax, at least if one adheres to the notion that a syntax must faithfully reflect the semantic structure. This anomaly is due to the facts that the syntax of expressions is defined regardless of type, that arithmetic operators are defined to be left-associative, and that logical operators are right-associative. For example, $x+y+z$ is understood to be equivalent with $(x+y)+z$; by contrast p&q&r is equivalent with p&(q&r). The logical connectives are defined in Modula in terms of conditional expressions, namely

$$p \& q \quad = \text{ if p then q else false}$$
$$p \text{ OR } q \quad = \text{ if p then true else q}$$

Consequently,

$$p \text{ OR } q \text{ OR } r \quad = \text{ if p then true else (if q then true else r)}$$

which is obviously right-associatve. The Boolean connectives are implemented not by logical operators, but by conditional jumps. And since Boolean expressions occur most frequently as constituents of if and while statements, a further complication arises: an efficient implementation must unify conditional jumps within expressions with those occurring in statements, thus effectively breaching the syntactic structure of the language. Fig. 5 indicates the structural transformations implied by the generated code sequences for the NS by showing a few simple examples. The resulting structures could not be expressed by a context-free syntax.

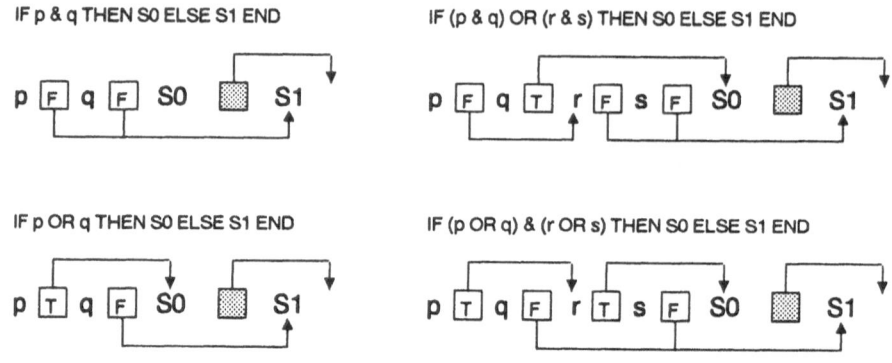

Fig. 5. Boolean connectives represented by conditional jumps

For Lilith, no structural transformations are necessary thanks to the existence of the and-jump and or-jump instructions which either cause a jump or the removal of the Boolean value on top of the stack. Whereas the compilation of Boolean expressions is straightforward, the resulting code is, however, less than optimal.

IF ((x < y) OR (z <= x)) & ((u < v) OR (w <= u)) THEN x := y ELSE u := v END

```
      LGW   x            CMPW   y x              MOVW   x(A5) D0
      LGW   y                                    CMPW   y(A5) D0
      LSS
      ORJ   L1            BGT    L1              BLT    L1
      LGW   z            CMPW   x z              MOVW   z(A5) D0
      LGW   x                                    CMPW   x(A5) D0
      LEQ
L1:   AJP   L2            BLS    L3              BGT    L3
      LGW   u      L1:   CMPW   v u       L1:   MOVW   u(A5) D0
      LGW   v                                    CMPW   v(A5) D0
      LSS
      ORJ   L2            BGT    L2              BLT    L2
      LGW   w            CMPW   u w              MOVW   w(A5) D0
      LGW   u                                    CMPW   u(A5) D0
      LEQ
L2:   JPC   L3            BLS    L3              BGT    L3
      LGW   y      L2:   MOVW   y x       L2:   MOVW   y(A5) x(A5)
      SGW   x
      JP    L4            BR     L4              BRA    L4
L3:   LGW   v      L3:   MOVW   v u       L3:   MOVW   v(A5) u(A5)
      SGW   u
L4:   ...          L4:   ...               L4:   ...
```

The considerable complications caused by the NS/MC architectures for handling expressions is modestly reflected by the introduction of a new item mode (cocMd) with the meaning "the item's value is represented by the condition code register". Its attributes are the mask value CC appropriately transforming the register value into a Boolean value, and two sequences of locations of branch instructions that require updating once their destination address is known. These sequences are labelled *Tjmp* and *Fjmp*, locating the branches taken for the Boolean result being TRUE or FALSE respectively.

In summary, we observe that - as expected - the NS/MC architectures lead to a smaller number of generated instructions compared to that of the pure stack architecture of Lilith. The gain is at the expense of more complicated compiling algorithms.

Measurements

The essential characteristics, source program length and object code size, of the investigated compilers are summarized in the following table.

Module	source text		object code (bytes)				
	lines	chars	Lilith	NS	MC	NS/L	MC/L
Scanner (M2S)	410	11200	2640	4180	5580	1.58	2.11
Parser (M3P)	1300	39520	8190	11340	18500	1.38	2.26
Table Handler (M3T)	270	8400	1350	2460	3500	1.82	3.29
Symbol File Gen. (M3R)	530	20850	3680	5730	9240	1.55	2.51
Code Gen. for Lilith	1490	50200	10190				
Code Gen. for NS	2050	69000	(15340)	22960		1.50	
Code Gen. for MC	3780	150000	(21550)		48630		2.26
Total			26050	46670	83450	1.79	3.28

Objectively considered, the results are not only against all intuitive expectations, but they are also *highly disappointing* with regard to the commercial microprocessors. Because of the complex instruction set, the hardware is considerably more intricate than that of Lilith, and the cost of it has been felt severely by long development delays. Another consequence is the need for more sophisticated code generators, if the "power" of the instruction set is to be fully tapped. The compiler program is 14% longer for NS and 56% longer for MC than for Lilith. If we consider the code generator parts only, the respective figures are 37% and 154%. But most disappointingly, the reward for all these efforts and expenses appears as *negative* : for the *same* programs, the compiled code is about 50% longer for NS and 130% longer for MC than for Lilith. The cumulative effect of having a more complicated compiling algorithm applied to a less effective architecture results in the compiler for the NS being 1.8 times more voluminous than that for Lilith, whereas the compiler for the MC is 3.3 times as long. Quite evidently, the value of a megabyte of memory strongly depends on the computer in which it is installed.

Translated into the time domain, we obtain similar results: measured over a reasonable number of diverse programs, execution times are in the average about 1.4 times as high on the NS32016 operated with a 10MHz clock as on a Lilith with a 7MHz clock. For the sake of fairness, it has to be mentioned that the commercial microprocessors offer a slightly richer set of data types (i.e. operand lenghts); Lilith does not offer operations for double precision integers and double precision floating point numbers. However, the influence on the instruction format or the compiling algorithm are rather marginal, and could not account for more than a few percent of additional length.

Naturally, one wonders where the architects have miscalculated. Measurements condensed into the following tables shed some light on this question. There is, of course, no single contributing factor to the poor result, but several; consequently, one sould not expect a single, simple answer.

The first table displays the relative frequencies of occurrence (in percents) of various instruction formats of Lilith. Of particular relevance is the distribution of *instruction lengths*. Hence we distinguish between operators with various lengths of their operand field: 0, 4, 8, 16, 24, 32. As objects of this investigation we again use the modules of the compilers themselves. Admittedly, this introduces some bias, e.g. against long operands (real numbers), but other measurements have largely confirmed these results.

	M2S	M3TL	M3RL	M3GL	M3PL	Total
length	2744	1356	3684	10274	8194	26252
no of instr.	1912	1026	2489	7001	4898	17326
bytes/instr.	1.44	1.32	1.48	1.48	1.67	1.52
operator	20.92	13.74	14.18	17.90	15.31	16.72
4bit opd	46.50	65.98	55.36	49.62	43.16	49.24
8bit opd	23.54	8.67	13.06	18.75	15.80	17.03
16bit opd	7.06	6.04	8.28	12.06	11.33	10.40
24bit opd	0.10	0.10	0	0.11	0	0.06
32bit opd	0.84	0	0.04	0.20	0	0.18
ext adr	1.05	5.46	9.08	1.36	14.39	6.36

The quotient of the length of the code and the number of generated instructions yields the average instruction length. Among the listed programs it varies between 1.3 and 1.7 bytes per instruction. About 15% of all instructions are operators without explicit operand fields, implicitly referring to operands on the expression stack. About 55% of all instructions have a single operand field 4 bits long; between 10 and 15% require a one-byte operand field. In the case of the 4-bit fields, it is packed together with the operator field into a single byte. This facility of short operand fields holds the key to Lilith's high code density. The idea stems from the Mesa instruction set of Xerox's D-machines [8, 9].

The next table shows the relative frequencies of occurrence (in percent) of instructions generated for the NS architecture, classified according to their *formats*. We emphasize that the pictures shown in Fig. 2 contain the operation code and one or two address mode indicators only. These one, two, or three bytes are usually followed by further bytes containing the addresses, operands, and indexing information. In rare cases, a single instruction may consist of a dozen bytes or even more. The average instruction length measured for the compiler is about 3.5 bytes for the NS versus 1.5 bytes for Lilith. The number of generated instructions, however, (the figure that looked so appealing in the examples above) is only 1.6 times higher for Lilith.

Module	M2S	M3TN	M3RN	M3CN	M3EN	M3HN	M3PN	Total
length	4184	2464	5732	5396	12216	5348	11344	46684
no of inst	1114	671	1512	1508	3288	1534	3251	12878
bytes/inst	3.76	3.67	3.79	3.58	3.72	3.49	3.49	3.63
F0 jump	17.24	10.58	14.02	13.06	23.48	17.93	21.04	18.66
F1 call	21.45	21.91	21.96	22.94	16.18	21.71	25.38	21.39
F2 quick	10.95	17.59	11.51	11.94	14.75	12.52	8.74	12.07
F3 case	0.09	0.60	0.40	0.33	0.18	0.07	0.03	0.19
F4 2-opd	29.62	32.49	46.36	35.21	30.93	33.57	32.17	33.84
F5	0	0.30	0	0	0.70	0.13	0.06	0.23
F6 1-opd	13.55	16.24	3.31	12.47	11.10	13.17	11.81	11.25
F8 check	3.59	0.30	2.45	4.05	1.06	0.91	0.74	1.65
F11 fltpt	3.50	0	0	0	1.61	0	0.03	0.72
1-byte	38.69	32.49	35.98	36.01	39.66	39.63	46.42	40.05
2-byte	40.66	50.67	58.27	47.48	45.86	46.15	40.94	46.10
3-byte	20.65	16.84	5.75	16.51	14.48	14.21	12.64	13.85

The NS and MC architectures feature a particularly rich set of *data addressing modes*, designed to reduce the number of instructions and to increase the density of code. The relative frequencies of their usage is tabulated below. Between 7 and 20% of references are to registers directly. This percentage roughly corresponds to the implicit stack references of Lilith. The stack mode of the NS is used exclusively for placing procedure parameters into the stack of activation records, and therefore has no relationship to Lilith's stack usage. The frequency of stack references is nevertheless surprisingly high (over 20%).

A noteworthy although not surprising result is that local objects are considerably more frequently accessed (via FP) than global ones (via SB). The ratio varies from 1.7 to 6.9, however. Surprisingly frequent are indirect accesses (up to 30%); this mode uses two displacements. This is a reflection of the preponderance of access to record fields via pointers. This addressing mode is present in the NS, but not so in the MC architecture.

Looking at constants, represented as immediate mode data placed in the instruction stream "immediately" following the instruction, one recognizes the predominance of 16-bit operands. Knowing the data size distribution measured for Lilith, one realizes that a major flaw of the NS/MC designs lies in the requirement that the length of an immediate operand be exactly as defined by the operator; no automatic lengthening (with either zero or sign extension) is provided, as it exists in the case of addresses (displacements).

Module	M2S	M3TN	M3RN	M3CN	M3EN	M3HN	M3PN	Total
no of inst	1114	671	1512	1508	3288	1534	3251	12878
no of addr	1322	824	1839	1891	3532	1736	3317	14461
addr/inst	1.19	1.23	1.22	1.25	1.07	1.13	1.02	1.12
register	16.87	7.89	10.55	21.89	14.89	15.67	9.41	13.87
reg indir	0.53	2.55	6.31	1.27	1.81	4.49	4.22	3.11

dir (FP)	18.00	12.86	16.91	17.72	11.44	13.94	18.24	15.50
dir (SB)	24.21	10.07	11.69	14.97	3.40	1.96	3.77	8.16
indir (FP)	0.23	11.17	18.71	12.37	30.27	21.37	12.00	17.36
indir (SB)	0	6.80	1.41	0	0	0	0.09	0.59
immed byte	12.78	6.80	3.53	7.09	4.53	5.70	13.14	7.74
immed word	8.17	9.59	5.06	9.52	9.31	10.43	6.84	8.28
immed dbl	0.98	0	0	0	0.45	0.06	0	0.21
absolute	0	0	0	0	0	0	0	0
external	0	5.22	3.10	0.53	2.80	3.28	10.49	4.25
indexed	3.18	0.49	2.01	3.23	0.99	0.81	0.75	1.51
stack	15.05	26.58	20.72	11.42	20.10	22.29	21.04	19.43

This brings us to a final investigation of the frequencies of the various *displacement sizes*. The NS architecture provides sizes of 1, 2, or 4 bytes. The length is not dictated by the operator code, but instead is encoded in the displacement value itself. This is a desirable solution also from the point of view of code generation. As expected, the 1-byte displacements dominate strongly.

Module	M2S	M3TN	M3RN	M3CN	M3EN	M3HN	M3PN	Total
no of disp	941	787	1979	1568	4160	1768	3808	15011
byte	50.37	79.16	59.47	78.51	74.50	78.22	70.14	71.00
word	45.59	20.71	40.47	18.18	23.73	21.72	28.94	27.65
double	4.04	0.13	0.05	3.32	1.78	0.06	0.92	1.35
bytes/disp	1.58	1.21	1.41	1.28	1.29	1.22	1.32	1.32

The frequency of instructions classified according to their length for the MC code is shown in the following table. These values hold for all compiler modules combined, and the avaerage instruction length is 3.46 bytes per instruction.

1-word	35.09%
2-word	57.36
3-word	6.85
4-word	0.70

The distribution of addressing modes indicates a dominance of register addressing with an offset. As in the tables above, PC relative addressing is not counted, as it is not being used for data access.

Addressing modes:	
D direct	12.72%
A direct	24.58
A indirect	5.67
A ind postincrement	5.62
A ind predecrement	12.12
A ind displacement	30.20
A ind index	0.49
immediate	8.53

An analysis of displacement sizes is only of academic interest, since there is only one size. However, it is noteworthy that 94% of all displacement values could be places into a single byte instead of a 16-bit word.

Conclusions

The NS and MC architectures have been compared with the Lilith architecture as a prototype of a regular, stack-oriented design. The increased complexity of their resources, instruction sets, and addressing modes fails to lead to a simpler, but rather requires a more complicated compiler. Regrettably, it also results in longer and often less efficient code. In the average, code for the NS is about 50% longer, code for the MC 130% longer than that for Lilith. Among the commercial products, this puts NS far ahead of MC. An improvement could be obtained only by substantial efforts for code "optimization", at the expense of much more complex compilers.

Although the two investigated microprocessors are by far the best architectures widely available, these results suggest that they also leave room for improvement. Among the two the NS yields markedly better results, particularly when judged by the compiler designer. In the author's oppinion, both designs could have avoided some serious miscalculations, if their compilers (for some high-level language) had been implemented *before* the designs of the processors were fixed.

The analysis presented here reveals the two main pinpointable causes of the low code density as being

1. the lack of short (less than 1 byte) address or operand fields, and
2. the use of explicitly addressed registers for intermediate results.

The principal underlying syndrome, however, is a misguided belief in complexity as a way towards better performance. Both the NS and MC architectures are marred by too baroque instruction sets and addressing modes. Obviously, they are compromises trying to satisfy many uncoordinated, imagined requirements, and results of an unbounded belief in the possibilities of VLSI. But not everything that *can* be done *should* be done.

This critizism appears to point in the direction of architectures featuring a simple structure, a small set of simple instructions, and only a few basic addressing modes. Such designs have become known as RISC architectures [2]. One should be cautious, however, not to rush from one extreme into the other. In fact, recent RISC schemes propose facilities - such as a register bank effectively implying a two-level store - which require complicated code generation algorithms to achieve optimal performance. Once again, the designers are primarily, if not exclusively, concerned with speed. But there is no reason why features could not be added to a design that cater to specific, genuine problems posed by the implementation of high-level languages. Under no circumstances, however, should such additions involve a complicated mechanism, or infringe on the regular structure of the existing scheme. *Regularity of design* emerges as the key. Features must *solve* problems, not *create* them. In order to promise genuine progress, the acronym RISC should stand for *Regular* (not Reduced) instruction set computers!

Regularity alone, however, is not sufficient. It must be accompanied by completeness. The instruction set must closely mirror the complete set of basic operators available in the language. In this respect the NS architecture represents a significant improvement over earlier products, whereas the MC design gives rise to innumerable grievances and is ill suited for effective compiler design. We emphasize that *regularity and completeness* have been pivotal concepts in mathematics for centuries. It is high time that they be recognized by engineers designing mathematical machines.

This analysis perhaps gave the expression of predominant concern with code density and code efficiency. But there is a much more profound reason to strive for regularity of design than efficiency. It is *reliability*. Reliability unquestionably becomes a victim whenever unnecessary complexity creeps in. Unreliability grows at least proportionally to the complexity of a device's specification, let alone its implementation. This "law" applies equally well to hardware as to software.

Reliability (and *not* convenience of programming) had also been the primary motivation behind the development of high-level, structured languages. They are supposed to provide suitable abstractions to formulate data definitions and algorithms in forms amenable to precise, mathematical reasoning. But these abstractions are useless unless they are properly supported by a correct, "water-tight" implementation. This postulate implies that all violations of the axioms governing an abstraction *must* be detected and reported. Checks against violations must be performed by the compiler whenever possible, and otherwise by additional instructions interpreted at run-time. It is therefore a primary characteristic of architectures designed for high-level languages that they support these abstractions by suitable facilities and efficient instructions in order to make the "overhead" minimal.

The consistent support of such checking is perhaps the most commendable characteristic of Lilith. The following violations lead to immediate termination of a computation:

1. Access to an array with an invalid index.
2. Access to a variable via a pointer variable with value NIL.
3. Overflow in integer, cardinal, and real number arithmetic.
4. Selection of an invalid case in a case statement.
5. Lack of data space on procedure call (stack overflow).

All these violations except the first are detected without the need for additional instructions. The checks are built into the address computation, arithmetic, case selection, and procedure entry instructions. Index values are validated by additional instructions inserted before the address calculation. These features - above anything else - have characterised Lilith as high-level language oriented. During five years of intensive use they have proven to be not only most valuable, but indispensible, and they have made possible a truly effective environment for program development. Protecting the validity of a language's abstractions is *not a luxury, but a necessity*. It is as vital to inspire confidence in a system as is the correctness of its arithmetic and its memory access. A processor must be designed such that the "overhead" caused by the guards is unnoticeable.

By these standards, both the NS and MC architectures can be called at best "half-heartedly high-level language oriented". But this already represents a tremendous improvement over all earlier commercial processor designs! Both processors feature convenient index bound checks. But unfortunately tests for invalid access via NIL values, or for stack overflow are available only at the cost of cumbersome instruction sequences which most programmers - too confident in their art - are unwilling to accept. Even tests for arithmetic overflow require additional instructions. It is simply incomprehensible that instructions specifically designed for reserving space for local variables upon procedure entry can be designed without the inclusion of a limit check.

The neglect of such essential properties is even harder to comprehend considering the sizeable efforts made towards easing other supposed problems. An example is the elimination of the requirement for address alignment. (A datum is said to be *aligned*, if its address a is a multiple of the computer's wordlength). A good compiler will align data even if the processor does not require this, thereby avoiding unnecessary memory accesses. The complex mechanism to handle non-aligned data will therefore rest unused all the time, whereas badly needed tests have to be implemented by additional, explicit instructions.

Acknowledgement

The author gratefully acknowledges the valuable contributions by W. Heiz and H. Seiler, who have ported the compiler to the MC 68000, have designed the new code generator, and have provided the data concerning that architecture.

References

1. N. Wirth. *Programming in Modula-2.* Springer-Verlag, New York, 1982.

2. D.A. Patterson. Reduced instruction set computers. *Comm ACM,* 28, 1 (Jan. 1985), 50-57.

3. N. Wirth. The personal computer Lilith. *Proc. 5th Int. Conf. on Software Engineering,* San Diego, March 1981, IEEE Computer Society Press.

4. R.S. Ohran. Lilith and Modula-2. *BYTE,* 9, 8 (Aug. 1984), 181-192.

5. Series 32000 instruction set reference manual. National Semiconductor Corporation, 1984.

6. MC68020 32-bit microprocessor user's manual. Prentice-Hall, Englewood Cliffs, 1984.

7. N. Wirth. A fast and compact Modula-2 compiler. (1985) To be published.

8. R.K. Johnsson and J.D. Wick. An overview of the Mesa processor architecture. *Proc. of the Symp. on Architectural Support for Programming Languages and Operating Systems;* Palo Alto, March 1982. (Also published in *SIGARCH Computer Architecture News* 10 (2) and in

SIGPLAN Notices 17 (4)).

9. R.E. Sweet and J.G. Sandman. Empirical Analysis of the Mesa Instruction Set. *Proc. of the Symp. on Architectural Support for Programming Languages and Operating Systems;* Palo Alto, March 1982. (Also published in *SIGARCH Computer Architecture News* 10 (2) and in *SIGPLAN Notices* 17 (4)).

15

FAULT TOLERANT VLSI MULTICOMPUTERS

Carlo H. Séquin and Yuval Tamir *

Computer Science Division
Electrical Engineering and Computer Sciences
University of California, Berkeley, CA 94720

ABSTRACT

An approach is presented to increasing the reliability of future high-end systems beyond what is possible with technological solutions alone. The system consists of computation nodes and communication nodes, interconnected by high-speed dedicated links. These components are relied upon to detect errors while system level protocols are used for error recovery and reconfiguration. The use of duplication and matching for implementing the self-checking nodes allows us to restrict a detailed analysis of the impact of all possible faults to the comparator, a circuit that can be implemented in a relatively straight-forward way in NMOS or CMOS technology.

1. INTRODUCTION

Certain computational problems such as weather forecasting, simulations of complex systems, or design optimizations, exceed the capabilities of current computers. They require a substantial increase in compute power and, since such computations may run for an extended amount of time, improved systems reliability. There are fundamental limitations to the gains in reliability and performance that can be obtained from advancing technology alone. A more important contribution will have to come from organizational improvements in such computing systems: performance can be enhanced by exploiting parallelism, while the limits on reliability can be overcome using fault tolerance techniques.

If the computational problem can be subdivided into a sufficient number of simultaneously executing tasks, then this inherent parallelism of the problem can be used to achieve high performance by a system that comprises many computational nodes. A possible systems architecture that is compatible with the constraints of VLSI interconnects these computation nodes using high-speed dedicated links, and *communication nodes* which provide hardware support for communication functions such as message routing.[23] Computation nodes my consist of a single processor chip and several memory chips surrounded with the associated glue

---

* Now with the Computer Science Dept., University of California, Los Angeles, CA 90024.

logic, forming a powerful self-contained computer. The communication node has several ports through which it is connected to computation nodes and other communication nodes. Such a system is called a *multicomputer*. Ideally, the two types of nodes and the links between them are building blocks that can be used to construct multicomputers with a wide range of organizations and performance.

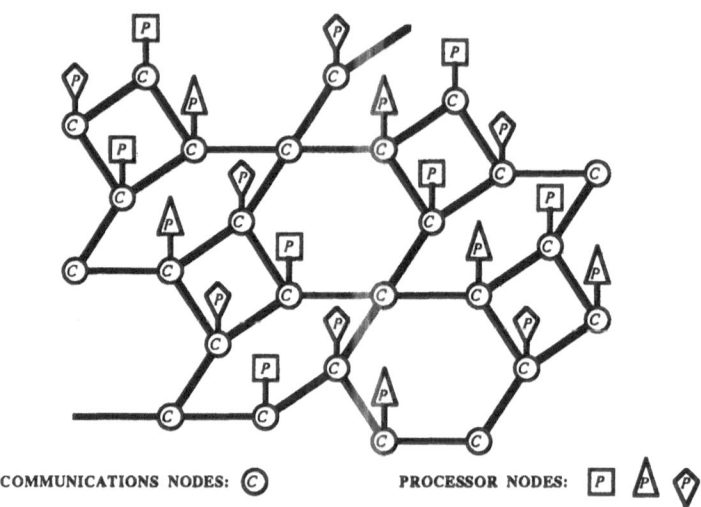

COMMUNICATIONS NODES: Ⓒ   PROCESSOR NODES: ▢ △ ⬠

Figure 1: *Conceptual View of a Multicomputer.*

When the behavior of a system deviates from its specification at the interface with the "outside" world, we say a *system failure* [1] has occurred. System failure is often the result of a failure of one of its components. However the failure of a component does not necessarily imply that a system failure must occur. The system is *fault-tolerant* if it can continue operating correctly despite the failure of some of its components. Various techniques can be employed to provide *fault tolerance* at the different levels of the system hierarchy.

A multicomputer is particularly well suited for reliability enhancement using fault tolerance techniques since it is naturally divided into fairly independent modules of substantial "intelligence" — the above mentioned nodes. Fault-free components can adjust their behavior to the changes in faulty components and continue their operation in such a way that the overall output of the system remains correct despite the occurrence of a fault.

The design of a VLSI chip or of a multicomputer system is in itself a very hard task. Adding the extra demands of fault tolerance may just make this task unmanageable, unless we simplify the task by using principles of regularity and repetition. Using the multicomputer system as an example, it will be demonstrated how the concerns of fault tolerance can be concentrated on a few critical components, and how, by a suitable modular approach, the whole system can become fault tolerant, without undue penalty to either system design time or system performance.

2. FAULT TOLERANCE

The reliability of any system can be enhanced by increasing the reliability of its components through *fault prevention*[1] techniques, such as specialized design methodologies, stringent quality control, and extensive validation and testing. These techniques typically result in more complex designs,[8] greater cost, and lower performance.[20] Furthermore, the effectiveness of these techniques is limited by our inability to exhaustively test complex VLSI chips.[19]

The reliability of components can also be increased by employing *fault tolerance* techniques at the component level. These techniques attempt to ensure that each component will continue to perform according to its specifications despite the failure of its subcomponents. Unfortunately, no component can tolerate an unbounded number of faults. Thus, the system must be able to handle component failure. The contamination of the system by incorrect output from a faulty component can be prevented only if, at some stage, other system components find out about the failure of the component and physically or logically isolate it from the rest of the system.

If a node fails due to a transient fault, it need not be removed from the system, but rather should be reset to a proper state, and then continue to be a useful part of the system. If the failure is detected by a neighboring node, then this node must have the authority to initiate some action that might eventually lead to a resetting of the node. However, the same authority also gives a *failed* node the potential to invoke the resetting of an operational neighbor, so that a single node failure could result in a total system failure. To prevent this undesirable situation, each node must be responsible for its own reset. Hence the node should include a mechanism to detect its own *erroneous states* and to initiate a reset.

System level fault tolerance techniques need not rely on the components to report faults themselves. Instead, system level protocols could be used for detection and recovery from the component failure. For example, each task may be performed in parallel on three nodes and a "majority vote" taken on the results. Such a system with triple modular redundancy[30,32] can continue to produce the correct output even if one of the nodes fails. The effectiveness of this approach is limited by the fact that after a single fault, a node may loose its ability to tolerate any additional faults. While the scheme does not make any assumptions about the nature of the individual subcomponents, it requires the system level protocols to ensure that the parallel task execute on different nodes and that the messages between themselves and the initiation node travel via independent paths. Hence this method leads to very high overhead in the use of the computation nodes as well as in message traffic. Additional problems concern locating failed components and effective handling of transient faults.

Many of the deficiencies of fault tolerance techniques that rely only on hardware or only on system-level protocols can be overcome by using a combination of hardware error detection in self-checking components and system-level protocols that perform error recovery and fault treatment. Errors caused by faults in the communication links are detected through the use of error-detecting codes. All nodes are self-checking and signal to the rest of the system when their output is incorrect. In addition, failed nodes attempt to reset themselves and reestablish a sane state. The immediate neighbors are informed whenever a node fails. If the node does not reset itself or fails too often, the neighbors can logically remove it from the system by refusing to communicate with it. The diagnostic status information is distributed throughout the system so that, eventually, no fault free node will attempt to use the faulty component.

On top of the self-checking hardware there is a low-overhead, application-transparent, distributed error recovery scheme. It involves periodic checkpointing of the entire system state and rolling back to the last checkpoint when an error is detected (Section 7).

3. SELF-CHECKING NODES

For all likely faults, a *self-checking* component must either produce the "correct" output (according to its specifications) or somehow indicate that its output is incorrect. A component that satisfies this requirement is said to be *fault secure*.[22] If the component does not produce an error indication immediately following the first fault, it is possible for several faults to exist in the component simultaneously without any indication to the rest of the system. Even if the component is fault secure with respect to any single fault, several faults together may lead to the failure of the self-check mechanism and, eventually, to incorrect output from the component. In order to prevent this situation, the component must also be *self-testing*.[22] In the presence of one or more faults, a self-testing component is guaranteed to produce an error indication before additional faults can occur that may lead to the failure of the self-check mechanism. Components which are fault-secure as well as self-testing are said to be *totally self-checking*[22] (*TSC*).

Error detecting or correcting codes can be used to implement *TSC* nodes. Redundant information is carried by busses, memories, and registers in order to detect (and possibly correct) errors.[22] Unfortunately, different coding schemes must be used for different parts of the node. This increases the complexity of the design task and makes design verification and testing more difficult. As a result, failure modes that are harder to predict and "tolerate" are more likely to occur.

An alternative is to construct the *TSC* computation or communication node using two identical, *independent* modules, each performing the function of the node. Inputs from neighbor nodes are fed to both modules. Except for the, hopefully nearly-impossible, case where both modules produce identical *incorrect* output (Section 6), if the modules operate synchronously, errors can be detected by simple comparison of the outputs of the modules. The comparator that performs this function is part of the node, and its output is connected to neighboring nodes through dedicated wires. The output from one of the two modules is the "functional" output from the node (Fig. 2). A "no-match" signal from the comparator is used locally as a reset signal and is also sent to all neighbors as a failure indicator. Similar failure indicators from the neighbors cause an interrupt and invoke system-level routines that handle the node failure.

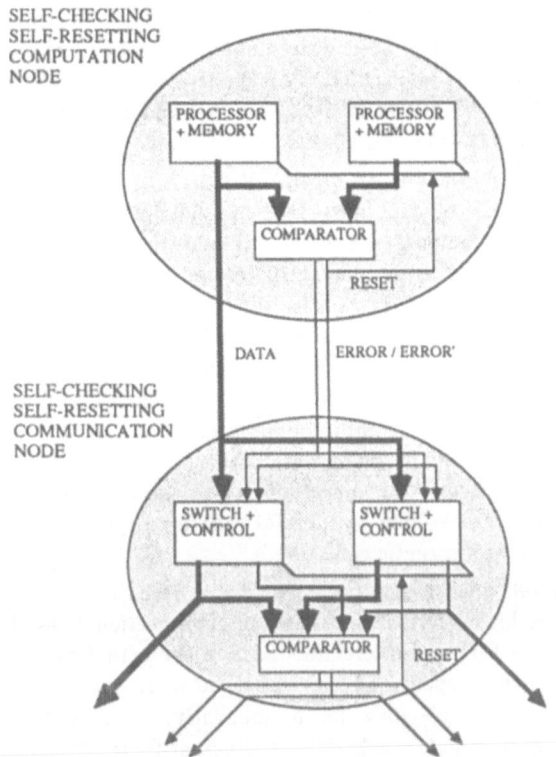

Figure 2: *Self-Checking Nodes for Multicomputers*

Implementing the *TSC* property in a component using duplication and comparison may appear wasteful since it more than doubles the required hardware. However, this scheme becomes more attractive when issues such as design complexity, fault coverage, reliability prediction, and the ability to recover from transient faults are taken into account.

Traditional fault models are not adequate for VLSI.[10, 29] As a result, low-cost error detection schemes, that are based on these models, may no longer be sufficient. With duplication and comparison, errors are detected as long as the comparator remains functional and the two modules produce different outputs the first time one or both of them fail. Since a faulty comparator can mask faulty functional modules, faults in the comparator must not go undetected, *i.e.*, the comparator must be self-testing. Thus a detailed analysis of the effects of all likely faults on the comparator is required.

4. DEFECTS AND FAULTS IN VLSI

The design of self-checking circuits requires an understanding of the physical defects that commonly occur in VLSI and of the resulting logical faults. In the past the stuck-at fault model has been widely used to model, at the logical level, the effects of physical defects in circuits. This model does not cover many of the possible defects in VLSI.[7,10,29] The fabrication flaws and physical processes that can cause malfunction of NMOS and CMOS VLSI circuits are summarized in this section.

VLSI chip failures may be caused by design or fabrication flaws, may be due entirely to environmental factors, or are the end result of a degenerative process invoked by operational and environmental stresses but often attributable to design or manufacturing flaws.[9,22] Fabrication defects in MOS chips consist mainly of shorts and opens in each interconnection level, (metallization, diffusion, and poly-silicon), shorts between different levels, and large imperfections such as scratches across the chip.[10] Other fabrication defects include incorrect dosage of ion implants, contact windows that fail to open, misplaced or defective bonds, and penetration of the package by humidity and other contaminants.[9] During the operation of the chip, faults may be caused by electromigration, corrosion, electrical breakdown of oxide, cracks due to thermal expansion, power supply fluctuation, and ionizing or electromagnetic radiation.[9]

At the logical level, most of the faults can be represented in a circuit model consisting of switches, loads (for NMOS), and interconnection lines that directly correspond to the transistors and interconnections in the actual circuit.[10] Most of the physical defects, such as opens and shorts, can be represented in this model in an obvious way.[7] A "switch" may be permanently on or permanently off, corresponding to a gate input stuck-at-1 or stuck-at-0, respectively. Shorted NMOS loads (pullups) are equivalent to an output line s-a-1. Disconnected gate inputs are usually equivalent to s-a-0 or s-a-1 faults.

Some physical defects have a more complex effect on the circuit. In NMOS, incorrect dosage of ion implants may cause a threshold shift in a load transistor. This can result in an output voltage that lies between the voltages assigned to logic 0 and logic 1. If the fanout from the gate is greater than one, some of the gates connected to its output may "interpret" it as logic 1 while others will interpret it as logic 0. If, at some point in time (clock cycle), the line is supposed to be a logic 1 but is interpreted by some of the gates as logic 0, we call it a *weak 1* fault. Conversely, if the line is supposed to be a logic 0 but is interpreted by some of the gates as logic 1, we call it a *weak 0* fault. A single physical defect, resulting in a single weak 0 or weak 1 fault, has the same effect as multiple s-a-1 or s-a-0 faults, respectively.

In CMOS, a transistor which is permanently off or a break in a line can result in a high impedance state where the output of a combinational logic gate is dependent on the previous output rather than the current input.[29] Such a fault (called a *stuck-open* fault) may escape detection even if all possible input vectors are used to test the circuit.[29]

5. SELF-TESTING COMPARATORS IN VLSI

The duplication and matching scheme relies entirely on a self-testing comparator to detect faults in the functional modules. Implementing such a comparator requires knowledge of how different faults will affect the circuit. Fortunately, a comparator is a simple circuit that can be implemented with a regular structure and is therefore amenable to thorough analysis. Hence, we can have confidence in our ability to predict the likely physical defects, develop a valid fault model, and prove that the implementation we propose is indeed self-testing.

We assume that physical defects in the node occur one at a time. A fault that is the result of a single physical defect is called a *single fault*. It is assumed that there is a negligible probability that the time interval between the occurrence of successive single defects in the comparator or between a single defect in the comparator and an arbitrary collection of defects in the functional modules, is less then some value T. In order to ensure that faults in the comparator will not mask future faults in the functional units, during normal operation, the comparator must "test itself" for any single fault in less than time T.

5.1. Single Stuck-At Faults

As a first step to constructing a comparator which is self-testing with respect to *any* single fault, we will discuss the implementation of a comparator which is self-testing with respect to any single *stuck-at* fault.

In this context "two-rail" codes prove useful. They consist of all words (bit vectors) such that a specified half of the word is the complement of the other half. If the output of one of the modules in a self-checking node is complemented, a two-rail code checker can serve as a "comparator" that checks the validity of the output (Fig. 3). Such a code checker, which is self-testing with respect to any single stuck-at fault, can be implemented as a two level NOR-NOR PLA (Fig. 4).[6,31] The output from the checker is a two-bit two-rail code that is 01 or 10 (*code output*) if the input is a two-rail code word (*code input*), and 00 or 11 (*noncode output*) otherwise (*noncode input*). It can be shown that if any single stuck-at fault exists in the checker, there is a two-rail code input word that results in a 00 or 11 output, thereby "detecting" the fault.[31]

The requirement that the checker must be self-testing with respect to any single stuck-at fault poses severe constraints on its implementation. It can be

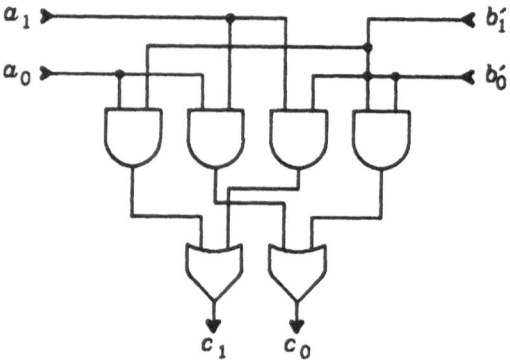

Figure 3: *Self-Testing Two-Rail Code Checker*

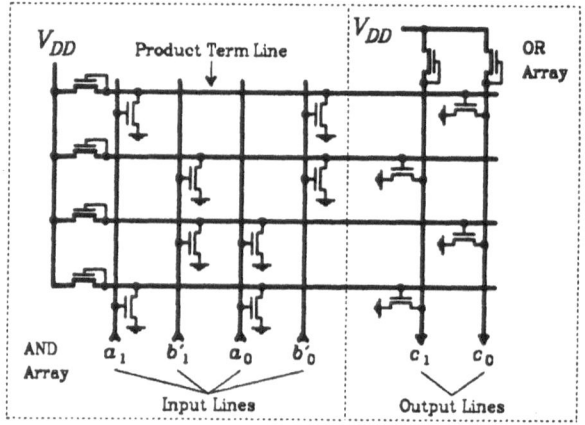

Figure 4: *NMOS Implementation of Code Checker*

shown that *any* two level AND-OR (or NOR-NOR) implementation for an input of $2n$ bits (n bits from each module) must use 2^n product terms, one for each code input.[28] If the output from each module is, say, 16 bits, this implementation is impractical since it requires $2^{16} = 65536$ product terms. Furthermore, all possible (2^n) code words must appear at the checker's inputs for it to perform a complete self-test.

Several small self-testing two-rail code checkers can be used as "cells" for constructing a self-testing checker for a wide input word (Fig. 5).[12, 22] While the self-testing property is preserved, the number of input patterns required for a complete self-test is dependent only on the size of the largest "cell".[12]

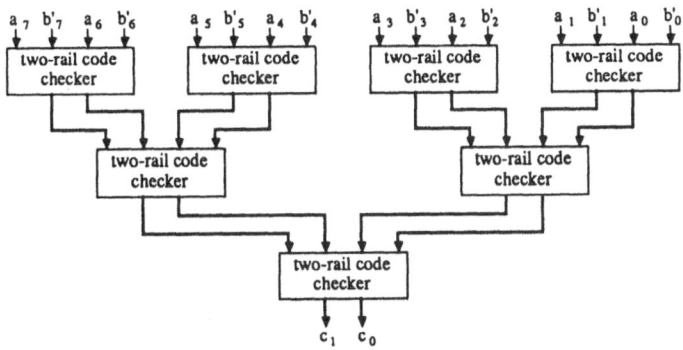

Figure 5: *A Self-Testing Two-Rail Code Checker Tree*

5.2. Other Single Faults

The faults that commonly occur in a MOS PLA are stuck-at faults, shorts between adjacent lines, breaks in lines, and contact faults that include missing or extra devices at crosspoints.[13,31] In addition, weak 0/1 faults can occur on the input or product term lines. Fortunately, it turns out that the straightforward NOR-NOR PLA implementation of the checker discussed above is self-testing with respect to any one of the aforementioned single faults. The rest of this section contains an informal "proof" of this claim; a more formal proof can be found elsewhere.[28] Faults in the input lines, product term lines, output lines, AND array crosspoints, and OR array crosspoints, are considered separately.

Any single stuck-at fault or short in the input lines will cause one or more 0's to change to 1's *or* one or more 1's to change to 0's (but not both) for some code input. It can be shown that such an error (called a *unidirectional error*[13]) on the input lines results in noncode output.[31] The effect of a break in an input line depends on its location. A break in the input line outside the AND array is equivalent to the line stuck-at-0 or stuck-at-1. A break in the middle of the AND array affects only some product terms. For an affected product term, if the break is equivalent to a stuck-at-1, the one code input that is supposed to select this product term won't, and a noncode output will result. If the break is equivalent to a stuck-at-0, there exists a code input that results in a noncode output since it selects two product term lines each of which is connected to a different output line.[28]

An extra device in the AND array is equivalent to the corresponding product term stuck-at-0. The code input that is supposed to select that product term results in a noncode output. If there is a missing device in the AND array, there exists a code input that produces a noncode output since it selects two product term lines, each of which is connected to a different output line.[28]

An extra device in the OR array means that one of the product terms is connected to both outputs. A missing device in the OR array is equivalent to the corresponding product term stuck-at-0. In either case, the code input that selects the relevant product term will result in a noncode output.

If the output lines are shorted, their values are equal and that is a noncode output. If one of the lines has a stuck-at fault, there exists a code input that causes the other line to have the same value, so the output is noncode. For some code input, a break in one of the output lines is equivalent to a stuck-at-1 or stuck-at-0 fault on that line.

A stuck-at-0 fault on a product term line will result in a noncode output if the input is the code word that is supposed to select that product term line. A stuck-at-1 fault on a product term line will result in a noncode output to any input that selects a product term line that is connected to the other output line. A break in a product term line is equivalent to a stuck-at fault on that line since each product term line is connected to only one output line. A short between two product term lines will result in a noncode output if the input selects either one of these lines.[28]

Product term lines are not susceptible to weak 0/1 faults since each product term line is connected to only one output line (fanout of one) so that a weak 0/1 fault is equivalent to a *single* stuck-at fault. Input lines have a fanout greater than one and are thus susceptible to weak 0/1 faults. A weak 1 fault on an input line is equivalent to one or more missing devices in the AND array. Each product term that is connected to a "missing device" will be selected by an input code word that also selects a product term line that is connected to the *other* output line.[28] Thus, a noncode output will result. A weak 0 fault on an input line is equivalent to one or more product term lines which are stuck-at-0. Any code input that is supposed to select one of these product terms will result in a noncode output.

In CMOS chips, PLAs are usually implemented in dynamic "pseudo NMOS".[29] All product term and output lines are precharged during every clock cycle before being selectively discharged according to the input. Therefore, no state is preserved from one cycle to the next, and the circuit is combinational despite any opens in the precharge or discharge paths.[28] Hence the PLA used in CMOS chips is only susceptible to the same faults as the traditional static PLA used in NMOS chips.

This analysis shows that for all single faults in our fault model, there exists a code input that results in a noncode output from the proposed two-rail code checker PLA. Thus, the checker is self-testing with respect to any likely single fault. Based on this result, it can be shown that the checker constructed as a tree of smaller self-testing checkers (Fig. 3) is also self-testing with respect to any likely single fault.[28]

6. IMPLEMENTATION ISSUES

The key to the fault tolerance technique presented in the previous chapters is the use of self-checking nodes implemented with duplication and comparison. As discussed in Section 3, one of the potential weaknesses of duplication and comparison is that if the two functional modules fail simultaneously in exactly the same way, the failure is not detected, and incorrect results are accepted as correct by the rest of the system. Thus we have to look at the causes of such *common mode failures* and at techniques for reducing their probability of occurrence. While it is not possible to entirely eliminated common mode failures, there are some practical implementation techniques for reducing the probability of these failures in the context of commonly used NMOS and CMOS circuits.

Common mode failures (henceforth, *CMFs*) may be caused by environmental factors such as power supply fluctuations, pulses of electromagnetic fields, or bursts of cosmic radiation, affecting both modules at the same time, triggering similar design weaknesses, and causing simultaneous identical failures of both modules. If the two modules to be matched are physical duplicates, then design weaknesses are a particular worrisome source of CMFs. Any pattern-sensitive marginal performance is likely to trigger the same erroneous output in both modules. Simultaneous module failures may also be caused by faults that occur at different times in parts of the modules that suffer from identical design weaknesses and are exercised only rarely.

Advancing VLSI technology will soon make it possible to implement an entire self-checking module, such as a computation or communication node in a multicomputer, on a single chip. This would provide nice logical building blocks[25] for the construction of powerful and reliable computer systems. Furthermore, such chips offer some advantages in production testing. Simplification of testing is achieved by eliminating the need to store the correct responses to long test sequences and compare them with the actual responses of the chip during testing. Testing can proceed at the normal system clock rate, and only the outputs of the comparator need to be monitored. However, the danger of CMFs masking design flaws may prohibit this approach for the case where the two modules are copies of the same physical design.

Unfortunately, if the two functional modules (and the comparator) are fabricated on the same chip, the probability of CMFs during normal operation is greater than if they are on separate chips. This increased probability of CMFs is due to the tighter electrical and physical coupling between the two modules and to similar weaknesses in the two modules that may be caused by fabrication flaws specific to the wafer containing the chip. Thus, having physical copies on the same chip enhances the possibility of CMFs to the point where it might defeat the overall purpose of fault tolerance. One must therefore consider to create different modules with the same desired behavior but with independent failure modes.

As noted in Section 3, one of the benefits of using duplication and comparison for self-checking subsystems is that relatively little extra design effort is required to implement the self-checking property. Creating two different implementations for every function clearly violates this goal. The question thus arises, how little extra effort is required to create two modules with the same function but with different enough implementation to reduce the chance of CMFs to an insignificant level.

How the two modules should differ to achieve independent failure modes depends on the implementation technology. In the following, approaches are outlined that are suitable for NMOS and CMOS VLSI.

6.1. Dual Implementations

For every combinational Boolean function $f(x) = f(x_1, x_2, \cdots, x_n)$ there is a corresponding *dual* function g such that $g(x) = \bar{f}(\bar{x})$ for every x. In the circuits C_f and C_g that implement the functions f and g, respectively, voltage levels represent the logic values. If the circuits are implemented using *positive-logic*, the "high" voltage level represents a logic 1 and the "low" level represents a logic 0. Because of the above relationship between the functions f and g, C_g is a *negative-logic* implementation of the function f, and C_f is a *negative-logic* implementation of the function g. The circuits C_f and C_g are said to be *dual implementations* of the function f, and C_f and C_g are said to be *dual circuits*.

Dual implementations of arbitrarily complex sequential logic circuits are also possible. If the inputs to the negative-logic implementation are complements of the inputs to the positive-logic implementation, the corresponding outputs from the two implementations are complements of each other.

Sedmak and Liebergot[21] have suggested that the probability of CMFs in a self-checking functional block can be reduced by using *dual* modules rather than pairs of identical modules. To make use of dual modules, the inputs to the self-checking block are passed unmodified to the positive-logic module (henceforth called the *P-module*), and are complemented for the negative-logic module (*N-module*). If the two modules are operating correctly, their outputs are complements of each other and can be "compared" using a two-rail code checker[6] (see Fig. 6).

There are some immediate advantages to the use of dual modules. The difference in the two modules forces the use of different masks, and thus it is not possible that a pattern defect gives rise to identical behavioral problems in both modules. Since one module is a *negative-logic* version of the other, electromagnetic pulses or noise on the power line will almost certainly produce different effects in the two modules. Finally, crosstalk problems within a module itself will typically appear at different times in the two modules because the sensitivity to electrical pickup at a particular circuit node is sensitive to the polarity of the voltage transition, and with dual circuits, the voltage transitions on corresponding lines in the two modules are in opposite directions.

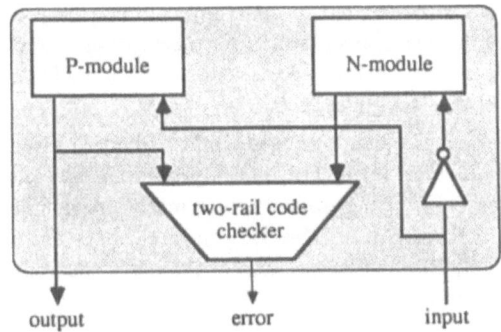

Figure 6: *Self-Checking Block Based on Dual Modules*

In SSI technology, the realization of a dual circuit is relatively straight-forward; the negative-logic module can readily be derived from the positive-logic module by a simple one-to-one replacement of gates and flipflops by their negative-logic equivalents. In VLSI technology, the implementation of dual circuits is more problematic since it is not possible to convert an existing positive-logic chip to negative-logic by a simple replacement of standard building blocks. Even the replacement of NOR gates with NAND gates is difficult. First, the different gates have different cell topologies and sizes, and the layout of the entire chip may have to be modified in order to accommodate the replacement gates. Second, the fan-in capability of the two gates may be different for example, in NMOS, it is possible to implement a NOR gate with a large number of inputs while NAND gates are limited to about four inputs. Finally, practical circuits are often not simply a collection of standard logic gates; they may contain transmission gates, precharged busses, register files, PLAs, dynamic logic subcircuits, etc. For some technologies, converting such circuits to negative-logic may require significantly more area and/or result in lower performance.

Thus, a practical conversion does not necessarily involve converting the entire module at the lowest level (i.e., individual FETs) to negative-logic. It may be preferable to design the N-module so that some of the subcircuits in the P-module have direct negative-logic equivalents in the N-module while other subcircuits are used unmodified in the N-module. The only critical requirement is that the N-module "behave" as the negative-logic equivalent of the P-module at the interface to the N-module.

6.2. Partial Conversions

Standard NMOS circuits are fundamentally asymmetrical. The available devices are enhancement mode transistors (EFETs) and depletion mode transistors (DFETs). There is no device that can perform the dual function of the EFET, i.e., be turned on by a low gate voltage and off by a high gate voltage. These constraints prevent a simple conversion of many common NMOS subcircuits into negative-logic. A more practical approach is to selectively convert only some of the circuits and keep others unchanged. If this is done judiciously, the sensitivity of the system can still be strongly reduced.[26]

Tamir proposes an approach[28] in which the N-module essentially stores and transfers all data in negated form, but where processing and control is done by positive-logic subcircuits. This approach avoids many problems with the conversion of control circuits: busses, multiplexers, and latches are not modified, and the transmission gates and pull-down transistors in them are controlled with signals of the same polarity in both modules.

Even though there are a lot of similarities between the two implementations of the modules, the probability of CMFs is greatly reduced. Shorts between data lines carrying complementary values usually pull both lines to the low voltage. Thus, both lines in the P-module change to logic 0 while similarly shorted lines in the N-module change to logic 1. Busses that fail to precharge in both modules will be interpreted as all zeroes in the P-module and all ones in the N-module. If timing is not properly designed and there is insufficient time to drive the bus from one of its sources, different lines on the bus will be affected (the ones that must be discharged), and the failure will be detected. The extra design effort with this approach is quite moderate.

CMOS technology offers switches of both polarities. Specifically, it can be shown that a positive-logic, ratioless CMOS circuit can be converted to a negative-logic circuit by simply replacing all NFETs with PFETs, replacing all PFETs with NFETs, connecting all V_{DD} lines to ground, and connecting all ground lines to V_{DD}. It thus appears that it should be simple to convert a P-module to negative logic.

Unfortunately, due to the different mobilities of the majority carriers in NFETs and PFETs, these devices are not completely symmetrical. The W/L ratio of a PFET has to be approximately twice the W/L ratio of an NFET in order to achieve similar drive capability. Therefore, a typical CMOS processor may employ many more NFETs than PFETs. In order to maintain similar performance and module area, the P-module cannot be converted to an N-module by simply complementing all FETs, and the difficulties in achieving an efficient conversion are often similar to the difficulties encountered for NMOS circuits. Thus, similar solutions and considerations apply, on the other hand, the availability of PFETs can simplify the conversion.[26]

6.3. Two Independent Implementations

Modules that are independently developed from the same specifications by two separate teams, are likely to fail in different ways. This approach is normally impractical because of the increased design costs. However, this situation may change for two reasons.

The generic modules needed to build VLSI multicomputers may become so popular, that different companies will develop the same product. Two modules fabricated by different companies can then be used to build self-checking nodes. Platteter[14] utilized this idea in constructing a fault-tolerant processor from three functionally identical microprocessors manufactured by different companies.

The other avenue to obtaining different implementations for a functional module[2] will come from the emergence of "silicon compilers". Before too long, design systems will get powerful enough to produce competitive macro modules or even whole chips in a fully automatic manner. The same set of specifications can then be run through two different compilers, or through the same compiler but with additional constraints that force two different implementations. At this stage most of the design effort will go into producing a full and unambiguous set of specifications. Once these specifications exist, obtaining different versions of the same module is only a matter of a few extra hours on a fast computer.

7. SYSTEM LEVEL PROTOCOLS

The *internal state* of a system is the ordered set of the *external states* (set of output values) of all of its components.[1] When a component fails, its external state is erroneous. Thus, *component failure* implies an erroneous internal system state, and an erroneous internal state can lead to *system failure*, i.e., incorrect output. Special measures, beyond simply detecting the error, must be taken in order to prevent system failure. In particular, in order to *recover* from the error, a *valid* internal system state must be restored, and the system must then be reconfigured so that it will not continue to use the faulty component. These actions require coordination between several (perhaps all) components. Hence, they involve system-level protocols.

7.1. Error Recovery

Most techniques for performing error recovery can be classified into two groups:[18] *Forward error recovery* techniques attempt to modify an erroneous system state so that it becomes a valid state. *Backward error recovery* techniques involve resetting (rolling back) the system to a previous valid state rather than trying to modify the current state.

Forward error recovery techniques are based on anticipating the types of errors that may occur and devising specific techniques for handling those errors. These techniques often involve special actions by the application program running on the system. On the other hand, backward error recovery techniques can cope with *unanticipated* errors. These techniques involve periodically recording the state of the system. When an erroneous state is detected, it is abandoned, and the system is reset to the previously recorded error-free state, called a *recovery point* or a *checkpoint*. The process of creating a recovery point is called *checkpointing*. No matter what type of error occurs, as long as it can be detected, some valid system state can be reinstated. Hence, a backward error recovery scheme can be totally independent of the application.

Many error recovery schemes are designed for a system where all communication is over a common bus or Ethernet.[4,15] This allows the implementation of a "recording node" that keeps a record of all inter-node messages transmitted in the system[15] and facilitates the implementation of an efficient *atomic* operation that transmits a message to a "primary" process and to its "backup" that resides on another node.[4]

On a *multicomputer*, communication is point-to-point between nodes. In order to keep track of messages that are transmitted throughout the system, they must be explicitly forwarded to the "recording node"[15] or to the "backup node".[4] This requires extra delays in processing: Before any action can be taken which counts on a message having been transmitted reliably, an acknowledgement from the destination and the backup node must be received.

Barigazzi and Strigini propose an error recovery procedure that involves periodic saving of the state of each process by storing it both on the node where it is executing and on another backup node.[3] The critical feature of this procedure is that all interacting processes are checkpointed together, so that their checkpointed states can be guaranteed to be consistent with each other. Therefore, the *domino effect* that may require backing up to successively older states[18] cannot occur. As a result, it is sufficient to store only one "generation" of checkpoints.

With the recovery scheme described in [3] a large percentage of the memory is used for backups rather than for active processes. The resulting increased paging activity leads to increases in the average memory access time and the load on the communication links. This load is increased further by the required acknowledgements of all messages and the transmission of redundant bits for error detection. The communication protocols, which are used to assure that the message "send" and "receive" operations are atomic, require additional memory and processing resources for the kernel. Thus, performance is significantly reduced relative to an identical system where no error recovery is implemented.

7.2. A Low-Overhead Error Recovery Scheme for Multicomputers

As we described previously,[27,28] the technique of simultaneously checkpointing the state of all processes belonging to the same "task" can be taken a step further: simultaneous checkpointing of the complete state of all the user and system processes on the system. A new global checkpoint is periodically stored on disks. When an error is detected, diagnostic information is distributed throughout the system. Normal operation is resumed after all the operational nodes are set to a consistent system state using the last checkpoint.

Creating and saving a global checkpoint is expensive; however, if the time between checkpoints is sufficiently large compared with the time it takes to establish a new checkpoint, the net system overhead for error recovery is still small. With the proposed scheme, in a large multicomputer the expected time to establish a new checkpoint is less than one minute. Thus, keeping the overhead low requires that a new checkpoint be established only once or twice an hour. It is clear that the loss of as much as an hour of processing when an error is detected is tolerable only for non-interactive applications.

The details of the proposed scheme are described elsewhere[27,28] and will not be repeated here. The technique consists of two major components: a scheme for saving a consistent global checkpoint of the entire system and a scheme for rolling back the system to a previously saved checkpoint once an error is detected. The technique relies heavily on the self-checking property of the nodes that ensures that faulty nodes are detected before erroneous information from them is allowed to spread throughout the system. As mentioned above, the technique is useful only for a system running non-interactive applications.

The scheme for saving a consistent global checkpoint is an adaptation of the standard two-phase commit protocol used for preserving consistency in distributed data base systems.[11] Initially, a designated node, say node 1, is assigned to serve as the *coordinator* for establishing global checkpoints. If the coordinator fails, all the other nodes are notified, and the next node, according to a total ordering between the nodes, takes over the task of being checkpointing coordinator. Every node includes a "timer" that can interrupt the node periodically. Checkpointing is initiated by the checkpointing coordinator when it is interrupted by its timer.[3]

The checkpointing coordinator initiates checkpointing by stopping all local normal processes and notifying all of its neighbors that checkpointing is in progress. Each node in the system, in turn, repeats this process. Once all the neighbors are informed, each node begins to send its state to a node with a disk where the state is saved. When the entire state of the node is saved, the checkpointing coordinator is informed. After all the nodes have saved their states, the checkpointing coordinator directs the entire system to resume normal operation.

Since the nodes are self-checking, the failure of a node is detected by its neighbors. The neighbors "spread the word" throughout the system, indicating which node has failed and that recovery is in progress. When a node with disk storage finds out that recovery is in progress, it begins sending the previously saved state to all nodes that used it for checkpointing. Each node that receives a complete previous state informs the coordinator. After all the nodes have obtained their previous states, the checkpointing coordinator directs the entire system to resume normal operation.

It is possible to obtain a rough estimate of the overhead of the proposed system by making several specific assumptions about such system based on the intended application environment and on current and near-future technology. We base our assumptions on the use of the INMOS Transputer chip as the node.[33] We assume a system with 1,000 nodes, each with 256,000 bytes of memory, connected in a network with a *diameter* of 15. With communication link bandwidth of 1.5×10^6 bytes/second, checkpointing or recovery are expected to take less than 20 seconds.[27] If the system has a mean time between failures of 10 hours and a checkpoint is saved twice an hour, the total overhead for checkpointing and recovery will be approximately 3.7 percent.

7.3. Reconfiguration

Following error recovery, it is easiest to resume normal system operation if no changes are made in the operation of the nodes or the interconnection between them. This is possible if the error was caused by a node that failed due to a transient fault. Following recovery the node can resume its previous role in the system if it is capable of resetting itself to a "sane state" at the same time it informs the neighbors of the failure (see Section 2). However, if the node fails due to a permanent fault, the system must be capable of continuing normal operation without this node.

One of the requirements for the interconnection topology of the system is that the failure of any one node does not partition the system into two independent networks that cannot communicate. More generally, the maximum number of nodes that can fail without the possibility of partitioning the system, is a critical parameter in determining system reliability.

Nodes that fail due to permanent faults are effectively removed from the system. The algorithms used to route messages between nodes in the system must adapt to such changes in the topology of the system. If the system uses table-driven routing, the routing tables throughout the system must be updated following error recovery.[5,24] If the system uses "algorithmic" routing that does not require routing tables, the interconnection topology must allow such routing even after some of the nodes are removed.[17]

If a node fails due to a permanent fault, processes that were executing on it must be moved to a different node and continue to execute there. Thus, following recovery, messages from processes that were communicating with processes on the failed node must somehow be redirected to the new node. This ability to transparently *migrate* processes between nodes is a critical requirement for the operating system of a fault-tolerant multicomputer. Powell and Miller propose one possible scheme for such process migration in multicomputers.[16]

8. CONCLUSIONS

As the number of switching elements in a VLSI system starts to exceed a few hundred millions, the reliability and thus the fault-tolerance of the system must become a major concern. The design of a VLSI system is in itself a very hard task, and adding fault-tolerance may just make it unmanageable, unless we use the principles of regularity and repetition to simplify the task.

At the example of a multicomputer system, consisting of hundreds or thousands of VLSI computation nodes interconnected by dedicated links, we have demonstrated how the concerns of fault-tolerance can be concentrated on a single critical component, and how, by a suitable modular approach, the whole system can become fault tolerant, without undue penalty to either system design time or system performance. The discussed scheme combines hardware that performs error detection with system-level protocols for error recovery and for fault treatment.

We have shown that a high probability of error detection can be achieved with self-checking nodes implemented using duplication and comparison. These nodes use two modules that perform identical functions but are not susceptible to simultaneous identical failures. The output of these modules is compared in a self-testing code checker that has been thoroughly analyzed for all likely defects in present-day VLSI circuits.

The proposed low-overhead, application-transparent error recovery scheme for the system involves periodic checkpointing of the entire system state using protocols that ensure that the saved states of all the nodes are consistent, and rolling back to the last checkpoint when an error is detected. No restrictions are placed on the actions of the application tasks, and the communication protocols used during normal computation are simpler than those required by most other schemes. A multicomputer system that follows the general principles outlined in this paper can provide a general-purpose, high-performance computing environment in which the fault tolerance features are completely transparent to the user.

Acknowledgements

This research was supported by the State of California MICRO program and the Defense Advance Research Projects Agency (DoD).

References

1. T. Anderson and P. A. Lee, "Fault Tolerance Terminology Proposals," *12th Fault-Tolerant Computing Symposium*, Santa Monica, CA, pp. 29-33 (June 1982).

2. A. Avizienis, "Design Diversity - The Challenge of the Eighties," *12th Fault-Tolerant Computing Symposium*, Santa Monica, CA, pp. 44-45 (June 1982).

3. G. Barigazzi and L. Strigini, "Application-Transparent Setting of Recovery Points," *13th Fault-Tolerant Computing Symposium*, Milano, Italy, pp. 48-55 (June 1983).

4. A. Borg, J. Baumbach, and S. Glazer, "A Message System Supporting Fault Tolerance," *Proc. 9th Symp. on Operating Systems Principles*, Bretton Woods, NH, pp. 90-99 (October 1983).

5. M. Bozyigit and Y. Paker, "A Topology Reconfiguration Mechanism for Distributed Computer Systems," *The Computer Journal* 25(1), pp. 87-92 (February 1982).

6. W. C. Carter and P. R. Schneider, "Design of Dynamically Checked Computers," *IFIPS Proceedings*, Edinburgh, Scotland, pp. 878-883 (August 1968).

7. B. Courtois, "Failure Mechanisms, Fault Hypotheses and Analytical Testing of LSI-NMOS (HMOS) Circuits," pp. 341-350 in *VLSI 81*, ed. J. P. Gray, Academic Press (1981).

8.   R. P. Davidson, M. L. Harrison, and R. L. Wadsack, "BELLMAC-32: A Testable 32 Bit Microprocessor," *1981 International Test Conference Proceedings*, Philadelphia, PA, pp. 15-20 (October 1981).

9.   E. A. Doyle, "How Parts Fail," *IEEE Spectrum* **18**(10), pp. 36-43 (October 1981).

10.   J. Galiay, Y. Crouzet, and M. Vergniault, "Physical Versus Logical Fault Models MOS LSI Circuits: Impact on Their Testability," *IEEE Transactions on Computers* **C-29**(6), pp. 527-531 (June 1980).

11.   J. N. Gray, "Notes on Data Base Operating Systems," pp. 393-481 in *Operating Systems: An Advanced Course*, ed. G. Goos and J. Hartmanis, Springer-Verlag, Berlin (1978). Lecture Notes in Computer Science 60.

12.   J. Khakbaz and E. J. McCluskey, "Concurrent Error Detection and Testing for Large PLA's," *IEEE Journal of Solid-State Circuits* **SC-17**(2), pp. 386-394 (April 1982).

13.   G. P. Mak, J. A. Abraham, and E. S. Davidson, "The Design of PLAs with Concurrent Error Detection," *12th Fault-Tolerant Computing Symposium*, Santa Monica, CA, pp. 303-310 (June 1982).

14.   D. G. Platteter, "Transparent Protection of Untestable LSI Microprocessors," *10th Fault-Tolerant Computing Symposium*, Kyoto, Japan, pp. 345-347 (October 1980).

15.   M. L. Powell and D. L. Presotto, "Publishing: A Reliable Broadcast Communication Mechanism," *Proc. 9th Symp. on Operating Systems Principles*, Bretton Woods, NH, pp. 100-109 (October 1983).

16.   M. L. Powell and B. P. Miller, "Process Migration in DEMOS/MP," *Proc. 9th Symp. on Operating Systems Principles*, Bretton Woods, NH, pp. 110-119 (October 1983).

17.   D. K. Pradhan, "Fault-Tolerant Architectures for Multiprocessors and VLSI Systems," *13th Fault-Tolerant Computing Symposium*, Milano, Italy, pp. 436-441 (June 1983).

18.   B. Randell, P. A. Lee, and P. C. Treleaven, "Reliability Issues in Computing System Design," *Computing Surveys* **10**(2), pp. 123-165 (June 1978).

19.   R. A. Rasmussen, "Automated Testing of LSI," *Computer* **15**(3), pp. 69-78 (March 1982).

20.   D. A. Rennels, "Architectures for Fault-Tolerant Spacecraft Computers," *Proceedings IEEE* **66**(10), pp. 1255-1268 (October 1978).

21.   R. M. Sedmak and H. L. Liebergot, "Fault Tolerance of a General Purpose Computer Implemented by Very Large Scale Integration," *IEEE Transactions on Computers* **C-29**(6), pp. 492-500 (June 1980).

22.   D. P. Siewiorek and R. S. Swarz, *The Theory and Practice of Reliable System Design*, Digital Press (1982).

23.   C. H. Séquin and R. M. Fujimoto, "X-Tree and Y-Components," pp. 299-326 in *VLSI Architecture*, ed. B. Randell and P.C. Treleaven, Prentice Hall, Englewood Cliffs, NJ (1983).

24.   W. D. Tajibnapis, "A Correctness Proof of a Topology Information Maintenance Protocol for a Distributed Computer Network," *Communications of the ACM*

20(7), pp. 477-485 (July 1977).

25.   Y. Tamir and C. H. Séquin, "Self-Checking VLSI Building Blocks for Fault-Tolerant Multicomputers," *International Conference on Computer Design*, Port Chester, NY, pp. 561-564 (November 1983).

26.   Y. Tamir and C. H. Séquin, "Reducing Common Mode Failures in Duplicate Modules," *International Conference on Computer Design*, Port Chester, NY, pp. 302-307 (October 1984).

27.   Y. Tamir and C. H. Séquin, "Error Recovery in Multicomputers Using Global Checkpoints," *13th International Conference on Parallel Processing*, Bellaire, MI, pp. 32-41 (August 1984).

28.   Y. Tamir, "Fault Tolerance for VLSI Multicomputers," Ph.D. Dissertation, CS Division Report No. UCB/CSD 86/256, Department of Electrical Engineering and Computer Sciences, University of California, Berkeley, CA (August 1985).

29.   R. L. Wadsack, "Fault Modeling and Logic Simulation of CMOS and MOS Integrated Circuits," *The Bell System Technical Journal* 57(5), pp. 1449-1474 (May-June 1978).

30.   J. F. Wakerly, "Microcomputer Reliability Improvement Using Triple-Modular Redundancy," *Proceedings of the IEEE* 64(6), pp. 889-895 (June 1976).

31.   S. L. Wang and A. Avizienis, "The Design of Totally Self Checking Circuits Using Programmable Logic Arrays," *9th Fault-Tolerant Computing Symposium*, Madison, WI, pp. 173-180 (June 1979).

32.   J. H. Wensley, L. Lamport, J. Golberg, M. W. Green, K. N. Levitt, P. M. Melliar-Smith, R. E. Shostak, and C. B. Weinstock, "SIFT: The Design and Analysis of a Fault-Tolerant Computer for Aircraft Control," *Proceedings IEEE* 66(10), pp. 1240-1255 (October 1978).

33.   C. Whitby-Strevens, "The Transputer," *12th Annual Symposium on Computer Architecture*, Boston, MA, pp. 292-300 (June 1985).

16

The VLSI Design Automation Assistant:
An IBM System/370 Design

T. J. Kowalski
AT&T Bell Laboratories
Murray Hill, New Jersey 07974

ABSTRACT

The Design Automation Assistant, DAA, is a knowledge-based expert-system that generates a technology-independent list of operators, registers, data paths and control signals from an algorithmic description of a VLSI system. This chapter shows the generality of design knowledge in the DAA by comparing and contrasting an IBM System/370 designed by an expert human designer, Claud Davis, against the design produced by the DAA. For each difference, possible changes in the CMU/DA system and the DAA are discussed. Davis himself felt the design produced by the DAA exhibited the quality he would expect from one of his better designers.

INTRODUCTION

Recent advances in integrated circuit fabrication technology have allowed larger and more complex designs to form complete systems[1] on single VLSI chips. These chips use one-micron to five-micron features to achieve complexities equivalent to 100,000 to 250,000 transistors. This level of design complexity has created a combinatorial explosion of details — a major limitation in realizing cost-effective, low-volume, special-purpose VLSI systems. To overcome this limitation, design tools and methodologies capable of automating more of the digital synthesis process must be built.

We have been developing just such synthesis tools[2,3] at AT&T Bell Laboratories and Carnegie-Mellon University. These tools help the designer develop the algorithmic description of the system and interactively add the details required to produce a finished design. This structured approach can decrease the time it takes to design a chip, automatically provide multi-level documentation for the finished design, and create reliable and testable designs.

This chapter focuses on the synthesis, or allocation, of the implementation-design space as it advances from an algorithmic description of a VLSI system to a list of technology-independent registers, operators, data paths and control signals. Our approach is aimed at aiding the designer by producing data paths and control sequences that implement the algorithmic system description within supplied constraints. Thus, the designer can consider many alternatives before deciding on a final design.

This task has inspired a variety of approaches, ranging from the most simplistic backtracking methods through the most complicated constraint propagation methods.[4-8] Owing to the complexity of design synthesis, simplistic backtracking schemes consume large amounts of CPU time, and the constraint propagation method is too cumbersome for large designs. Because of the combinatorial explosion of details and implicit dynamic constraints involved in choosing an implementation, this problem does not lend itself to these algorithmic solutions. An alternate approach to design synthesis uses a large amount of design knowledge to eliminate backtracking; whenever possible, the focus is on specific design details and constraints. Artificial intelligence researchers have called systems developed under this heuristic approach knowledge-based expert systems, KBESs.[9]

1. CONCEPTION

KBESs are generally developed in several stages. First, "book knowledge" of the problem is codified as a set of situation-action rules; interviews with experts then fill in knowledge gaps and refine current knowledge. Then, many example problems are given to the KBES, and experts closely examine and validate the results. Often, errors are found through the examples, and new rules are added to the system to correct the error situations.

This iterative process is necessary because experts are often unaware of exactly how they go about designing a chip and are inexperienced at articulating the procedure. Furthermore, the knowledge base is not an exact codification of the expert's knowledge, but a compilation of what is understood by the knowledge engineer.

After gathering current book knowledge about synthesis of the architectural design space,[4-6] we interviewed four designers of varied experience: one was a novice, two were moderately experienced, and one was an expert. The interviews, which lasted about an hour each, started with a determination of the designer's background, including years of experience, logic families used, and designs created. Most of the time was spent discussing the design process, with some time given to a discussion of the DAA system. Our interview method was designed to allow the interviewees as much freedom as possible in generating ideas; we emphasized such questions as "What do you do next?" and "Could you elaborate?"

The designers discussed the global picture, partitioning, selection, and allocation tasks. They began with a high-level overview of the hardware, which listed inputs and outputs to the outside world, the functions the hardware should provide, general constraints, and design feasibility with consideration of the target technology. They generally partitioned the global picture into smaller blocks and emphasized minimizing connections among blocks, selecting blocks that operate as parallel or serial units, and grouping according to similarity of function. Partitions were chosen for allocation in a decreasing order of difficulty or degree of constraint. The designers reasoned that if the most difficult part could be designed, the rest of the design was feasible.

Once a partition was selected for allocation, it was carried out either in parallel or in series. A parallel design made thinking of the control logic much simpler, while a serial design minimized the design area. The constraints of the parallel design were examined for size violations to determine the parts to be serialized by adding data paths, registers, and control logic to the initial parallel design. The constraints of the serial design were examined for speed violations to determine the parts to be reimplemented in parallel. If the designers recognized a part of the design as similar to a part of a previous design, they used what they knew had worked in the past. Within each partition, designers allocated clock phases, operators, registers, data paths, and control logic. The order was interesting because once registers and data paths were allocated, they were not changed. The control was changed because it was the hardest thing to think about and because it depended on a constant structure for the data path elements.

The designers described the iteration process as a step-by-step refinement to meet violated constraints. They looked for a technology change to meet a constraint before making a design change. This could be as simple as finding a new chip in the TTL data book or as complicated as a design-rule shrink. Next, they would sacrifice functionality to meet a constraint. One designer summed it up best by saying, "An engineer's training teaches him when constraints can be swept under the rug."

The relative importance of constraints is application dependent. The designers mentioned the constraints of speed, area, power, schedule, cost, drive capabilities, and bit width. Other design changes consisted of global improvements not recognized until the design neared completion. This suggests that the general choice of partitions and the initial design style selections approached optimum and that designers do not seem to use much backtracking in their designs.

2. BIRTH

Even though many details were missing, enough book knowledge had been gathered to put together a prototype version of the DAA system using the OPS5[10] KBES writing system. While the DAA system was far from perfect at this point, it stimulated further elicitation sessions with expert designers. We now turn our attention to the flow of control in the prototype system and how the KBES approach formulate the problem.

2.1 Flow of Control

The DAA starts with a data flow representation extracted from the algorithmic description. This representation resembles the internal description used by most optimizing compilers,[11] but computer programs manipulate it more easily, and it is felt to be less sensitive when the same algorithm appears in a variety of writing styles.

The DAA produces a technology-independent hardware network description. This description is composed of modules, ports, links, and a symbolic microcode. The modules can be registers, operators, memories, and buses or multiplexers with input, output, and bidirectional ports. The ports are connected by links and are controlled by the symbolic microcode.

The DAA uses a set of temporally ordered subtasks to perform the synthesis task. It begins by allocating the base-variable storage elements — constants, architectural registers, and memories with their input, output, and address registers — to hardware modules and ports. Then a data flow BEGIN/END block is picked, and the synthesis operation assigns minimum delay information to develop a parallel design. Next, it maps all data flow operator outputs not bound to base-variable storage elements to register modules. Last, it maps each data flow operator, with its inputs and outputs to modules, ports, and links. In doing so, the DAA avoids multiple assignments of hardware links; it supplies multiplexers where necessary. The last two mapping steps place the algorithmic description in a uniform notation for the expert analysis phase that follows.

The expert analysis subtask first removes registers from those data-flow outputs where the sources of the data-flow operator are stable. Operators are combined, according to cost and partitioning information across the allocated design, to create ALUs. The DAA also examines the possibility of sharing non-architectural registers. Where possible, it performs increment, decrement, and shift operations in existing registers. Where appropriate, it places registers, memories, and ALUs on buses. Throughout this subtask, constraint violations require trade-offs between the number of modules and the partitioning of control steps. The process is repeated for the next data flow BEGIN/END block.

2.2 The OPS5 Writing System

The DAA is implemented as a production system via the OPS5 KBES writing system. The KBES tool is based on the premise that humans solve problems by recognizing familiar patterns and by applying their knowledge in the current situation. The tool formulated a problem by using three major components: a working memory, a rule memory, and a rule interpreter.

2.2.1 Working memory. The working memory is a collection of elements that describe the current situation. The elements resemble the records in conventional programming languages:

> literalize module
>   id: adder.0
>   type: operator
>   atype: two's complement
>   bit-left: 17
>   bit-right: 0
>   attribute: +

This working-memory element describes an operator module *adder.0*, which can perform two's complement addition on 18 bits of binary data.

2.2.2 Rule memory. The rule memory is a collection of conditional statements that operate on elements stored in the working memory. The statements resemble the conditional statements of conventional programming languages:

IF:

  the most current active context is to create a link
  and the link should go from a source port to a destination port
  and the module of the source port is not a multiplexer
  and there is a link from another module to the same destination port
  and this other module is not a multiplexer

THEN:

  create a multiplexer module
  and connect the multiplexer to the destination port
  and connect the source port and destination port link to the multiplexer
  and move the other link from the destination port to the multiplexer

This rule recognizes situations in which a multiplexer needs to be created to connect one port to another.

Each subtask in the DAA is associated with a set of rules for carrying out the subtask. An example of a rule for the fourth subtask appeared above. Most of the rules, like the example above, define situations in which a partial design should be extended in some particular way. These rules enable the DAA to synthesize an acceptable design by determining, at each step, whether a certain design extension respects constraints.

2.2.3 The rule interpreter. The rule interpreter pattern matches the working-memory elements against the rule memory, to decide what rules apply to the given situation. The rule-selection process is data driven; the rule interpreter looks through the rule memory for a rule whose antecedents match elements in the working memory. This is also called forward chaining or antecedent reasoning. The consequences of the rule are applied, and the process is repeated until no more rules apply or until a rule explicitly stops the process. If more than one rule applies, the rule dealing with the most current working memory is selected first. If multiple rules are still applicable, the most specific rule is selected. This selection mimics following a train of thought, as far as possible, and uses special-case knowledge before general-purpose knowledge. The separation of expert knowledge from the reasoning mechanism makes the incremental addition of new rules and the refinement of old ones easy because the rules have minimal interaction with one another.

3. FIRST STEPS

The prototype DAA system had about 70 rules and could design a MOS Technology Incorporated MCS6502 microcomputer in about three hours of VAX 11/750 CPU time. We asked many expert designers at INTEL and AT&T Bell Laboratories to critique the design by explaining what was wrong, why it was wrong, and how to fix it. After each critique, rules were modified, new rules were added, and the MCS6502 was re-designed. Based on the critiques, the development DAA system now has over 300 rules, and has designed a much better MCS6502 microcomputer in about five hours of VAX 11/750 CPU time. In retrospect, clearly much of what we learned was common-sense design knowledge, the same things human designers learn through apprenticeship. The DAA has undergone many improvements and produced many designs of the MCS6502 microcomputer. Below, we illustrate a few of these changes.

Each knowledge acquisition interview began by giving the designer a drawing of the design with a sheet of clear plastic over it. Before the designer started the critique, pieces of cardboard were placed over the design. As the designer proceeded, a piece of cardboard had to be lifted, the plastic written on, and covered by new plastic to correct the design. This provided a complete record of where the designer was focusing attention and what was corrected. The designers found this elicitation procedure compatible with their normal spatial mode of operation.

The first prototype DAA system was used to produce the design summarized in Column 1 of Table 1. Each row shows the bits of the specified operator or register type found in the design.[†] The expert criticism is summarized in four points:
- Operators of different types and sizes should be combined into ALUs.
- One-bit operators within the same block should not be combined, because multiplexers are more expensive than most one-bit modules.
- Registers should increment, decrement, and shift their values internally, where possible.
- Temporary registers to the controller should be eliminated, and one latched register should be placed in front of the controller.

The rules were changed to produce the design summarized in Table 1, Column 2, and illustrated in Figure 1. To produce this design, partitioning information was added, based on connectivity of data paths and similarity of operators among blocks. This simplified the decision about which modules to combine when hardware operators are shared among abstract operations detailed in the algorithmic description. Rules were also added to combine modules of different sizes

---

† A figure is not provided because it is totally inscrutable.

Table 1. MCS6502 — THREE DESIGNS

Designs	1	2	3
And	20	20	20
Cmp	177	1	1
Minus	64	0	0
Or	9	9	9
Not	21	21	21
Plus	540	0	0
Shifts	35	1	1
Xor	9	9	9
Alu	0	35	35
Dreg	450	281	210
Treg	1227	0	62
Mux In	2122	2657	473
Mux Out	293	377	84
Bus In	0	0	769
Bus Out	0	0	210

and types. As Column 2 shows, the ALU number increased, decreasing the plus, minus, shift, and compare numbers. Rules were also added to decrease the amount of temporary register storage.

Figure 1 shows the eight-bit data paths of the MCS6502. The one-bit and 16-bit data paths were omitted for clarity. Each of the symbols represents a module. The circles are single function ALU modules to AND, SHIFT, NOT, XOR, and OR data, the small trapezoids are multiplexers that gate one of their inputs to the output, the small rectangles are registers, the large rectangle is the memory, the large trapezoid is a multi-function ALU, and each of the lines represents a link between the modules. Where the links join with the modules, a port is defined. An obvious problem, pointed out by our experts, was overuse of multiplexers. They suggested ways of distributing the multiplexer hardware to form buses.

The rules were changed in response to these critiques, resulting in the design of Table 1, Column 3, and Figure 2. To produce this design, rules were added to

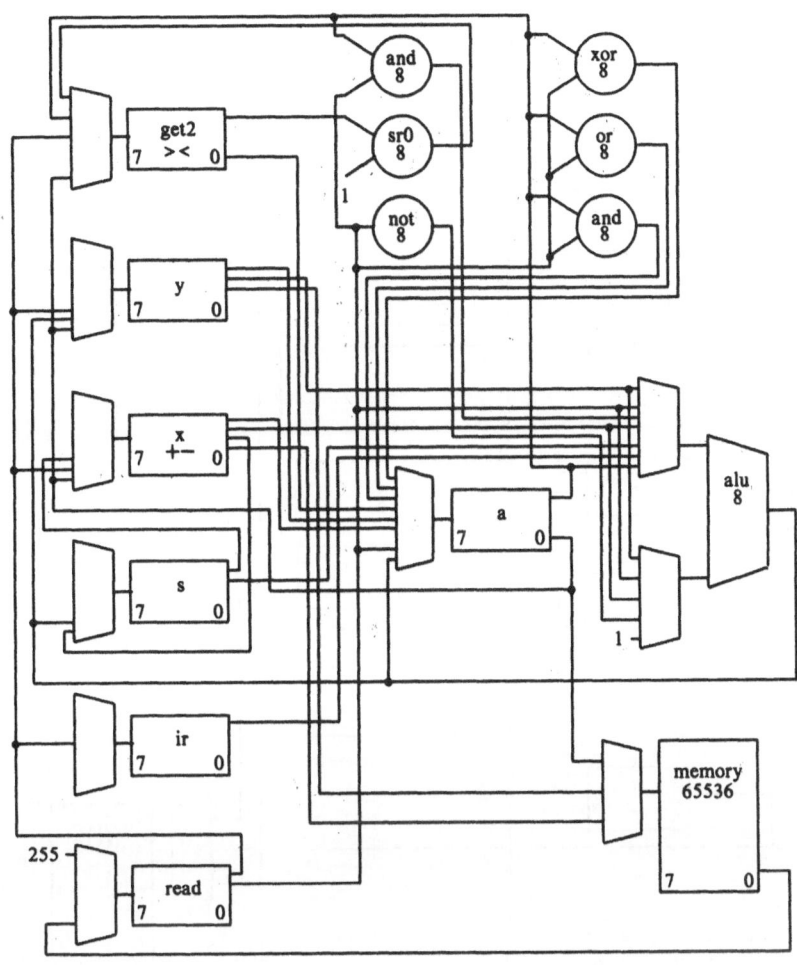

Figure 1. MCS6502 — 8 BIT MUX DATA PATHS

recognize when a multiplexer should be converted into a bus and how to share that bus with other distributed multiplexers. In addition, new rules decreased the amount of declared register storage. Specifically, registers were not needed to multiplex information into the data flow BEGIN/END blocks. As Column 3 shows, the multiplexer numbers decreased, increasing the bus numbers. The declared register

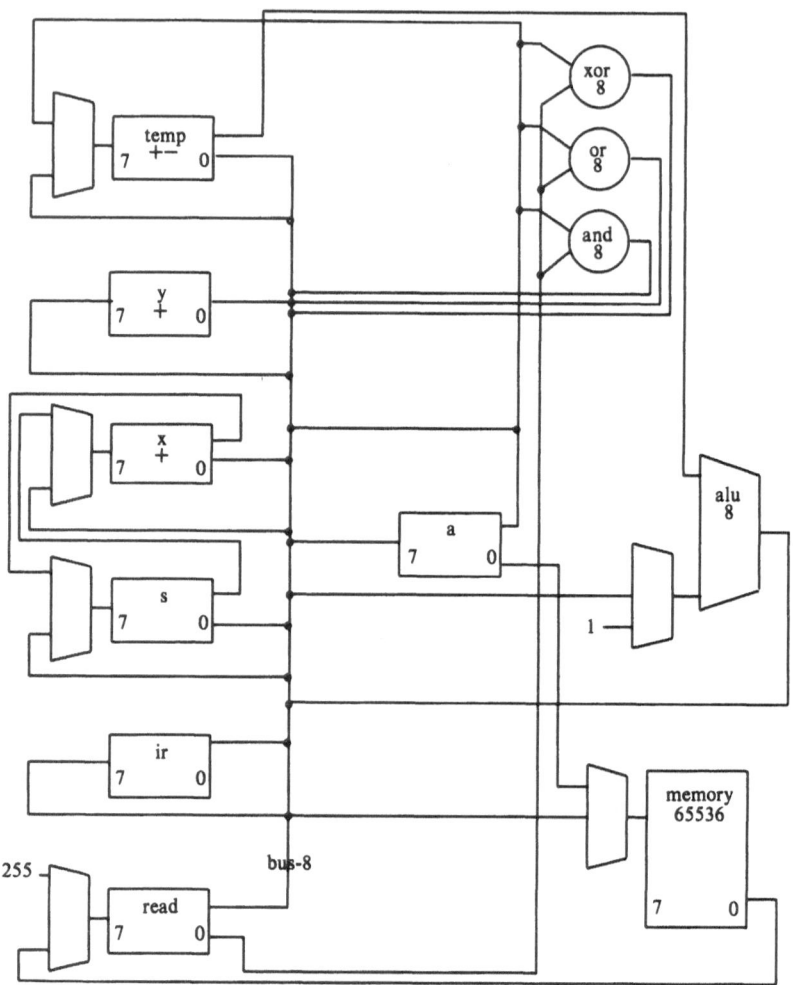

Figure 2. MCS6502 — 8 BIT BUS DATA PATHS

number also dropped.

Though this design was acceptable to our experts, it was not perfect. Further changes led to improvements such as multiple buses of different widths. However, these changes did not affect the MCS6502 because it did not require multiple buses. This brings up an interesting point about expert systems: they are never totally finished. Like human designers, the DAA becomes a better designer as its rule memory expands. Until all possible world knowledge about designing microprocessors has been codified in the DAA's rules, there will always be room for improvement in its designs.

4. THE IBM SYSTEM/370 EXPERIMENT

After the DAA successfully designed a MCS6502 microprocessor, it had to be determined whether the system had also acquired knowledge about processor design in general. In this regard, an experiment was designed to see whether the DAA could design a processor substantially different and more complex than the MCS6502.

An ISPS description was chosen for the complete IBM System/370 from the descriptions maintained at Carnegie-Mellon University. This description included memory-management operations, channel controller I/O instructions, and all the 370 instructions, except the extended-precision floating point, the characters under mask, the edit and mark, and the packed-decimal instructions. The unmodified System/370 description, missing only a small percentage of the total 370, is more than 10 times larger than that of the MCS6502,[†] and it had not been used to build the DAA. Important benefits of this choice are that a single-chip design of the 370 had been made at IBM, information is publically available, and Claud Davis, the design team manager and a key designer, was willing to critique the design. Thus, the experiment was a fair and convenient way to test the generality of the DAA's design knowledge.

4.1 The DAA 370 Design

The DAA designed the D370, its version of the System/370, in 47 hours of CPU time on a VAX 11/780 with six megabytes of memory and two memory controllers. The D370 was designed without rule modifications or design iterations of any type.

The D370 is an IBM System/370 data-flow design using a $50x$ clock, where x is some scaled unit of time like μseconds, with multiplexer and bus style data paths. The DAA's constraints were set to produce a high-performance machine — that is, it could use as much hardware as required to allocate the data paths and retain maximum parallel operator usage. To meet this performance constraint, the D370

[†] The next bigger description, the Digital Equipment Corporation VAX 11/780, wouldn't compile through the VT compiler in a six megabyte address space

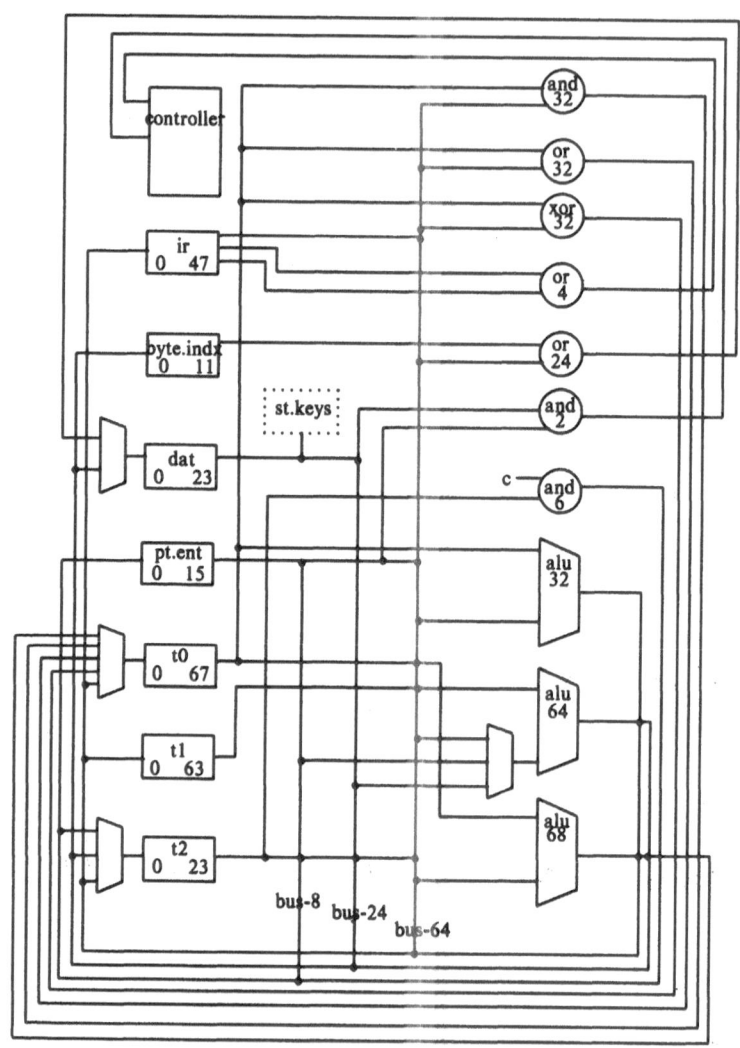

Figure 3. THE D370 DESIGN — PART 1

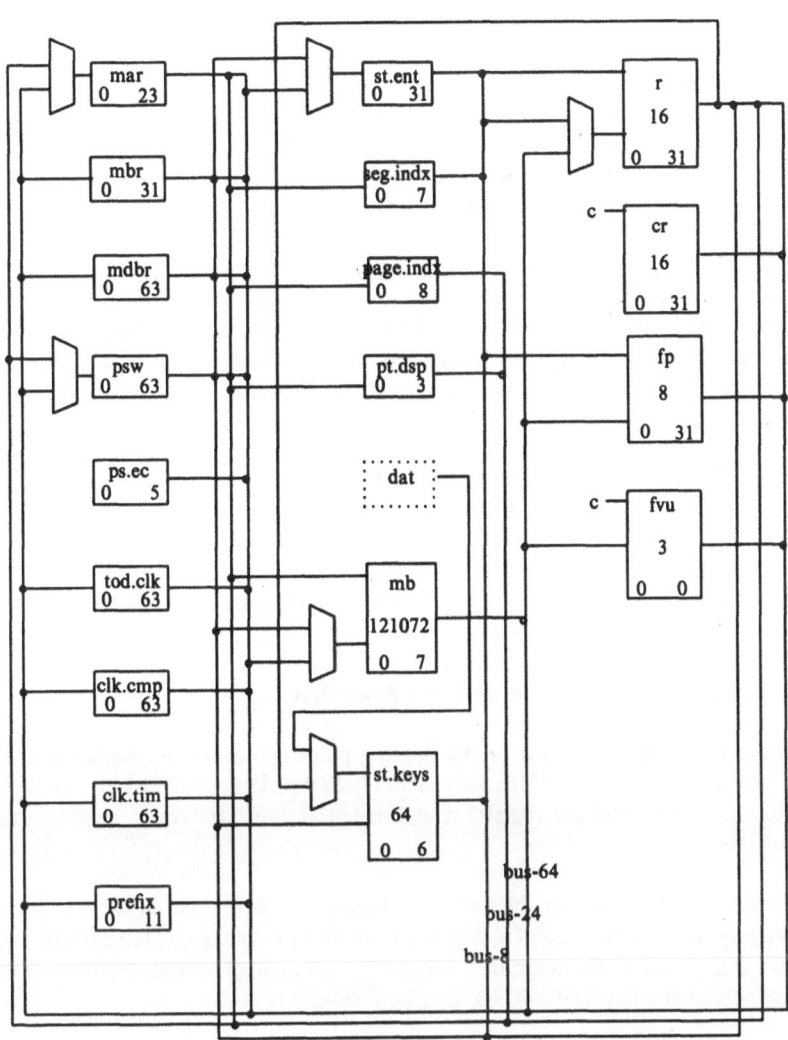

Figure 4. THE D370 DESIGN — PART 2

has eight-bit, 24-bit and 64-bit buses, 32-bit, 64-bit and 68-bit ALUs, a few discrete components, six memory arrays, and a great many architectural registers.

4.1.1 The three ALUs. The three ALUs and bus sizes arose from three different groups of data operations and major transfer widths in the IBM System/370. The basic busing style placed a temporary register before an input to each ALU and assumed the ALU latches its result so it can be read on the next clock phase transition. Thus, a two-phase clock set up the inputs to an ALU in one clock cycle and stored the result on the next clock cycle.

For clarity, the data path of the D370 is drawn in two separate figures, Figures 3 and 4. The connection between the two figures is through the eight-bit, 24-bit and 64-bit data buses. Figure 3 contains the arithmetic portion, the temporary registers, and the controller; Figure 4 shows the architectural registers, including the register arrays, such as the 16 general purpose registers R.

The 32-bit ALU is used for most of the arithmetic operations in the System/370 architecture. It can ADD, SUBTRACT, and COMPARE two binary numbers from the $T0$ temporary register and the 64-bit bus. It gates a result out on this bus.

The 64-bit ALU is used for most of the address calculation operations and a few low-frequency operations, such as MULTIPLY and MODULUS in the System/370 architecture. It can ADD, SUBTRACT, COMPARE, MULTIPLY, MODULUS, and SHIFT RIGHT two binary numbers from the $T1$ temporary register and a bus. It can gate a result out on either the 24-bit or 64-bit bus.

The 68-bit ALU is used for most of the floating-point operations in the System/370 architecture. This ALU can ADD, SUBTRACT, COMPARE, and SHIFT LEFT two binary numbers from the $T0$ register and the 64-bit bus. Its result is gated onto the 64-bit bus.

4.1.2 The discrete components. Not all data manipulation is done in the ALUs. To aid debugging, one of out expert designers from INTEL keeps single-function logic outside the ALU. Thus, the logic instructions are implemented with separate distributed logic elements: the 32-bit AND, OR, and XOR.

Smaller logic elements are provided for a variety of functions. The four-bit OR takes input from two fields of the instruction register IR and feeds the result to the microcontroller. This aids in instruction decoding. The virtual storage system uses three discrete components. The 24-bit OR takes input from the byte index $BYTE.INDX$ register and the 64-bit bus and places the resultin the dynamic address translation DAT register. The two-bit AND takes input from the 24-bit bus and the page table entries *PT.ENT* and feeds the result to the microcontroller. The six-bit AND takes input from a group of constants and *T2* and places the result on the eight-bit bus.

4.1.3 The memory arrays. The D370 architecturally defines the primary memory *MB*, the storage keys *ST.KEYS*, the general purpose registers *R*, the control regis-

Table 2. MEMORY ARRAYS IN THE D370 DESIGN

Abbreviation	Bits	Words	Address	Inputs	Outputs
MB	8	121072	24bb	8bb, 64bb	64bb
ST.KEYS	7	64	R, DAT	8bb	8bb
R	32	16	8bb	8bb, 64bb	ST.KEYS, 8bb, 24bb, 64bb
CR	32	16	Constants		64bb
FP	32	8	8bb	64bb	64bb
FVU	1	3	Constants	64bb	64bb

ters *CR*, the floating point registers *FP*, and the floating point error registers *FVU*. Table 2 lists the bit width of each memory array, the number of words in the array, and what buses *bb* or registers connect to the address, input, and output ports. Thus, there are 64 storage keys, each seven bits wide, with their address port connected to the general purpose registers and the dynamic address translation register. The input and output ports are connected to the eight-bit bus.

4.1.4 The architectural and temporary registers. The D370 architecturally defines many registers. Table 3 lists the bit width of each register and tells what buses or registers connect to the input and output ports. Thus, the instruction register is 48 bits wide with its input connected to the 64-bit bus and its output connected to the 64-bit bus and the four-bit OR described above.

4.1.5 The control specification. A symbolic microcode word controls the D370. A microcode word is required for each cycle of the machine. The generation of either PLA or ROM based micro-engine is possible in later phases of the design synthesis task. A sample sequence from the dynamic address translation BEGIN/END has the 24-bit bus gating the *MAR* to the *MB* address port, the 64-bit ALU adding the *MAR* from the 24-bit bus and the temporary register *T1* and gating the result on the 64-bit bus to the *MB* input port, with the six-bit AND anding the temporary

Table 3. REGISTERS IN THE D370 DESIGN

Abbreviation	Name	Bits	Inputs	Outputs
MAR	memory address	24	24bb, 64bb	24bb, 64bb
MBR	memory buffer	32	64bb	8bb, 64bb
MDBR	memory double buffer	64	64bb	8bb, 64bb
PSW	processor status word	64	24bb, 64bb	8bb, 24bb, 64bb
PS.EC	extended code	6		64bb
TOD.CLK	clock	64	64bb	64bb
CLK.CMP	clock comparator	64	64bb	64bb
CPU.TIM	CPU timer	64	64bb	64bb
PREFIX	prefix	12	64bb	64bb
IR	instruction register	48	64bb	64bb, 4-bit OR
ST.ENT	segment table entry	32	8bb, 64bb	8bb
PT.ENT	page table entry	16	8bb	8bb, 64bb, 2-bit AND
SEG.INDX	segment index	8	24bb	8bb
PAGE.INDX	page index	9	24bb	24bb
PT.DSP	page table displacement	4	24bb	24bb
BYTE.INDX	byte index	12	24bb	24-bit OR
DAT	dynamic address translation	24	24bb, 24-bit OR	24bb, ST.KEYS
T0	temporary 0	64	64bb, 32-bit AND, OR, XOR	64bb, 32-bit ALU, 68-bit ALU, 32-bit AND, OR, XOR
T1	temporary 1	64	64bb	64bb, 64-bit ALU
T2	temporary 2	24	8bb, 24bb, 64bb	8bb, 24bb, 64bb, 6-bit AND

register *T2* and a constant, gating the result on the eight-bit bus to a field in the *PT.ENT*. This illustrates the high degree of parallelism possible in the D370.

4.2 The μ370 Design

The μ370 is an IBM System/370 micro-processor data flow on a single bipolar gate-array masterslice chip. It uses a 100-nanosecond cycle clock and is capable of executing 200,000 instructions per second. The physical chip is 7x7 mm and dissipates 2.3 watts. The plan was to use no more 5000 wired circuits, 3 watts of power, and 200 pins. To meet these size and power constraints, the problem was divided into on-chip and off-chip sections. This section discusses the functional blocks of the μ370.

4.2.1 The on-chip functional block. The on-chip functional block has an eight-bit ALU, a 24-bit incrementer/decrementer, I/D, a 24-bit shifter, two nine-bit parity generators, 17 eight-bit working registers, two eight-bit buffer registers, a 16-bit status register, a 24-bit register, and hardware to calculate the next microcode address. These components are wired together with two fan-in eight-bit buses, a fan-out eight-bit bus, a bidirectional 16-bit bus, a fan-in 24-bit bus and a fan-out 24-bit bus. These are shown in Figure 5.

The eight-bit ALU can ADD, SUBTRACT, OR, AND, and XOR either binary or packed-decimal numbers. The arithmetic operations of the ALU can be controlled directly from a microcode field or indirectly through a status bit located in register *S*. This indirect control feature allows sharing of microprogramming routines for the ADD and SUBTRACT operations. Two eight-bit buses feed two eight-bit buffer registers, *A* and *B*, which feed the ALU. These two registers can selectively gate groups of four bits that correspond to hex digits within the byte or pass the complete byte to the ALU. This gating is used for decimal operations. The *A* register can also pass its eight bits, rotated by four bits to reverse its two hex digits. This is used by the pack and unpack instruction. The output of the ALU is placed on an eight-bit bus gated to all the working registers.

The 24-bit I/D is a special purpose adder that can add a 24-bit binary number with the constants 0, 1, 2, 3, −1, −2, or −3. The constant input is directly controlled from a microcode field. This I/D is dedicated to address calculations. An address is gated from a set of three working registers onto the 24-bit bus feeding the memory address register *MAR*. Any value present in this register is gated to the memory address bus MAB, the input of the I/D, and a shifter (discussed below). The output of the I/D is placed on the 24-bit fan-out bus and gated back to the same three registers that fed the *MAR*.

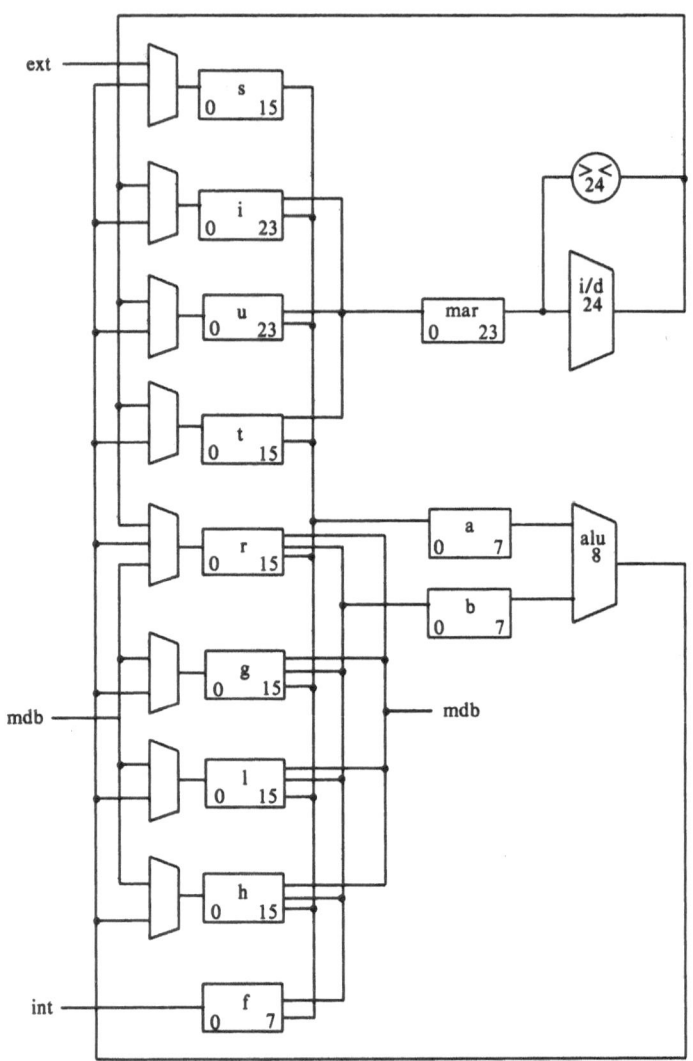

Figure 5. THE μ370 DESIGN

The 24-bit shifter can do a few complex shift operations to produce a 16-bit result. This shifter is dedicated to handling the 12-bit address field used for page addressing in the virtual storage system. Input is gated from the *MAR*, and output is gated onto the 24-bit fan-out bus.

Two nine-bit parity generators check the parity of each data byte arriving at the chip and place it on the 16-bit memory data bus, MDB. They can also affix a parity bit to each data byte leaving the chip from the MDB.

The μ370 has 17 eight-bit registers, two eight-bit ALU buffer registers, a 16-bit status register, and a 24-bit *MAR*. The 17 eight-bit registers are grouped in functional pairs and triples. The *R*, *G*, *L*, and *H* register pairs are primarily the memory data registers, MDRs. Because the *G* register pair has special microcode branching capabilities, it is the OP code register. The *I* and *U* register triples are the program counter and operand register, respectively. The *T* register pair is the local-store address. The *I*, *U*, *T*, and *R* register groups can pass two bytes of data to one another with or without a displacement. This feature is part of the virtual storage management of the IBM System/370 architecture. The *S* register pair is the CPU status register and serves as input to the microcontroller. *S1* can be set and reset by external inputs. The interrupt register *F* is also settable by external conditions.

A 54-bit microcode word controls the μ370. A microcode word is required for each cycle of the machine and is fetched during the last 75 nanoseconds of each cycle from the read only store, ROS. To select the next ROS word, a 16-bit address is generated in the first 25 nanoseconds of each cycle. Six bits of the ROS address are taken directly from the ROS word. The low order two bits are extracted from conditions within the chip. ROS fields dictate the internal conditions to be examined. The remaining bits are taken from the previous ROS address. However, if an external-trap bit is raised, the ROS address is forced to a specific value for a trap handler. Possible traps are parity errors, IPL request, page overflow, storage wrap, memory protect violation, stop request, and I/O control.

4.2.2 The off-chip functional block. The off-chip functional block has the architectural registers, two external memories, an I/O port, and the ROS. The μ370 chip uses 512 bytes of architectural registers. They are kept in a local store that can be accessed in 60 nanoseconds. The local store, which is limited to 64K bytes, is addressed by the *T0* and *T1* registers. The μ370 uses up to 16 megabytes of memory, which is addressed by the MAB, while data is gated on the MDB. *Read*, *write*, *memory-1*, *memory-2*, and *ready* signals allow up to two memories of any speed to be interfaced to the chip. If neither memory line is asserted, the MDB is connected to an I/O bus that uses the MAB to choose the I/O device being serviced.

4.3 The D370 and μ370 Design Comparison

The previous two sections have discussed the individual attributes of the two designs. This section brings those designs together by comparing and contrasting their differences. Claud Davis compared the two designs at IBM Poughkeepsie. During his career of over 25 years at IBM, Davis has worked on designs and managed teams of designers for the 701, 702, 7074 MA, 360/50, FAA, 360/67, and the μ370. His vast experience with the higher-performance processors and the μ370 made his critique valuable in two ways. First, we could determine what is needed for a single-chip IBM System/370 architecture; second, we could determine what is needed for a higher performance processor. Davis summarized his comparison thusly:[12]

> "The 370 data-flow we reviewed exhibited the quality I would expect from one of our better designers. The level of detail was what we call second level design. This encompasses all 'architected' registers, status latches and sufficient working registers to implement the functions defined by the instruction set. This level of design is independent of implementing technology.
>
> The review included a test for 'architected' registers, data path widths, latches for exceptional conditions, signs, and latches for temporary information in multi-cycle instructions.
>
> The assumptions for clocking and controls were examined and found to be consistent."

The complete transcript[13] is also available.

In the following discussion, differences are grouped by objectives and functional blocks including: ALUs, buses, memories, and registers, which are summarized in Table 4. For each difference possible changes in the CMU/DA and the DAA systems are discussed.

4.3.1 The objectives. The objectives and testing of the two designs differed. The μ370's objective was to place a fully functional System/370 on a single chip, while observing such technology constraints as number of wired circuits, power, and I/O pins. The D370's objective was to design a high-performance System/370 sensitive to technology constraints, but independent of power and number of I/O pins. The μ370 was produced as a working chip, whereas the D370 is only a paper design.

Table 4. IBM SYSTEM/370 — DESIGN DIFFERENCES

Design	D370	μ370
Objectives	High performance, technology sensitive, independent of power and I/O pins; paper design	Strict observance of technology criteria such as number of wired circuits, power, and I/O pins; working chips
ALUs	32-bit, 64-bit, and 68-bit; Binary numbers; hardware for virtual memory, floating point and multiply	Eight-bit and 24-bit; Binary and packed-decimal numbers; microcode for virtual memory, and multiply
Buses	Eight-bit, 24-bit, and 64-bit; bidirectional	Three 8-bit, a 16-bit, two 24-bit; fan-in, fan-out and bidirectional
Memories	12-byte buffer; single ported	Eight-byte buffer; single ported
Registers	Discrete	Memory array

4.3.2 The ALUs. The number, size, and type of functions supported by the ALUs in the two designs differed. The D370's design had one extra ALU that can be directly traced to the implementation of floating-point operations, while the μ370 planned a separate floating-point chip. In addition, the D370 implemented the dynamic address translation hardware, while the μ370 supplied a shifter that had a few complex shift patterns to aid in calculating the virtual address by using microcode.

The DAA does a high-level floor layout to help decide how to partition the algorithmic description, but this does not currently allow exclusion of functionality; the whole algorithmic description is implemented. Changes that modify the algorithmic description by including or excluding functionality are best made by changing the initial description or by having a postprocessor feed size constraints to the

DAA.

The ALUs also differed in size. The μ370 serialized the 32-bit and 64-bit operations of the System/370 architecture into four or eight cycles through an eight-bit ALU. The DAA's constraints were set to design a high-performance processor, and thus the data paths were not serialized. Less than a dozen rules could be added to the DAA to allow it to serialize on ALU width. However, this change would be better made by adding a transformation to the CMU/DA system that removes a single-abstract data flow operation, and replaces it with several smaller ones.

Finally, the ALUs differed in the functions they provided. The μ370 has an ALU that can ADD and SUBTRACT packed-decimal numbers; the D370 performed these operations by adding hardware and microcode. The D370 has an ALU that can MULTIPLY, while the μ370 MULTIPLIED by SHIFTING and ADDING. Davis felt the choice of an ALU that could MULTIPLY was reasonable and consistent with the constraints used by the DAA.

4.3.3 The buses. The designs differed in the number, size and type of buses used. The μ370's eight-bit buses and 16-bit bus serve the same purpose as D370's eight-bit and 64-bit buses. The size difference is accounted for by the μ370 serializing the 32-bit and 64-bit operations of the System/370 architecture down to eight-bit operations, as discussed above. Also, the D370 uses bidirectional buses, where the μ370 used separate fan-in and fan-out buses. Davis felt the design choices made by the DAA were reasonable for the higher-performance D370 design, citing the IBM System/370 model 158 as an example of this style of busing.

4.3.4 The memories. The memory functional blocks differed only slightly. The D370 uses an eight-byte buffer with a four-byte memory data register, while the μ370 uses eight bytes of memory data registers. Davis felt this and even more elaborate cache schemes suited the higher-performance processors. He suggested the D370 use dual-ported memories for its general-purpose registers, to allow the use of two registers during the same cycle. Dual-ported memories would require a few rule changes, but would allow up to two memory array accesses during the same clock cycle. The rules for finding the address, input and output ports of memories would have to be enhanced to check for idle ports. All told, about 20 rules would have to be modified or extended to effect this change.

4.3.5 The registers. Both descriptions have about the same number of bytes of architectural and temporary registers. However, the μ370 groups all the architectural registers off chip in a fast local store, which can be thought of as memory. This would be a major change in the structure of the DAA. However, it could be accomplished simply, as a post-processor pass by the CMU/DA system as other technology-specific hardware is bound to the modules.

5. SUMMARY

We are exploring the allocation problems of operators, registers, data paths, and control paths from an algorithmic representation of a VLSI system. We are using a KBES to test the knowledge gathered from interviews with experts; from this knowledge, we are attempting to create interesting and usable designs.

This chapter has shown the generality of design knowledge in the DAA by comparing and contrasting an IBM System/370 designed by an expert human designer, Claud Davis, against the design produced by the DAA. The differences were either explained and shown to be unimportant, or changes to the system were discussed. Davis felt the design produced by DAA exhibited the quality he would expect from one of his better designers.

This chapter has shown the first large implementation design, automatically generated from an algorithmic description and constraints, that has been favorably critiqued by an expert designer. Furthermore, the design required 47 hours of CPU time, which with some work can be reduced by a factor of 12 to about 4 hours of CPU time. This clearly shows the dramatic improvement in CPU time for large designs obtained using methods that replace backtracking by match techniques. Finally, because this design was generated using synthesis techniques, it is possible to verify its operation by construction[14] and link it to the rest of the CMU/DA design environment.

ACKNOWLEDGMENTS

We would like to thank K. Chong, C. Davis, D. Ditzel, M. Maul, G. Mowery, A. Ross, C. Schneider, G. Williams, and A. Wilson for donating time to critique the various designs.

REFERENCES

[1] Mead, C. and Conway, L., *Introduction to VLSI systems*, Addison-Wesley Publishing Company, Reading, Massachusetts, (1980).

[2] Director, S. W., Parker, A. C., Siewiorek, D. P., and Thomas, D. E., "A design methodology and computer aids for digital VLSI systems," *IEEE*

Transactions on Circuits and Systems, Vol. cas-28, No. 7, (July, 1981).

[3] Thomas, D. E., Hitchcock, C. Y. III, Kowalski, T. J., Rajan, J. V., and Walker, R., "Automatic Data Path Synthesis," *Computer*, Vol. 16, No. 12, (December, 1983), pp. 59-70.

[4] Marwedel, P. and Zimmermann, G., *MIMOLA Software System User Manual*, Vol. 1, Institut Fur Informatik und Praktische Mathematik, Christian-Albrechts-Universitat Kiel, (1979).

[5] Hafer, L. J., *Automated data-memory synthesis: A Format Method for the Specification, Analysis, and Design of Register-Transfer Level Digital Logic*, PhD Thesis, Department of Electrical Engineering, Carnegie-Mellon University, (June, 1981). Also in Design Research Center DRC-02-05-81.

[6] Hafer, L., *Data — Memory Allocation in the Distributed Logic Design Style*, Master's Thesis, Carnegie-Mellon University, (December 21, 1977).

[7] Hitchcock, C. Y. III, Automated Synthesis of Data Paths, CMUCAD-83-4, SRC-CMU Center for Computer-Aided Design, Carnegie-Mellon University, (January, 1983).

[8] Tseng, C. J. and Siewiorek, D. P., "Facet: A Procedure for the Automated Synthesis of Digital Systems," *Proceedings of the Twentieth Design Automation Conference*, (June 27, 1983), pp. 490-496.

[9] Feigenbaum, E. A., *Knowledge Engineering: The Applied Side of Artificial Intelligence*, Computer Science Department, Stanford University, (1980).

[10] Forgy, C. L., *OPS5 User's Manual*, Department of Computer Science, Carnegie-Mellon University, (1981).

[11] Aho, A. V. and Ullman, J. D., *Principles of Compiler Design*, Addison-Wesley, Reading, Mass., (1979).

[12] Davis, C., Personal letter to Dr. D. E. Thomas, (August 12, 1983).

[13] Kowalski, T. J., *The VLSI Design Automation Assistant: The IBM 370 Critique*, Department of Electrical and Computer Engineering, Carnegie-Mellon University, (1983).

[14] McFarland, M. C., *Mathematical Models for Verification in a Design Automation System*, Phd thesis, Carnegie-Mellon University, (1981). Also in Design Research Center DRC-02-06-81.

17

Higher Level Simulation and CHDLs
(Computer Hardware Description Languages)

Reiner W. Hartenstein, Udo Welters
Kaiserslautern University
Department of Informatics
Postfach 3049, D-675 Kaiserslautern, F.R.G.
phone: (xx49 - 631) 205 - 2606, or: (xx49 - 7251) 3575

Abstract

This paper gives an introduction to Computer Hardware Description Languages (CHDLs) and their application in early phases of the VLSI design process. It first gives a survey on objectives of its use in simulation. Then it briefly introduces a subset of a modern register transfer language (RT language). Finally it gives a survey on various CHDL-based CAD tools and its linkage to physical design, as well as its integration into CAD environments.

Contents

1. Introduction: why CHDLs ?
2. Early phases of the Design Process
    2.1 Specification and Design Problem Capture
    2.2 Experimenting with alternative Architectures
    2.3 Design for Testability
    2.4 Structured VLSI Design
3. Introducing the Use of a CHDL
    3.1 Language Primitives
    3.2 Cell Modules with Floor Plan Capability
    3.3 Description of Wiring Patterns
    3.4 Step-wise refinement using a CHDL
    3.5 CHDL Use as a Design Calculus
4. CHDL-based Design Environments
    4.1 Using an Interactive Graphic CHDL Editor
    4.2 CHDL-based CAD Tools
    4.3 Interfacing to other CAD Tools
5. Conclusions
6. Acknowledgements
7. Literature

INTRODUCTION: WHY CHDLs ?

The next higher abstraction level above gate level is called register transfer level (RT level), since its primitives are registers, register arrays, and data transfer paths, such as e. g. operators, buses, multiplexers, and others. Hardware descriptive notations at RT level (and sometimes above) are called *Computer Hardware Description Languages (CHDLs)*. Some of them [HLW86] have a mnemonics which is similar to that of the Pascal programming language. CHDLs may be used to feed higher level simulators, to feed silicon compilers (at least future ones), for automatic generation of *go / no go* test patterns [HaWo85], for hardware specification, and for documentation, for more concise teaching the principles of digital hardware, use as a design calculus, and for many other applications.

All modern CHDLs are hierarchical and thus have structural description capabilities, so that this should not be used as a classification criterium. We may distinguish two major classes of CHDLs: non-procedural languages and algorithmic languages. Non-procedural languages may be used only for description of input/output behaviour of systems and modules. Algorithmic CHDLs, have additional programming features to describe sequences of mircoinstructions and other sequences. So the same language may be used e. g. to describe and simulate the data paths of a microcomputer and the microprograms running on it. The question only is, which solution has more advantages: using an algorithmic CHDL, or, using a non-procedural CHDL system, which is interfaced to a separate microprogram compiler.

Substantial Reduction of Complexity. What are the benefits in using such CHDLs, compared to traditional abstraction levels, such as e. g. the gate level ? The most important benefit is the reduction of notational complexity. Gate level notations, such as e. g. Boolean equations, do not yield a substantial reduction of complexity, compared to circuit diagrams. The average number of transistors per gate is the quotient of complexity reduction: this is only about 3 to 5, in CMOS and some other circuit techniques only about 4 to 8. In using CHDLs this quotient may by much higher, sometimes up to several hundreds.

One reason for reduced complexity is the fact, that CHDLs use to bundle a bit vectors of bits into words, like in high level programming languages. In describing a 32 bit data path, for instance, its data values are kept in a single word to be processed at once within the simulator and other tools. A second reason is the fact, that at RT level more powerful operators are available, such as e. g. multiplication, an equivalent up to hundreds of gates, and many others. A third reason for reduced complexity is found in the flexibility of RT language use. Modern CHDLs feature capabilities to describe a particular hardware in different levels of abstraction. That's why in using the same language a design process may start with a very high level specification of very low complexity, and after several steps of refinement it may end up with a more complex and more detailed description of a solution concept.

Training. So it is sure, that the complexity problem of VLSI design can be solved only by using CHDLs. Later in this paper a number of additional benefits will be illustrated, such as support of design for testability, early test pattern development, structured VLSI design, experimenting with alternative architectures before starting logic design, and many others. Why does a majority in industry hesitate to introduce CHDLs? Some quite interesting tools are available. However, most Universities do not teach using CHDLs, which would be an effort taking about less than half of the time needed to introduce Pascal. Another problem is the lack of methodology. At gate level a very elaborate and formal design methodology has grown, mainly within the last 30 years, due to contributions of thousands of scientists throughout the world. At register transfer level, however, most contributions are more of narrative character and of analytical nature, rather, than being a design methodology. A generally and widely accepted formal notation - comparable to Boolean algebra at gate level - has not yet been established in most application areas at RT level. So design tends to be more a trial and error procedure, or, to use one of a few popular concepts, such as e. g. systolic arrays and others.

EARLY PHASES OF THE DESIGN PROCESS

CHDLs are an important opportunity to avoid expensive redesigns needed to correct errors, such as e. g. bad testability, bad topology and bad structure of the circuit, to much area comsumption, or, bad (VLSI-) architecture, missing the requirements, and others. Many of such errors could be avoided, if design concepts would be decided at a very early phase of the design process, definitely before the costly logic design procedure has been started.

Specification and Design Problem Capture

One important role of CHDL use could be design problem capture. To be sure to meet the requirements the design problem has to be pinned down the correct way. A concise notation has to be used to express the design problem. A description of a design problem using such a notation is called a specification. To be sure to capture the design problem correctly the specification has to be checked against the requirements.

Specification Verification by Simulation. If a CHDL imple- mentation including a simulator is available, the requirements could be simulated, after the specification has been accepted by the tool. So the CHDL system may serve for design problem capture. Such a CHDL system could be also used as a communication medium between customer and design center, or, if within the same company: between product planning division and design division. The customer uses the CHDL system, such as e. g. a KARL compiler and simulator, for design problem capture. Reacting to simulation results the customer successively debugs the specifications. Finally the verified and debugged specifications (for instance, a

KARL description of the design problem) are handed over to the design center.

Experimenting with alternative Architectures

Bugs in specifications are not the only possible reasons for missing the requirements. Sometimes the principles of a design concept are critical with respect to real-time performance, to design cost, or other important aspects. Often several possible solutions have to be considered and analyzed, so that experimenting with alternative architectures is needed. Of course such experiments should be carried out at the highest possible level of abstraction to avoid incomprehensibility of descriptions and high labour cost because of high complexity. For more detailed discussion see [GHW85]. For a survey on automated optimization support see[Wod86].

Design for Testability

Testing VLSI circuits currently is a major desaster area in industry, since the technology of testing and test pattern generation is far behind the possibilities of manufacturing technology and design capabilities. For mass production very often the time needed for testing is too long. Desirable would be around a second or less. For automatic test pattern development often an excessive amount of CPU time is needed. A very critical aspect is the fact, that for a given set of test patterns often the test coverage is much too low. This means, that the percentage of circuit faults, which will be detected by using a given set of test patterns, is far below 100%. This issue is critical, since it severely affects product quality. In some applications, such as where malfunction of circuits could be a danger to human life (process control, aerospace, some modern automotive, medical applications etc.), or, could make the entire mission fail (aerospace applications etc.) this quality aspect is one of the most important objectives at all.

Very early Test Pattern Developoment. In many cases the design is the reason, why a circuit's fault coverage is low. In such a case the best possible test patterns could not achieve high fault coverage. That's because of properties of the design important inner subcircuits cannot be reached by a sufficient percentage of stimuli. Nor a sufficiently high percentage of its responses could be observed from outside the circuit. Only an expensive redesign of the circuit, which takes testability aspects into account, could solve such a problem. The product development schedule could slip for months or more.

All this illustrates that design for testability is an important ingredient of the VLSI design process. Not only testability per se, but also the length of the test needed is a very important objective in design for testability. The best solution is to carry out test patern development in very early phases of the design process, at least before logic design has been started. The most desirable time would be, when the specifications are ready. Instead of being part of the logical design, and thus being highly expensive, testability would be a subject of early design planning and partitioning definition. The designer could fully concentrate on the testability architecture of a circuit and could experiment with alternative architectures. However, this would require, that the test patterns are available at such an early time, so that testability data and test length data of different version architectures are available. Otherwise the designer would not know, which alternative to decide, and, whether the design for testability efforts have to be continued or not.

Functional Testing. All this is feasible, since fortunately for production testing of integrated circuits only a *go / no go* test (sometimes called a *functional test*) is needed. That's because integrated circuits are not repaired, so that fault locating is not required. A functional test can also be developed without any structural knowledge about the circuit. (A test also exhibiting fault location diagnostics would be called a *structural test.*) That's why a functional test can already be developed from the functional description of a circuit, i. e. from its specification.

Integrating Simulation and Test Development. Although the area of functional test pattern generators currently is rather immature, it is useful to have it available along with the circuit specificaiton for designing for testability. For instance, the output of the KARATE test pattern generator (currently being implemented in Kaiserslautern [HaWo85]) uses the same language SCIL [HHa86], which is also accepted by the KARL simulator. Problems yet to be solved, are the following ones. For large circuits an exhaustive simulation is not possible, unless an accelerator is available which runs the simulator, or, a physical model extension is used. So the user will have to select subsets of the test patterns in a clever way to run the simulator. (This would be supported by the good readability of the SCIL language, and the KARL simulator's dialogue mode capability.) Currently there is also no way to an optimization of functional test patterns with respect to those errors which have an extremely low probability for technological reasons. However, it is not known, how many of such errors could be expected. Also the automatic testability analysis area is rather immature. Some of the more widely known analyzers having been published yield quite obscure results. About a more recent one [SM85, SMP86] this paper's authors have only limited information.

Structured VLSI Design

The term of *structured VLSI design* has been coined by the Mead-and-Conway scene. It stands for a method to implement algorithms directly onto the planar surface of silicon in a way, which attempts that most cells of the design are connected by abutment. This means, that by means of port matching between neighbour cells no routing area between these cells is needed. This in many application problems is a highly efficient way to save chip area, since routing areas tend to eat up very much more chip area (sometimes up to about 95% of the chip) than active cells. The best way to use this method is it, to try to plan the chip in a way, that most of it is made up by arrays of abuttable cells. The success of such a solution highly depends on the cleverness in planning the shape and the topology of *key cells,* being efficiently abuttable. Often a successful key cell design is possible only, when a clever partitioning and placement strategy has been used in chip floor planning. All this means, that layout considerations are needed at very early phases of the design process, about when the specification is formulated.

Innovative Power of VLSI Design. Structured VLSI Design as a design style has an innovative power. The success of structured VLSI design efforts depends on selection of the best possible task realization algorithm for a VLSI solution. Sometimes the smart memory approach is a good solution (this is shown in tutorial-like explaining the design of a simple sorter chip example in [BBad85]). Also this illustrates the benefit of very early chip planning. To provide means for design plan verification at this early phase, a simulator input language (a CHDLS) is needed, which can express such partitioning and topological features already at specification level. Such language features will be shown later.

INTRODUCING A CHDL AND ITS USE

To get a more illustrative presentation a particular CHDL will introduced. It is the KARL-III non-procedural language which is the most familiar one to the author [Har77, NN85, HLW86, HHa86]. KARL-III is a multi-level language which includes the **RT level, gate level,** and the **switch level.** This has good reasons: sometimes small pieces of a description cannot be expressed at RT level, so that this 'remainder' can only be presented in using gate level primitives. Another important reason is the bus, being an important architectural resource at high levels of system description: it is a switch level concept [Ha77]. This multi-level paradigm has more advantages: in using the same language as a top-down design medium a specification can be successively refined to a more hardware-near detailed concept.

KARL-III is a multi-paradigm language. It is not only multi-level, but also strongly typed (for diagnosability), it features structural description as well as functional description, it also features topological description including a cell abutment expression sublanguage (for floor plan capability). We believe, that all this is a nearly optimum mix to fulfil the requirements of design problem capture, early design for testability, and, structured VLSI design. Nevertheless, KARL is not a baroque language, like Ada, for instance.

Comprehensibility. Although being multi-level, KARL is substantially more easy to learn than Pascal. For instruction and documentation we believe, that the simultaneous use of two versions of a RT language is quite useful for better comprehensibility. This fact has been recognized quite earlier, so that in fact at gate

Structural primitives

group	primitives
modules (user-defined)	**cell** declaration topology: front, back, left, right, ports: in, out, bi **function** declaration
abutting chip floor planning	**make** expr.: @, :, mirx, miry, rotr, rotl, rotu
inter-connect	simple: .= := bus, terminal formatting: •, | (catenate) [...] (subscript) *also see wiring functions*
user-defined cells	

KARL functional primitives

level	combinational	w. memory
RT level	**arithmetic:** +, -, *, /, mod **relational:** >, <, =, =<, >=, <> **multiplexers:** if........ case.......	register, array reg. RAM ROM constant
gate level	**logical:** not, and, nand, or, nor, exor, coin	delays
switch level	**bus drivers:** oeo, oco, enables **buses:** upbus, downbus tribus	dynamic memory
	clocking: at, on, wile	

Utilities (standard functions):

operative standard functions	wiring standard functions	
comment	word format	array format
code conversion fcts: decode encode decount encount **test functions:** equ, odd, even testunary, testsingulary **miscellaneous:** inc *increment* dec *decrement* pril *priority left* prir *priority right*	**shift functions:** shr shl dshr dshl cshr cshl nshr nshl eshr eshl cirshr cirshl **shuffle functions:** fold **butterfly fcts.** (not y.released) **mirror functions:** feflect **field select functions** msb, lsb	push pop dpush dpop cpush cpop npush npop epush epop cirpush cirpop | merge | reverse | msw, lsw

Fig. 3.1. Survey on KARL-III language primitives and utilities

Language Primitives

The language, its power, and its flexibility is determined by the repertory of its primitives. We may distinguish structural primitives from functional ones. Structural primitives within KARL-III are uniform throughout all abstraction levels: they are all the same, no matter whether being used at RT level, gate level, or, at switch level, and even at circuit, and symbolic layout levels [Wel86]. KARL-III structural primitives are much more easy to use than those of KARL-II (which is no more supported).

Structural primitives may be subdivided into module definition features, and, into notations to specify interconnect. KARL III provides two different module declaration facilities, the func declaration for user-defined function modules, and the cell declaration also including topological features, which are explained in section 3.2 (for a survey see fig. 3.1). There are two kinds of interconnect descriptions: implicite descriptions within expressions and explicite descriptions in terminal, bus, assign- ment statements, as well as by means of actual parameters (connections to cell ports) in cell instantiations (for details see [HLE86, NN85]).

Functional primitives may subdivided into RT level, gate level, and switch level primitives (see fig. 3.1). RT level primitives of KARL-III are: arithmetic and relational operators, multiplexers, and all lements with permanent memory (register, RAM, ROM, etc.). This supply is extended by RT level utilities (standard functions), such as e. g. decode, encode, priority functions etc. The group of wiring operators
also considered to be RT level operators, are subject of sect.3.3. Gate level primitives of KARL-III include the usual logical operators, and two kinds of delay elements which may be used to model propargation delays at all levels. At switch level a simple but flexible technology-independant bus modelling scheme has been developed [Ha77, NN85, HaEIS86], providing an upbus, downbus, and, tribus (for three-state bus) declaration for the 3 basic bus types. Three technology-independant bus driver primitives oco ('open collector output'), oeo ('open emitter output'), and enables (having a separate control input) provide a method for modelling bus systems, and to model the circuit principles of of MOL (matrix-oriented logic) using personality matrix specifications (see section 4.0).

Cell Modules with Floor Plan Capability

The concepts underlying the KARL cell definition and instantiation features efficiently supports structured VLSI design and the integration of KARL-based CAD tools into physical design. Relative to its declaration orientation a KARL cell distinguishes four different sides (left, right, front, and, back) of port location (fig.

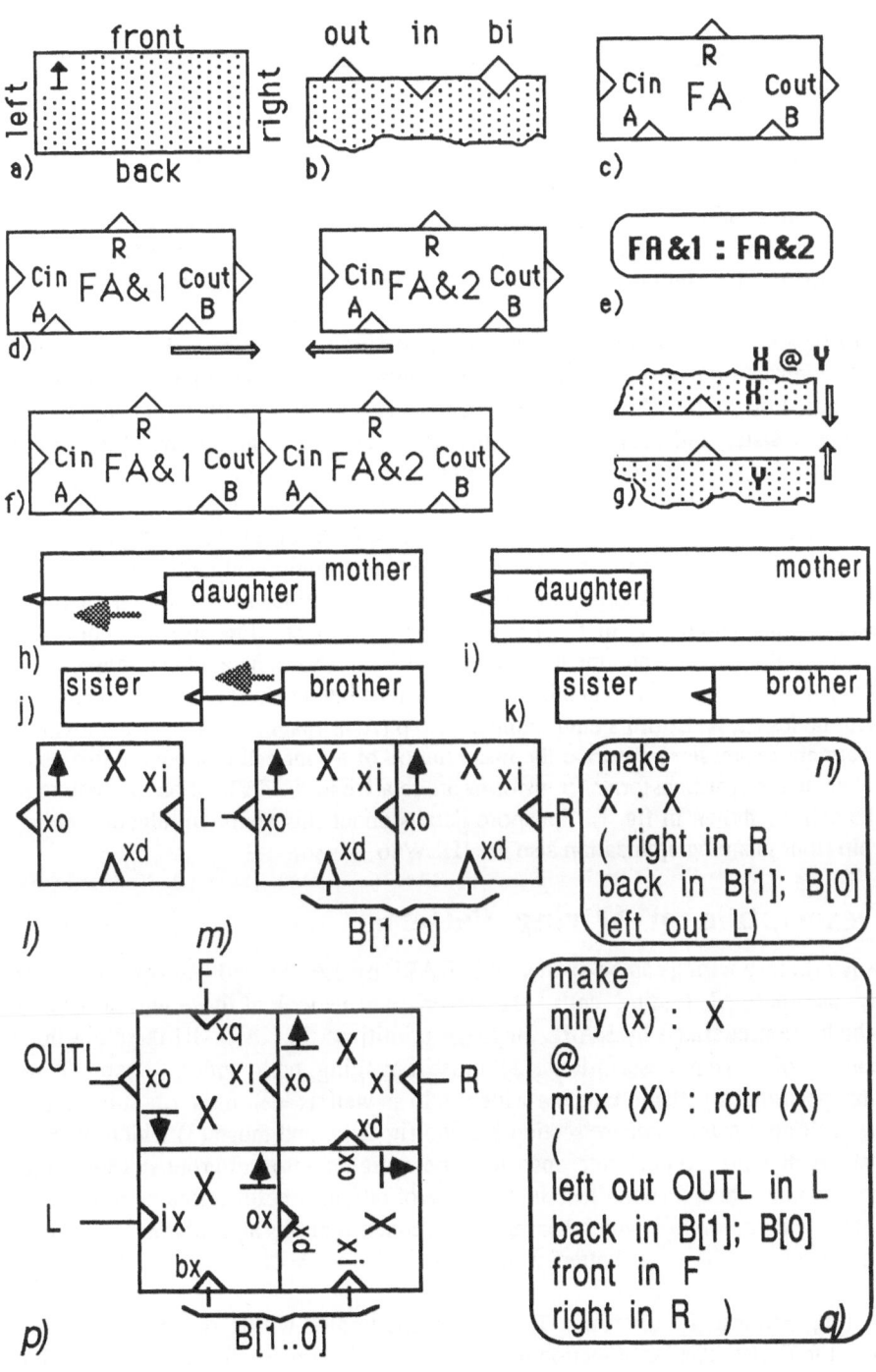

Fig. 3.2.

3.2 a), as well as 3 different port types in, out, and bi (fig. 3.2 b). These declaration attributes are used for automatic interconnect generation for cell abutments. Fig. 3.2 c shows the ABL diagram of the external view of a full adder cell example named FA. Fig. 3.2 d shows two instances FA&1 and FA&2 of this cell, due to be abutted ('&' is the separator between cell name and instant number). Fig. 3.2 e shows the abutment expression which is the notation for this instruction, where ':' stands for horizontal abutment, i. e. for abutment of slices. Fig. 3.2 f shows the result of the abutment operation. Fig. 3.2 g illustrates vertical abutment of two cells X and Y, described by the abutment expression X @ Y, using the '@' connective.

Abutment Expressions. Figures 3.2 h through k illustrate the alternatives between routing connection (h and j) and abutment (with automatic interconnect , generated by the KARL compiler, see figs.i and k). Figures j/k illustrate interconnect between sister and brother cell (same level within the hierarchy). Firuges h/i illustrate daughter-to-mother interconnect (the daughter cell is an internal component within the mother cell). Also complex abutment expressions may be formulated for the synthesis of complex supercells including several levels of cell hierarchy, as well as rotate and mirror transforms on single cells and compound cells. Figures 3.2, l through q show two examples of compound cells described by non-trivial abutment expressions. The cell X in fig. l is used as a component. Fig. m shows the ABL diagram of a two cell abutment example, and fig. n shows its abutment expression. The parameter list within parentheses describes the interconnect of the compound cell with nodes (L, R, B) of its environment. Fig. p (ABL diagram) and q (its equivalent abutment expression) describe the instantiation of a four cell compound also using rotate and mirror transforms (transforms are relative to the declaration orientation of the cell X, shown in fig. l). For more details about this abutment algebra and its chip floor planning application also see [HLW86, NN86].

Description of Wiring Patterns

Any arbitrary wiring can be expressed in KARL by user-defined descriptions, and, by user-defined routing cells. However, let us look at those wiring patterns which are predefined by KARL language primitives. In KARL-III there are three classes of wiring descriptions: 1) those changing path width ('•' and '|' for juxtaposition of paths to create a wider path, as well as (2) uses of subscripting to split up a path; compare section 3.1 and fig. 3.1), and those (3) which preserve data path width. The latter ones may be split up into subgroups: (3a) direct connections which do not affect the sequence of bits, user-defined routing boxes, and, (3b) *wiring operators* which rearrange the sequence of bits algorithmically, such as shift, shuffle, reflect, and butterfly operators.

Wiring Standard Functions. The implementation of wiring operators in KARL uses the KARL standard function format. Fig. 3.3 illustrates a few examples, 16 bits wide: a) cirshr&3 a circular shift right by 3 bits, b) fold&2, the *'perfect shuffle'*,

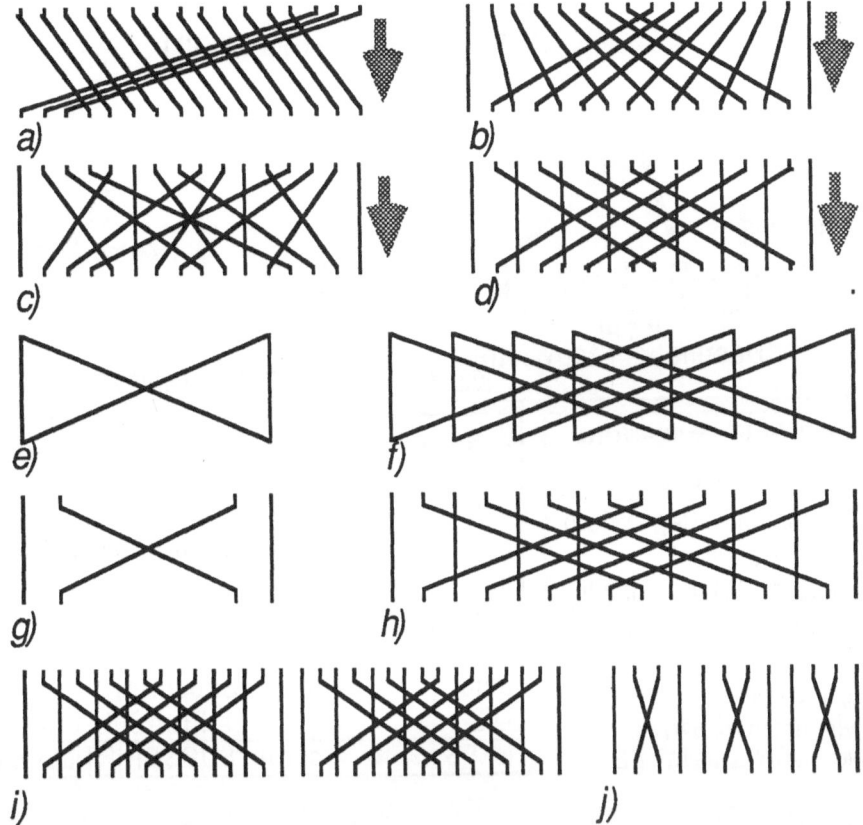

Fig. 3.3.

c) fold&4, a shuffle with destination step width of 4, d) a butterfly example. The constant number behind the '&' separator is the function parameter. For shift operators the wiring pattern is well known. The algorithm generating the shuffle wiring pattern is illustrated for a particular path width (8) and parameter (4) by fig. 3.4/3.5. By the KARL compiler the wiring patterns are automatically adapted to both parameters. Illegal parameter combinations (possible with shuffle and butterfly patterns) are detected and diagnosed by the compiler. Fig. 3.4 illustrates the meaning of the destination step width parameter in shuffle patterns.

Butterfly Wiring Patterns. Butterfly patterns have many applications, such as for example in digital signal processing, micro processor and micro computer interconnect networks and many other areas [Bat76, GoL73, MGN79, TYF74, TYF81, etc.]. Fig. 3.3. e illustrates the butterfly shape. The precise wiring pattern of the elementary butterfly, however, is that shown in fig. 3.3. g. For wider data

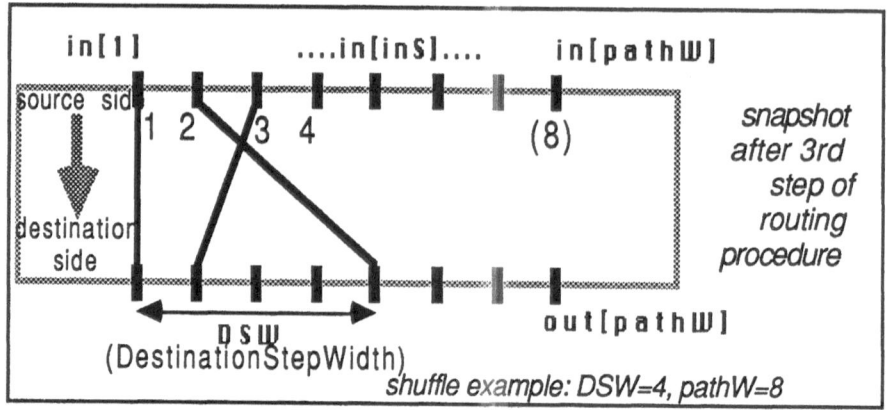

Fig. 3.4.

SHUFFLE procedure	Variables used in fig. 3.5

parameters:	*name*	*type*	*meaning*
Destination Step Width DSW (explicit)	pathW		DATApathWidth
	DSW		DestinationStepWidth
Datapath Width pathW	in		input
(automatically adapted)	out	*integer*	output
	inS		inputSubscript
legal parameter ratio: iff	outS		outputSubscript
Rem (pathW/DSW)=0	LFS		LeftmostFreeSubscript
and DSW ≤ pathW / 2	isLFS	*Boole*	currentSUBSCRIPTisLFS

paths the butterfly pattern usually is shaped by the superposition of several elemantary butterflies, such as illustrated by the 4-layer example in fig. 3.3 f. This figure is only an illustration of the pattern generation principle. The precise wiring pattern of it, however, is shown by fig. 3.3. h. The butterfly function parameter indicates the number of segments. Figures 3.3 d through h only show examples with the parameter value i = 1 (butterfly default parameter value within KARL). Figures 3.3 i, or, j, show examples with i = 2, or, with i = 3, respectively.

Within the KARL System the wiring operators illustrated above are provided in two versions: a *word format* version rearranging the bit sequence, and, an *array format* version reordering the word sequence in an array. For a survey see fig. 3.1. The array format versions of these wiring operators are useful for concise description of interconnect patterns in switch boxes like bunyan networks etc. (e. g. see [LaMa86, BaV86]), for parallel signal processing circuits, such as e. g. for fast fourier transform. Shuffle operators are also useful for mixed mode (word format) RT level and (single-bit format) logic simulation in using KARL, where the shuffle pattern may be used as an adapting interface between both kinds of the hardware description. For more details you may request [HLW86, NN86].

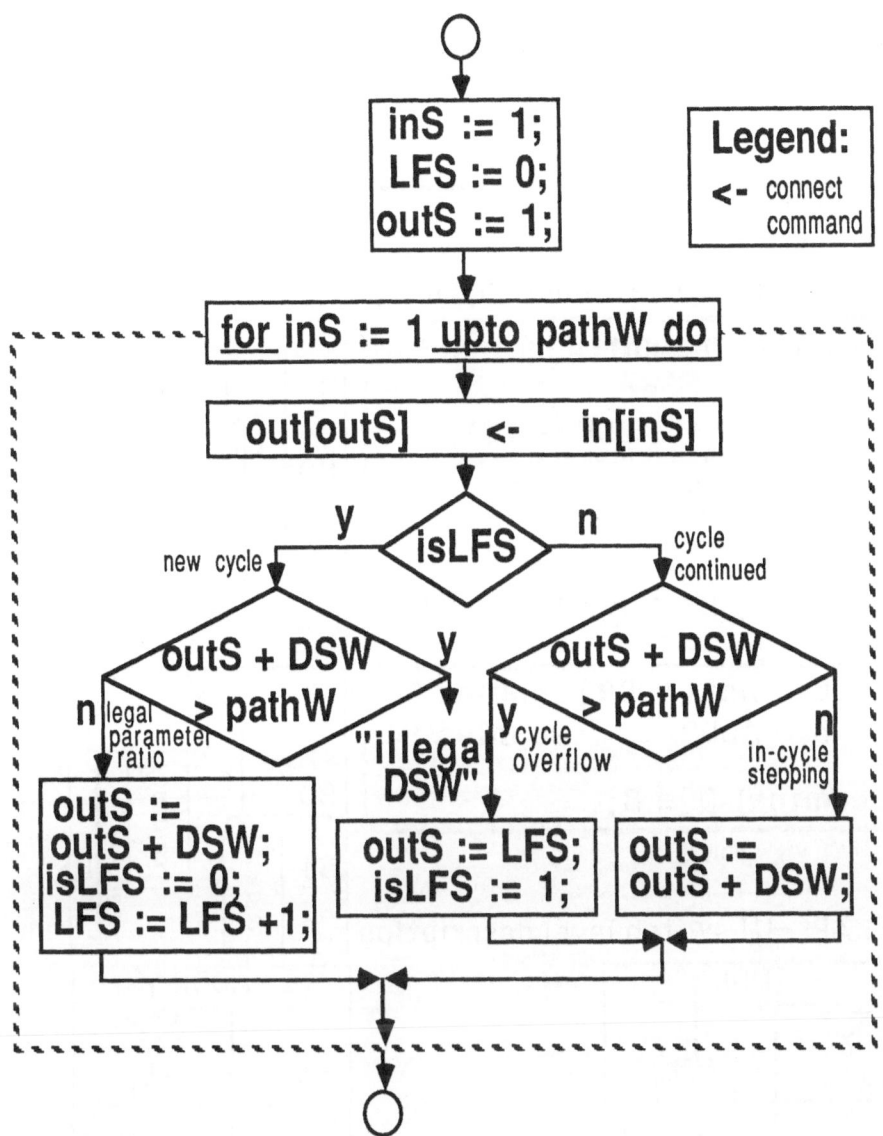

Fig. 3.5

Step-wise refinement using a CHDL

Refinement capability is an important language property to achieve its use as a design language. So you do not need to change the language in top-down planning from a purely functional specification to a more detailed description which then may be used to enter logic design and physical design. So you may use the same language and the same tools to analyze and to synthesize a conceptual description being a

488 *VLSI CAD Tools and Applications*

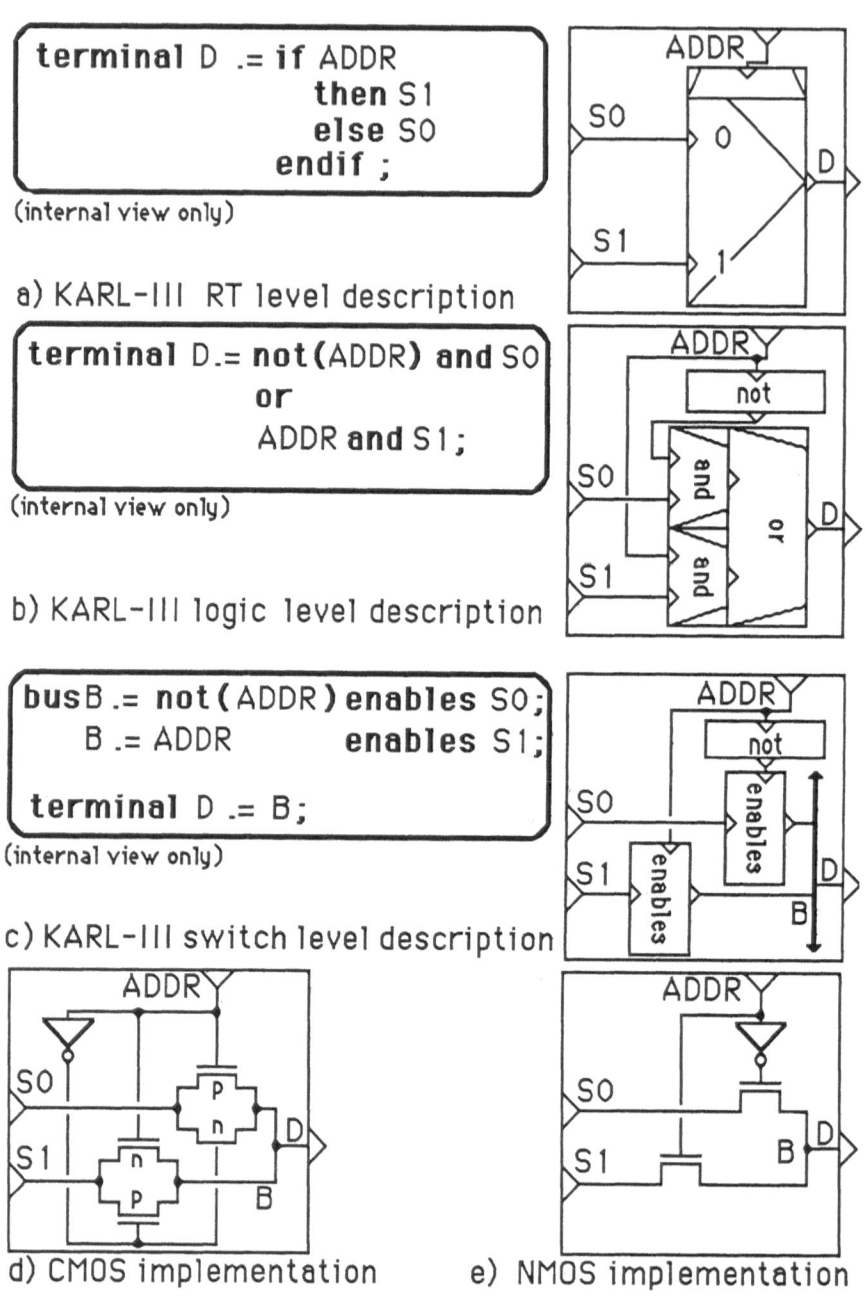

a) KARL-III RT level description

b) KARL-III logic level description

c) KARL-III switch level description

d) CMOS implementation

e) NMOS implementation

Fig. 3.6

structural / topological / functional notation which carries along all the clever architectural ideas for testability, for structured design, etc. over to the silicon implementation team.

Fig. 3.6 illustrates stepwise refinement in KARL and ABL by an example circuit, a simple 2-way multiplexer. It always shows two notations: the textual (KARL) notation at the left side, and the ABL notation (such as produced by the ABLED editor [GHW85, Far86]) at the right side. It shows the circuit in different degrees of detailedness, in different abstraction levels: a) at RT level, b) at gate level, c) at switch level. So this also illustrates, that the KARL descriptions a) thru c) are technology-independant. By the way: the simulator gives the same response for all three of them. So this is required to have a consistent implementation of the language KARL.

Figures 3.6 d) and e) are no more KARL descriptions, but circuit diagrams of different implementations: d) in CMOS technology, e) in NMOS technology. Figures d) and e) have been produced bei another graphic editor MLED [Wel86]. This is a mixed-level editor including all levels from RT down to layout. Bei menue guidance any mixed-mode representation can be arranged, such as e. g. showing one cell at ciruit level, another one at layout level, a third one at RT level (here using ABL) etc.

CHDL Use as a Design Calculus

Section 3.4 has illustrated the consistency of KARL and its capability for step-wise refinement, so that top-down design can be carried out without changing the language. This is a requirement if it is desired to use the language as a design calculus in order to reach design goals by means of a sequence of algebraic manipulations. However, such a language can only be a medium to express such algebraic rules, however, it cannot be this algebra itself. This medium, however, is powerful, suitable for effective exploration of many areas of application by experimenting with alternative architectures and structures. In [NN85, Lem86] an example is described, where a regularly structured integer multiplier layout has been developed from the algorithm description by a sequ3nce of successive algabraic manipulations. In [Ha77, Lem 86] approaches to a general algebraic schematics development are illustrated in using binary-to-BCD, BCD-to-binary code converter, and universal shifter examples.

CHDL-BASED DESIGN ENVIRONMENTS

The KARL core system (fig. 4.1 a) as well as the KARL environment are modular systems. The KARL core system uses three different languages: the hardware description language *KARL* as a description source, the simulator activation and test description language *SCIL* [HHa86], and the executable intermediate form called

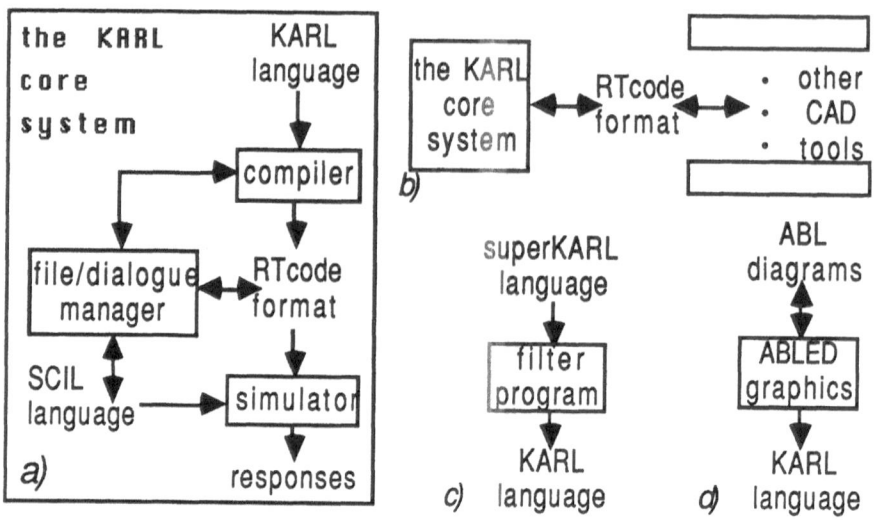

Fig.   4.1.

RTcode. RTcode [HaMa86] has turned quasi-standard interface to a number of tools within and outside the CVT project [NN86] (fig. 4.1 b). For instance, three different simulators have been implemented, which accept RTcode: the original KARL simulator [Web81, HHa86], the fault simulator of the CVT CAT system [SMP86, SM85], and an event-driven fast simulator having been implemented at CSELT [Per86].

Figures 4.1 c) and d) illustrate the implementation of other languages by means of precompilers or interpreters generating KARL source descriptions. The language *superKARL* [GHHO86] is a KARL extension featuring parametric cell descriptions using array size and path width parameters [HaHa86]. Fig. 4.6 illustrates the generation of cell arrays under control of superKARL array size parameters and data path width parameters: on-dimensional cell arrays (a), bit node arrays (e), and word node arrays (f); two-dimensional cell arrays with linear growth(c), exponential growth (d), node arrays with exponential growth (g), as well as recursively definde two- dimenstional cell arrays (b). superKARL also features a rule-driven, and thus technology-adaptable algorithm for translation of personality matrixes into KARL functional descriptions for a wide variety of matrix-oriented logic (MOL) circuit techniques, such as e. g. PLAs, folded PLAs, Weinberger arrays, folded Weinberger arrays, Lopez/Law dense gate matrix layout, KOLTE arrays, and others [GHHO85]. There are also other CAD tools having interfaces to the KARL system [NN86], not described here because of lack of space.

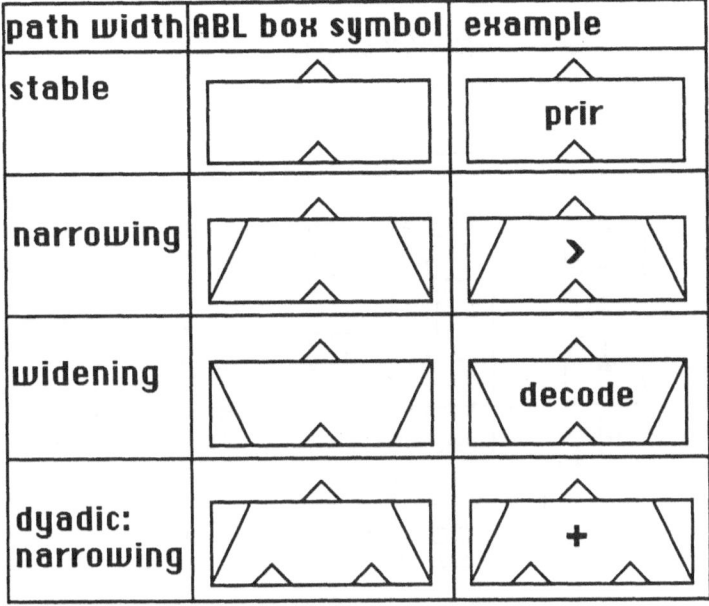

path width	ABL box symbol	example
stable		prir
narrowing		>
widening		decode
dyadic: narrowing		+

Fig. 4.2.

Using an Interactive Graphic CHDL Editor

This section briefly illustrates a new interactive graphic RT level editor ABLED, from a user's point of view. ABLED has been developed within the CVT project bei CSELT (Torino, Italy) and Kaiserslautern University [GHW85]. It is an interactive graphic interface to the KARL system (see fig. 4.1 d). Its typical diagram symbols are illustrated by the right side of figures 3.6 a thru c, as well as by Fig. 4.2. The arrows to indicate ports are placed on the edge of cell boundary boxes to allow the DOMINO notation [Ha77] to show symbolic abutment (of abstract boxes) in architectural diagrams, and, to show physical abutment (when boxes reflect the shape of real cells) in case of a partitioning derived from a chip floor plan. (This DOMINO feature is especially useful in MLED [Wel86], which combines the features of ABLED and those of editors for layout, circuit diagrams, and logic diagrams (also see section 3.4).) The symbols are automatically created by picking via menue. The editor also includes an on-line graphical syntax check, which immediately diagnoses illegal matings. This accelerates working with KARL considerably, since most of the diagnostics is already interactively available, before the KARL compiler has been called to parse the derivative of the ABL data structure (fig. 4.1 d). ABLED (and MLED) uses a tightly guiding menue technique, so that working with KARL is much more easy to the user via ABLED, than directly at the KARL textual interface. For illustration fig. 4.3 shows a more complex diagram example having been plotted

INT (* *)
CK
SIGN
ZERO[7..0]
LEVIND
CMX
INT

Fig. 4.3

KARATE
SCIL
language
Test
assembler
Tester
the KARL
core
system
a)
SCIL
language
the KARL
core
system
simulator
physical
model
Tester
b)

Fig. 4.4.

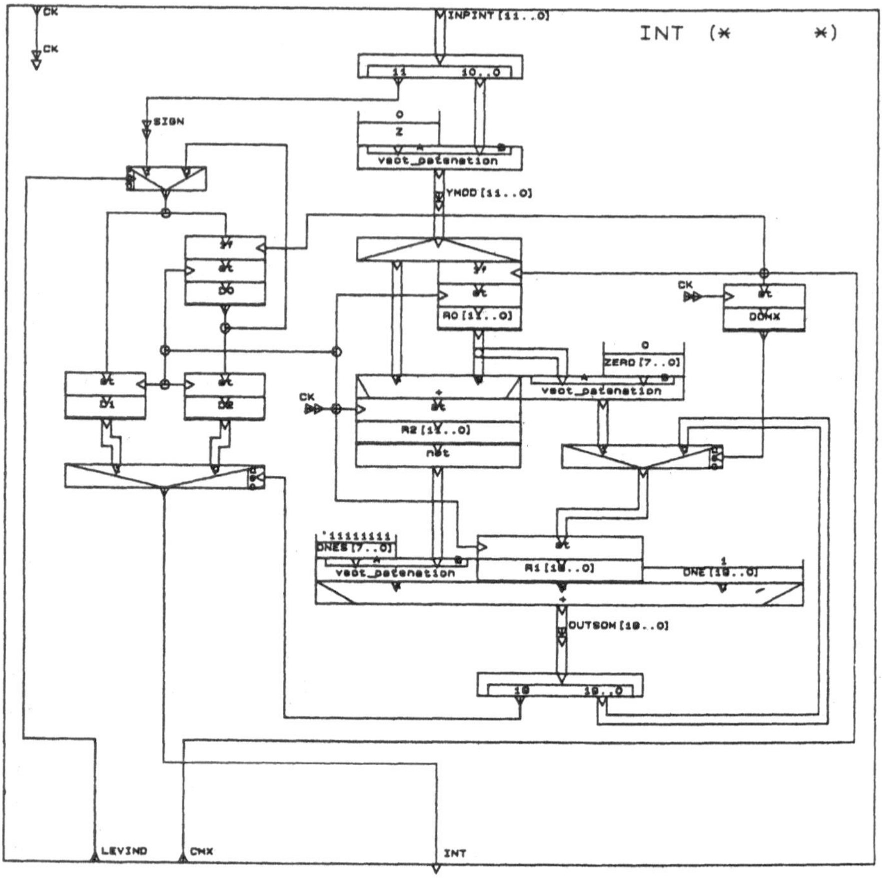

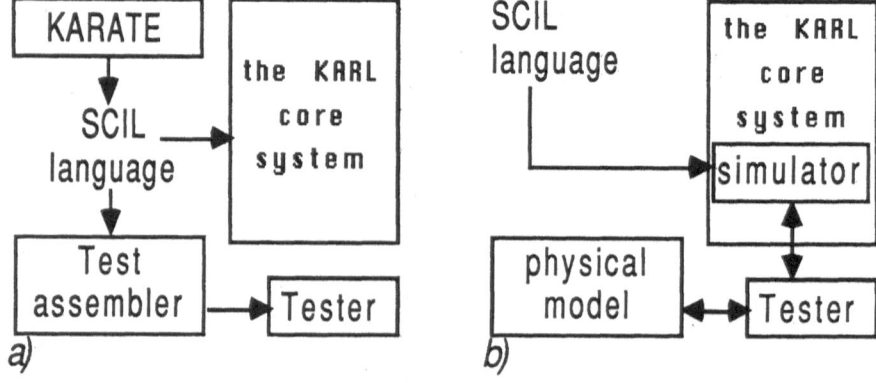

by ABLED. A detailed description of ABLED, from a users point of view, gives [GHW85].

CHDL-based CAD Tools

A number of KARL-related CAD tools [NN86] have been developed within research projects: within the CVT project having been funded by the Commission of the European Communities, by the multi-university E.I.S. project being funded by the German Federal Minister of Research and Technology, and also outside these projects. Those programs are tools for automatic synthesis (silicon compilers using KARL source input [ArMa86, EvP85]), for interactive synthesis (microprogram transformation [RaGr86], interfacing to RTcode of the KARL core system, compare fig. 4.1 b) for RT level verification [GSch85, SchG86], for test pattern development and testability analysis (the CVT CAT Environment [SM85, SMP86], the KARATE program [HaWo85]). Fig. 4.4 a.) shows how test development and simulation are integrated by using the same language SCIL (Simulator Command and I/O Language [HHs86]), which at the same time is the test description language used as an output language by the test generator KARATE, and, the stimuli and command language used as an input to the KARL simulator inside the KARL core system (compare fig. 4.4 b). A test assembler program is used to assemble

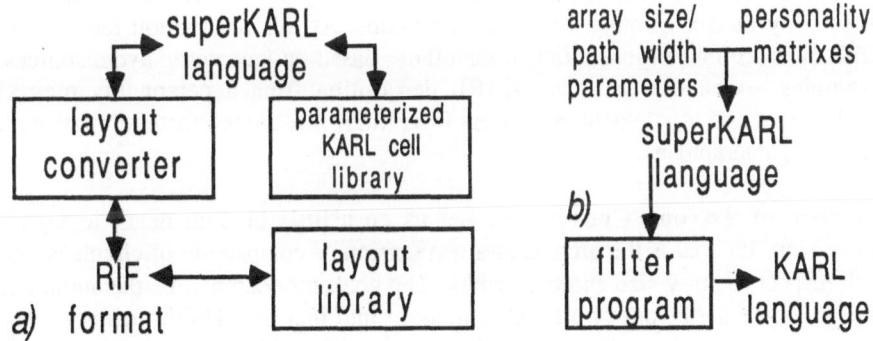

Fig. 4.5

a device-specific test program for the test device in use (fig. 4.4 a). Only this assembler has to be changed, when an other test device will be used.

Fig. 4.4 b shows, how also a physical model extension feature (PMX [HaHi86]) is integrated into the same set-up. (A PMX allows the simulator to communicate with real hardware *'physical model'* , such as e. g. an already existing prototype hardware module of the system to be developed. This requires by far less CPU time, than simulating the whole system.) Since the tester anyway is directly coupled to the computer which is hosting the KARL core system, as well as the test program assembler, the physical model may be just plugged into the adapter of the testing device. The only difference is, that the tester is talking to another piece of software: to the simulator.

Interfacing to other CAD Tools

This section briefly illustrates interfacing to layout level by means of a few example configurations. The most direct connection from CHDL down to layout is the silicon compiler, having been mentioned above [LaMa86, EvP85].

KARL Extraction from Layout. The other way around is behavioural extraction from layout. The REX system [Neb86] directly generates KARL descriptions from Layout (so the interface is simple). It is rule-driven, and thus technology-adaptable. REX is useful, but wasting CPU time for large circuit design verification, however, would be more efficient for verification of critical cells and library
cells. So such an extractor could be very useful to verify the basic circuit library needed for a much more efficient methodologies based on personality matrix notations of MOL (matrix-oriented logic; compare section 4.)

KARL Extraction from Personality Matrixes. The MOL extractor [GHHO85] accepts description sources at higher symbolic layout levels, using personality matrix formats, or similar notations. At this abstraction level many CAD tools are much more efficient than those based on lgeometric ayout sources. Examples: extractors (generate KARL description from a personality matrix) synthesizers and topological optimizers [HaEIS86] layout architectural converters (see next paragraph) etc.

Support of Layout Conversion. Let us go a little bit into detail in layout conversion. Its goal is the architectural exptension or compaction of circuit layout with respect to array size and path width. The goal, for example, be the automatic conversion of the layout of a 64 kbit memory into that of a 1Mbit memory. The memory array size has to be extended, and its peripheral logic has to be extended linearly and to be relocated geometrically, where a modified layout format RIF (relative intermediate form) may be helpful (fig. 4.5 a). (Also fan-out changes etc. have to be considered, of course). However, the organizational aspects are of much higher level, than layout. Such a layout converter has to be guided by RT level data. A KARL extension like *superKARL* (also see section 4.) or its forerunner version *hyperKARL* [Borr85] uses such array size and path width parameters. The actual cell hierarchy tree is produced after parameter assignment and translation into KARL (fig. 4.5.b). The layout converter (fig. 4.5 a) has to be partly implemented in a similar way, or, has to be part of, or, interfaced to, the filter program. This again shows the

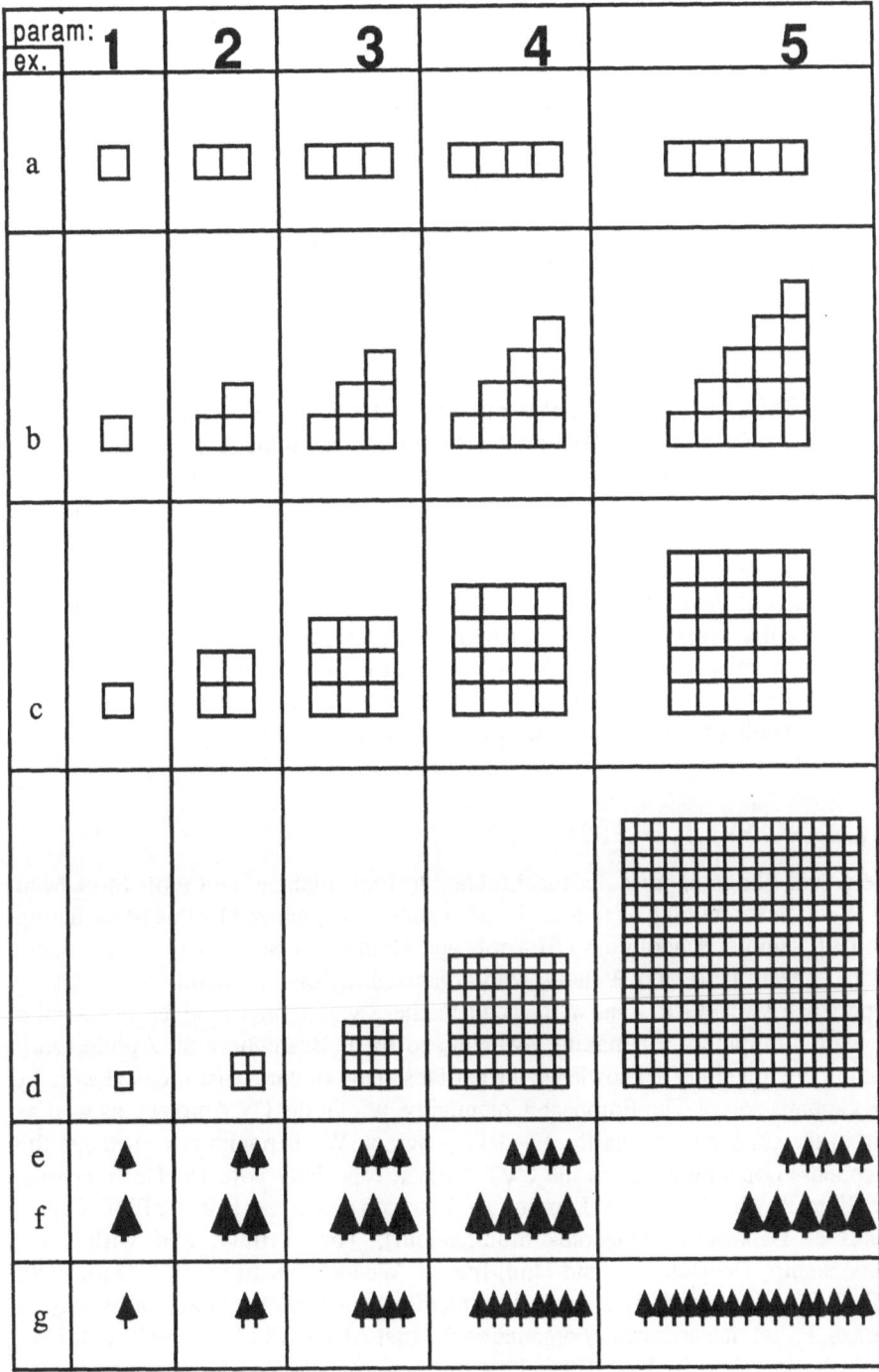

Fig. 4.6

usefulness of interfacing RT level tools to low level CAD tools.

KARL interfaces to other CAD Tools. The following tools, having been interfaces to KARL, or, beeing KARL-based, have not yet been mentioned in this paper. The VERENA verifier, based on theorem proving techniques, compares two different KARL descriptions for equivalence [GSch85, SchG86]. Also the ARIANNA chip planner has interfaces to KARL [ARA84]. The CVT CAT (Computer-Aided Test Development) environment uses 5 different tools, part of it beeing KARL-based [MHM84, SMP86]. For more literature and other information on KARL and its applications see [NN86, LM86].

CONCLUSIONS

We have illustrated the usefulness of using CHDLs and related tools. We shave shown, that contemporary CHDLs are a better front end of the ASIC design process, that CHDLs are very useful in concise design problem capture, in design for testability and, in efficient structured VLSI design in directly casting algorithms onto silicon. We tried to illustrate the substantially improved acceptance of CHDLs by provided an interactive graphic user interface. We also tried to show the substantial acceleration of the design process by using such a graphic interface including an on-line graphic syntax check. We hope to creasonalble onvince the readership of the VLSI community that in an ASIC-oriented and USIC-oriented (User-specific IC) design environment there is no way to cope with complexity of design planning and product planning without using such high level tools.

ACKNOWLEDGMENTS

We would like to express our thankfulness to Prof. Fichtner and Prof. Morf from ETH Zurich for inviting us to give the presentation summarized by this paper during the IFIP summer school on VLSI Tools and Methods at Beatenberg, Switzerland, 1986, as well as to Udo Welters, who organized the on-site facility for hands-on experience with KARL and ABL, and to the Swiss representatives of Apollo Domain Computers for making available to us at Beatenberg an Apollo work station. We gratefully acknowledge the partial support of our KARL- related work by the Commission of The European Community, within the CVT project, as well as within the CVS project via the ESPRIT program. We herewith acknowledge the excellent cooperation within the CVT project, especially with Dr. DeVincentiis, Dr.Giorcelli, Dr. Girardi, Dr. Leproni, Dr. Patrucco and others from CSELT (Centro Studi et Laboratori Telecom- municazioni), Torino Italy, and with Prof. Gallenkamp, Dr. Delafera, and Dipl.-Ing. H. Wendt from FI at FTZ, Darmstadt, F.R.G. We also gratefully acknowlege the efforts of Prof. W. Graß, University of Passau, F.R.G. in successfully organizing the first ABAKUS (ABL and KARL User Group) Workshop in June 1986. Last but not least R. Hauck, K. Lemmert, A. Mavridis, W. Nebel, and A. Wodtko of my group at Kaiserslautern, as well as all my students involved in our work should be mentioned. Without them all these results would not have been possible.

LITERATURE

[ABA86]   ABAKUS: List of KARL-related literature; ABAKUS, Kaiserslautern University, Kaiserslautern, F.R.G., 1986

[AGA84]   G. Arato, O. Gaiotto, P. Antognetti, A. de Gloria; ARIANNA: Floor Planning Tool; CVT/CSELT technical report, Genova /Torino, 1984

[ArMa86]   G. Arato, R. Manione: PSICO: A System for Automatic Layout Generation; report, CSELT, Torino, 1986

[Bat76]   K. E. Batcher: The Flip Network in STARAN; Proc. 1976 Conf. on Parallel Processing; IEEE 1976

[BaV86]   G. Balboni, V. Vercellone: Experiences in Using KARL-III in Designing a CMOS Circuit for Packet- switches Networks; ABAKUS (ABL and KARL user group) workshop, Passau, F.R.G., 1986

[BBad85]   K. Bastian, R. Hartenstein, W. Nebel: VLSI-Algorithmen: innovative Schaltungstechnik statt Software; VDI/VDE-GME-Tagung "Mikroelektronik in der Automatisierungs-technik", Baden-Baden, 1985, VDI- BerichtNr. 550, VDI-Verlag, Düsseldorf, 1985

[BBG86]   A. M. Biraghi, A. Bonomo, G. Girardi: ABLEDitor: User Manual; CSELT, Torino, Italy, July 1986

[Borr85]   H. Borrmann: hyperKARL-III Language Reference Manual, report, Fachbereich Informatik, Kaiserslautern University, Kaiserslautern, F.R.G., 1986

[EvP85]   E. von Puttkamer: Das SIC-System als Kern eines Silicon-Compilers, report SFB-124, No. 18/85, Fachbereich Informatik, Univ. Kaisaerslautern, 1985

[Far86]   D. Farelly: The Practical Application of ABLED and   KARL in Designing a Speech Synthesizer; ABAKUS (ABL&KARL user group) worksh., Passau, F.R.G., 1986

[GHHO85]   J. Gebhard, R. Hartenstein, R. Hauck, D. Oelke: Behavioural
             Extraction from Personality Matrixes of Matrix-Oriented Logic
             (MOL) Circuits; (submitted for publication)

[GHHO86]   J. Gebhard, R. Hartenstein, R. Hauck, D. Oelke: The
             superKARL-III Language Specification; report, Fachbereich
             Informatik, Kaiserslautern University, Kaiserslautern, F.R.G.,
             1986

[GHW85]   G. Girardi, R. Hartenstein, U. Welters: ABLED: a RT   level
             Schematics Editor and Simulator Interface; Proc. EUROMICRO
             Symposium Brussels, Belgium, 1985,(ed.: K. Waldschmidt),
             North Holland Publ. Co., Amsterdam/New York, 1985

[GoL73]   R. Goke, J. Lipovski: Banyan Networks for Partitioning
             Multiprocessor Systems; Proc. 1st Ann. Symp. on Computer
             Architecture, 1973; IEEE 1973

[GSch85]   W. Graß, M. Schielow: VERENA: a Program for Automatic
             Verification of the Register Transfer Description; IFIP Int'l
             Symp. on CHDLs and their Applications, Tokyo, Japan, 1085;
             North Holland Publ. Co., Amsterdam/New York, 1985

[HaEIS86]   R. Hartenstein, R. Hauck: Entwurf von Symbolischem   Layout
             mit Werkzeugen der Register-Transfer-Ebene; in: (Hrsg.: H.
             Heckl, A. Kaesser, K. Koller, K. Woelcken) Entwurf Integrierter
             Schaltungen, 2. EIS-Workshop im Wissenschaftzentrum, Bonn,
             1986

[HaHa86]   R. Hartenstein, R. Hauck: Functional Extraction from
             Personality Matrixes of MOL (Matrix-oriented Logic) Circuits;
             ABAKUS (ABL and KARL user group) workshop, Passau,
             F.R.G., 1986

[HHa86]   R. Hartenstein, R. Hauck: The KARL System User Guide; CVT
             report, Kaiserslautern University, Kaiserslautern, F.R.G., 1986

[HaHi86]   R. Hauck, A. Hirschbiel: A Physical Model Extension for
             KARL Simulators; ABAKUS (ABL and KARL user group)
             workshop, Passau, F.R.G., 1986

[HaMa86]   R. Hartenstein, A. Mavridis: RTcode Instant for KARL-III,
             second edition; report, Fachbereich Informatik, Kaiserslautern
             University, Kaiserslautern, F.R.G., 1986

[Har77]   R. Hartenstein: Fundamentals of Structured Hardware Design - A Design Language Approach at Register Transfer Level; North Holland Publ. Co., Amsterdam /New York 1977

[Har86]   R. Hartenstein (ed.): Hardware Description Languages; in: (ed.: T. Ohtsuki) North Holland Book Series on Advances in CAD for VLSI; Elsevier Scientific, Amsterdam/New York, 1986

[HaWo85]   R. Hartenstein, A. Wodtko: Automatic Generation of Functional Test Patterns from an RT Language Source; Proc. EURO-MICRO Symposium, Brussels Belgium, 1985, (ed.: K. Waldschmidt), North Holland Publ. Co., Amsterdam/New york, 1985

[HLW86]   R. Hartenstein, K. Lemmert, A. Wodtko: KARL-III Language Reference Manual, CVT report, Kaiserslautern University, Kaiserslautern, F.R.G., 1986

[LaMa86]   L. Lavagno, R. Manione: Automatic Layout Generation from (KARL) RT level descriptions; ABAKUS (ABL and KARL user group) workshop, Passau, F.R.G., 1986

[Lem86]   K. Lemmert: Steps Towards KARL Use as a Design Calculus; ABAKUS workshop, Passau, F.R.G., 1986

[LM86]   L. Müller: KARL-related literature: reference list and order form; ABAKUS, Bau 12, Universität Kaiserslautern; 1986 (regularly updated)

[MHM84]   S. Morpurgo, A. Hunger, M. Melgara, S. Segre: RTL Test Generation and Validation for VLSI: an integrated set of Tools for KARL; IFIP CHDL'85, Tokyo, Japan, Sept 1985, North Holland Publ. Co., Amsterdam, 1985

[MGN79]   G. M. Masson, G. C. Gingher, S. Nakamura: A Sampler of Circuit Switching Networks; Computer, June 1979

[Neb86]   W. Nebel: KARL Extraction from Layout; ABAKUS (ABL / KARL user group) workshop, Passau, F.R.G., 1986

[NN85]   NN: KARL Primer; CVT report, Kaiserslautern University, Kaiserslautern, F.R.G., 1986

[NN86]   NN: KARL-related Software Catalogue; ABAKUS, Kaisers-
              lautern University, Kaiserslautern, F.R.G., 1986

[Per86]   G. Girardi: personal communication

[RaGr86]   M. Rauscher, W. Graß: Experiences in using KARL in
              PRIMITIVE (Program for Interactive Microprogram
              Transformation and its Verification; ABAKUS (ABL and KARL
              user group) workshop, Passau, F.R.G., 1986

[SchG86]   M. Schielow, W. Graß: KARL use in VERENA (Verification
              of RT Structures); ABAKUS (ABL and KARL user
              group)workshop, Passau, F.R.G., 1986

[SM85]   I. Stamelos, M. Melgara: CVT TIGER: a RT level Test Pattern
              Generation and Validation Environment; CVT report, CSELT,
              Torino, Italy, 1985

[SMP86]   I. Stamelos, M. Melgara, M. Paolini, S. Morpurgo, C. Segre:
              A Multi-Level Test Pattern Generation and Validation
              Environment; International Test Conference, 1986

[TYF74]   Tse-yun Feng: Data Manipulationg Functions in Parallel
              Processors ant Their Implementations; IEEE-Trans. C-23
              (1974), no. 3

[TYF81]   Tse-yun Feng: A Survey of Interconnection Networks;
              Computer, Dec. 1981

[Web81]   B. Weber: PASCAL-Implementierung eines Simulators auf
              KARL-II-Basis; Diplomarbeit, Fachbereich Informatik, Kaisers-
              lautern University, Kaiserslautern, F.R.G., 1981

[Wel86]   U. Welters: Specification of the MLED Multi-Level Editor;
              internal report, Fachbereich Informatik, Universität Kaisers-
              lautern, Kaiserslautern, F.R.G., 1986

[Wod86]   A. Wodtko: RT Languages in Goal-oriented CAD Algorithms;
              in: [Har86]

18

NEW TRENDS IN VLSI TESTING

G. SAUCIER   C. BELLON   M. CRASTES DE PAULET

Laboratory Circuits & Systems - IMAG
46 Av. F. Viallet 38031 GRENOBLE cedex. FRANCE

INTRODUCTION

Test data generation for complex integrated circuits or today printed circuits requires an hierarchical approach including the possibility of using diversified types of descriptions (behavioral, functional, structural) as well as diversified test generation methods for elementary blocks.

Moreover, new techniques, mainly Artificial Intelligence ones, seem to be useful to store test experiences, to provide test advices and to solve some problems encountered during the test generation.

The scope of this paper is to present an integrated approach of the test generation problem fullfilling these requirements (figure 1).

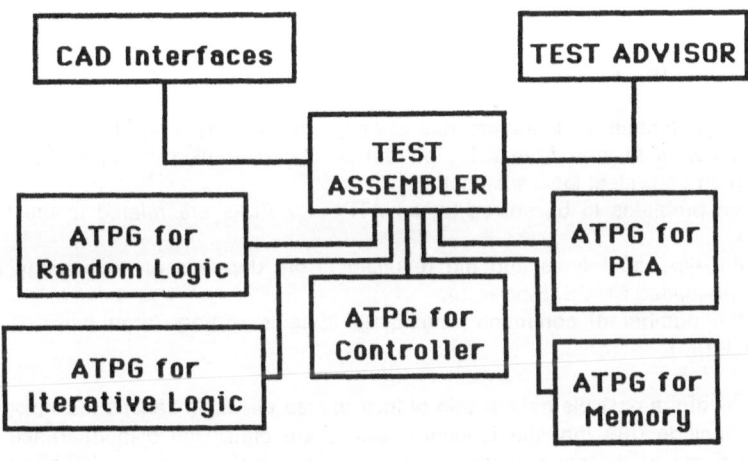

Figure 1 : a system for an integrated approach of test generation

Automatic test programs generators (ATPG) are used as low level servers. They are specific to different types of logic (random logic, structured or iterative logic, PLA, memory, processor,...). The most original ones (dealing with controllers, PLA, and memories) will be illustrated in section 1.

Section 2 defines the task of the test assembler which receives informations about the test patterns generated for the elementary blocks and which helps to construct the global test program. This global test program has to exercize each elementary block according to the local test patterns.

Section 3 outpoints the help which may be expected from a test advisor : the test advisor starts from a high level description and aims at providing the test assembler with an efficient help or advices about the global optimization of the test.

Finally, the interface with the CAD data base is discussed in section 4 ; in order to organize the different informations related to the test problem in an efficient way, an object oriented description based upon frames will be used.

1 - THE LOW LEVEL ATPG SERVERS
1.1 - ATPG for random and semi random logic

ATPG for random logic detecting stuck-at faults on gate level models are of common use [1] ; a lot of progress deal with algorithmic improvments [2] and research concerns mainly the extensions to complex MOS gates [3]. By semi random logic, we mean mainly iterative or bit sliced logic. Iterative logic test methods have been extensively studied [4] but their implementation in a convenient ATPG is less common. As random logic is, in most systems, computation logic thus iterative logic, this point is of high importance.

1.2 - ATPG for PLA

A huge amount of litterature has already covered this topic ([5]) ; the regularity of PLAs allows to start from a list of realistic defects which is better than a list of faults on an equivalent logic schemata.

The main problems to be solved by an ATPG for PLAs are related to the following remarks :

- the type of defects and the resulting errors depends on technology (NMOS, CMOS, preloaded CMOS, Bipolar,....),

- the number of commonly considered defects is very large (shorts at every crosspoint,....).

The study of the possible defects and of their related effects in various technologies lead us to conclude ([6]) that the functional errors are either the disappearance of PLA product terms or the appearance of other product terms which can be characterized (product terms "adjacent" to the implemented ones,...).

In addition, memorization errors must be considered in CMOS technologies (stuck open faults).

Thus the ATPG implemented in our system is composed of 2 modules (figure 2) : the first one deduces the set of functional errors equivalent to the set of defects selected by the user ; the second one generates the test itself, starting from a classical description of the PLA and the previous functional errors list.

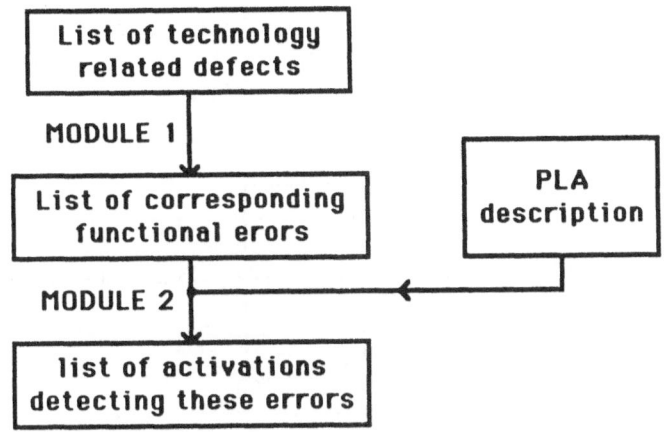

Figure 2 : ATPG for PLAs

The related software is 2000 Pascal lines long and is running on VAX-VMS.

1.3 - ATPG for memories

Algorithms for memory testing are well known and are implemented on any test equipments. The assistance given in our system consists in selecting a test algorithm according to the following criteria :
- the error hypotheses given by the user,
- the internal organization of the memory cell array, in order to reduce the test length without loss of the efficiency (for example, low probability cell couplings are eliminated),
- the suitable test length and the memory environment ; the assistance leads to define a balanced test for complex circuits : the test of a local memory, for instance, is generated with respect to the coverage rate in the other blocks of the circuit.

1.4 - ATPG for controllers

A controller implements next state and output equations, the set of states being stored in memory points. It is well known that a structural approach to test controllers is inadequate and inefficient.
The good way to approach its test is to start from its functional specification in term either of a classical state graph or a flowchart. In our system, the state graph of a controller is described using the high level language CADOC.LD ([7]).
In CADOC.LD, outputs signals are associated with the nodes and timed input conditions with the transitions (an example is given on figure 3) ; a special effort is made to allow an easy description of timing diagrams for both input and output signals.

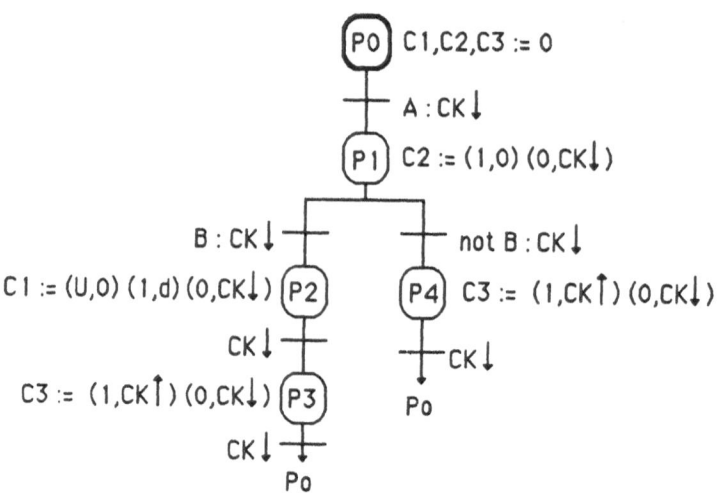

Figure 3 : Example of a CADOC.LD description

Remark :

$C1 := (U,0) (1,d) (0,CK\downarrow)$

will be interpreted as :

CK

C1

d

        The ATPG for controllers aims at giving the timing diagrams of required inputs to go through the pathes of the graph. These test cases may be found automatically by "timed symbolic execution" of the state graph ([8],[9]). Nowadays, only guided assistance is given to find the path conditions required to test the controller. These path conditions are stored as test results associated with the controller and will be used for determination of the global test involving this controller. These path conditions express time-symbolic timing diagrams and are illustrated on figure 4 ; the outputs results are stored under the same form.

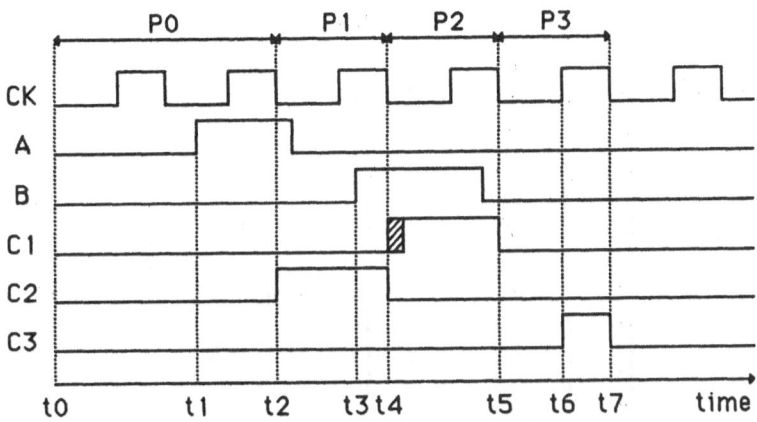

Figure 4a : Timing diagrams for path P0,P1,P2,P3

path	A = (0,0) (1,t1)	t0 < t1
condition	B = (0,0) (1,t2)	t2 = fex (Ck,t1,1)
	C1 = (0,0) (U,t4) (1,t4+d) (0,t5)	t4 = fex (CK,t1,2)
output	C2 = (0,0) (1,t2) (0,t4)	t5 = fex (CK,t1,3)
sequences	C3 = (0,0) (1,t6) (0,t7)	t6 = rex (CK,t5,1)
	t7 = fex (CK,t6,1)	
	t2 < t3 < t4	

where
- fex (CK,ti,j) represents the jth falling edge on variable CK after time ti
- rex (CK,ti,j) represents the jth rising edge on variable CK after time ti

Figure 4b : time-symbolic timing diagrams

2 - THE TEST ASSEMBLER

The test assembler is the key point of the system ; its goal is to generate a test program for a complex circuit or a complex printed board starting from the knowledge of the local test patterns for each elementary block. In other words, the system has to find functional global test activations exercizing the different blocks with the expected test patterns (figure 5).

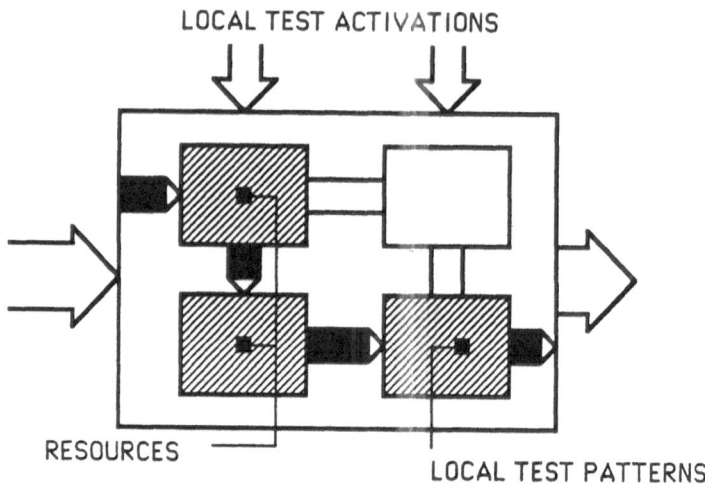

LOCAL TEST ACTIVATIONS

RESOURCES

LOCAL TEST PATTERNS

Figure 5 : multilevel approach of the test generation problem

2.1 - Description language and knowledge base for elementary blocks

The description language has to, model a circuit as an interconnection or a cooperation of elementary blocks or resources. A block may be described at a behavioral, functional or structural level ; the language used in practice in this approach is still CADOC ([7],[8]). A resource is described by its timed IO specifications and its internal functioning which may be described by a classical bipartite state graph (nodes and transitions). Let us notice here that the same language is used for describing both control and data path parts of circuits.

Two types of knowledges have to be stored for each elementary resource, one is made up of the global informations about the functional activations of the block, the other one is made up of the test activations suitable for the block.

2.1.1 - General knowledge of a resource

With a resource, must be associated the set of possible functional pathes labelled by the corresponding IO timing diagrams. These informations may be stored using a rule-like representation :

IF {input sequence } THEN {ouput sequence}   (figure 6).

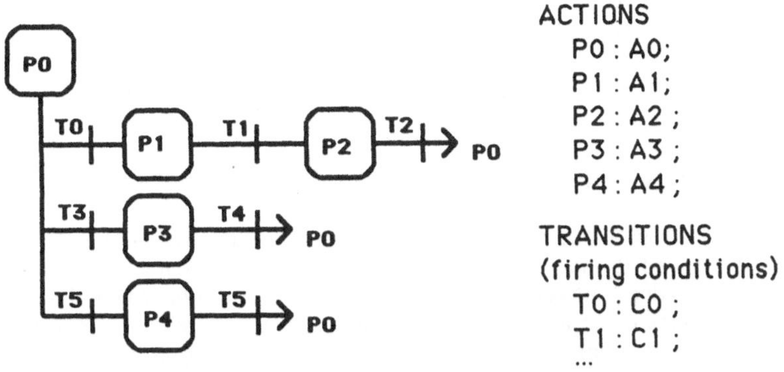
P0
T0
P1
T1
P2
T2
P0
T3
P3
T4
P0
T5
P4
T5
P0
ACTIONS
P0 : A0;
P1 : A1;
P2 : A2 ;
P3 : A3 ;
P4 : A4 ;
TRANSITIONS
(firing conditions)
T0 : C0 ;
T1 : C1 ;
...

CADOC description of a resource

RULES :
1 : **IF** (C0,C1,C2) **THEN** (A0,A1,A2,A0)
path condition expected outputs
2 : **IF** (C3,C4) **THEN** (A0,A3,A0)
3 : **IF** (C5,C6) **THEN** (A0,A4,A0)

Figure 6 : knowledge base of a resource

The automatic generation of this information is equivalent to process the timed symbolic execution of the functional model of the resource ; this has been extensively studied in ([9],[10]). The difference with controllers (cf. 1.4) is that the execution is symbolic with respect to time and value.
In practice, an assistance is given to the designer to create this knowledge; this assistance generates the set of pathes and gives explicitely the sequences of inputs appearing in path conditions.

2.1.2 - Test oriented knowledge of a resource

With a resource is also associated the test sequences and/or the test data which are the target test of the resource. These data may be generated either by ATPG tools or by the designer. These informations are declared in the same format as the global informations associated to complex blocks (IO timing diagrams or rule based descriptions). In the case of combinatorial blocks, the test oriented knowledge is reduced to a list of vectors.

2.2 - Top-down interactive and semi-interactive mode of the test assembler

In the interactive mode, the test assembler works like a classical fault simulator; but it works at a functional level instead of the logic level and the notion of fault coverage is replaced by the notion of functional test activations covering. The designer chooses himself test inputs and for every activation of the global circuit, the system updates the list of covered activations for each resource ; it analyses the completness with regards to the local test target ; it gives, for instance, the list of not yet reached test cases (pathes or data within a path).

The semi-interactive mode leads to a guided generation of the inputs. This guidance is based either on control flow or data-flow, according to the type of the circuit.

2.2.1 - Control guided generation

This mode allows a "semi-symbolic" simulation : the designer determines some inputs but is allowed to leave other ones undetermined. The simulator performs the simulation by declaring "unassigned" all variables depending on unassigned values. As long as no simulation choice (path selection or activated resource selection) is conditionned by such a variable, the simulation and analysis go on. If a choice raises up, it is submitted to the designer jointly with the current state of the test. To complete its test, the designer may choose a path rather than another one or activate a block rather than another one by assigning some unassigned variables.

2.2.2 - Data-flow guided generation

This mode is based on the result of a previous analysis of the data flows in the circuit ; this analysis is performed by the test advisor presented in next section. This test advisor suggests a general organization of the test which is followed by the designer helped by the functional simulator.

2.3 - Bottom-up test generation : backward and forward chaining of a test case

This mode is dedicated to difficult cases ; when a local test case (a critical point of the circuit) does not show up easily using the first two modes, an automated test generation may be started. The test case is locally identified and consistency to the inputs as well as propagation to the outputs are looked up for ([10]). As previously, this is very similar to a classical approach (path sensitization) but at a different level and on different types of descriptions. The system refers to the designer and an interactive research is performed if an impossibility or a loop is detected.

The two steps are illustrated on figure 7.

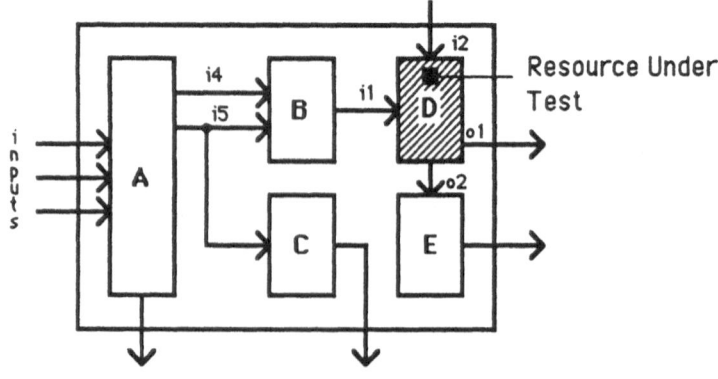

Figure 7

STEP 1 : consistency

Goal : activate a path P_i of the Resource Under Test (RUT)
Procedure :

1 : Determine all the resources connected to the inputs of the RUT.
2 : Find the set $\{R_k\}$ of rules of these resources the right part of which satisfies
the left   part of the rule
related to path P_i. **(backward chaining)**
3 : Do step 2 until reaching the primary outputs.

Example (The RUT is D)

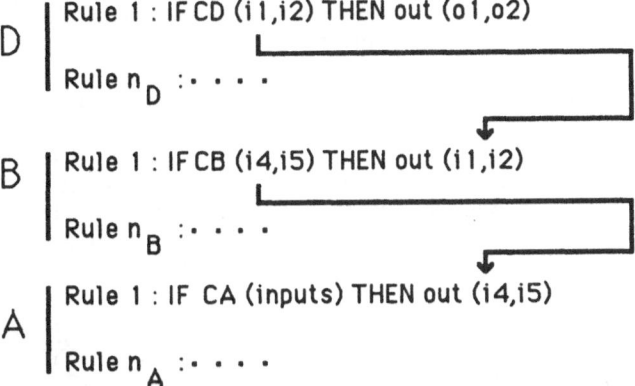

$$D \quad \begin{array}{l} \text{Rule 1 : IF CD (i1,i2) THEN out (o1,o2)} \\ \\ \text{Rule } n_D : \cdots \end{array}$$

$$B \quad \begin{array}{l} \text{Rule 1 : IF CB (i4,i5) THEN out (i1,i2)} \\ \\ \text{Rule } n_B : \cdots \end{array}$$

$$A \quad \begin{array}{l} \text{Rule 1 : IF CA (inputs) THEN out (i4,i5)} \\ \\ \text{Rule } n_A : \cdots \end{array}$$

STEP 2 : forward propagation

Goal : observe the test results at the primary outputs
Procedure :

1: Determine all the resources connected to the outputs of the RUT.
2: Find the rules $\{R_j\}$ of these resources the left part of which is satisfied by the
right part of the rule related to path P_i. **(Forward chaining)**
3: Do step 2 until reaching the primary outputs.

An automatic system realizing the backward and forward propagation is not yet available ; but it is possible to give a very efficient interactive help which, for a given test case in a resource, processes parts 1 and 2 of the consistency and propagation steps.

2.4 - Conclusion on the test assembler

This test assembler is original on many aspects :
- logic and structural levels are definitively left,
- functional activation covering replaces fault covering,
-propagation and consistency problems for test generation are
    studied as backward and forward chaining on a rule based
functional model.

3 - THE TEST ADVISOR

The test advisor gives advices about the general organization of the test, the adequacy of the test to its objectives (type of test and equipment), insertion of test facilities, ...

3.1 - Advices on the test organization

Such an advice has been implemented successfully in the CATA system ([11]) starting from a data flow analysis of the circuit or the board. A simplified abstract model describes how the information is propagated through the different blocks of the circuit and the system calculates the set of data flows and the set of activated blocks associated with each data flow (figure 8a,b). The designer chooses a strategy (presently a startsmall one) and the systems suggests an ordering on the set of activations and gives the resulting diagnosis.

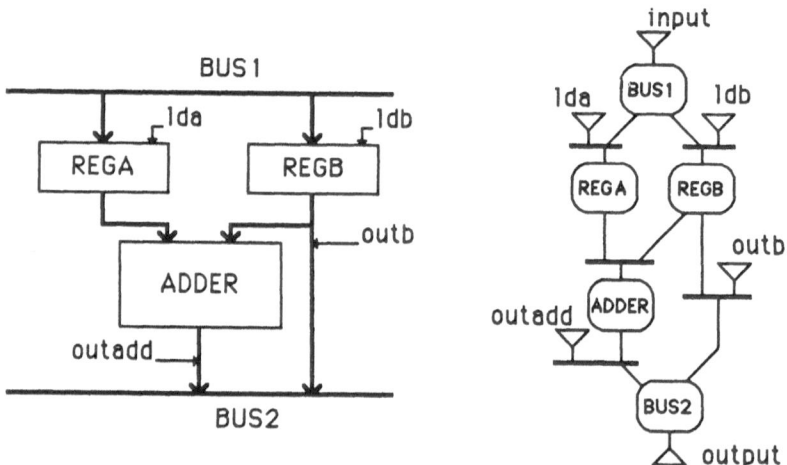

Figure 8a : CATA model

data flows :	Diagnosis blocks
BUS1, REGB, BUS2, LDB, OUTB	(BUS1,REGB, BUS2)
BUS1, REGA, REGB, BUS2, LDA, LDB, OUTADD	(REG1, ADDER)

Figure 8b : CATA results

3.2 - Advices on the complexity and the test coverage

According to the goal of the test, the advisor gives several options; for instance, for a first debugging test, each block has to be activated once in order to have a rough idea of the circuit functioning (correct power supply, ...). A complete debugging test requires the activation of all the pathes within the resources (this leads to a huge test program), end of manufacturing test has to exercize each block with shorts optimized patterns.

3.3 - Insertion of test points or BIST facilities

From a previous global expertise (using CATA system for instance) or according to special difficulties in the test assembly process, the advisor suggests the insertion of new test points. These points may be either connected to IO pads or stored serially through a shift register or compacted through a signature analysis device. It suggests also the use of local test generators if the controlability problem is critical.

3.4 - Advice on diagnosis

In the case where a fault is detected, the designer must find which blocks have to be suspected, according to the test organization. This may be prepared by the test organization advisor. For a required diagnosis, the advisor suggests new activations in order to distinguish between the suspected blocks.

4 - INTERFACE WITH THE GENERAL DATA BASE

In order to facilitate test data generation, all informations related to a circuit and belonging to different domains (functional and structural descriptions, test methodology, ...) must appear within the same data base.

The proposed solution is to consider a circuit as an abstract object with a list of attributes related to each domain and a set of methods indicating either how to calculate the values of attributes or how CAD tools can interact with a given object. Methods will also be useful to check consistency between informations related to different domains.

Such needs will be easily implemented using an object oriented paradigm ; a complete description of this implementation is out of the scope of this paper and we will just briefly introduce its main characteristics.

In a first time we will define a set of classes according to their characteristics in the test domain :

- Class "random logic",
- Class "memory element",
- Class "controller",
- Class "memory".
- ...

The following attributes will then be defined for the previous classes :
- associated test data (these data will either be given by the designer or calculated by an ATPG linked to this attribute),

- complexity and efficiency of the test algorithm,
- Built In Self Test facility,
- ....

A global circuit will be defined hierarchically as a cooperation of objects belonging to one of the previous classes, mechanisms to check data consistency will be activated according to the methods defined within the classes.

CONCLUSION

Test generation is presently one of the most acute problem for all people involved in the ICs area (designer, manufacturers, etc...). The   three main features outpointed in this paper are, according to us, required to solve it.

First, a hierarchical approach makes a clear distinction between low level servers (ATPG for blocks of the circuit) and the test assembler guided by test advisors.

Secondly the levels of representation of blocks and of the whole circuit or system have to be diversified and in any cases, a functional normalized representation is needed.

Finally test coverage is evalued with respect to functional covering (covering of functional pathes) ; test pattern generation is replaced by test case generation ; backward and forward chaining are performed on functional models.

REFERENCES

[1] : D.J. WHARTON, G.D. ROBINSON
      "THE HITEST test generation system. Overview"
      1983 Int. Test conference, Philadelphia, october 83,
      pp. 302-323.

[2] : P. GOEL
      "An implicit enumeration algorithm to generate tests for combinational logic circuits"
      10th Fault-Tolerant Computing Symp., Kyoto, Japan, ctober 80, pp.145-151.

[3] : K.W. CHIANG, Z.C. VRANASIC
      "On fault detection in CMOS logic networks"
      20th Design Automation Conference, Miami Beach, June 83, pp.50-56.

[4] : A. VERGIS, K. STEIGLITZ
      "Testability conditions for bilateral arrays of combinatorial cells"
      IEEE Trans. on Comp., vol C35 n°1, january 86, pp.13-22.

[5] : R.S. WEI, A. SANGIOVANNI-VINCENTELLI
      "PLATYPUS : a PLA test pattern generation tool"
      22nd Design Automation Conference, Las Vegas, june 1185, pp.197-203.

[6] : Z. LOTFI
"Etude et analyse des schémas logiques combinatoires à aide de représentations numérique et graphique"
PhD Thesis, Université de Nancy, France, february 85.

[7] : P. AMBLARD, M . CRASTES DE PAULET, G. SAUCIER
"CADOC ; a functional specification and simulation tool"
IEEE Int. Conf. on Computer Aided Design (ICCAD 83),
Santa Clara, USA, Sept. 83, pp; 65-66.

[8] : C. BELLON, M. CRASTES DE PAULET, G. SAUCIER
"CADOC system : a tool for multilevel description and test generation for VLSI circuits"
7th Int. Conf. on Computer Hardware Description Languages (CHDL85),
pp. 364-380, Tokyo, Japan, Aug. 85.

[9] : C. BELLON
"Le test fonctionnel de circuits intégrés complexes"
PhD thesis, Lab. Circuits & Systems, INPG/USMG, Grenoble,
France, Oct. 84.

[10] : J. RARIVOMANANA
"Système CADOC : génération fonctionnelle de test pour les circuits complexes"
PhD Thesis, Lab. Circuits & Systems, INPG/USMG, Grenoble,
France, Nov. 85.

[11] : Ch. ROBACH, P. MALECHA, G. MICHEL
"CATA : a Computer-Aided Test Analysis system"
IEEE Design & Test, vol1, n°2, may 84, pp 68-79.

19

VLSI Testing: DFT Strategies and CAD Tools

M. Gerner/ M. Johansson

Siemens AG

Otto-HahnRing 6

D-8000 München 83

West-Germany

Introduction

The increasing complexity of the design primitives used and the higher degree of integration now possible are two factors that impede testability. This is due to the higher number of gates which is not matched by an adequate increase in pin count. Using CAD tools, the cost of designing such "more complicated chips" can of course be kept within reasonable limits, but the cost of test preparation will explode due to the level of complexity. Another fact is that semicustom design is on the increase. Bearing in mind that the intention underlying semicustom design is to achieve low-volume production of a great variety of circuits in a very short turn-around time, it is obvious that the factors of high cost and long test preparation time are becoming more critical, as compared with universal chips produced in large quantities. Thus it is essential to automate test preparation by using adequate CAD tools, such as automatic test pattern generation (ATPG). The basis for the effectiveness of these tools is a strict design for testability (DFT), even if the chip area becomes somewhat larger.

This paper presents several DFT strategies, which can be selected as a function of the internal structure of the chip to be tested. Another main topic is the presentation of state-of-the-art CAD tools.

DFT Strategies

Design for testability is necessary for achieving a high quality standard and supporting the automation of test preparation. What all the presented methods have in common is that additional circuits, called test aids, have to be implemented by the chip designer. An essential condition for the designer is to keep in mind certain design rules.

The main idea is a fully synchronous design. The result is circuits whose behavior can be completely described as a function of the external inputs and the internal states at discrete times ($tn-1, tn, tn + 1....$). The change from one state tn to the next state $tn + 1$ is initiated by a special signal, called the clock signal. This can be achieved by using master slave flipflops, strictly partitioning between clock and data signals and by forbidding loops in combinatorial circuits. Most of the dynamic faults can thus be avoided.

Another rule is the limitation of sequential depth S during test mode, i.e. the number of clock steps necessary to make a signal change at an input visible at the outputs. This can be realized by using additional input/outputs and multiplexers, controlled by a test mode signal (Fig. 1). A more systematic way is using a scan architecture. A further step in DFT is the built-in self-test techniques.

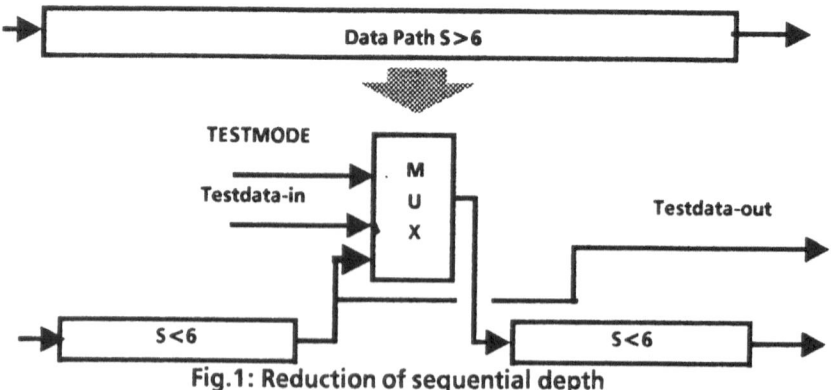

Fig.1: Reduction of sequential depth

Scan Architecture

The first and foremost method of obtaining an automatically testable design is to reduce the sequential depth, because most automatic pattern generators perform best for the purely combinatorial part. Most of the well-known ad-hoc methods to achieve this, such as partitioning, are within the province of the user of a design system. But since they involve additional investigation effort on the part of the designer to keep test preparation cost low, they are only of limited value. Another disadvantage is the lack of standardization, which is necessary for automatically deriving a circuit model for ATPG. The scan technique, on the other hand, can mostly be implemented automatically by a CAD system. There are two representatives of this technique: -   **scan path**

-     **random access scan.**

Both have the aim of transforming a sequential circuit into a purely combinatorial one. In this paper we will only discuss the scan-path strategy, and refer, with respect to the random-access scan, to the literature.

The introduction of a scan path into a circuit means providing a possibility of concatenating the flipflops into a shift register (Fig. 2). An extension of the basic flipflop is necessary. Well-known is the LSSD concept from IBM (Level-Sensitive Scan Design). The basic level-sensitive latch is here extended by an additional one, (Fig. 3) to implement the shift function. In this paper, however, we will discuss edge-triggered flipflops. The resulting model for ATPG is the same.

The general flipflop consists of a D-flipflop and an appendix A defining the type of the flipflop (Fig. 4). For realizing the shift function, we only have to consider the basic D- flipflop. This can be done by inserting a multiplexer, which switches as a function of the TESTMODE signal, the flipflop input from function to scan (Fig. 5).Every flipflop gets its input data from, and passes its output data to a combinatorial part of the circuit. Data coming from, or

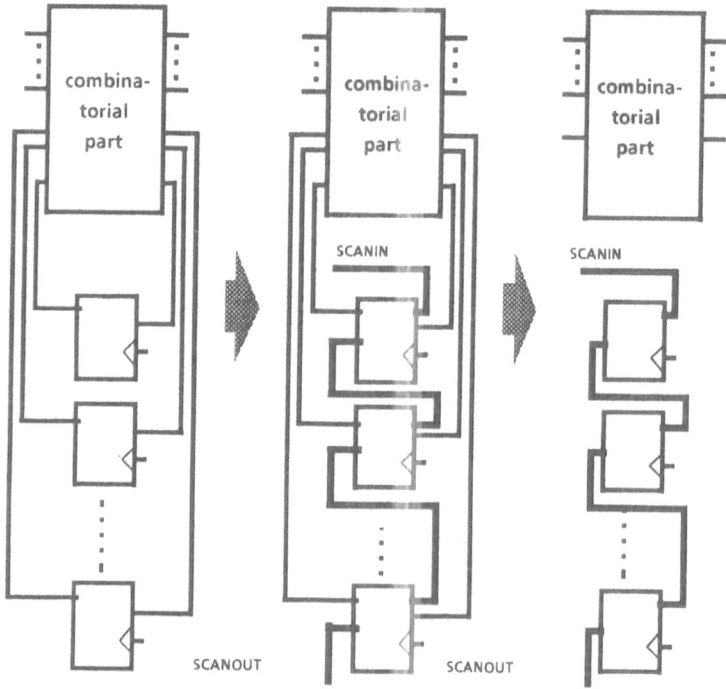

Fig.2:General principle of scan path

going to, the combinatorial part can now be observed and set by means of
the scan path. One test sequence consists of three steps:

- shifting the test pattern into the scan flipflops
- storing the test answer in the scan flipflops after the delay
  time of the combinatorial part by using a clock signal
- observing the test answer by shifting it out.

The model of each flipflop presented to ATPG can be substituted by a
fictitious outpout and a fictitious input (Fig. 6). Thus every sequential circuit is
transformed into a purely combinatorial one, allowing the use of a very
effective ATPG. Besides , the sequential part can also be tested separately by
shifting certain test patterns through the scan path.

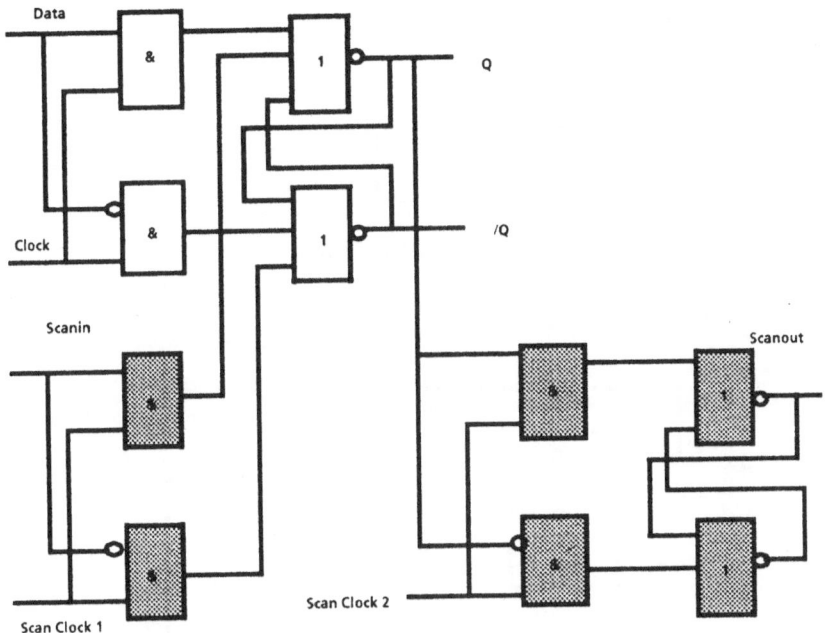

Fig.3:LSSD storage element

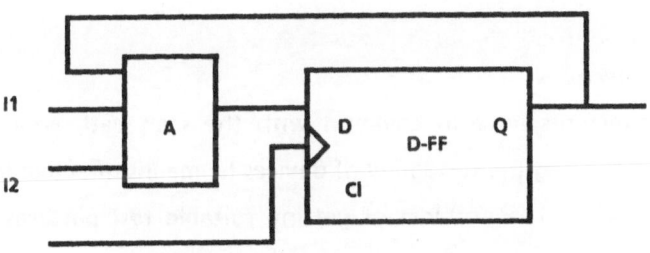

Fig.4: Universal master slave flipflop

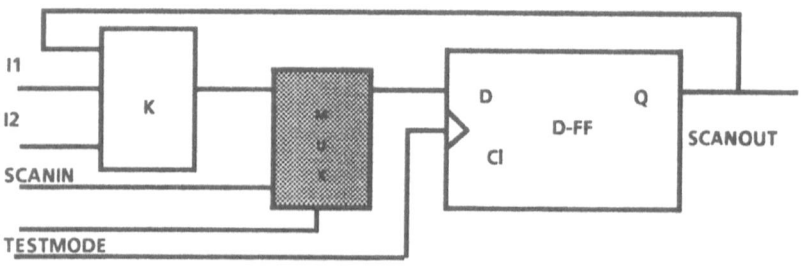

Fig.5: Universal master slave scan flipflop

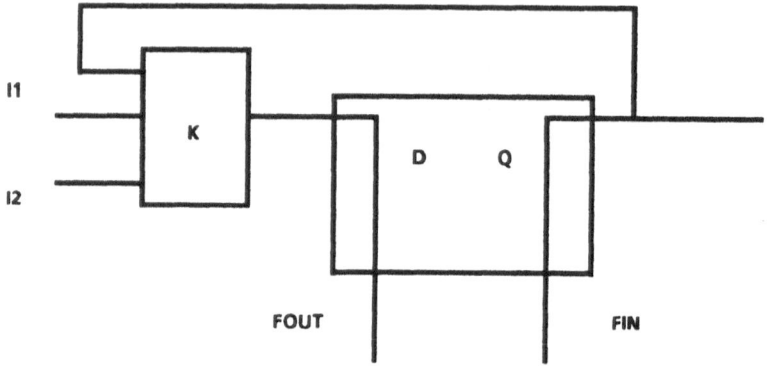

Fig.6:generator model

Self-Test Architectures

Self-test architectures have in common with the scan path concept the objective of enhancing the testability of devices by means of active test aids applied to the chip. The problem of getting suitable test patterns to the module to be tested and of transporting the test answers to the circuit outputs is solved by generating the test patterns locally and evaluating the test answers in the module itself. There are two alternative selftest methods: the first (incomplete self-test) is limited to hard-to-get-at circuit sections (e.g. embedded RAMs), while the second tests the entire device. The self-test can be controlled either on the chip or externally.

When automatic testing equipment is used, the test objects are examined sequentially, i.e. one device after the other. This is associated with a relatively long test time for circuits of high complexity. In contrast, the self-test permits parallel testing. This is an advantage especially for the wafer test in which no preselection has yet been made. The precondition for parallel testing is that all devices are supplied with the operating voltage and provided with a few signal lines required for the self-test. Suitable test adapters must be made available for this purpose.

The self-test can be used not only for production testing but also for the field test of the system. For this purpose, the possibility of self-testing must be offered to the user of an integrated circuit. In a number of semicustom circuits, the self-test option is already contained in the device specification. In order to simplify the field test, a suitable interface must be made available.

Like all integrated testing aids, the self-test results in a hardware overhead in the form of additional chip area and, possibly, a higher pin count. As integration density continues to increase however, this point may lose its significance. The modification and inclusion of components which are in any case required for the device function also permit the outlay to be reduced.

Extensive automation is necessary for minimizing the design overhead required by the self-test equipment. It relies on suitable CAD tools which allow for, or implement, self-test integration at all design levels. Design reglementation is a prerequisite for design automation, involving a certain restriction of the designer's leeway in favor of enhanced testability.

Besides the principles of self-test a number of conceivable self-test architectures will now be presented in more detail. The individual concepts are organized in line with the structure of the circuit to be tested. Since only the most important distinctions are to be indicated, a high degree of abstraction was chosen for the presentation. In real architectures, the rule is a mixed form of the outline architectures shown. Naturally, other concepts are also conceivable in addition to the procedures shown here.

Principle: In traditional tests with an automatic test equipment, the device is stimulated by precalculated test patterns and an anticipated-vs.-actual comparison of the individual test answers is made by the test equipment. In a self-test, these functions are performed with the aid of supplementary logic present on the device itself. In an externally activated test mode , test pattern generators (TPGs) produce the stimuli, and test answer evaluators (TAEs) assess the answers at the circuit outputs (Fig. 7). A control logic ensures the interaction of these test circuits.

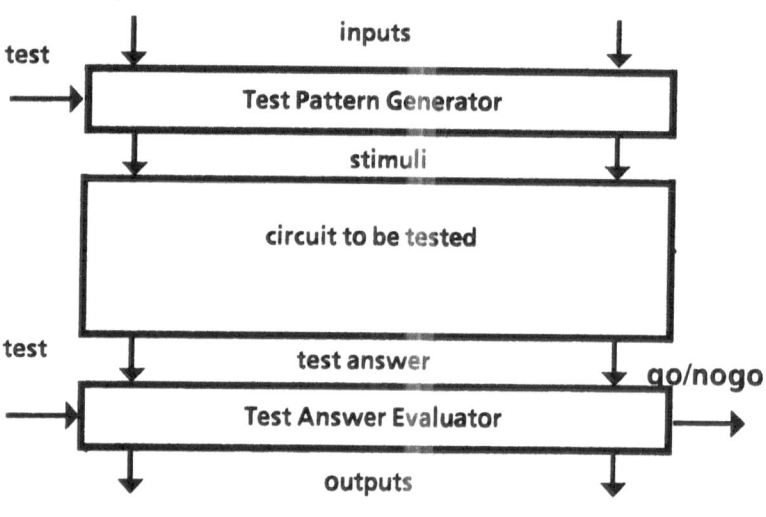

Fig.7: Self test scheme

An effective self-test architecture selects the test pattern generators and test answer evaluators most suitable for a given type of device. The efficient generation of test stimuli and the evaluation of the test answers are of critical importance for a self-test. The storage of deterministic test patterns in a ROM is no economic proposition unless a processor structure is present. As a rule, test patterns are generated by linear-feedback shift registers (LFSR). Fig. 8a shows a test pattern generator of this kind. By suitable selection of the feedback function, the most diverse pseudorandom numerical sequences may be generated. Of particular importance are sequences which contain all 2^n (n: width of the register) vectors with the exception of the zero vector. They can be used to perform a test which is complete with respect to the stuck-at-fault model in a relatively simple way.

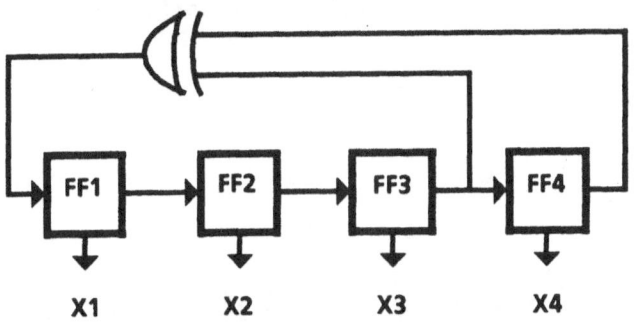

Fig 8a: Test pattern generator (LFSR)

Nonlinear-feedback shift registers (NFSR) are more useful for a number of self-test methods (Fig. 8b). In them, the feedback function is not performed by linear circuit elements (exclusive OR) but rather by a network of AND or OR gates. Deterministic test patterns, in particular, such as are required for a RAM test (walking one/zero, marching one/zero etc.), may thus be simply generated.

Compression of the individual test answers to a signature is the rule for evaluating the test answers. This task is performed by signature registers. As is shown in Fig. 8c, these are built out of linear-feedback shift registers with additional parallel inputs (MISR: Multiple Input Shift Registers).Due to this data compression, the fault detection rate for the occurence of multiple faults is less than 100% and depends on the register width. Apart from this, a comparison of the actual signature with an anticipated signature allows only

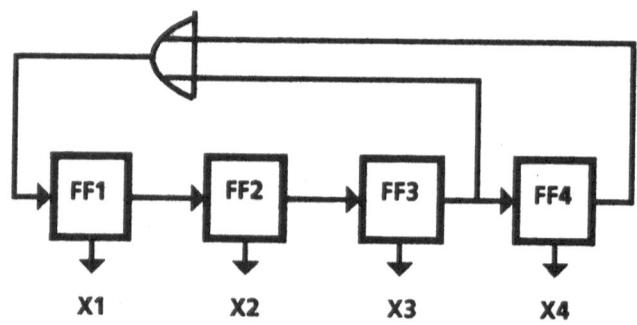

Fig 8b: Test pattern generator (NFSR)

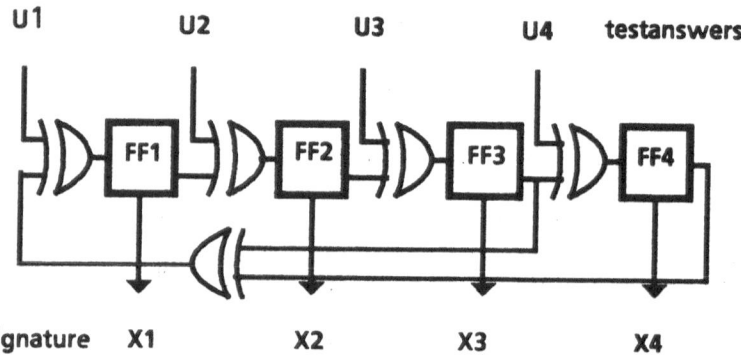

Fig. 8c: Test answer evaluator

a Go/Nogo statement to be derived. Fault location is not possible, but usually not necessary either.

Test pattern generators, test answer evaluators and control units are the elementary modules of a self-test architecture. With their aid, a self-test can be performed at module level. Combination of single module tests leads to a device test concept. Over and above this, a convenient self-test architecture should also take into account the possible system environment of the individual device.

Random structure: The most general application of a self-test architecture is in testing random logic. For this function, a self-test procedure is recommended which does not presuppose any specific circuit structure. Random logic may consist of combinatorial and sequential parts. It was shown in the chapter on scan architecture that the scan path principle is suitable for testing this class of devices. In a self-test, the scan path must be extended by a test pattern generator for the primary and fictitious inputs and by a signature register for the outputs. Fig.9 shows this concept schematically.

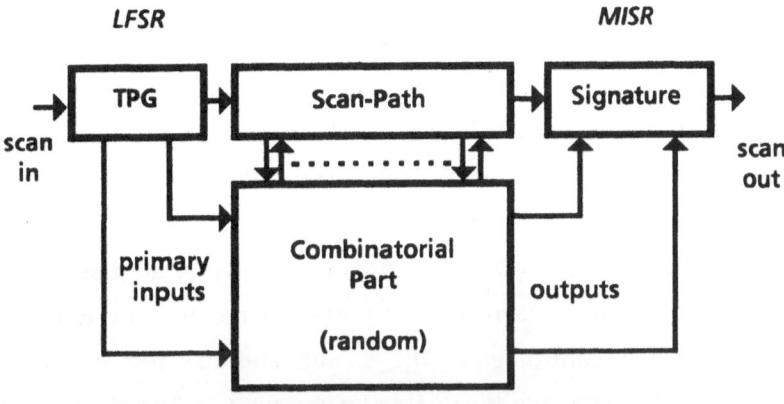

Fig. 9: Selftest architecture for random logic

By combining the scan path register, test pattern generator and signature register into a BILBO (Built-in Logic Block Observer), the hardware overhead can be reduced. This may be done by using the test answer evaluator for one module as the test pattern generator for a second module in the subsequent test. But this involves the disadvantages of an increasing control overhead.

A serious disadvantage of this self-test procedure is the greatly increasing scan-path length for a large number of primary inputs and of storage elements. Since the rule is the exhaustive test, very long test times result. This can be at least partly relieved by the use of pseudorandom test patterns. This involves the generation not of all possible test patterns but only of a number sufficient to obtain the required fault coverage. Another approach aims at Partitioning the circuit into several smaller units. An important point is to determine the optimal subunits to ensure minimal mutual coupling.

Bus structure: When a bus structure is present, the device can be separated relatively easily into single modules to be tested, with several modules being connected by common buses. Enable control lines can then be used to link the individual modules to the bus or uncouple them from it. This permits relatively simple partitioning into single submodules.

In the self-test of bus structures, a rough division can be made into two types of test architecture: **global self-test**

local self-test

In case of global self-test a common test module exists for all submodules to be tested (Fig. 10). It contains the test pattern generator, the test answer evaluator and the control unit. The stimuli and the test answers are transported on the common bus. At any given time, only a single module (test specimen) on the bus is activated. The advantage of this procedure lies in the low extra expenditure it involves, since TPG, TAE and the control unit can be used in common by several modules. However, since all modules must be tested in sequence, the test time can become unreasonable long under certain circumstances.

This disadvantage is avoided by the local self-test (Fig. 11). Every module to be tested contains its own test pattern generator, test answer evaluator and control unit. The bus is needed here only to decouple the individual modules from each other. All modules perform the self-test simultanously, the individual go/nogo statements being linked to a common go/nogo flag. It is obvious that the reduction of the test time must be traded off with a higher

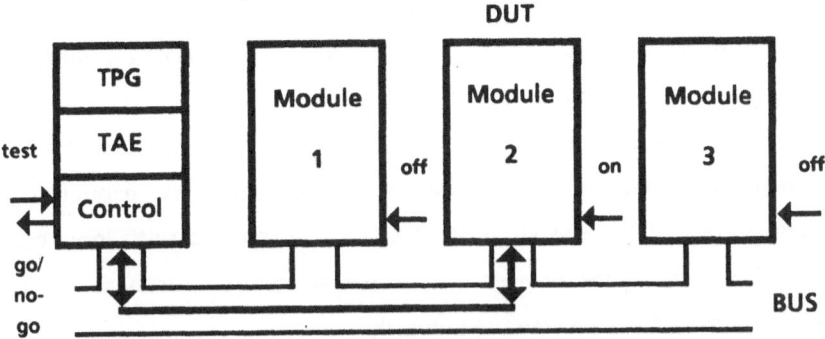

Fig. 10: Bus structure with global selftest

hardware overhead. A conceivable and effective mix of both variants could be realized with common test pattern generation and local answer evaluation for all modules.

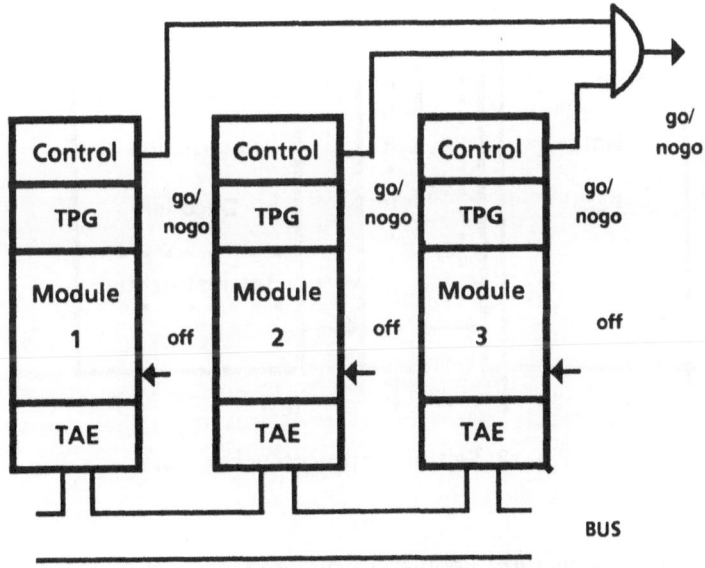

Fig 11: Bus structure with local selftest

Processor Structure: Fig. 12 shows a simplified form of a possible self-test architecture for processor devices. Use is made of the bootstrap method. In the first step, the command decoder is tested. This fault-free component can subsequently be used to test the data path. In this way, an increasingly powerful test hardware is available step by step. By exploiting the "intelligence" of the device, a self-test can be realized with low control expenditure. Processor devices with self-test components have therefore been on the market for some time already.

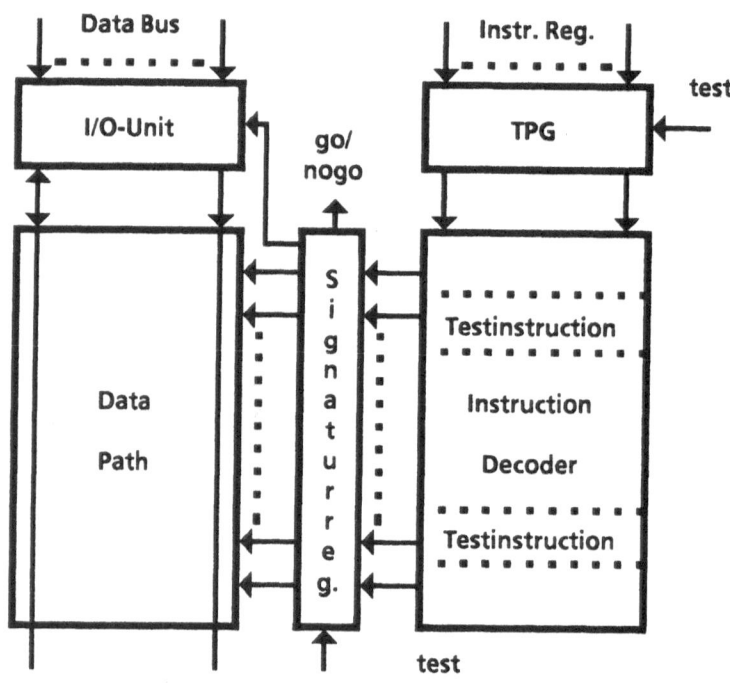

Fig. 12: Selftest of processor structure

Regular Module Structure: Regularly structured modules like PLA, ROM and RAM are used in many complex circuits, because they are easily designed. For PLAs there exist minimization programs and layout generators. Memories are parameterizable in address space and word length. However, the increasing complexity leads to great difficulties in testing these circuits. Accordingly self-test may be the solution for the types of structure described before .In addition, self-test has the advantage of exploiting the topological structures of these circuits. Typical faults in the PLAs are shorts between adjacent bitlines or product-term lines. These faults are detectable by efficient self-test methods.

For PLAs, many self-test methods have been published. For small PLAs (number of inputs <20), an exhausting test with an LFSR and a signature register is possible. For larger ones, it is better to separate the test for the AND and the OR plane. In Fig. 13 the structure of a typical MOS PLA is

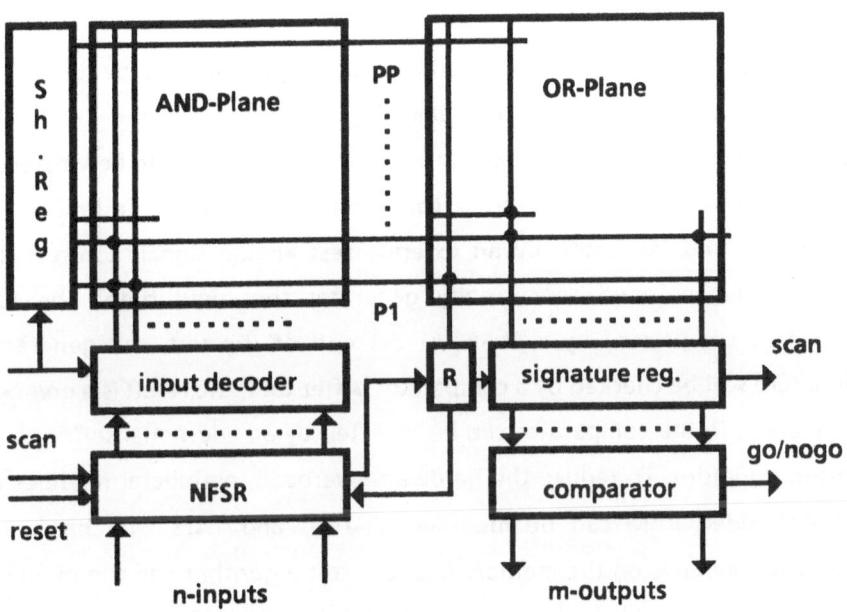

Fig 13: PLA selftest

depicted. Both planes consist of NOR gates. Nor gates can be easily tested. The problem is to observe the outputs (product terms) of the AND plane and to control the inputs of the OR plane. In Fig. 14 a shift register disables all product terms except one. Therefore the selected one can be observed at the output without interference from the other product terms. The input stimuli are generated by an NFSR, which produces a walking-one sequence. The product term shift register also supplies the test patterns needed for testing the OR plane. The number of test cycles is $\approx$ n p (p = number of product terms), which is far less than 2^n.

The difference between PLAs and ROMs ist that ROMs are fully coded. This means that there exists one output for every input combination. For testing a ROM, all input combinations must be applied. Therefore, a counter can be used as a pattern generator. One may prefer an LFSR, because the order of the test pattern sequence is of no importance.

For testing static RAMs, simple and linear algorithms can be used. For example, the "marching one/marching zero" algorithm may be implemented. Both units of information, one and zero, have to be written in all cells. Therefore, one needs a counter to generate all addresses. An additional unit, activated by an external test enable signal, controls the input/output unit.and concurrently generates the input data. The test answer is compressed by a MISR. At the end of the test, the generated signature will be checked by a comparator. After that, the result is a go/nogo flag. Errors in the comparator can be detected by an additional shifting-out of the signature. To reduce the hardware overhead, peripheral modules of an embedded RAM can be modified in TPGs and TAEs. The hardware overhead depends on the memory size, the test algorithm and the modules of the circuit available in the periphery (Fig. 14).For large RAMs the test time increases enormously. Parallel testing of sections will reduce the test time.

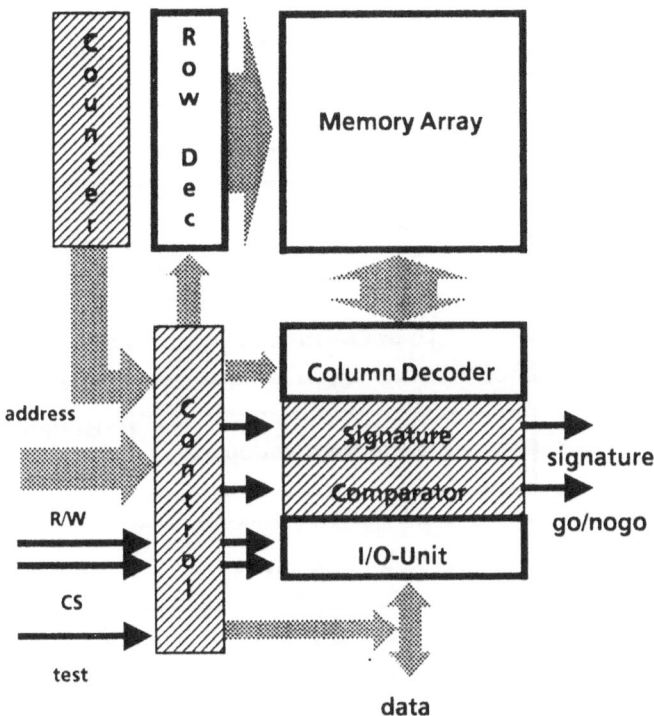

Fig 14: RAM selftest

CAD-Tools

The test problems have to be considered at every stage of the CAD process. As part of the logical design process it must be verified that the circuit is testable, and that it is possible to generate testvectors automatically and with a high fault coverage. As part of the physical design process, the test program is generated, which contains all necessary tests.

A typical system with the most important functions for automatic test preparation in a CAD environment is shown in figure 15. Input to the system is a description of the circuit, output is a tape with a test program, written in the language of the automatic test equipment.

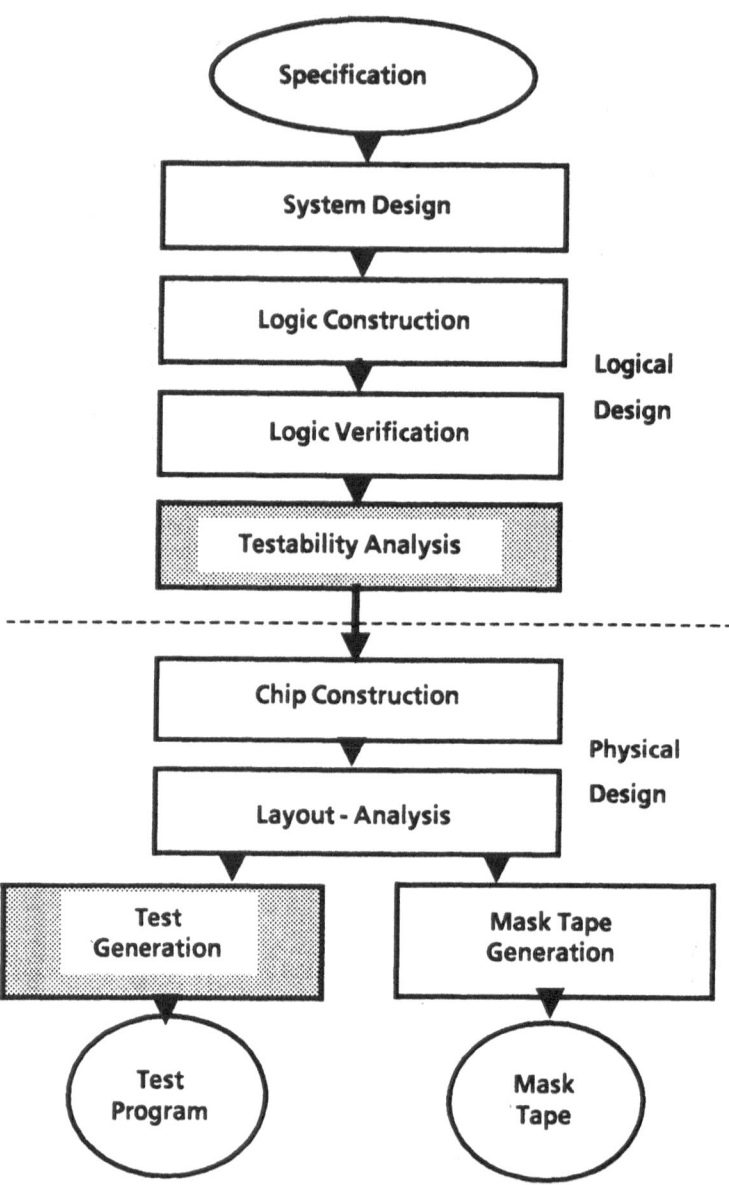

Fig 15: Testing in the CAD process

We shall here discuss principles and algorithms for testability analysis, automatic test pattern generation, fault simulation and test program generation. However, the test problems often have to be considered also by other CAD tools, for example if test aids are to be integrated automatically.

Testability Analysis

It is important to be able to verify at an early stage of the design process that the circuit will be easily testable and that the tests can be generated automatically. One important part of a "Design for Testability" is the application of the test aids discussed earlier, another is the testability analysis, for which we shall here present two approaches:

- Design rule audits which investigate whether or not the circuit obeys a set of testability-related design rules.
- Algorithms which calculate values for controllability and observability for different nodes of the circuit.

Rule Audit: In a CAD system, a set of testability-related design rules are defined as a basis for the automation of test preparation. This means that rules are given, which will create a design standard that allows for testing. The main goals of the rules are

- to achieve testability of the circuits
- to avoid certain kinds of faults (e.g. dynamic ones) or to make them detectable as static faults
- to reduce the costs for test preparation
- to guide the designer with respect to the integration of test aids.

A fundamental principle of the design rules is the construction of strictly synchronous circuits. These are usually more tolerant of variation in production parameters, and in them dynamic faults can be avoided or detected as static faults.

On the basis of the circuit structure and of information about the elements, a "rule audit" can investigate whether the defined design rules were obeyed or not. The number and character of the rule violations are a measure for testability.

A rule audit uses algorithms from the field of graph theory. Violations of structural rules can be detected by tracing paths through the circuit with due regard to the different types of elements. Rule violations involving timing problems, however, cannot always be detected through this kind of structural analysis. The result of the rule audit is a listing of the violated rules and the sites of the violations. This provides the designer with information on what must be changed in the circuit to make it more easily testable. A further development of this approach is to have the rule audit automatically give advice as to how the violations can be corrected or how test aids can be integrated.

A rule audit can be used with advantage for hierarchical designs where the blocks can be analyzed individually.

Controllability and observability: To quantify the testability of a circuit, measures aiming to cope with the difficulty of controlling and observing the logical values of internal nodes from inputs and outputs are calculated. The result of this analysis is statistical information giving values for all nodes. Nodes, which are difficult to observe or control can thus be found, and the circuit can be modified to make it more easily testable.

This analysis can support the integration of different test aids. The calculated measures can also provide guidance for the test pattern generator.

Six functions are defined for each node (N):

   CC0/1 (N):   Combinatorial 0/1 controllability

CO (N):   Combinatorial oberservability

SC0/1 (N):   Sequential 0/1 controllability

SO (N):   Sequential observability

The combinatorial controllability of a node is related to the minimum number of node assignments required to justify a 0 or a 1.

The combinatorial observability of a node is related both to the number of elements between the node and an observable output and to the minimum number of node assignments required to propagate the logical value from the node to the observable output.

The sequential controllability and observability, on the other hand, estimate the number of sequential nodes that must be set to put the desired value on N or to propagate the value of N to an output. The sequential values are a measure of the number of time frames required to control or observe an internal node of a sequential circuit.

For an external input node I , the controllability functions are defined as:

$$CC0\,(\,I\,) = 1$$
$$CC1\,(\,I\,) = 1$$
$$SC0\,(\,I\,) = 0$$
$$SC1\,(\,I\,) = 0$$

Starting with the external inputs, the controllabilities are calculated for all nodes. For a 3-input ($I\,1$, $I\,2$, $I\,3$) NOR gate, the controllabilities for the output (O) are defined as:

$$CC0\,(O) = \min\,[CC1\,(\,I\,1), CC1\,(\,I\,2), CC1\,(\,I\,3)] + 1$$
$$CC1\,(O) = CC0\,(\,I\,1) + CC0\,(\,I\,2) + CC0\,(\,I\,3) + 1$$
$$SC0\,(O) = \min\,[\,SC1\,(\,I\,1), SC\,1\,(\,I\,2), SC1\,(\,I\,3)]$$
$$SC1\,(O) = SC0\,(\,I\,1) + SC0\,(\,I\,2) + SC0\,(\,I\,3)$$

To set the output to 0, one of the inputs has to be set to 1; thus the most easily controllable input is chosen. To set the output to 1, all inputs have to be set to 0.

The NOR gate has the combinatorial depth 1 and the sequential depth 0, so the combinatorial controllabilities are increased by 1.

For an external output node O, the observability functions are defined as:

$$CO(O) = 0$$
$$SO(O) = 0$$

Starting from the external outputs, the observabilities can now be calculated for all nodes. For one input (e.g. the first one) of the 3-input NOR gate, the observabilities are defined as:

$$CO(I\,1) = CO(O) + CCO(I\,2) + CCO(I\,3) + 1$$
$$SO(I\,1) = SO(O) + SCO(I\,2) + SCO(I\,3)$$

To observe the value of I 1, it is necessary to set I 2 and I 3 to 0 and to observe O.

Automatic Test Pattern Generation

Generally speaking, to test a circuit means to apply some values to the inputs of the circuit, to observe the output values, and to compare these with the expected output values. If there is a discrepancy between the observed and the expected values, the circuit is said to have a fault.

There are many different approaches and ideas how to find test patterns for a circuit. The easiest one would be to apply all possible input vectors and observe the outputs (exhaustive test). This is not normally possible because of the amount of necessary test patterns. For a combinatorial circuit with n inputs, we would have 2^n test patterns. For a sequential circuit, the number would be even larger by several orders of magnitude.

There are different methods of generating a smaller number of test patterns for a given circuit:

- functional test pattern generation, in which a functional circuit description is used and the faults are in some way defined as deviations from this function.

- Random test pattern generation, in which a randomly chosen subset of the exhaustive test is used. The method can be improved if a suitable distribution of zeros and ones can be calculated for the inputs.

- Structural test pattern generation, a structural circuit description is used and the considered faults are directly related to the physical elements or signals.

The state of the art is to use structural and/or random test pattern generators. We shall here discuss the D algorithm as an example of structural test pattern generation. Most of the structural ATPG algorithms are extensions of the D algorithm.

Before we go on to discuss the algorithms for automatic test pattern generation (ATPG), we have to define what faults we shall test for, i.e. a fault model.

Faults and fault models: There are many different types of fault:
A static (logical) fault changes the function of the circuit. A commonly considered type of static fault is a fault through which the value of a signal is stuck at logical 0 or 1. (The stuck-at-fault model).

A <u>dynamic fault</u> changes the function of the circuit at a certain frequency or under certain timing conditions. A typical dynamic fault is an excessive time delay for an element.

A <u>parametric fault</u> changes the magnitudes of circuit parameters, such as voltage or current.

These faults may, further, be permanent or intermittent, single or multiple; they many be caused by shorts, open connections or other physical defects, or their occurrence may depend on temperature, aging and so on.

To be able to generate test patterns automatically, it is necessary to specify what faults shall be considered, that is, to define a <u>fault model</u> which describes the effect of the faults on the logical behavior of the circuit. Such a fault model will necessarily regard only certain types of fault.

Usually, test patterns are generated automatically only for single stuck-at-0/1 faults. Most static faults are detected by using test patterns generated for this fault model. There are however static faults, e.g. CMOS-stuck-open-faults, which are not always detectable with this fault model.

There are different approaches to the automatic generation of tests for some dynamic faults. One possibility is the so called clock rate test, i.e. a static test executed at a high frequency.

Tests for parametric faults are usually added to the other test patterns, depending on the technology involved.

Structural test pattern generation: As discussed above, test aids such as the scan path technique transform a sequential circuit into a combinatorial circuit during test. Hence, we shall only consider combinatorial circuits here. Extensions for sequential ones are described in the literature, but they are much more sensitive to the size and the complexity of the circuits.

To generate a test pattern for a certain fault, the behavior of the faulty circuit is compared with that of the faultfree one. A test pattern is successfully generated, when an observable value differs in the faulty case from that of the faultfree case. To facilitate the comparison, the two circuits are considered in parallel using composite generation values:

Generation value	Fault free case	Faulty case
0	0	0
1	1	1
D	1	0
$\overline{D}$	0	1

To describe an element for test pattern generation, different aspects have to be considered. They will here be shortly explained by using a two-input NAND gate as an example.

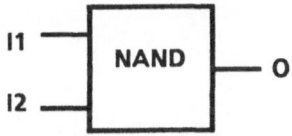

Fig 16: A two-input NAND-gate

- The <u>primitive cubes</u> describe the normal, logical behavior of the faultfree element.

- The <u>propagation- D cubes</u> describe how a faulty signal with the value D or $\overline{D}$ can be propagated from an input to an output of the

	I1	I2	O
a	0	x	1
b	x	0	1
c	1	1	0

element,i.e. how the other input values have to be set, to make the output value change if one input value changes. Normally, there should also be vectors with more than one input $D/\overline{D}$. This means that all of these inputs have to change to make the output value change.

	I1	I2	O
d	$\overline{D}$	1	D
e	1	$\overline{D}$	D
f	D	1	$\overline{D}$
g	1	D	$\overline{D}$

• The <u>D cubes of a logic fault</u> describe how the element is to be tested, i.e. what fault model was chosen. They contain the input vectors that must be applied to the element in order to test it, and the expected output values (embryonic test).

We shall here consider the single stuck-at-fault model, which for the NAND gate implies six different faults:

    I 1:   s-a-0   s-a-1

	I1	I2	O	I1	I2	O
h	0	1	D	s-a-1	-	s-a-0
i	1	0	D	-	s-a-1	s-a-0
j	1	1	$\overline{D}$	s-a-0	s-a-0	s-a-1

I 2:	s-a-0	s-a-1
O :	s-a-0	s-a-1

To explain the main steps of the algorithm, we shall study a simple circuit consisting of four NAND gates (figure 17). We want to find a test pattern for the fault : input I1 of gate G2 is stuck-at-1 .

1. Embryonic test. Select a D-cube for the fault. This defines the faultfree values of the inputs and the possibly faulty one (D/$\overline{D}$) of the output.

For G2 we choose h, which defines L6: = 0, L3: = 1 and L7: = D. (Figure 17a)

2. D drive . Select propagation D-cubes for a chain of elements from the considered one to an observable output, so that the output value will change depending on whether the fault exists or not.

For G4 we choose f, which defines L8: = 1 and L9: = $\overline{D}$.(Figure 17b)
Since L9 is an observable output, the D-drive is finished.

3. Line justification. During the first two steps, many elements got defined input values. These have to be set via controllable inputs. To do so, the primitive cubes will be used.

Injected fault: G2-I1: s-a-1

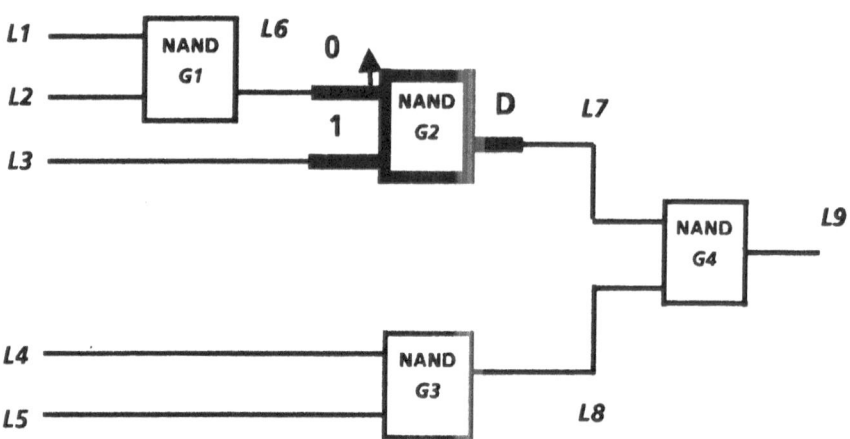

Fig 17a: Embryonic test

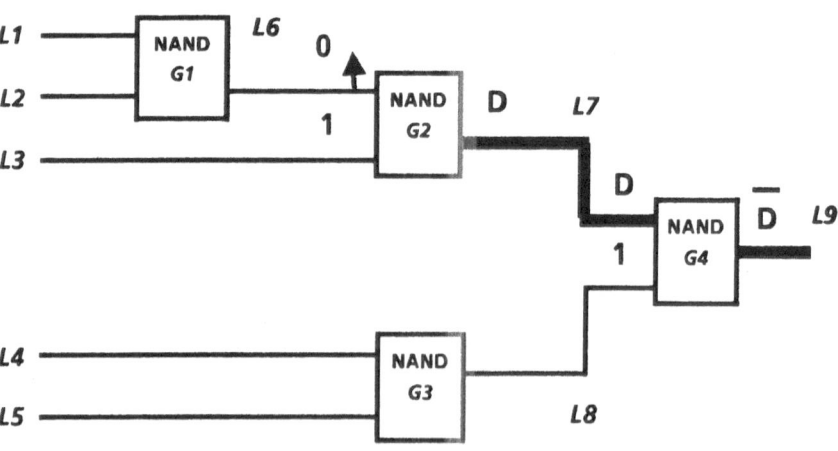

Fig. 17b: D-drive

We have defined values for *L6, L7* and *L8*. We now have to make sure that the signals *L6* and *L8*, which are not directly controllable, are set properly. To set *L6* to 0, we use **C** for *G1*, which defines *L1*: = 1 and *L2*: = 1. To set *L8* to 1 we choose **a**, which gives *L4*: = 0.(Figure 17c)

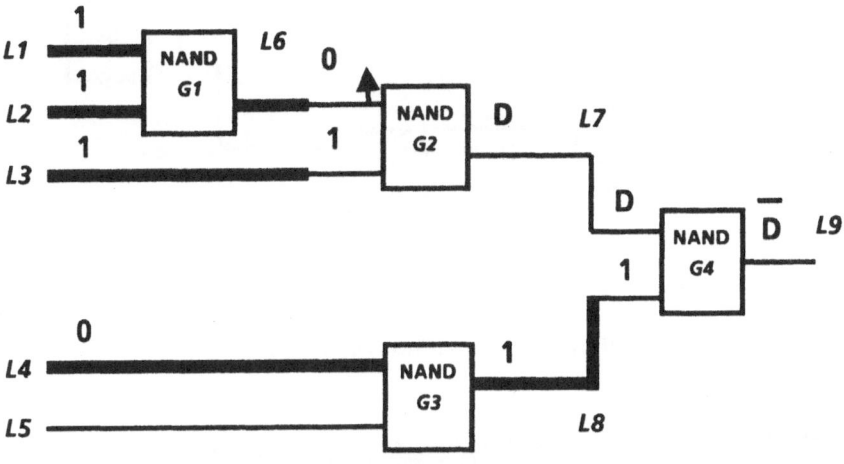

Fig. 17c: Line justification

We now have a test pattern for the studied fault:

$$(L1, L2, L3, L4, L5, L9) = (1, 1, 1, 0, X, \overline{D}).$$

If the fault exists, the output value will be a 1, otherwise a 0.

In all steps of pattern generation there may exist many alternative choices. It must be kept record of what alternative was chosen so that, if necessary, the other ones can be considered later on. The reason is the common case that a signal is set to a certain value over one path and must be set to the opposite value over another path (figure 17d). To resolve such an inconsistency, it is necessary to go back to the last step, where a not yet used alternative exists (backtracking), and make another choice. For *G3* in the example of figure 17d

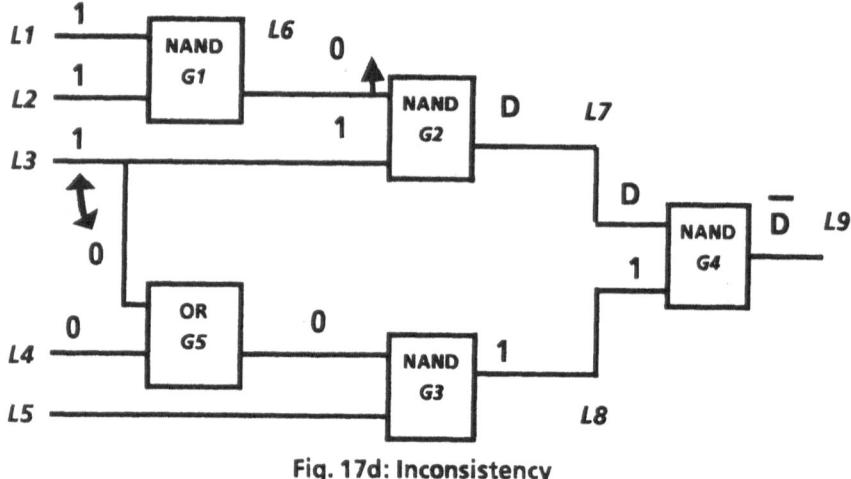

Fig. 17d: Inconsistency

we choose **b** instead of **a**, which defines *L5* : = 1.

Fault simulation

Fault simulation means that a circuit is simulated under certain fault assumptions and the results are compared with a simulation of the faultfree circuit. The fundamental goal is to find out if certain faults can be detected with the given test patterns.

In testing, fault simulation is used for the following tasks:

●   Calculation of exactly which of the considered faults were detected by the given patterns. This information is used as input to the ATPG tool, which then shall generate test patterns explicitly for the not yet detected faults. Since the ATPG tools do not normally exactly know all the faults that are detected with the generated patterns, an iterative strategy is often used, where ATPG and fault simulation are executed alternatingly.

- Calculation of the fault coverage of a given test pattern set. The fault coverage is the ratio of the number of detected faults to the total number of faults possible with the given fault model. The fault coverage is a measure of the quality of the test patterns.

- Creation of a fault catalog with information on which faults are detected through which patterns. This is important if a fault diagnosis is to be carried out. For ICs a fault diagnosis is not normally made, since there are no components which could be exchanged or repaired.

Test Program Generator

A test program contains parts for the following functions:

- Parametric tests. Here the magnitudes of some circuit parameters, such as voltage or current, are measured.

- Functional tests for logical faults.
  Here the logical function of the circuit is tested. This is done by using the generated test patterns, which can be extended by manually written patterns.
  Tests should be carried out both for static and for dynamic faults.
  The functional test for static faults can be executed once more but at a high frequency (clockrate test), to detect certain dynamic faults.

- Statistics and shmooplots. Statistics are used to calculate the yield and to detect design weaknesses. The shmooplot describes within what area (e.g. of voltage or frequency) the circuit functions correctly.

As part of a CAD system, the test program should be generated automatically for the automatic test equipment (ATE) used. To do this manually would consume too much time. The generation of the test program can be done by a program with the following functions:

- The parts for the parametric tests are generated. On behalf of technological parameters and a general test program for the actual technology and ATE.

- The test patterns from the ATPG and/or the manually generated ones are translated from some neutral format into the language of the actual ATE.

- The instructions for the desired statistics and shmooplots are integrated.

Conclusions

To test an arbitrary complex and highly integrated circuit for all possible faults may be an almost impossible task. To obtain the tests automatically as a part of a CAD process is even more difficult. In order to be able to guarantee a high quality and a highly automatic design process a test strategy and a standardization are necessary, which means that testability aspects have to be considered in all stages. A high degree of testability may be achieved by observing certain design rules and/or by integrating extra test aids.

Whether or not the testability aspects have been considered, can be checked by CAD-tools for testability analysis in an early stage of the design process. If the results from this analysis are good, it can be guaranteed that the circuit is easily testable and that the necessary tests can be obtained automatically.

Literature

(AbGa81)   Abraham, J. A.; Gajski, D:D.: Design of testable structures defined by simple loops. IEEE Trans. CAS-28, 1079 (1981)

(AgMe82)   Agraval, V.D., Mercer, M.R.: Testability measures -- What do they tell us? Proc. Int. Test Conf., Philadelphia 1982, p. 391

(Benn84)   Bennetts, R.G. Design of testable logic circuits. London: Addosin-Wesley 1984

(Bott77)   Bottorff, P.S.; France, R.E.; Garages, N.H.; Orosz, E.J.: Test generation for large logic networks. Proc. 14th Design Autom. Conf., New Orleans 1977, p. 479

(Bowd75)   K.R. Bowden: A Technique for Automatic Test Generation for Digital Circuits. Proc. IEEE Intercon, 1975, Session 15

(BuSi82)   Buehler, M.G.; Sievers, M.W.: Off-line, built- in test techniques for VLSI circuits. Computer 15, 69 (1982)

(BrFr84)   Breuer, M.A.; Friedman, A.D.: Diagnosis & reliable design of digital systems. Computer Science Press, Inc. (1984)

(BMR81)   Bennetts, R.G.; Maunder, C.M.; Robinson, G.D.: CAMELOT: a Computer-Aided MEasure for LOgic Testability. in: Dal Ciin, M.; Dilger, E. (eds.) Selfdiagnoses and Faulttolerance. Tübingen: Attempto Verlag 1981

(CeAb83)   Cerny, E.; Aboulhamid, E.M.: Built-in-testing of pI-testable iterative arrays. Proc. FTCS-13, Milano 1983, p.33

(DaMu81a)   Daehn, W.; Mucha, J.: A hardware approach to self-testing of
              large programmable logic arrays. IEEE Trans. CAS-28, 1033
              (1981)

(Daeh83)   Daehn, W.: Deterministische Testmustergenerierung für den
              eingebauten Selbsttest von integrierten Schaltungen. NTG-
              Fachbericht 82 (1983)

(EiLi80)   Eichelberger, E.B.; Lindbloom, E.: A heuristic test-pattern
              generator for programmable logic arrays. IBM Journal of Res.
              and Dev., 24, 15 (1980)

(EiWi77)   Eichelberger, E.; Williams, T.W.: A logic design structure for
              LSI testability 14th Design Autom. Conf. 1977, 462- 468

(ElZi83)   El-Zig, Y.M.: S**3: VLSI self-test using signature analysis and
              scan path techniques. Proc. IEEE ICCAD-83, Santa Clara 1983,
              p. 73

(EGM83)   Echtle, K.; Görke, W.; Marhöfer, M.: Zur Begriffsbildung bei
              der Beschreibung von Fehlertoleranz-Verfahren. Universität
              Karlsruhe, Fakultät für Informatik, Interner Bericht Nr. 6/83
              (1983)

(Frie73)   Friedman, A.D.: Easily testable iterative systems. IEEE Trans. C-
              22, 1061 (1973)

(FuAb84)   Fuchs, W.K.; Abraham, J.A.: A unified approach to concurrent
              error detection in highly structured logic arrays. Proc. FTCS-
              14, Kissimmee 1984, p.4

(FuKi81)   Fujiwara, H.; Kinoshita, K.: A design of programmable logic arrays with universal tests. IEEE Trans. CAS-28, 1027 (1981)

(FuTa83)   Fujiwara, H.; Takesh, S.: On the Acceleration of Test Generation Algorithms. IEEE Trans. C-32, 1137 (1983)

Fuji85)   Fujiwara H.: Logic Testing and Design for Testability. The Massachusets Institute of Technology (1985)

(GaKu83)   Gajski, D.D.; Kuhn, R.H.: New VLSI tools - guest editors introduction. Computer 16, 11 (1983)

(GeNe84)   Gerner, M.; Nertinger, H.: Scan-Path in CMOS-Semicustoms LSI Chips, IEEE Test Conference 1984

(Goel81)   Goel, P.: An implicit enumeration algorithm to generate tests for combinational logic circuits. IEEE Trans. C-30, 215 (1981)

(Gold79)   Goldstein, L.H.: Controllability/Observability Analysis of Digital Circuits. IEEE-Trans.Cas-26, 685 (1979)

(Gras82)   Grassl, G.: Built-in test for a 32-Bit ALU. IEEE Workshop on Design for Testability, Vail, Co., April 1982 (presentation only)

(GrPf83)   Grassl, G.; Pfleiderer, H.-J.: A function-Independent sself-test for large programmable logic arrays. INTEGRATION, the VLSI journal 1, 71 (1983)

(HaMc83)   Hassan, S.Z.; McCluskey, E.J.: Testing PLAs using multiple parallel signature analysers. Proc. FTCS-13, Milano 1983, p. 422

(HoOs80)   Hong, S.J.; Ostapko, D.L.: FITPLA: A programmable logic
             array for function independent testing. Proc. FTCS-10, Kyotot
             1980, p.131

Joha83)   Johansson, M.: The GENESYS-Algorithm for ATPG without
             Fault Simulation 1983, IEEE Test Conference pp. 333 - 337

(Kaut67)   Kautz, W.H.: Testing faults in cellular logic arrays. Proc. IEEE
             8th Ann. Symp. on Switching & Automata Theory, 1967, p.161

(Khak83)   Khakbaz, J.: A testable PLA design with low overhead and
             high fault coverage. Proc. FTCS-13, Milano 1983, p. 426

(Köne80)   Könemann,B.; Mucha, J.: Built-in-test for Complex Digital
             Inetgrated Circuits. IEEE Journal of Solid State Circuits, Vol.
             15, 3, 1980

(LeWa83)   Leisengang, D.; Wagner, M.: Signaturanalyse in der
             Datenverarbeitung , Elektronik 21, 10/83, 67-72

(Levi81)   Levi, M.L.: CMOS is most testable. Proc. Int. Test Conf.,
             Philadelphia 1981, p. 217

(LoGl81)   Lord, D.H.; Gleason, D.: Design & evaluation methology for
             built-in test. IEEE Trans. R-30, p.222 (1981)

(Marh84a)   Marhöfer, M.: Entwurf von Testbarkeit (Systematische
             Übersicht und Literaturschau). Arbeitspapier, Mai 1984

(Marh84b)   Marhöfer, M.: Entwurfs von Testbarkeit bei der Entwicklung
             hochintegrierter Schaltungen. Arbeitspapier, Juni 1984

(MoAb83)   Mantoye, R.K.; Abraham, J.A.: Built-in tests for arbitrarily structured VLSI carray-lookahead adders. in: Anceau, F.; Aas, E.J. (eds.): VLSI '83, 361 (1983)

(Moor56)   Moore, E.F.: Gedanken-experiments on sequential machines. Automata Studies, Princeton University Press 1956, p. 129 - 153

(Much83)   Mucha, J.: Grundprinzipien· und Entwicklungstendenzen beim Testen digitaler integrierter Schaltungen. in: Zimmer, G. (Hrsg.): Praxis der Grossintegration, Seminar, Uni Dortmund 1983

(RSP82)   Ratiu, I.M.; Sangiovanni-Vincentelli, A.; Pederson, D.O.: VICTOR: A fast VLSI testability analysis program. Proc. Int. Test Conf., Philadelphia 1982, p. 397

(Roth67)   Roth, J.P.: Diagnosis of Automata Failures: A Calculus and a Method. IBM Journal, Vol. 10, No. 7, July 1967

(Savi80)   Savir, J.: Syndrome-testable design of combinational circuits. IEEE Trans. <u>C-29</u>, 442 (1980)

(Savi83)   Savir, J.: Good controllability and observability do not guarantee good testbility. IEEE Trans. <u>C-32</u>, 1198 (1983)

(SaBa83)   Savir, J.; Bardewll, P.H.: On random pattern test lenght. Proc. Test Conf., Philadelphia 1983, p. 95

(ShFe83a)   Shen, J-.P.; Ferguson, J.: Easily-testable array multipliers. Proc. FTCS-13, Milano 1983, p. 37

(Smit79)   Smith, J.E.: Detection of faults in programmable logic arrays. IEEE Trans. C-28, 845 (1979)

Some83)   Somezi, F.; Gai, S.; Mezzalama, M.; Prinetto, P.: PART: Programmable ARray Testing based on a PARTitioning algorithm. Proc. FTCS-13, Milano 1983, 430

(SrHa81)   Sridhar, T., Hayes, J-.P.: A functional approach to testing bit-sliced microprocessors. IEEE Trans. C-30, 563 (1981)

(SDB83)   Savir, J.; Ditlow, G.; Bardell, P.H.: Random pattern testability. Proc. FTCS-14, Milano 1983, p. 73

(Tris84)   Trischler, E.: An integrated design for testability and automatic test pattern generation system: an overview. Proc. 21th Des. Autom. Conf., Albuquerque 1984, p. 209

(Wads78)   Wadsack, R.L.: Fault Modeling and Logic Simulation of CMOS and MOS integrated circuits. The Bell System Technical Journal, Vol. 57, June 1978

(Walt80)   Waltrich, J.B.: A Method of estimating the effect of design for testability on PC board test costs. Proc. Int. Conf., Philadelphia 1980, p. 176

(Yama83)   Yamanda, T.: Syndrome-testable design of programmable logic arrays. Proc. Int. Test Conf., Philadelphia 1983, p. 453

(YaAr81)   Yajima, S.; Aramaki, T.: Autonomously testable programmable logic arrays. Proc. FTCS-11, Portland 1981, p. 41